T0317478

Chemistry and Technology of Emulsion Polymerisation Second Edition

Chemistry and Technology of Emulsion Polymerisation

Second Edition

Editor

A.M. van Herk

Institute of Chemical and Engineering Sciences, Singapore

WILEY

This edition first published 2013

Registered office
John Wiley & Sons Ltd, The Atrium, Southern Gate, Chichester, West Sussex, PO19 8SQ, United Kingdom

For details of our global editorial offices, for customer services and for information about how to apply for permission to reuse the copyright material in this book please see our website at www.wiley.com.

Library of Congress Cataloging-in-Publication Data applied for.

A catalogue record for this book is available from the British Library.

ISBN: 9781119953722

Set in 10/12pt Times by Aptara Inc., New Delhi, India.

Contents

List of Contributors

Elena Akhmastkaya Basque Center for Applied Mathematics (BCAM), Spain

Jose M. Asua POLYMAT, University of the Basque Country UPV/EHU, Spain

Bernadette Charleux Chemistry, Catalysis, Polymers & Processes, Université de Lyon, France

Thierry Delair Laboratoire des Matériaux Polymères et des Biomatériaux, Université Claude Bernard Lyon 1, France

R. G. Gilbert Centre for Nutrition & Food Science, University of Queensland, Australia, *and* Tongji School of Pharmacy, Huazhong University of Science and Technology, China

Finn Knut Hansen Department of Chemistry, University of Oslo, Norway

Ola Karlsson Division of Physical Chemistry, Lund University, Sweden

Haruma Kawaguchi Graduate School of Engineering, Kanagawa University, Japan

Hans Heuts Department of Chemical Engineering & Chemistry, Eindhoven University of Technology, The Netherlands

Jose Ramon Leiza POLYMAT, University of the Basque Country UPV/EHU, Spain

Yuri Reyes Mercado POLYMAT, University of the Basque Country UPV/EHU, Spain

Jan Meuldijk Department of Chemical Engineering & Chemistry, Eindhoven University of Technology, The Netherlands

Michael J. Monteiro Australian Institute for Bioengineering and Nanotechnology, The University of Queensland, Australia

Christian Pichot Saint-Priest, France

Brigitte E.H. Schade Particle Sizing Systems,Waterman, Holland

Jürgen Schmidt-Thümmes BASF AG, GMD, Germany

Peter Schoenmakers Department of Chemical Engineering, University of Amsterdam, The Netherlands

Bernhard Schuler BASF AG, ED/DC, Germany

Dieter Urban BASF AG, GMD, Germany

A.M. van Herk Institute of Chemical and Engineering Sciences, Jurong Island, Singapore *and* Department of Chemical Engineering & Chemistry, Eindhoven University of Technology, The Netherlands

Brian Vincent School of Chemistry, University of Bristol, UK

Abbreviations

AA	Acrylic acid
ABS	Acrylonitrile-butadiene-styrene
Aerosol MA	AMA, sodium di-hexyl sulfosuccinate
Aerosol OT	AOT, sodium di(2-ethylhexyl)sulfosuccinate
AFM	Atomic force microscopy
AIBN	Azobisisobutyronitrile
ATRP	Atom transfer radical polymerization
B	Butadiene
BA	n-Butyl acrylate
BPO	Benzoyl peroxyde
Buna N	Butadiene-acrylonitrile copolymer
Buna S	Butadiene-styrene copolymer
CCA	Colloidal crystalline array
CCD	Chemical composition distribution
CDB	Cumyl dithiobenzoate
CFM	Chemical force microscopy
CFT	Critical flocculation temperature
CMC	Critical Micelle Concentration
CMMD	Control molar mass distribution
CPVC	Critical pigment volume concentration
CRP	Controlled radical polymerization techniques
CTA	Chain transfer agents
CVP	Colloid vibration potential
Cyclam	Tetrazacyclotetradecane
DLVO	Derjaguin-Landau-Verwey-Overbeek
DMA	Dynamic mechanical analysis
DNA	Desoxy nucleic acid
DSC	Differential scanning calorimetry
EDTA	Ethylene diamino tetraacetic acid
EHMA	2-Ethylhexyl methacrylate
EPA	Environmental Protection Agency
ESA	Electrokinetic sonic amplitude
ESEM	Environmental scanning electron microscopy
FESEM	Field emission scanning electron microscopy
FIB-SEM	focused ion beam SEM

FFF	Field-flow fractionation
FLGN	Feeney, Lichti, Gilbert and Napper
HASE	Hydrophobically modified alkali-swellable emulsions
HDPE	High density polyethylene
HEC	Hydroxy ethyl cellulose
HEMA	2-Hydroxyethyl methacrylate
HEUR	Hydrophobically modified ethylene oxide urethanes
HIC	Hydrophobic interaction chromatography
HPLC	High performance liquid chromatography
HUFT	Hansen, Ugelstad, Fitch, and Tsai
IR	Infrared
K	Kelvin
LV-SEM	low voltage SEM
LRP	Living radical polymerisation
MA	Methyl acrylate
MFFT	Minimum film forming temperature
MMA	Methyl methacrylate
MMD	Molar Mass Distribution
MONAMS A5	1-(methoxycarbonyl)eth-1-yl initiating radical
NMP	Nitroxide-mediated living radical polymerisation
NMR	Nuclear magnetic resonance
NR	Natural rubber
OEM	Original Equipment Manufacturer
OM	Optical microscopy
PCH	Phenyl-cyclohexene
PCS	Photon correlation spectroscopy
PDI	Polydispersity index
PDMS	Poly(dimethylsiloxane)
PEO	Poly(ethylene oxide)
PGA	Poly(glycolic acid)
PHS	Poly(hydroxystearic acid)
PLA	Poly(D, L-lactic acid)
PLGA	Poly(glycolic-co-lactic acid)
PMMA	Poly(methylmethacrylate)
PNIPAM	Poly(N-isopropylacrylamide)
PPO	Polypropylene oxide
PRE	Persistent radical effect
PSA	Pressure sensitive adhesives
PSD	Particle size distribution
PTA	Phosphotungstic acid
PTFE	Poly tetrafluorethylene
PVAc	Poly(vinyl acetate)
PVC	Pigment volume concentration
QCM-D	quartz crystal micro-balance with dissipation monitoring
RAFT	Reversible addition fragmentation transfer
RCTA	Reversible chain transfer agents

S	Styrene
SAM	Self-assembled monolayer
SANS	Small angle neutron scattering
SAXS	Small angle X-ray scattering
SB	Styrene butadiene
SBLC	Styrene butadiene latex council
SBR	Styrene butadiene rubber
SDS	Sodium dodecyl sulphate
Sed-FFF	Sedimentation field-flow fractionation
SEM	Scanning electron microscopy
SFM	Scanning force microscopy
SPM	Scanning probe microscopy
SRNI	Simultaneous reverse and normal initiation
SSIMS	Static secondary ion mass spectrometry
STM	Scanning tunneling microscopy
TEM	Transmission electron microscopy
TEMPO	2,2,6,6-Tetramethylpiperidine-l-oxyl
Texanol®c	2,2,4-Trimethyl-1,3-pentanediol-diisobutyrat
UAc	Uranyl acetate
UV	Ultraviolet
Vac	Vinyl acetate
VCH	Vinyl-cyclohexene
VOC	Volatile organic compound
W	Watt
Wet-SEM	wet scanning transmission electron microscopy
XPS	X-ray photoelectron spectroscopy
XSB	Carboxylated styrene-butadiene dispersions

List of Frequently Used Symbols

a_e	Specific surface area for a emulsifier molecule on a polymeric surface
A	Arrhenius constant of the initiation (A_i), propagation (A_p), termination (A_t) and transfer (A_{tr})
$\bar{d}$	average particle diameter d_n, number average diameter, d_s surface average diameter, d_w weight average diameter, d_v volume average diameter
d_w/d_n	particle diameter non-uniformity factor
E	energy of activation for initiation (E_i), propagation (E_p), termination (E_t) and transfer (E_{tr})
f	Initiator efficiency
F	Efficiency factor for adsorption
ΔG	Partial molar free energy of droplets ΔG_d, ΔG_a of the aqueous phase and of the latex particles ΔG_l
H	enthalpy
ΔH	change in enthalpy
j_{crit}	Critical length of an oligomer at which precipitation from the aqueous phase occurs
k	exit frequency
k	rate constant of the initiation (k_i), propagation (k_p), termination (k_t) and transfer reaction (k_{tr})
[M]	concentration of monomer, $[M]_p$ concentration of monomer in the polymer particles. If this depends on quantities such as radius r, time t, etc., the recommended notation is $[M(r,t,\ldots)]_p$. $[M]_a$ for the monomer concentration in the aqueous phase, $[M]_{a,sat}$ for the saturation concentration in the aqueous phase.
M	average molar mass: number-average molar mass (M_n). weight-average molar mass (M_w),
N	number of latex particles per unit volume of latex
N_n	Number of particles with n radicals per particle
N_A	Avogadro constant
n	number of radicals in a latex particle
$\bar{n}$	average number of radicals per particle
n_{m0}	initially added number of moles of monomer per unit volume
$\overline{P}_n$	number average degree of polymerisation
R	gas constant
$r_{1,2}$	reactivity parameters in copolymerisation
r_p	rate of polymerisation per particle

r_e	rate of entry of radicals per particle
r_t	rate of termination per particle
r_o	the radius of the unswollen micelles, vesicles and/or latex particles.
R_p	Rate of polymerisation
S	entropy
ΔS	change in Entropy
T	temperature
T_g	glass transition temperature
t	time
V	volume of monomer swollen latex particles
V_m	molar volume of the monomer
v_p	volume fraction of polymer
W	stability ratio
w_p	mass fraction of polymer in the particle phase
x	fraction conversion of monomer to polymer
x_n	number-average degree of polymerisation, x_w weight-average degree of polymerisation
z-mer	The length of an oligomer in the aqueous phase at which surface activity occurs
α	fate parameter (fate of excited radicals)
χ	Flory-Huggins interaction parameter
δ	solubility parameter or chemical shift
ε	permittivity
γ	interfacial tension
η	viscosity
$[\eta]$	intrinsic viscosity
ν	kinetic chain length
π	osmotic pressure
ρ	entry frequency
ρ_i	radical flux or rate of initiation (2 kd f [I])
μ	Volume growth factor
τ_g	time of growth of a polymer chain

Introduction to the Second Edition

The increasing need for environmentally benign production methods for polymers has resulted in a further development and implementation of the emulsion polymerisation technique. More and more companies switch from solvent based polymer production methods to emulsion polymerisation.

Since the introduction of the first edition in 2005 the experience gained with using this book in a teaching environment, led us to this second improved edition. Besides some of the new developments we added a new chapter on latex particle morphology development as especially in this area much progress has been made and a lot of research efforts, both in academia and in industry, has been devoted to this important area. Furthermore the chapter on the use of controlled radical polymerization in latex production has been substantially updated as most of the other chapters.

Powerpoint slides of figures in this book for teaching purposes can be downloaded from http://booksupport.wiley.com by entering the book title, author or isbn.

Introduction to the First Edition

New polymerisation mechanisms like controlled radical polymerisation are combined with the emulsion polymerisation technique, encountering specific problems but also leading to interesting new possibilities in achieving special nanoscale morphologies with special properties. In the past years many people have been trained in the use of the emulsion polymerisation technique. Many courses on the BSc, MSc and the PhD level as well as special trainings for people in industry are given all over the world. Despite this no recent book exists with the purpose of supporting courses in emulsion polymerisation.

This book is aiming at MSc students, PhD students and reasonably experienced chemists in university, government or industrial laboratories, but not necessarily experts in emulsion polymerisation or the properties and applications of emulsion polymers. For this audience, which is often struggling with the theory of emulsion polymerisation kinetics, this book will explain how theory came about from well-designed experiments, making equations plausible and intuitive. Another issue experienced, especially in industry, is that coupling theory and everyday practice in latex production is really hard. This is another aim of the book; showing how theory works out in real life.

The basis for the contents of this book can be found in the course emulsion polymerisation taught for many years at the Eindhoven University of Technology in the framework of the Foundation Emulsion Polymerisation. Many people have contributed to shaping the aforementioned course and therefore laying a basis for this book: Ian Maxwell, Jenci Kurja, Janet Eleveld, Joop Ammerdorffer, Annemieke Aerdts, Bert Klumperman, Jos van der Loos and last but not least Ton German. Most of the contributors to the chapters are member of the International Polymer Colloids Group, a group of experts around the world that meet on a regular basis and form a unique platform for sharing knowledge in the field.

The book is focussing on emulsion polymerisation in combination with both conventional and controlled radical polymerisation. Except for miniemulsion polymerisation, more exotic techniques like inverse emulsion polymerisation, microemulsion polymerisation and dispersion polymerisation are not covered.

The first chapter is giving a historic overview of the understanding of emulsion polymerisation, also focussing on the solution of the kinetic equations. In the second chapter an introduction is given in the radical (co)polymerisation mechanism, explaining kinetics and the development of molecular weight and chemical composition. In chapter three the basic element of emulsion polymerisation are explained, again focussing on rate of reaction and molecular mass distributions. In chapter four, emulsion copolymerisation, process strategies are explained. In chapter five the implementation of controlled radical polymerisation mechanisms in emulsion polymerisation is discussed. In Chapter 6 the development of morphology in latex production is discussed. Colloidal aspects of emulsion polymerisation are discussed in chapter seven. In chapter eight an overview of the molecular characterization techniques of (emulsion) polymers is given whereas in chapter nine the characterization techniques available for particle size, shape and morphology are reviewed. In Chapter ten and eleven bulk and specialty applications are discussed. As much as possible the nomenclature for polymer dispersions according to IUPAC has been followed (Slomkowski, 2011).

We hope that this book will become a standard textbook in courses in emulsion polymerisation.

1

Historic Overview

Finn Knut Hansen
Department of Chemistry, University of Oslo, Norway

1.1 The Early Stages

Polymers are composed of very large molecules, each of which includes a large number of repeating structural units. The oldest and most abundant group of polymers consists of the natural polymers, such as cellulose, proteins, rubbers, and so on. One of these, natural rubber, occurs in the form of a latex, that is defined as the "viscid, milky juice secreted by the laticiferous vessels of several seed-bearing plants, notably *Castillia elastica*," and so on (Bovey *et al.*, 1955). By far the most important natural latex is that obtained from the rubber tree *Hevea brasiliensis*. This tree, originally from Brazil, as may be deduced from its name, was transplanted to Malaya, Sri Lanka and the East Indies (Hauser, 1930) in 1876, and eventually has made this area the most important source of natural rubber. The latex that is obtained from the tree is usually denoted as "natural latex" and is a colloidal suspension of rubber particles stabilized by protein. The rubber content of the latex is between 32 and 38% by weight, the protein 1 to 2%, different natural sugars about 2% and about 0.5% of inorganic salts (Hauser, 1930). The rubber particles vary largely in size from quite small, circa 50 nm, up to 1–2 micrometres. The rubber latex is coagulated, washed and worked into sheets that form the basis for further industrial use.

In view of the latex origin of natural rubber, it was not surprising that, when the need for a synthetic equivalent arose, the mimicking of natural rubber latex was an obvious starting point. The effort, and great success, of making synthetic rubber by emulsion polymerisation has led to the word "latex" eventually being used to refer to a colloidal suspension of *synthetic* polymers, as prepared by emulsion or suspension polymerisation. Such *synthetic latexes* are to be distinguished from dispersion of polymers prepared by grinding the polymer with water and a dispersing agent. This chapter will treat the early

Chemistry and Technology of Emulsion Polymerisation, Second Edition. Edited by A.M. van Herk.

stages of the "invention" and production of synthetic latexes by emulsion polymerisation from the beginning and up to the middle of the twentieth century. Several reviews and book chapters on the early developments in emulsion polymerisation have already been written, and have been a natural starting point for this text. One of the first reviews is that of Hohenstein and Mark from 1946 (Hohenstein and Mark, 1946). The following is a direct quotation from their work (reprinted from J. Polymer Sci., by permission):

> The earliest observations on polymerisation of olefins and diolefins as far back as 1838 (Mark and Rafft, 1941; Regnault, 1838) refer almost entirely to the pure liquid phase and describe the gradual transition from a liquid monomer to a viscous or solid polymer under the influence of heat, light, or a catalytically active substance. The idea of using a finely divided monomer in an aqueous suspension or emulsion seems to have been first conceived, about 1910, by Hofman and Delbrück (Hofman and Delbrück, 1909, 1912) and Gottlob (Gottlob, 1913). There were two main reasons for the desire to carry out the polymerisation of various simple dienes in the presence of a diluting agent: one, the fact that the use of metallic sodium as catalyst, which was common practice at that time, led to highly heterogeneous materials and posed a rather difficult problem regarding the complete removal of the alkali metal from the final polymer. The more important incentive for the use of an aqueous system, however, were the facts that all native rubbers occur in the form of latexes and that, obviously, polymerisation in the plant takes place under mild conditions in an aqueous phase without the application of elevated temperatures and high pressures, and certainly without the use of such catalysts as metallic sodium or alkali alkyls.
>
> The aim of reproducing the physiological conditions occurring in the plant is mentioned in some of the earlier disclosures (Gottlob, 1913; Hofman and Delbrück, 1909, 1912), and led to the preparation and stabilization of the "emulsions" as described in these patents *not* with the aid of soap or other surface-active agents, but by application of hydrophilic protective colloids such as gelatin, egg albumin, starch, milk, and blood serum. Certain remarks in the text of these patents indicate that these protective colloids not only emulsify the hydrocarbon monomer but may also act as catalysts during the polymerisation. We have carried out a number of polymerisations, following closely the methods given as examples in two of these patents and have substantially confirmed the results of the claims. In these experiments we observed a *very slow, partial* conversion of the monomer (isoprene, dimethylbutadiene) into a polymer latex. The total amount of polymer formed varied between 40% and 80%; the duration of the reaction was in certain cases as much as six weeks. The results, in general very erratic and almost irreproducible, create the impression that the reaction under such conditions could be considered a *suspension polymerisation* catalyzed by the oxygen of the air, which was never specifically excluded in any of the examples. In order to check this conclusion we repeated a few experiments of this type with deaerated monomer and deaerated water under nitrogen and found that under these conditions only *extremely slow* polymerisation can be observed. In some instances conversion was not achieved at all.
>
> It seems, therefore, that the early practice, as disclosed in the above-mentioned patents, is substantially different from what is known today as emulsion polymerisation, and is essentially a suspension polymerisation in which the protective colloids act as suspension stabilizers and which is catalyzed by the presence of small amounts of oxygen.
>
> In 1915 and 1916, Ostromislensky (Ostromislensky, 1915, 1916; Talalay and Magat, 1945) carried out similar experiments with vinyl halides and discussed the advantages of the presence of an inert diluent. However, since there is no mention of the use of soap or other micelle-forming substances in his articles either, it seems that his observations also refer to "uncatalyzed" or photocatalyzed polymerisation in solution and suspension.

It was only in 1927 that the use of *soap* and similar substances (ammonium, sodium, and potassium oleates, sodium butylnaphthalene sulphonate) was disclosed in patents by Dinsmore (Dinsmore, 1927) and Luther and Heuck (Luther and Heuck, 1927). The examples cited in these disclosures approach present practice to a considerable degree; they specify the simultaneous use of emulsifiers and catalyst (water- or monomer-soluble peroxides) and describe conversions and reaction times of the same order of magnitude as reported in more recent scientific articles. It seems, therefore, that the use of catalyzed emulsion polymerisation is about twenty years old (in 1946, Ed. note).

In the years following a large number of additional patents accumulated, with an almost confusing multitude of disclosures and claims (compare references (Hoseh, 1940, 1941; Scheiber, 1943; Talalay and Magat, 1945)). On the other hand, during this same period (1930–1940) only very few articles were published in scientific journals. Dogadkin (1936) and his collaborators (Balandina *et al.*, 1936a, 1936b; Berezan, Dobromyslowa, and Dogadkin, 1936) studied the polymerisation of butadiene in the presence of soap, peroxides, and other catalysts at different temperatures and investigated the kinetics of this reaction. Fikentscher (Fikentscher, 1934), at a meeting of the Verein Deutscher Chemiker in 1938, gave a general description of the course of emulsion polymerisation of dienes and advanced, for the first time, the hypothesis that polymerisation takes place essentially in the aqueous phase and not inside the monomer droplets. In 1939, Gee, Davies, and Melville (Gee, Davies, and Melville, 1939) investigated the polymerisation of butadiene vapour on the surface of water containing a small amount of hydrogen peroxide and came to certain conclusions about the kinetics of this process. While the mechanism of emulsion polymerisation was thus only infrequently and briefly discussed in the scientific literature between 1930 and 1940, much work was carried out during this same period in the research departments of various industrial organizations, as shown by the large number of patents filed and issued in many countries.

One of the authors (H. M.) had an opportunity to discuss the problem of emulsion polymerisation in the period between 1935 and 1938 with Drs. Fikentscher, H. Hopff, and E. Valko in Ludwigshafen am Rhine. At that time they offered several arguments in favour of polymerisation taking place preponderantly in the aqueous phase. Valko even considered it as highly probable that the monomer, solubilised in the micelles of the soap solution, was most favourably exposed to the action of a water-soluble catalyst and, therefore, might be considered as the principal site of the reaction. At a seminar on high polymers in Kansas City in September, 1945, Dr. F. C. Fryling told us that he had, at the same time, independently arrived at very similar conclusions on the basis of his own observations. It appears, therefore, that some of the more recent developments were anticipated to a certain extent in the unpublished work between 1930 and 1940.

No work in emulsion polymerisation was published in the next three years, except for brief references in the books of Mark and Raft (Mark and Rafft, 1941) and of Scheibler (Scheiber, 1943). In 1941, Fryling (Fryling, 1944) described a very useful method for carrying out emulsion polymerisation experiments in 10-gram systems and, together. with Harrington (Fryling and Harrington, 1944), investigated the pH of mixtures of aqueous soap solutions and substituted ethylenes, such as acrylonitrile, styrene, etc.; they concluded that the monomer which was solubilized in the McBain layer micelles (McBain, 1942; McBain and Soldate, 1944) was very likely to be the most important site for initiation of polymerisation. Hohenstein, Mark, Siggia, and Vingiello (Hohenstein, 1945; Hohenstein, Siggia, and Mark, 1944a; Hohenstein, Vigniello, and Mark, 1944b) studied the polymerisation of styrene in aqueous solutions without soap and in aqueous emulsions in the presence of soap. At the New York meeting of the American Chemical Society in September, 1944, Vinograd delivered three excellent lectures (Vinograd, Fong, and Sawyer, 1944) on the polymerisation of styrene in aqueous suspension and emulsion. At the same meeting, Frilette (Frilette, 1944) reported on experiments on the polymerisation of styrene in very dilute aqueous systems.

In 1945, Hohenstein, Siggia, and Mark (Siggia, Hohenstein, and Mark, 1945) published an article on the polymerisation of styrene in agitated soap emulsions, and Hughes, Sawyer, and Vinograd (Huges, Sawyer, Vinograd, 1945), Harkins (Harkins, 1945a), and Harkins with a number of collaborators (Harkins 1945b) contributed very valuable x-ray data on the McBain micelles (McBain, 1942) before, during, and after polymerisation. In the same year, two very interesting articles appeared, by Kolthoff and Dale (Kolthoff and Dale, 1945) and Price and Adams (Price and Adams, 1945), on the influence of catalyst concentration on the initial rate of polymerisation; and Montroll (Montroll, 1945) developed a general phenomenological theory of processes during which diffusion and chemical reaction cooperate in the formation of large molecules.

A large amount of basic research was carried out on all phases of emulsion polymerisation as part of the government rubber program, most of which has not yet (1946, ed. note) been released for publication. [The paper of Kolthoff and Dale (Kolthoff and Dale, 1945) was part of this program and was published with the permission of the Rubber Reserve Company, Washington, D. C.] One can, therefore, look forward in the not too distant future to many informative articles in this field.

As far as our present knowledge goes, it seems appropriate to distinguish between the following three types of vinyl polymerisation of diluted monomers:

1. Polymerisation in *homogeneous solution* in which the monomer, all species of the polymer molecules, and the initiator (catalyst) are soluble in the diluting liquid (e.g., styrene polymerisation in toluene with benzoyl peroxide). If the solution is sufficiently dilute, such a process begins and ends in a completely homogeneous system with a dilute molecular solution of the monomer at the beginning and a dilute molecular solution of the various species of the polymer at the conclusion of the reaction. A number of recent papers (see original publication) describe studies on olefin polymerisations under such conditions. If the system is not sufficiently dilute, toward the end of the reaction a concentrated polymer solution is obtained containing aggregations and entanglements of the macromolecules which represent a certain deviation from molecularly homogeneous dispersion. A particularly interesting case of solution polymerisation occurs if the monomer is soluble in the liquid, whereas certain species of the polymer, namely, those of higher degrees of polymerisation, are insoluble in it. The polymerisation of styrene, the copolymerisation of vinyl chloride and vinyl acetate in methanol, and the polymerisation of acrylonitrile in water are examples of reactions that start in a molecularly homogeneous phase but continue and end in a system consisting of a swollen gel and a supernatant liquid solution.
2. Polymerisation in heterogeneous suspension, in which the monomer is mechanically dispersed in a liquid, not a solvent for it and for all species of polymer molecules. The initiator is soluble in the monomer. In such cases polymerisation takes place in each monomer globule and converts it gradually into a polymer "bead" or "pearl"; the liquid plays only the role of a carrier, which favours heat transfer and agitation but does not interfere with the reaction as such. The polymerisation of styrene or dichlorostyrene in aqueous dispersion is an example of such a process. It must, however, be noted that the monomer is never completely insoluble in any carrier liquid and, in certain cases, such as bead polymerisation of vinyl acetate in water, is even fairly soluble in it. These reactions are, then, processes in which solution polymerisation and suspension polymerisation occur simultaneously in the different phases of the heterogeneous system-the former in the aqueous, the latter in the monomer, phase. The amount of polymer formed in each phase depends upon the solubility of the monomer in water, and upon the distribution of the catalyst or catalysts in the two phases. If the monomer is only moderately soluble in water, the amount of polymer formed in the aqueous phase is not considerable but its degree of polymerisation is low, because of the small monomer

concentration, and one obtains a polymer containing a noticeable amount of low molecular weight species. In fact, polymers prepared under such conditions occasionally show a molecular weight distribution curve with two distinct peaks, the smaller of which corresponds to the lower molecular weight. This effect is exaggerated if, for some reason, one increases the solubility of the monomer in the aqueous phase by the addition of organic solvents like methanol, alcohol, or acetone. This consideration shows that suspension polymerisation can be a fairly complex process the complete elucidation of which is rather difficult. In the articles which attempt to contribute quantitative results (Hohenstein, 1945; Hohenstein, Vigniello, and Mark, 1944b; Vinograd, Fong, and Sawyer, 1944), monomers and catalysts were selected which are only very slightly soluble in water and probably approach the case of a heterogeneous suspension polymerisation to a fair degree. Another factor which may complicate the elucidation of suspension polymerisation is the use of suspension stabilizers, which may solubilize part of the monomer and, therefore, create an intermediate case between solution and suspension polymerisation.

3. Polymerisation in emulsion, in which the monomer is: (a) *dispersed in monomer droplets* stabilized by an adsorbed layer of soap molecules (Fryling and Harrington, 1944; Vinograd, Fong, and Sawyer, 1944; Kolthoff and Dale, 1945; Price and Adams, 1945; Siggia, Hohenstein, and Mark, 1945); (b) solubilised in the soap micelles (McBain, 1942; McBain and Soldate, 1944; Harkins, 1945a) which exist in an aqueous soap solution of sufficient concentration; and (c) molecularly dissolved in the water. The amount of polymer formed in the droplets, in the micelles, and in solution will depend upon the way in which the monomer and catalyst are distributed in the *three* existing phases: the monomer phase, the soap micelle phase, and the water phase – and possibly also upon the accessibility and reactivity of the monomer in these three phases. In certain aqueous soap emulsions, such as styrene, dichlorostyrene, or isoprene, the amount of molecularly dissolved monomer is small and, therefore, the reaction will occur preponderantly either in the monomer droplets or in the soap micelles. If the polymer formation occurs preponderantly in the micellar phase, one is inclined to speak of a typical *emulsion* polymerisation. If, however, polymerisation takes place to a considerable extent both in the monomer droplets and the soap micelles, the case is intermediate between *suspension* and *emulsion* polymerisation. There also exist emulsion polymerisations (vinyl acetate, acrylonitrile) in which the *monomer* is substantially soluble in water and a reaction which is a superposition of solution, suspension, and emulsion polymerisation is expected.

These brief remarks suffice to show that one must select the system for investigation with care if complications and overlapping between different types of reactions are to be avoided.

This citation tells much about the early start of our understanding of the emulsion polymerisation mechanisms, even though, at that time, a quantitative theory was not yet developed. Basic understanding of the relative importance of the aqueous, organic and micellar phases was also somewhat lacking. These topics will be treated thoroughly throughout this book. At this point, however, the very important, so-called, GR-S recipe for synthetic rubber must be mentioned. Even if the production of synthetic latexes was known in the 1930s, the cost was higher that that of natural rubber. However, the need for large amounts of synthetic rubber arose as a result of World War II, after the Japanese conquests in South East Asia. The secret United States Synthetic Rubber Program (1939–1945) resulted in the famous GR-S rubber recipe, the so-called "Mutual" recipe that was used for the first time by the Firestone and Goodrich companies in 1942 and adopted for large-scale production in early 1943 (Bovey *et al.*, 1955):

Table 1.1 *A typical recipe for a styrene-butadiene latex.*

Ingredients	Parts by weight
Butadiene	75
Styrene	25
Water	180
Soap	5.0
n-Dodecyl mercaptan	0.50
Potassium persulfate	0.30

The American Chemical Society has declared this program as one of their "historic chemical landmarks". By 1945, the United States was producing about 920 000 tons per year of synthetic rubber, 85% of which was GR-S rubber. As we see (Table 1.1), the recipe is quite simple, and each ingredient has it specific function. The 3 : 1 ratio (5.8 : 1 molar) of butadiene and styrene gives the polymer its useful physical properties. In addition, butadiene does not homopolymerise readily, and the copolymerisation with styrene gives the process a "normal" rate. The soap controls the nucleation and stabilization of the particles, whereas the potassium persulfate acts as initiator. The traditional soap used was a commercial fatty acid soap containing mainly C_{16} and C_{18} soaps, but the effect of different soaps from C_{10} to C_{18} was investigated. The role of the mercaptan has been debated, and it has been frequently stated that the mercaptan and persulfate form a redox couple. However, the most accepted role of the mercaptan is as an inhibitor and chain transfer agent: to inhibit the formation of cross-linked, *microgel*, particles during the polymerisation. When the rubber is used in end products, such as car tyres, and so on, it is cross-linked in its final shape, a process that is called vulcanisation. This utilizes the tetra-functionality of the butadiene (two double bonds), but this cross-linking is, naturally, not wanted during the emulsion polymerisation. Adding (amongst others) mercaptan to avoid this cross-linking action thus controls the process. The process is also stopped at 60–80% conversion and the monomers removed by flash distillation. The GR-S rubber recipe has been modified from the "Mutual" recipe over the years. Especially, the lowering of the polymerisation temperature to 5 °C has improved the process by increasing the achievable molecular weight. That again makes it possible to "extend" the polymer by adding inexpensive petroleum oils and rosin derivatives. Because persulfate is too slow as an initiator at such low temperatures, the development of more active (redox) initiator systems was required.

In Germany, production of synthetic rubber had also been developed during the War. These products were named Buna S (a butadiene-styrene copolymer) and Buna N (a butadiene-acrylonitrile copolymer) and these products were patented by the I.G. Farbenindustrie in the 1930s. In 1937 the annual German production of Buna S was 5000 tons. Though these were much more expensive than natural rubber, production was pushed ahead for the very same reasons the American synthetic rubber program was accelerated – the uncertain access to natural rubber under war conditions. After the war, the know-how that had been developed both in Germany and in the US was used in many other industrial emulsion polymerisation systems that began their development both before and after the war. Another example of this is neoprene rubber, poly(chloroprene) [poly(2-chloro-1,3,butadiene)]. Because neoprene is more resistant to water, oils, heat and solvents than natural rubber, it was ideal for industrial uses, such as telephone wire insulation and gasket and hose material in automobile engines. Neoprene was developed at DuPont's research laboratory for the development of artificial materials; founded in 1928, the laboratory was being led by the famous chemist Wallace Hume Carothers. DuPont started production of this polymer in 1931, but improved both the manufacturing process and the end product throughout the 1930s. Elimination of the disagreeable odour that had plagued earlier varieties of neoprene made it popular in consumer goods like gloves and shoe soles. World War II removed neoprene from the commercial market, however, and although production at the Deepwater plant was stepped up, the military claimed it all. DuPont purchased a government-owned neoprene plant in Louisville, Kentucky, to keep up with increasing demand after the war.

The emulsion polymerisation of PVC was patented by Fikentscher and coworkers at the I.G. Farben already in 1931 (Fikentscher, 1931). PVC is a polymer that has many useful properties, among others very low permeability of small molecules such as air (oxygen) and water. In many examples, the use of water-soluble initiators and a range of emulsifiers including sulfonated organic derivatives, such as the sodium salts of Turkey Red Oil and diisobutylnaphtalene sulfonic acid, were described. This was the birth of the modern PVC emulsion polymerisation process and further development work continued both in Germany and in the US during the 1930s and eventually in the UK in the late 1930s. Because of Germany's lead in this field, the plants there continued with the emulsion process for most applications for a longer period after World War II, whereas in the US and the UK production methods changed from emulsion to suspension polymerisation for all but the plastisols and special applications. Polymerisation of PVC was also started as an emulsion process in Sweden in 1945 by (what became) KemaNord and in Norway in 1950 by Norsk Hydro. This was the origin of the Norwegian work with emulsion polymerisation (and also that of the present author).

We see from the citation above that Mark and Hohenstein mention the monomers styrene, dichlorostyrene, isoprene, vinyl acetate and acrylonitrile. After the invention of emulsion polymerisation, many monomers were investigated, but not all of these were of commercial interest. Further development of emulsion polymerisation of vinyl acetate and the acrylates, especially for paint and binder applications, first speeded up after the War, when more advanced copolymers were developed. This development is described further in Chapter 2.

In academia, these developments were closely paralleled by increasing understanding of the mechanistic and, subsequently, kinetic theories. Among these, the Harkins and Smith–Ewart theories are the most prominent and important. The Harkins theory has already been mentioned in the citation from Hohenstein and Mark above (Hohenstein and Mark, 1946). It appeared in a series of publications between 1945 and 1950 (Harkins, 1945a, 1945b, 1946, 1947, 1950). Harkins' interest was chiefly the role of the surface-active substances in emulsion polymerisation. The Harkins theory is, therefore, a qualitative theory, but it is often looked upon as the starting-point of all "modern" theories of emulsion polymerisation, see Figure 1.1. The essential features of the theory are (Blackley, 1975):

1. The main function of the monomer droplets is to act as a reservoir of monomer.
2. The principal locus of initiation of polymer particles is monomer swollen emulsifier micelles

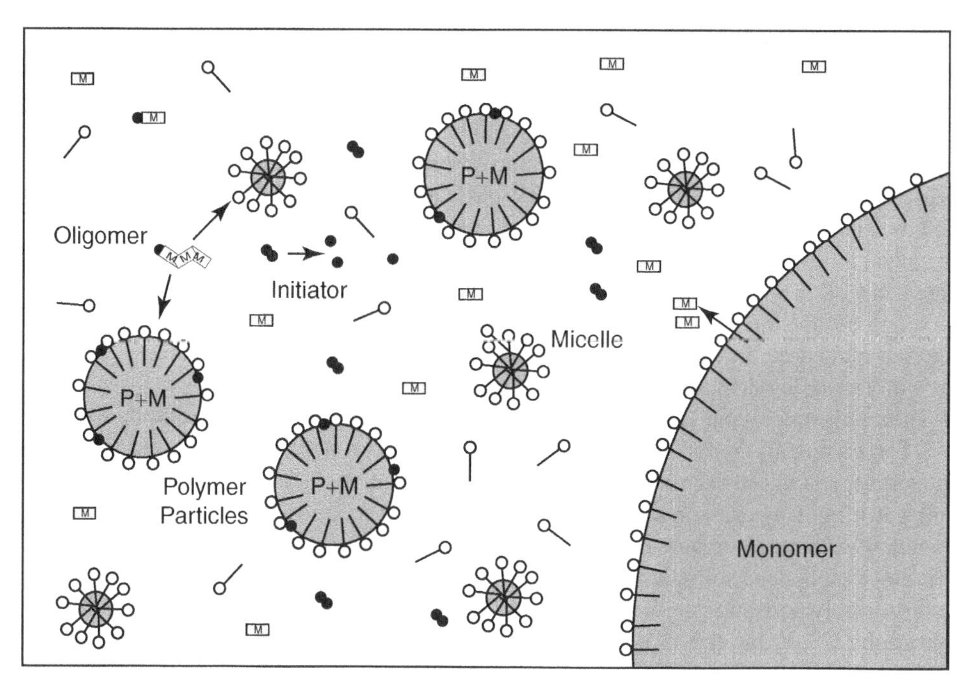

Figure 1.1 *Cartoon of an emulsion polymerisation based on the Harkins theory. Ingredients are monomer, surfactant, and initiator. The surfactant forms micelles and the initiator is soluble in water. This "snapshot" is taken during* Interval I, *when particles are being formed and monomer is present both as free droplets, in aqueous solutions, in micelles and in already formed polymer particles. The surfactant is distributed as dissolved molecules, in micelles, adsorbed on polymer particles and on monomer droplets (to a lesser degree).*

3. The main locus of polymerisation is the initiated polymer particles. During polymerisation monomer diffuses through the continuous phase and particles grow by this adsorption and subsequent polymerisation.
4. A small amount of particle nucleation can occur within the true aqueous phase. The significance of this nucleation is considered less and less important as the amount of soap increases.
5. Growth of the polymer particles leads to an increase in surface area. This increase leads to the adsorption of soap from the aqueous phase, which again leads to dissolution of micelles.
6. Nucleation stops when no more micelles are present and the major part of polymerisation takes place in the polymer particles.
7. Continual absorption of monomer into growing polymer–monomer (swollen) particles leads to the disappearance of the monomer droplets as a separate phase. This happens after micellar soap has disappeared, and the system therefore only consists of monomer–swollen polymer particles.

Harkins did not state explicitly how the water soluble initiator would be able to initiate the swollen monomer, and therefore "oil-rich" soap micelles. This detailed mechanism was somewhat unclear at the time (maybe still is), but it has been assumed that the initial polymerisation takes place within the aqueous phase. How these polymers (oligomers) would be able to go into the micelles was not discussed. Harkins based his theory both on earlier opinions, as described above, and on experimental evidence. Building on the Harkins theory, the Smith–Ewart theory, which appeared in 1948, was the major leap forward in emulsion polymerisation. This is described further in Section 1.2.2 below.

1.2 The Second Half of the Twentieth Century

Following the pioneering work on synthetic rubber, and also other earlier patents such as that for neoprene and PVC, several new industrial processes were developed utilizing emulsion polymerisation. In the second half of the twentieth century emulsion polymerisation was developed to high sophistication, both experimentally and theoretically. It has indeed reached such a level of sophistication that it is called by many a "ripe" technology. This means that the major problems, both experimental and theoretical, have been solved and that current activities are concerned with reaping the profits and refining both products and theories. However, new developments are still possible, and theories are maybe not as solidified as may be imagined. In this section, the stages leading up to the present situation will be considered.

1.2.1 Product Development

As a part of the interest in more advanced applications of emulsion polymers, many have investigated the different ingredients in the polymerisation. In the beginning, different emulsifiers and different initiators were being developed. The GR-S recipe was, as mentioned, eventually modified with different soaps and with redox initiator systems in order to make it possible to run the process at lower temperatures. Earlier, other emulsifier systems, especially natural resins, had been tested. In the second part of the century, non-ionic emulsifiers were becominging more important. Among the most popular of these were the "Triton" and similar emulsifiers. These are nonyl- or octyl-phenols modified by ethylene oxide to give poly(ethylene glycol)-based emulsifiers. (Because of their toxicity these are now replaced by fatty alcohol-based polymers). It was shown early that these were not efficient for particle nucleation (see below), but excellent as emulsion stabilizers. They therefore became very popular as co-emulsifiers in addition to sulfates, sulfonates, and similar. In the same class come surface-active polymers, *protective colloids*. Many types of these have been developed, and they are used extensively in industrial production, either during polymerisation, or as post-additives to improve storage stability and other properties. There have also been efforts to develop surface-active initiators ("ini-surfs") and copolymerisable emulsifiers ("surf-mers"). The idea behind this is to anchor the stabilizing groups better to the particle surface in order to improve stability. Many research groups have been working on this during the last quarter of the twentieth century, but successful commercial products are not abundant. The reason is probably that the combination of functionalities makes the emulsion polymerisation process more difficult to control and have several unexpected side effects. For instance will surface active initiators and/or monomers influence the nucleation process and make this more difficult to control.

A post-war outgrowth of the synthetic rubber work found tremendous interest in the US for styrene butadiene rubber (SBR) dispersions for their utility in water-based latex paint. The first SBR was sold into architectural coatings application in 1948. Consumer desire for easy clean-up and new roller technology combined to make a rapid market shift. Sales of SBR latex increased extremely quickly, with 33% of solvent-based interior paints replaced by latex paint within four years of its introduction in 1947. Other polymers during the 1950s and 60s gradually replaced SBR. The deficiencies of SBR that account for this shift include colour stability and chalking. Styrene acrylics were introduced in 1953 to address some of these issues; current styrene acrylics are often sold as "modified acrylics" even though they may contain as much as 50% by weight styrene. The technology that is incorporated into acrylic gloss paints is based on over 40 years history of research and development by the world's major polymer manufacturers. The first 100% acrylic emulsion polymer developed for use as a paint binder was introduced by the Rohm and Haas Company in 1953. This company had its early business in the production and sales of Plexiglas (PMMA homopolymer), and the introduction of emulsion polymers based on PMMA (and other co-monomers) was therefore a natural development. During the last 50 years these polymers have been developed into a much diversified class of binders for all kinds of applications, including inks, industrial and maintenance finishes, floor polishes, cement modifiers, roof mastics and adhesives.

In a similar way other polymers, like PVC and poly(vinyl acetate) (PVAc) homo- and copolymers have been developed further into the wide range of products seen today. PVAc-based polymers are also used in paint binders, as well as in the very popular carpenter's glue. As paint binders, they compete with acrylates, but are less hydrolytically stable and, therefore, not as durable in moist environments and less scrub resistant. They are, however, often used in less expensive paints because of their lower cost. PVAc homopolymer emulsions began to be used in paints before the war, with one British company founded in 1939 for PVAc manufacture. After the war, development of vinyl acetate-based resins continued in Western Europe. The high T_g of PVAc homopolymer made the use of plasticizer necessary. The superior colourfastness and yellowing resistance of vinyl acetate-based resins helped drive the market in Europe away from SBR. Copolymers of vinyl acetate with acrylates, versatate, and ethylene reduced the necessity for plasticizer and enhanced performance in terms of alkali resistance, scrub, and so on. In both acrylics and PVAc-based products, development has been much concentrated on finding copolymer compositions with good application properties at the same time as giving a stable polymer latex and a controllable process. Surfactants and other additives have played a major role in this development. One example is the introduction of amino functionality in latex paints in order to improve wet adhesion properties.

In academia, as well as in some companies, new advanced types of emulsion polymer particles have been developed during the last quarter of the twentieth century. Among these are, for instance, core-and-shell particles for paint and binder applications. In order to obtain a continuous film in a dry paint, film-forming agents in the form of high boiling glycols or hydrocarbons (volatile organic compounds, VOCs) are often added. These are, however, not so environmentally friendly and also unwanted for technical reasons. Poly(acrylate) copolymers have, therefore, been developed with a soft shell polymer on top of a hard core. The technical requirements for producing and controlling such a particle structure have been the object of many scientific papers (Sundberg *et al.*, 1990; Lee and Rudin, 1992, Gonzáles-Ortiz and Asua, 1996a, 1996b, 1996c), but a predictive theory for the structure–property relationship of this type of emulsion polymer is still missing, probably because of its extremely complex nature. This has not, however, hindered industrial products based on this type of latex. Another similar product is the hollow latex particles, produced by the Rohm and Haas company (Kowalski, Vogel, and Blankenship, 1981). These are based on core-and-shell particles in which the core is an originally water-swollen polymer that is later collapsed into a void. The application of these particles is for pigment substitutes and other additives. The same company also has developed very advanced multi-lobe particles by means of multistage addition of co-monomers with subsequent phase separation into separate, but still connected spheres. They show that this type of latex gives the product especially useful rheological properties.

Core-and-shell composite particles based on inorganic cores with a polymer shell have also been investigated by several researchers, but do not seem to have reached industrial products. The reason for this is probably the high cost and possibly limited benefits of this type of latex compared to existing products. A similar type of product is composite particles based on pre-emulsified polymers, like epoxies or polyesters (alkyds), with a subsequent addition of new monomers and polymerisation. This technique is partly connected to the process of "miniemulsion" polymerisation described in the next section. A type of core-and-shell particle or at least multi-phase particle may be obtained in this type of process. However, industrial applications of this type of product are not yet found on a large scale. Applications of polymer particles, mainly made by emulsion polymerisation, in the biomedical field were concentrated initially in the areas of blood flow determination and *in vitro* immunoassays. Microspheres have been employed for the determination of myocardial, cerebral and other blood flow and perfusion rates. Polymer particles and latexes in particular have been extensively used in immunoassays, starting in 1956 with the development of the Latex Agglutination Test (Singer and Plotz, 1956). Later a significant number of additional applications of polymer particles in the biomedical field emerged. These applications exploit advances in polymer chemistry in combination with new developments in the field of biotechnology. Some of these applications are solid-phase immunoassays, labelling and identification of lymphocytes, extracorporeal and haemoperfusion systems, and drug delivery systems. Magnetic microspheres have also been introduced by several companies for cell separation and other therapeutic as well as diagnostic applications. This technology has obtained enormous popularity since around 1990 (see also Chapter 10).

1.2.2 Kinetic Theory

Definitely the most important theory in emulsion polymerisation is the Smith–Ewart theory. This theory was first published in 1948 (Smith and Ewart, 1948) and since then has been the subject of continuing discussion and refinement. The theory is based on the Harkins mechanisms and then tries to predict the rate of reaction and its dependence upon the concentrations of the main components of the system. The rate of reaction is considered to be equal to the total rate of polymerisation in the nucleated soap micelles, which then have been converted to polymer particles. There is no polymerisation in the aqueous phase or in the monomer drops. The total rate can then be set equal to the rate in each polymer particle, multiplied by the number of particles:

$$R_p = -\frac{d[M]}{dt} = k_p[M]_p\bar{n}\frac{N}{N_A} \tag{1.1}$$

Here [M] is the total amount of monomer in the system, k_p is the propagation rate constant, $[M]_p$ is the concentration of monomer in the latex particles, $\bar{n}$ the average number of radicals in the particles, N the total number of particles, and N_A is Avogadro's number.

The quantitative theory is therefore centred on predicting (i) the number of particles nucleated and (ii) the rate of polymerisation in each particle. The Smith–Ewart theory operates in the three *intervals* of the polymerisation process, and defines three *cases* for the kinetics. The intervals correspond to the three stages in the Harkins theory: *Interval I* is the nucleation stage where micelles are present and the particle number increases; *Interval II* corresponds to the stage when the particle number is constant and free monomer drops are also present, and *Interval III* is the last part of the polymerisation when the monomer drops have disappeared. Smith and Ewart developed an expression for the particle number created by nucleation in the soap micelles that is still considered essentially correct, within its limits (meaning that monomers, surfactants and generally conditions can be found when the S–E theory is *not* correct and that our understanding today is more detailed). The expression for the particle number, N, is

$$N = k(\rho_i/\mu)^{2/5}(a_s[S])^{3/5} \tag{1.2}$$

Here ρ_i is the rate of initiation, μ is the volumetric growth rate, $\mu = dv/dt$, a_s is the specific surface area of the emulsifier ("soap") and [S] is the concentration of emulsifier (also denoted as [E]). The constant k has a value between 0.37 in the *lower limit* and 0.53 in the *upper limit*. The two *limits* are obtained by deriving the particle number under slightly different suppositions: In the *upper limit* the rate of nucleation is constant and equal to the rate of radical generation, ρ_i, up to the point where there are no micelles left. This means that the particles implicitly are not assumed to absorb any radicals during the nucleation period, or that at least this rate is negligible. This may or may not be true, as discussed later (Chapter 3). On the other hand, in the *lower limit* the particles adsorb radicals at a rate according to their surface area. This, naturally, leads to a lower particle number, but the two limits, surprisingly enough, only differ by the constant k and are otherwise equal! The mathematics involved in deriving these equations is quite straightforward in the case of the upper limit, but somewhat more involved in the case of the lower limit. Smith and Ewart did this derivation very elegantly and later work, both analytical and numerical has shown Equation 1.2 to be a limiting case of a more general solution for the particle number.

The second part of the Smith–Ewart theory concentrates on calculating the average number of radicals per particle. As long as the monomer concentration in particles is constant, as may often be the case in *Interval II*, this number then yields the rate of polymerisation. Smith and Ewart did this by means of a recursion equation that is valid for the situation prevailing after particle formation is finished,

$$\begin{aligned}&\rho_A N_{n-1}/N + (n+1)(k_s a_s[S]/v)N_{n+1} + (n+2)(n+1)(k_t^*/v)N_{n+2}\\&= \rho_A N_n/N + n(k_s a_s[S]/v)N_n + n(n-1)(k_t^*/v)N_n\end{aligned} \tag{1.3}$$

where ρ_A is the total rate of radical absorption or entry in the particles (in molecules per unit volume), k_s is the rate "constant" for desorption or exit of radicals from the particles, a_s the specific surface area and k_t^* the termination constant in latex particles. The particle number N_n denotes the number of particles with n-occupancy of radicals. Smith and Ewart then discussed three *limiting cases:* Case 1: $\bar{n} \ll 0.5$, Case II: $\bar{n} = 0.5$, and Case 3: $\bar{n} \gg 0.5$. Case 2 is that which has later been most generally known as the Smith–Ewart theory and is the only case that has been given a complete treatment by Smith and Ewart. The solution for this case is also obvious from simple consideration of the situation in a randomly selected particle. The condition for this case is

$$k_s a_s[S]/v \ll \rho_A/N \ll k_t^*/v \tag{1.4}$$

This means that the rate of adsorption of radicals in polymer particles is much larger than the rate of desorption (so the latter can be neglected) and much lower than the rate of termination. The kinetic conditions may, for this case, be easily deduced by regarding the adsorption and termination processes in a single particle. When a radical enters a "dead" particle ($n = 0$), it becomes a "living" particle ($n = 1$), and polymerisation proceeds with the present monomer. This situation is maintained until another radical enters ($n = 2$). Because the rate of termination is high, the two radicals terminate immediately, and the particle is again "dead". Due to the random nature of the adsorption process (diffusion), the particle is switched on and off at random intervals, but as a time average each of the two states are present half of the time, or the half is present all the time, that is, $\bar{n} = 0.5$. This number has become more or less synonymous with the Smith–Ewart theory, but is only a special case.

The two other cases occur when the left side (Case 1) or the right side (Case 3) of Equation 1.4 are not fulfilled, giving negative or positive deviations from the 0.5 value. Smith and Ewart did not treat these cases completely, re-absorption of radicals was only included for the case when termination in the particles is dominating (their *Case 1B*) and particles with more than one radical (Case 3) were only considered when desorption is negligible. Also they did not give the full solution of the recursion equation (1.3). This was not solved until 1957 by Stockmayer (Stockmayer, 1957). If desorption is neglected, the solution is

$$\bar{n} = \frac{I_0(a)}{I_1(a)} \tag{1.5}$$

where I_0 and I_1 are Bessel functions of the first kind, and

$$a = \sqrt{8\alpha}, \quad \alpha = \frac{\rho_A}{N k_t^*/v} \tag{1.6}$$

Stockmayer also presented solutions for the case that takes into account desorption of radicals. This solution, however, is wrong for the most important range in desorption rates. But Stockmayer's solution(s) led the way for the possibility of exact mathematical solution of emulsion polymerisation kinetics at a time when digital computers were not yet very important in chemical computations. The general solution when desorption is taken into account was presented by O'Toole (O'Toole, 1965). He applied a modified form of the

Smith–Ewart recursion equation that gave the solution

$$\bar{n} = \frac{a}{4}\frac{I_m(a)}{I_{m-1}(a)} \tag{1.7}$$

where the dimensionless parameter m is given by $m = k_d/(k_t^*/v)$, that is the ratio between the desorption and termination rates. Here, the Smith–Ewart desorption "constant" $k_s S/v$ has been replaced by k_d, signifying that the desorption rate must not necessarily be proportional to the particle surface area. In addition, desorption also normally would only happen to monomer (or other small) radicals produced by chain transfer; k_d will, therefore, also include the chain transfer constant. We see that when desorption is zero, $m = 0$, and O'Toole's solution is equivalent to Stockmayer's. O'Toole used radical occupancy probabilities in the modified recursion equation, and was thus able to compute the probability distribution functions that have importance for computing the molecular weight distribution. However, neither Stockmayer nor O'Toole took into consideration the fate of the desorbed radicals. This was the main objection of Ugelstad and coworkers (Ugelstad, Mørk, and Aasen, 1967; Ugelstad and Mørk, 1970) when presenting their theory in 1967. Their main incentive was that the kinetics of PVC emulsion polymerisation did not fit the Smith–Ewart theory. First, they found a very low value of $\bar{n} < 0.5$ and secondly, the Smith–Ewart Case 1 kinetics did not fit either. Ugelstad's argument was that ρ_A and ρ_i cannot be treated as *independent* parameters as in both Stockmayer's and O'Toole's solutions, but that they are connected by processes in the continuous (water) phase. The desorbed radicals may be re-absorbed, either before or after having polymerised to some degree in the continuous phase, or they may terminate there. Ugelstad therefore introduced an additional equation for taking these processes into account in a simplified fashion:

$$\rho_A = \rho_i + \sum k_d N_n n - 2k_{tw}^* [R\cdot]_w^{*2} \tag{1.8}$$

Here, k_{tw}^* is the termination constant and $[R\cdot]_w^*$ is the radical concentration in the water phase. This equation is brought into dimensionless form by dividing by Nk_t^*/v and by realizing that $\bar{n} = (\Sigma nN)/N$, so that resultant equation is

$$\alpha = \alpha' + m\bar{n} - Y\alpha^2 \tag{1.9}$$

This treatment only introduced one additional dimensionless parameter, Y, which is a measure of the degree of water phase termination. $Y = 0$ therefore represents the case when all desorbed radicals are re-absorbed. The disadvantage of this treatment is that a general solution cannot be made without the use of numerical methods, that is, computers. Aasen also simplified O'Toole's Bessel function expression for $\bar{n}$ to a simple converging continued fraction (Ugelstad, Mørk, and Aasen, 1967) that can be solved simultaneously with Equation 1.9. This equation is

$$\bar{n} = \cfrac{\alpha}{m + \cfrac{2\alpha}{1 + m + \cfrac{2\alpha}{2 + m + \cfrac{2\alpha}{3 + m + \dots}}}} \tag{1.10}$$

If $m = 0$ and α is small ($\ll 1$), this equation is seen to give the famous $\bar{n} = 0.5$. For a given system the rate of initiation and thus ρ_i, and correspondingly α', is an independent

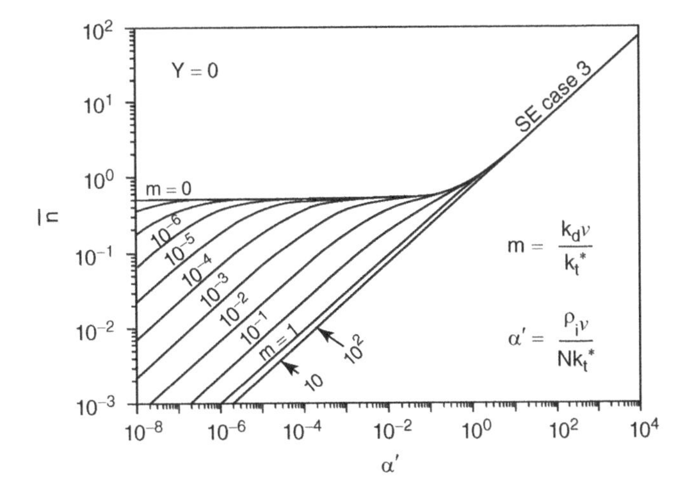

Figure 1.2 *Average number of radicals per particle, $\bar{n}$, calculated from the theory of Ugelstad and coworkers. $\bar{n}$ is given as a function of the dimensionless parameters α' and m when there is no termination in the aqueous phase ($Y = 0$). Case 2 of the Smith–Ewart theory, $\bar{n} = 0.5$, is described by the horizontal line where $m = 0$.*

variable; and Equations 1.9 and 1.10 can be solved by successive approximations to given$\bar{n}$. These equations give the famous curves for $\bar{n}$ as a function of α', as shown in Figure 1.2 for the most simple case when $Y = 0$.

Solutions for other values of Y are given by Ugelstad and Hansen (Ugelstad and Hansen, 1976) in their 1976 review of emulsion polymerisation. In this way, the complete solution to the steady-state Smith–Ewart based theory is available. Ugelstad and coworkers (Ugelstad *et al.*, 1969; Ugelstad and Mørk, 1970) found the theory to fit the emulsion polymerisation kinetics of PVC to a very high precision and later found this also to be the case for bulk polymerisation of the same monomer because of PVC's low solubility in its monomer. One important factor in these calculations was the particle size dependence of the desorption constant. The surface area/volume dependence assumed by Smith and Ewart was discussed by Nomura, Harada and coworkers in 1971 (Nomura *et al.*, 1971; Harada *et al.*, 1972). They concluded that the desorption constant should be proportional to the particle radius/volume, that is, inversely proportional to the square of the particle size. This dependence was used successfully by Ugelstad and coworkers in their calculations.

Around the same time as Ugelstad and coworkers introduced their theoretical and experimental results, Gardon also published in a series of papers (Gardon, 1968, 1970a, 1970b) a re-examination and recalculation of the Smith–Ewart theory. Some of the results that were obtained were more special solutions of the more general solutions developed by Stockmayer–O'Toole and Ugelstad and some assumptions have later been disputed. One of these assumptions is that the rate of adsorption of radicals in micelles and particles is proportional to their surface area. This is the same assumption that was made by Smith and Ewart, and was derived by Gardon from geometric considerations assuming the radicals move in straight lines to collide with the surface. For that reason this model is also called the *collision model*. However, this has been shown to be correct for only a limited range of conditions, especially because Gardon did not take the concentration gradient necessary for mutual diffusion into consideration, the so-called *diffusion model* that gives proportionality with the particle radius rather than with its surface area. Also Gardon did not include desorption and re-absorption of radicals. Much of Gardon's semi-analytical computations has later been made needless by numerical computer technology. It may be said then that the Gardon theory has not been applied very much in the later years.

During the last quarter of the twentieth century several groups have been occupied with the kinetic theory of emulsion polymerisation, bringing it to still higher degrees of sophistication by investigating different details that had not earlier been considered. Among the most well-known of these groups are Nomura and coworkers in Japan and Gilbert, Napper and coworkers in Australia. One of their main contributions has been the independent measuremens and estimation of many of the rate constants involved in initiation, propagation and termination, in addition to producing advanced models with computer simulation. Among these are non-steady state reaction kinetics, and the development of particle size and molecular weight distributions. In order to have a realistic model that can be used for prediction and/or process control it is necessary to have good independent estimates for the constants in order to avoid what is popularly referred to as "curve fitting". Through a series of publications they have investigated many aspects of these problems. Also Asua and his coworkers in Spain have contributed to more detailed descriptions of the mechanisms. Among other contributions, primarily in process control and reaction engineering, they have published a more detailed description of the desorption mechanism, taking also the reactions in the aqueous phase into consideration (Asua, Sudol, and El-Aasser, 1989). They have also published work on general parameter estimation (De la Cal, Adams, and Asua, 1990a; De la Cal *et al.*, 1990b). More thorough descriptions of the more recent work are given in Chapter 3.

The first part of the Smith–Ewart theory, the nucleation part (Interval I) was not in the beginning debated to the same degree as the rate of polymerisation. This may be because people found that observations agreed with the theory, or maybe rather that they did not. Observations have not always agreed with the exponents 0.4 and 0.6 predicted by the theory in Equation 1.2 nor with the absolute particle number, but this has been found to be very dependent on the specific system studied. Experiments with more water-soluble monomers, such as those by Priest (Priest, 1952) and Patsiga (Patsiga, Litt, Stannett, 1960) with vinyl acetate and Giskehaug (Giskehaug, 1965) with PVC did not fit this theory. In the beginning there were some researchers who performed modifications and recalculations of the Smith–Ewart theory (Gardon, 1968; Parts, Moore, and Watterson, 1965; Harada *et al.*, 1972), and found that some of the details of the theory had to be modified. Parts *et al.* (Parts, Moore, Watterson, 1965) proposed, for instance, that in order to explain the experimental particle numbers, the absorption efficiency of radicals in micelles is lower than in particles. The particle numbers (or more correctly, particle sizes) calculated by Gardon were found to describe some experimental results for styrene and methyl methacrylate fairly well, whereas other data on particle numbers were 2–3 times lower than predicted. Another feature of the Smith–Ewart theory is that the reaction rate at the end of Interval I is expected to be higher than the steady-state value of 0.5, but there is little evidence for such a maximum in rate. There was, therefore, need for a more detailed description of the conditions during Interval I. Objections to Harkins and subsequently the SE theory also appeared for other reasons: Particles can be formed and stabilized even in systems with no micelles (below the critical micelle concentration, CMC) and even in systems completely without emulsifier. Roe (Roe, 1968), in a well-known article, showed experimental evidence that in a mixture of ionic (sodium dodecyl sulfate, SDS) and non-ionic emulsifiers the particle number is not dependent on the total number of micelles, but rather only on the number of SDS micelles. That means that the non-ionic micelles do not participate in particle formation, at least not to the same degree as the ionic. Roe then went forward and re-derived the Smith–Ewart expression for the particle number, Equation 1.2, on a pure non-micellar basis. The quantity S was then redefined to be the total surface area of emulsifier available for particle stabilization, but apart from that, all parameters and conditions were the same. Roe proposed to use "some sort of adsorption isotherm" to better describe the role of the emulsifier, but did not propose any quantitative equations for such isotherms. Roe's considerations could, therefore, explain some of the controversies of the Harkins theory, but still many questions were left, because a detailed description of the process of so-called homogenous nucleation was not given.

The derivation of a separate theory for homogenous nucleation was started by Fitch and coworkers (Fitch, Prenosil, and Sprick 1969; Fitch and Tsai, 1971) who worked with methyl methacrylate. They based their qualitative description on that of Priest (Priest, 1952) where a growing (oligomeric) radical in the aqueous phase can self-nucleate when reaching a certain chain length, the so-called critical chain length. Fitch and coworkers both determined this chain length for MMA and derived an expression for the particle number. This expression was based on the finding that the rate of polymerisation of MMA in Interval I could be described by homogenous polymerisation in the aqueous phase during the whole nucleation period for initiator concentrations below 10^{-3} M. In their model they used non-steady state homogenous kinetics. They also based the rate of adsorption in particles on Gardon's collision model. They found that the expression they derived gave a good prediction of the particle number when high amounts of emulsifier were used. In many systems, however, it has been shown that the collision model is incorrect, and Fitch and Shih (Fitch and Shih, 1975) found later that the diffusion model was more correct for seeded nucleation experiments (it was shown later by the present author, however, that both models may be correct, depending on the conditions). The work on the theory for homogenous nucleation was continued at that time (1975) by Hansen and Ugelstad (Hansen and Ugelstad, 1978, 1979a, 1979b, 1979c) based on Fitch and Tsai's ideas. They derived an expression for the rate of adsorption of radicals in micelles and particles that can take into consideration both reversible diffusion and electrostatic repulsion. By means of this expression the low capture efficiency of micelles that was postulated by Parts and coworkers (Parts, Moore, and Watterson, 1965) could be explained, as well as many other special cases, such as the possible validity of both the diffusion and collision theories under different conditions. They also developed an expression for the particle number in the case where all nucleated particles are stable and found this to fit well the observed data for styrene. They solved their model by numerical integrations by means of digital computers that were beginning at that time to become useful for advanced simulations. They also formulated expressions to calculate the so-called limited coagulation in order to explain the much lower particle numbers formed in systems with low or zero emulsifier concentration. Because of the computational requirements of their model, however, they were not at that time able to follow this model to any equilibrium situation. Fitch later named this combined model the HUFT (Hansen, Ugelstad, Fitch, and Tsai) model, which acronym has obtained some popularity. Hansen has in later publications (Hansen and Ugelstad, 1982, Hansen, 1992a, 1992b) described the consequences of the model in more detail.

Fitch and coworkers (Fitch and Watson, 1979a; Fitch *et al.*, 1984) later investigated the limited coagulation process. They performed coagulation experiments with MMA, using photo-initiation of homogeneous solutions and light scattering detection. Fitch and Watson used flash initiation and investigated the subsequent coagulation process. They clearly showed that coagulation takes place below the CMC and they could calculate the stability ratio as a function of surfactant (SDS) concentration. The Australian group, Feeney, Lichti, Gilbert and Napper (FLGN) initiated and continued work on particle nucleation during the 1980s. Especially, they have contributed with new experimental work, and this has been followed up partly by new theoretical ideas. Traditionally the comparison between theory and experiment with respect to particle nucleation is done by comparing (final) particle numbers and/or the rate of polymerisation. FLGN argue that several other parameters provide additional and more sensitive information about the nucleation mechanism. Such parameters are the particle size distribution, molecular weight distribution (also in the aqueous phase), and the rate parameters for absorption (entry) and desorption (exit). By measuring the rate constants explicitly, they were aiming to avoid the "curve fitting" dilemmas that are inherently present in the theoretical calculations cited above. They measured the particle size distribution as a function of time (Lichti, Gilbert, and Napper, 1983; Feeney, Napper, and Gilbert, 1984), and from the observation that these distributions are positively skewed, they concluded that the particle formation rate must be an increasing (or at least not decreasing) function of time, and that this may only be explained by a limited coagulation mechanism (they named this *coagulative nucleation*). That such a mechanism is active below the CMC comes as no big surprise, while it seems contradictory to other experimental and theoretical work that this should also be a governing mechanism above the CMC, especially for monomers such as styrene that adsorbs surfactants well, and emulsifiers such as SDS that form gaseous/liquid expanded layers when used alone and, therefore, have very fast adsorption/desorption kinetics. The theories of Gilbert and coworkers are further described in Chapter 3.

More recently, Tauer (Tauer and Küehn, 1995, 1997; Tauer and Deckwer, 1998) proposed an alternative framework for modelling particle nucleation in an emulsion on the basis of a combination of classical nucleation theory and the Flory–Huggins theory of polymer solutions. The basic assumption is that water borne oligomers form stable nuclei under critical conditions. The only adjustable model parameter is the activation energy of nucleation. The model allows calculation of the chain length of the nucleating oligomers, the number of chains forming one nucleus, the diameter of the nucleus, the total number of nuclei formed and the rate of nucleation. Based on the kinetic constants and model parameters, numerical results characterizing particle nucleation were calculated for the polymerisation of styrene, methyl methacrylate, and vinyl acetate as model systems. Still, this model has not been thoroughly tested, and several objections may also be raised to itsvalidity. It will remain to be seen to what degree this model will be adopted in the future. There is, however, another aspect of nucleation and kinetics that was discovered in the early 1970s: the role of the monomer droplets was reconsidered. This is described in the next section.

Interval III of the Smith–Ewart theory has maybe not been the object of the same attention as Intervals I and II. This stage, when monomer drops have disappeared, is more like a suspension or bulk polymerisation, and some of the special features of emulsion polymerisation are not so essential. However, the compartmentalization effect on the kinetics is still present, and this interval also has its own special problems when the monomer concentration and the termination constant decrease. The Smith–Ewart theory, the Stockmayer–O'Toole solution and the work of Ugelstad and coworkers mentioned above describes the kinetics in this interval as well, as long as the monomer concentration and the termination constant are accounted for. The connection between these two and their effect on the rate, and also possibly on nucleation, has been the source of separate research work. The so-called gel effect was investigated already by Gerrens in 1956 (Gerrens, 1956). He showed that the rate increase due to this effect varies with the particle size of the latex; the strongest increase is obtained with the largest particle sizes. This is a natural consequence of the rate of termination being the lowest for high particle volumes and thus the possibility for $\bar{n}$ to increase beyond 0.5 is most probable for these. Comprehensive treatments of this interval were done by Nomura and coworkers (Nomura *et al.*, 1971, Nomura *et al.*, 1975) and Friis and coworkers (Friis and Hamielec, 1973a; Friis and Nyhagen, 1973b, Friis and Hamielec, 1974a; Friis *et al.*, 1974b) in the early 1970s. Friis and Hamielec made use of kinetic results from bulk polymerisation from which they found k_t as a function of conversion. By modelling k_t versus conversion by a mathematical expression, it was possible to calculate the rate in Interval III by computer simulation. This methodology has also been the way in later work by others, where different mathematical expressions have been proposed for the termination constant.

1.2.3 Emulsion Polymerisation in Monomer Droplets

As mentioned in Section 1.2.2 the Harkins theory states that no, or at least very little, polymerisation takes place in the monomer droplets. This is essentially correct, and the reason is that the *number* of monomer droplets compared to the particles nucleated from micelles is many orders of magnitude lower. This does not mean that the monomer droplets are not initiated, however, and in many processes a few extra large particles may be observed. Also, monomer suspension polymerisation is often the source of reactor fouling. Many believe these large particles are the left-over of the monomer drops that are probably all initiated, but contribute very little to the over-all conversion because of the peculiar compartmentalization kinetics. It might be thought then, that if the monomer drops could be made smaller and thus more numerous, they might be more important in the nucleation process. This has indeed been shown to be the case.

In the late 1960s Ugelstad and coworkers were investigating an industrial PVC emulsion process that used a fatty alcohol in addition to the ordinary emulsifier in order to obtain especially large polymer particles. These large particles have advantages when used in some PVC paste products. The thought behind the process was that the fatty alcohol was causing limited flocculation of the latex and thus larger particles. The problem was, however, that the process, and especially particle size, was difficult to control. Unknown factors sometimes caused the particles to become very small, like an ordinary emulsion polymer, but it proved very difficult to discover which factors exactly were causing the problem. Every imaginable analysis was done on the ingredients, but there was no clue! It had been observed that the use of the fatty alcohol produced a much "better" monomer emulsion, but this was not connected to anything special. It was not until 1972 that Ugelstad, at that time on sabbatical at Lehigh University, proposed that the reason for the large particles could be initiation in the monomer droplets, because these were much smaller in these systems. Experiments done more or less simultaneously in Norway and the US confirmed this theory (Ugelstad, El-Aasser, and Vanderhoff, 1973; Ugelstad, Hansen, and Lange, 1974). The fine monomer emulsion has two effects: first it increases the number of monomer drops to an extent where they become comparable to (but still larger than) ordinary latex particles, secondly the greatly increased surface area causes adsorption of most of the emulsifier and leaves little left in the aqueous phase for "ordinary" nucleation. It was also shown that the reason for the reproducibility problems was the instability of the monomer emulsion (Hansen, Baumann Ofstad, and Ugelstad, 1974). The initial emulsion is produced by spontaneous emulsification by a diffusion process into small fatty alcohol/emulsifier aggregates (drops), but the emulsion is destabilized with time by Ostwald ripening because the fatty alcohol is slightly water-soluble. When the monomer emulsion is destabilized, the emulsifier concentration in the aqueous phase increases and will cause more "ordinary" nucleation, especially if the concentration exceeds the critical miceller concentration.

The conditions for droplet and ordinary nucleation were later investigated in more detail, using styrene as the monomer (Hansen and Ugelstad, 1979c). In these experiments, the monomer emulsions were produced by homogenizing the monomer with a high pressure homogeniser, rather than using a fatty alcohol and spontaneous emulsification. In order to stabilize the emulsion against Ostwald ripening, a water-insoluble substance (hexadecane) was used instead of the fatty alcohol. The advantage of using hexadecane or other paraffins is that the emulsion is much more stable because of the much lower water solubility, and the emulsifier concentration can be controlled more independently of the drop size. Another advantage is that other polymers, such as polyesters, polyamines, and so on, can also be included in the emulsified drops, and subsequently copolymerised with added monomers. It is also possible to add monomers to a homogenized emulsion of hexadecane or other substances so the monomer will swell the preformed emulsion like in a seeded emulsion polymerisation. This process was named "Method #2" by Ugelstad and was patented in 1978. The process of emulsification of the monomer and subsequently droplet initiation has been called the "miniemulsion" process by El-Aasser and has been the object of thorough investigation and numerous publications from the Lehigh group. Lately, it has also been taken up by others.

In this emulsion polymerisation process a water soluble initiator was originally used, giving the process its characteristic kinetic properties. Dependent on the type of monomer and on the drop size, all types of kinetic behaviour may be observed, but usually the drops are rather large (>1 μm), and SE Case 3 kinetics is often observed. Especially if SE Case 2 kinetics is present ($\bar{n} = 0.5$), but even in the case where $\bar{n} \gg 0.5$, there will be a narrowing of the particle size distribution for most monomers (Hansen and Ugelstad, 1979c), and this is thus a characteristic feature of the miniemulsion process. However, oil-soluble initiators may also be used in this process, and the process might then rather be named "minisupension" (or maybe "microsuspension"). Method #2 was shortly after further developed by Ugelstad and coworkers into Method #3, which has later been better known as the Ugelstad Process. This is the so-called two-step swelling process based on polymer seed particles. The intention is to get the seed particles to take up much more monomer than they would otherwise do, because of the limited free energy of mixing of monomer and polymer (mostly entropy driven). In the first step the seed particles are "activated" (swollen) by a relatively low molecular weight water-insoluble substance (for instance hexadecane) by adding a water-soluble solvent (acetone, methanol, etc.). Afterwards the solvent is removed, effectively trapping the water-insoluble substance in the seed particles. These are now able to take up much more monomer (up to ca 1000 times their volume) because of the increased entropy of mixing in the particles. By using "ordinary" monodisperse seed particles (diameter < 1 μm), much larger monodisperse particles can be produced in one polymerisation process. By repeating the process, extremely large monodisperse particles can be produced (> 100 μm). These particles have, by some, been given the name "Ugelstad particles" or "Ugelstad beads". The process has been reputed to produce large, monodisperse particles, but in itself it has nothing to do with monodispersity. It is not an emulsion polymerisation process either, because oil-soluble initiators have to be used to avoid new particle nucleation, so it is rather a peculiar suspension polymerisation. The particles have been given several additional properties, like macroporosity, magnetism, different surface coatings and so on, and have become very successful products, especially in the biomedical field. Because of this several groups have developed similar emulsion polymers, based on a variety of modifications of the process.

1.2.4 Industrial Process Control and Simulation

From the earlier days, an objective (the major objective?) of making theories of particle nucleation and growth was to use these for process development, prediction, and finally process control. With the advent of modern digital computer technology, modelling for process control was becoming more realistic. This is a relatively new technology, and has emerged during the last part of the twentieth century. Many companies that do emulsion polymerisation now have developed their own technology in this field that by its nature is regarded as confidential. The public scientific exchange of new developments, and especially clever computer control procedures and modelling, are therefore limited. Another source of separation of the industrial processes from the scientific community working in emulsion polymerisation kinetics is the difference in objectives. Kinetic models do, to a large extent, only predict the molecular level properties and not the macroscopic properties that are important for the users ("customers") – the so-called end-use properties.

The traditional emulsion polymerisation processes were run in batch reactors and even today, the majority of products are still produced in batch reactors. This is due to both the nature of the nucleation process, that cannot easily be controlled in continuous reactors without some sort of seeding to avoid oscillatory behaviour in particle size and/or molecular weight, and to such factors as co-monomer composition (in copolymerisation processes), fouling, temperature control, sensor technology, and so on. Models for emulsion polymerisation reactors have been published by several researchers. The simplest reactors to model are batch reactors that closely resemble lab reactors. One important design criterion for industrial reactors that differ from lab reactors is temperature control. When the reactor increases in size, the decreased surface/volume ratio of the vessel makes heat transfer an increasing problem, especially because the reaction rate, for reasons of economy, should be as high as possible. Hamielec (Hamielec and MacGregor, 1982) concludes that "the results from simple calculations indicate that for reactor volumes greater than 5000 gal additional cooling capacity would likely be required to achieve commercial production rates".

A model for a continuous stirred tank reactors (CSTR) was first presented by Gershberg and Longfield (Gerschberg and Longfield, 1961) in 1961. It is based on the SE Case 2 model, and has been further described and elaborated by Poehlein (Poehlein, 1981). A pioneering modelling framework was presented by Min and Ray (Min and Ray, 1974, 1976a, 1976b, 1978) in 1974–1978. After the work of Min and Ray others have concentrated on practical solutions of the model, testing different numerical techniques and comparing predictions to experimental data for specific polymerisation systems. The main challenge is to do appropriate model simplifications while at the same time producing a sufficiently accurate model. When a model is available, it may be used for predictive process control of the reactor. Several researchers in chemical engineering are now working on this topic, and much of it has appeared only during the last 10–15 years. The group of Asua has published many works in this field and is presently one of the most active in emulsion polymerisation process control. The details of these processes will be treated further in Chapter 4.

2

Introduction to Radical (Co)Polymerisation

A.M. van Herk
Institute of Chemical and Engineering Sciences, Jurong Island, Singapore and *Department of Chemical Engineering & Chemistry, Eindhoven University of Technology, The Netherlands*

In this chapter the basics of free radical polymerisation are described in a concise way with an emphasis on development of molecular mass and rate of polymerisation. For a more extensive discussion of all aspects of free radical polymerisation and controlled radical polymerisation the reader can resort to two excellent books; Moad and Solomon (1995) and Matyjaszewski and Davis (2002).

Some important new insights on transfer to polymer reactions are included in this chapter because of their relevance to emulsion polymerisation.

2.1 Mechanism of Free Radical Polymerisation

The mechanism of free radical polymerisation belongs to the class of so-called chain reactions. Chain reactions are characterized by the fast subsequent addition of monomers to an active centre at the chain end. The activity of the growing chain is transferred to the adding unit. The active centres are present in very low concentrations (10^{-5}–10^{-8} mol l^{-1}). The rate of addition is very high (10^3–10^4 units per second) and the time of growth of a chain (time between initiation and termination of a chain) is quite short (0.1–10 s) relative to the total reaction time, which can be of the order of several hours.

This means that the composition of the chain and the chain length are determined in seconds. Terminated chains, in principle, do not take part in further reactions (except when

Chemistry and Technology of Emulsion Polymerisation, Second Edition. Edited by A.M. van Herk.

transfer to polymer events occur, Section 2.3). The final chemical composition distribution and molecular mass distribution is determined by the accumulation of rapidly produced dead chains (chains without an active centre). In free radical polymerisation the active centre is a free radical. In controlled or living radical polymerisation (Section 2.5) the radical is protected against termination and continues to grow during the complete reaction time.

During the formation of a polymer chain a number of subsequent kinetic events take place: (1 radical formation, (2) initiation, (3) propagation and 4) termination. Transfer of the radical activity to another molecule is a complication that will be dealt with in Section 2.3:

1. Radical formation

   The formation of free radicals can take place in a number of ways. Radicals can be produced by photo-initiation, radiation (γ-radiation or electron beams), electrochemically and by thermal initiation. Well-known examples of the thermal decomposition of initiators are:

$$\phi\text{–C–O–O–C–}\phi \xrightarrow{80-90^\circ\text{C}} 2\ \phi\text{C–O}^\bullet \rightarrow 2\phi^\bullet + \text{CO}_2$$

benzoyl peroxide (BPO)

$$\text{NC–C(CH}_3)_2\text{–N=N–C(CH}_3)_2\text{–CN} \xrightarrow{60-70^\circ\text{C}} 2\text{NC–C(CH}_3)_2^\bullet + \text{N}_2$$

α–α' azobis(isobutyronitrile) (AIBN)

   A schematic representation of the decomposition of the initiator (I) into two radicals ($R^\bullet$), with a decomposition rate coefficient (k_d) and an expression for the rate of decomposition of the initiator (R_d) is given below:

$$\text{I} \xrightarrow{k_d} 2\text{R}^\bullet \quad \text{with rate} \quad R_d = \frac{d[\text{R}^\bullet]}{dt} = 2k_d[\text{I}] \tag{2.1}$$

   This reaction has a high activation energy (140–160 kJ mol^{-1} depending on the initiator) so k_d depends strongly on temperature.

   In fact not all radicals will initiate a polymeric chain; some of the radicals are lost in side reactions (like recombination of the initiator fragments). For this reason the efficiency factor f is introduced. f is the fraction of radicals that actually initiate a polymeric chain, the rate of radical production (leading to an actual initiation step) ρ_i equals then:

$$\rho_i = 2k_d\, f\, [\text{I}] \tag{2.2}$$

2. Initiation

   This is the addition of the first monomeric unit to the initially formed free radical:

$$\text{R}^\bullet + \text{M} \xrightarrow{k_i} \text{RM}^\bullet \quad \text{with rate} \quad R_i = k_i[\text{R}^\bullet][\text{M}] \tag{2.3}$$

3. Propagation

   This is the process for the growth of the chains:

$$\text{RM}^\bullet + \text{M} \xrightarrow{k_p} -\text{M}_2^\bullet$$

$$-\text{M}_2^\bullet + \text{M} \xrightarrow{k_p} -\text{M}_3^\bullet$$

$$-\text{M}_i^\bullet + \text{M} \xrightarrow{k_p} -\text{M}_{i+1}^\bullet \quad \text{with rate} \quad R_p = k_p[\text{M}^\bullet][\text{M}] \tag{2.4}$$

It is assumed in this approach that the rate coefficient of propagation k_p for equivalent active centres does not depend on chain length (principle of equal reactivity, P.J. Flory, 1953). Recent investigations have shown that the first propagation steps are faster, but this effect quickly levels off at approximately 10 monomeric units. After reaching 10 units the propagation rate coefficient is no longer a function of chain length.

4. Termination
   Termination takes place via two types of bimolecular free radical reactions:
   In combination the two radicals form a new bond, connecting the two growing chains to form one dead chain with the combined length of the two growing chains.
   – combination: $-M_i^{\bullet} + -M_j^{\bullet} \xrightarrow{k_{tc}} -M_{i+j}-$
   In disproportionation a radical abstracts a proton from the chain end of another growing chain, leading to two dead chains, one with a double bond and one with a saturated chain end.
   – disproportionation: $-M_i^{\bullet} + -M_j^{\bullet} \xrightarrow{k_{td}} -M_i = + - M_j.$

$$\text{at a rate } R_t = 2k_t[M^{\bullet}]^2 \tag{2.5}$$

   The rate of termination is usually determined by the rate of diffusion of the polymer chains. Because rates of diffusion are dependent on the viscosity of the medium and the size of the diffusing species this means that the rate of termination is dependent on conversion and on the chain length of the polymer chains (see also Section 2.2.4.).

Because k_t cannot be regarded as a constant during a polymerisation reaction sometimes an average value of k_t is introduced in the expression for the termination rate ($<k_t>$).

2.2 Rate of Polymerisation and Development of Molecular Mass Distribution

2.2.1 Rate of Polymerisation

When the radical polymerisation reaction proceeds (usually after a so-called induction period, see Section 2.2.4.) we are interested in the rate of polymerisation and the produced molecular mass of the polymer (Section 2.2.2.).

In determining the rate of polymerisation we usually look at the rate of consumption of the monomer. There are two reactions that consume monomer; the initiation reaction and the propagation reaction.

For the rate of polymerisation (R_{pol}) the following equation holds:

$$R_{pol} = -\frac{d[M]}{dt} = R_i + R_p \cong R_p \approx k_p[M^{\bullet}][M] \tag{2.6}$$

in a normal polymerisation reaction high molecular mass material would be formed, therefore $R_p \gg R_i$, each initiation reaction forms a new growing chain, each propagation reaction extends the chain with one more monomeric unit, so the ratio R_p over R_i is a measure of the number of monomeric units per chain and thus for the average molecular mass of the formed polymer chains.

$[M^\bullet]$ cannot be measured easily and therefore it is impractical to use. However, we know that the free radical concentration will increase initially, but will attain a constant value when the termination reactions start taking place. In other words, the steady state for free radicals is assumed (both in $R^\bullet$ and $M^\bullet$). This is frequently done in the case when highly reactive species are present in low concentrations. What is actually done is setting the rate of formation (R_i) and the rate of disappearance (R_t) of a radical to be equal, which means that the actual rate of change of that radical concentration (the sum of the rate of production and rate of disappearance) equals zero (steady state), for $M^\bullet$ this leads to:

$$d[M^\bullet]/dt = k_i[R^\bullet][M] - 2k_t[M^\bullet]^2 \cong 0 \tag{2.7}$$

The same can be done for the radical $R^\bullet$:

$$d[R^\bullet]/dt = 2k_d\, f\,[I] - k_i[R^\bullet][M] \cong 0 \tag{2.8}$$

From these equations we can solve for $[M^\bullet]$; hence

$$[M^\bullet] = (k_d\, f\,[I]/k_t)^{1/2} \tag{2.9}$$

so that

$$R_p = k_p \left\{ \frac{k_d\, f\,[I]}{k_t} \right\}^{1/2} [M] \tag{2.10}$$

In some cases the conditions are chosen as that the initiator decomposes relatively slowly. At the beginning of the reaction, [I] will be approximately constant, so that

$$R_p \cong k'[M]; \quad [M] \cong [M]_o e^{-k't} \tag{2.11}$$

where k' is a composed "rate coefficient". The polymerisation is an apparent first order reaction with respect to monomer concentration.

Changing the initiator concentration will have an effect on R_p according to the square root of the change (so a fourfold increase in [I] will lead to a doubling of R_p).

2.2.2 Kinetic Chain Length

With a chain reaction polymerisation the *kinetic chain length* ν is defined as the average number of monomeric units that is added per initiating species (in this case a free radical). The kinetic chain length is defined independently from the mode of termination (disproportionation or combination), therefore, to get to the actual average molecular mass a correction has to be made.

Remember that in the steady state $R_i = R_t$;

$$\nu = \frac{R_p}{R_i} = \frac{R_p}{R_t} = \frac{k_p[M^\bullet][M]}{2k_t[M^\bullet]^2} = \frac{k_p[M]}{2k_t[M^\bullet]}$$
$$\nu = \frac{k_p[M]}{2(k_d\, f\,[I]k_t)^{1/2}} \tag{2.12}$$

The average lifetime (or growth time) of a chain radical in the steady state is:

$$\tau_g = \frac{[M^\bullet]}{2k_t[M^\bullet]^2} = \frac{1}{2k_t[M^\bullet]} \approx 10^{-1}\text{–}10\ \text{s} \tag{2.13}$$

Another way of identifying the chain length is to multiply the rate of monomer addition at the growing chain end (k_p[M]) by the average growth time of a chain radical (τ_g);

$$\nu = k_p[M]\tau_g = \frac{k_p[M]}{2k_t[M^\bullet]} \tag{2.14}$$

Equation 2.14 can easily be used to predict the effect of changes in a recipe on the kinetic chain length:

- with increasing conversion the monomer concentration decreases and therefore the kinetic chain length decreases approximately exponentially:

$$\nu \cong \nu_o e^{-k't} \tag{2.15}$$

- when increasing the initiator concentration we increase the rate of termination (Equation 2.5) and therefore the average growth time of a chain (Equation 2.13) decreases, which in turn will decrease the kinetic chain length
- when comparing monomers with different k_p values it is obvious that, under otherwise similar conditions, a monomer with a higher k_p value will produce chains with a higher kinetic chain length
- another process that will affect the average growth time of a chain is a transfer reaction (Section 2.3), terminating a growing chain sooner than without transfer, decreasing τ_g and thus creating a shorter chain
- with increasing conversion the viscosity of the medium increases and, therefore, the rate of termination decreases, this will increase the average growth time of the chain and increase the chain length (note: the steady state might no longer be applicable in this regime.

For the number averaged degree of polymerisation $\bar{P}_n$ of the chains that are formed at a particular point in time, the following holds: $\bar{P}_n$ = number of monomer units added in the time interval dt over the number of dead chains formed in dt. Therefore, $\bar{P}_n = 2\nu$ for termination by combination, because two growing chains produce one dead chain and $P_n = \nu$ for termination by disproportionation, because two growing chains will produce two dead chains.

The order of magnitude for several kinetic parameters that control molecular weight is given in Table 2.1.

2.2.3 Chain Length Distribution

The chain length distribution of a polymer has a great influence on the mechanical properties and the processability. From the mechanism of polymerisation, it should be possible to calculate the chain length distribution of the formed polymer chains.

Table 2.1 *The order of magnitude for a number of important parameters.*

$k_d = 10^{-4}$–10^{-6} s^{-1}	[M] = 10–10^{-1} mol l^{-1}
$k_i \sim k_p = 10^2$–10^4 l mol^{-1} s^{-1}	[M$^\bullet$] = 10^{-7}–10^{-9} mol l^{-1}
$k_t = 10^6$–10^8 l mol^{-1} s^{-1}	[I] = 10^{-2}–10^{-4} mol l^{-1}

In order to discuss the chain length distribution the probability of formation of an i-mer must be considered, or in other words, we must calculate the mole fractions of 1-, 2-, ..., i-mer. Therefore, we define the probability p of chain growth:

$$p = \frac{R_p}{R_p + R_t} = \frac{\text{number of growth events in } \Delta t}{\text{number of growth events} + \text{number of stopping events in } \Delta t} \quad (2.16)$$

The probability of termination is then $(1 - p)$.

In chain reactions the time of growth of a polymeric chain is of the order of 0.1–10 s. In the time interval Δt all conditions will remain constant (v_p and v_t are constant during Δt) and the probability p will be the same within the time interval. With conversion the value of p changes because R_p and R_t will change.

In order to predict the chance of obtaining, for example, a trimer we need to have two propagation steps followed by one termination (by disproportionation) step, multiplying the probabilities of these individual steps will give the probability of obtaining a trimer:

$$x_3 = \underbrace{p \quad p}_{\text{probability of two propagation steps}} \quad \underbrace{(1 - p)}_{\text{probability of a termination step}} \quad (2.17)$$

2.2.3.1 *Chain Termination by Disproportionation*

In this case or with chain transfer (not to polymer!) the probability of an i-mer = the mole fraction of i-mer is given by the *Flory distribution*:

$$x_i = p^{i-1}(1 - p) \quad (2.18)$$

Also the mass distribution of the Flory distribution can be calculated:

$$w_i = \frac{x_i i}{\sum x_i i} = i p^{i-1}(1 - p)^2 \quad (2.19)$$

Flory distributions at different probabilities p are shown in Figure 2.1.

For $\bar{P}_n$ of the *instantaneously formed* product the following is valid:

$$\bar{P}_n = \sum x_i i = \frac{1}{1 - p} \quad (2.20)$$

For $\bar{P}_w$ of the instantaneously formed product we find:

$$\bar{P}_w = \sum w_i i = \frac{1 + p}{1 - p} \cong \frac{2}{1 - p} \quad (2.21)$$

It appears that for $p \to 1$ P_n is the maximum in the mass distribution curve and $\bar{P}_w$ the inflection point on the high degree of polymerisation side.

To produce high molecular mass material p has to be close to 1 ($p > 0.99$) so that we have for the degree of dispersity:

$$D = \frac{\bar{P}_w}{\bar{P}_n} = 1 + p \cong 2 \quad \text{for instantaneously formed product !} \quad (2.22)$$

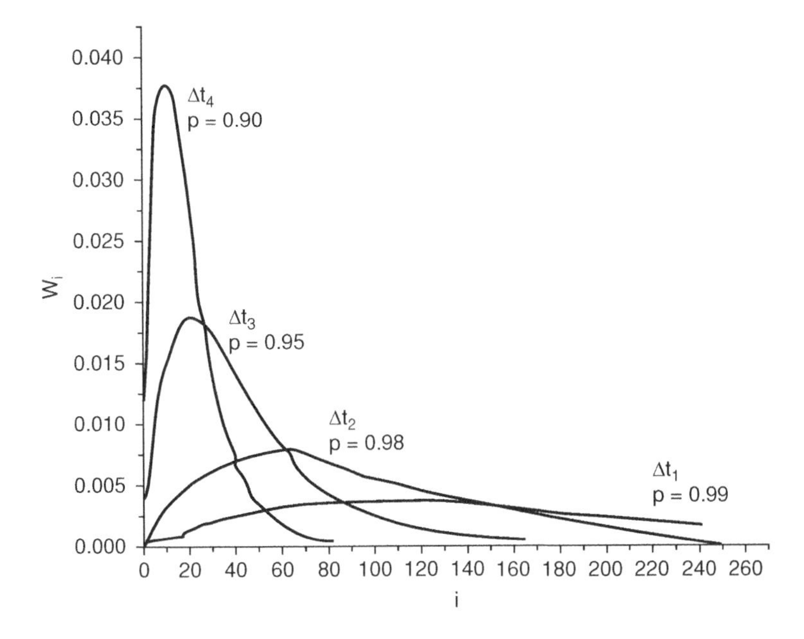

Figure 2.1 *Flory distributions at different values of the probability p.*

2.2.3.2 *Chain Termination by Combination*

In the case of chain termination by combination the chains will be longer, and two Flory distributions (from the two combining chains) make the overall distribution sharper. It can be derived that the following holds:

$$\bar{P}_{\mathrm{n}} = \frac{2}{1-p}, \quad \bar{P}_{\mathrm{w}} = \frac{2 \mid p}{1-p}, \text{ so that } D = 1 + 1/2p \cong 1\ 1/2 \tag{2.23}$$

By bifunctional combination the resultant distribution is sharper than in the case of disproportionation.

2.2.3.3 *High Conversion*

Several effects can make the chain length distribution broader than is described above:

- high or complete conversion ($D \sim 2$–5), especially if the reaction involves:
- gel effect ($D \sim 5$–10) and/or
- side-chains ($D \sim 20$–50).

$$p = \frac{R_{\mathrm{p}}}{R_{\mathrm{p}} + R_t} = \frac{1}{1 + 1/\upsilon} = \frac{k_{\mathrm{p}}[\mathrm{M}]}{k_{\mathrm{p}}[\mathrm{M}] + 2k_{\mathrm{t}}[\mathrm{M}^\bullet]} = f(\text{time}) \quad \text{with } [\mathrm{M}^\bullet] = (k_{\mathrm{d}}[\mathrm{I}]/k_{\mathrm{t}})^{1/2} \tag{2.24}$$

Since $[\mathrm{M}^\bullet]$ depends linearly on $[\mathrm{I}]^{1/2}$ and is therefore almost constant, and [M] decreases continuously as the reaction proceeds, the probability of chain growth (p) decreases in time and will be 0 at the end of the reaction. Consequently $\bar{P}_{\mathrm{n}}$ decreases in time.

The decreasing $\bar{P}_n$ of the instantaneously formed product occurs at different points in time resulting in differing "dead" distributions. The sum of these dead distributions (Flory distributions) in the complete product is a very broad distribution, and is no longer a Flory distribution.

2.2.4 Temperature and Conversion Effects

Temperature dependence. If the pertaining reaction constants can be described by the Arrhenius equation with E = energy of activation, then it follows from Equations 2.10 and 2.12, respectively, that:

$$\frac{\partial \ln R_p}{\partial T} = \frac{E_p - 1/2E_t + 1/2E_d}{RT^2} \tag{2.25}$$

$$\frac{\partial \ln \bar{P}_n}{\partial T} = \frac{E_p - 1/2E_t - 1/2E_d}{RT^2} \tag{2.26}$$

For most monomers $E_p - 1/2E_t \sim 20\,\text{kJ mol}^{-1}$ and for many initiators $E_d \sim 130\,\text{kJ mol}^{-1}$, so that:

$$\frac{\partial \ln R_p}{\partial T} \cong \frac{85}{RT^2} \quad \text{i.e. } R_p \textit{ increases} \text{ by a factor of 2 or 3 per 10 °C}$$

$$\frac{\partial \ln \bar{P}_n}{\partial T} \cong \frac{-45}{RT^2} \quad \text{i.e. } \bar{P}_n \textit{ decreases} \text{ by a factor of factor 2 or 3 per 10 °C}$$

Radical polymerisations are often characterised by a sudden increase in R_p when a certain conversion is reached.

The higher the initial concentration $[M]_0$, the earlier this effect occurs.

Only at low $[M]_0$ ($\approx$10%) is v_{pol} as expected. $\bar{P}$ increases during the occurrence of the *gel effect* (also called the Trommsdorff–Norish or auto-acceleration effect). This is evidence for the explanation: As conversion increases the viscosity of the system increases (especially at high $[M]_0$). The chains cannot diffuse fast enough and the apparent termination rate constant decreases faster and faster. In this way both R_p and $\bar{P}$ increase (see Figure 2.2 for R_p)!

During the gel effect the steady state in radicals is no longer valid; the radical concentration will increase. This means that substituting Equation 2.9 for the radical concentration is not allowed.

Although the propagation reaction can be hindered as well, this effect is not significant: termination involves a reaction between two large molecules, whereas propagation involves a reaction between one large molecule and one small molecule. Only at very high $[M]_0$ (see 100% curve) can the propagation reaction be hindered considerably.

With the occurrence of the gel effect both R_p and $\bar{P}$ are higher than were predicted. Although the great increase in heat production with increasing viscosity can be dangerous with respect to reaction runaway, in practice proper use can be made of the gel effect.

Induction period. In many cases the reaction does not start immediately but after minutes or sometimes even hours. This induction period is caused by preferential reaction of radicals with impurities in the reaction mixture, for example oxygen. When the radicals have reacted with all the impurities then the normal polymerisation reaction starts. Sometimes both the reaction with impurities and the normal polymerisation reaction proceed simultaneously; in that case there is a *retardation effect* (slower overall polymerisation rate).

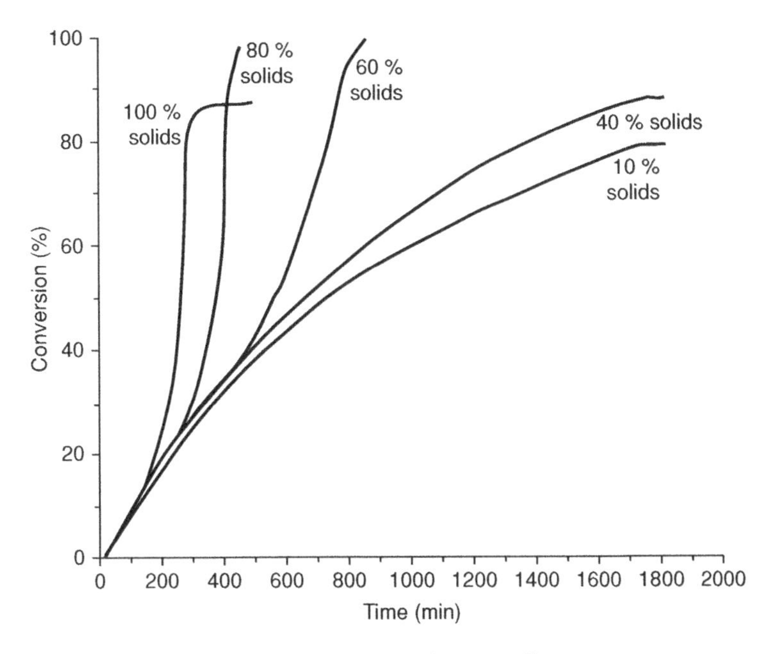

Figure 2.2 *Autoacceleration effect.*

2.3 Radical Transfer Reactions

2.3.1 Radical Transfer Reactions to Low Molecular Mass Species

For many polymerisations it is observed that $\bar{P}$ is lower than is predicted on the basis of termination by combination or disproportionation. This is caused by the occurrence of chain transfer reactions to low molecular mass species. The chain transfer agent, T, can be solvent, monomer, initiator, polymer or "modifier".

$$-\mathrm{M}_n^{\bullet} + \mathrm{T} \xrightarrow{k_{\mathrm{tr}}} -\mathrm{M}_n + \mathrm{T}^{\bullet} \tag{2.27}$$

$$\text{e.g. } -\mathrm{M}_n^{\bullet} + \mathrm{CCl_4} \xrightarrow{k_{\mathrm{tr}}} -\mathrm{M}_n\mathrm{Cl} + \mathrm{CCl}_3^{\bullet} \tag{2.28}$$

The growing chain is stopped, but re-initiation with rate constant k_{i}' can take place:

$$\mathrm{T}^{\bullet} + \mathrm{M} \xrightarrow{k_{\mathrm{i}}'} \mathrm{M}_1^{\bullet} \quad \text{after which} \tag{2.29}$$

$$\mathrm{M}_1^{\bullet} + \mathrm{M} \xrightarrow{k_{\mathrm{p}}} \mathrm{M}_2^{\bullet} \quad \text{etc.} \tag{2.30}$$

for example $\mathrm{CCl}_3^{\bullet} + \mathrm{M} \xrightarrow{k_{\mathrm{i}}'} \mathrm{CCl_3M}_1^{\bullet}$ followed by

$$\mathrm{CCl_3M}_1^{\bullet} + \mathrm{M} \xrightarrow{k_{\mathrm{p}}} \mathrm{CCl_3M}_2^{\bullet} \text{ or}$$

$$\mathrm{CCl_3M}_1^{\bullet} + \mathrm{CCl_4} \xrightarrow{k_{\mathrm{tr}}} \mathrm{CCl_3M_1Cl} + \mathrm{CCl}_3^{\bullet}$$

Table 2.2 Effect of transfer on rate of polymerisation and kinetic chain length.

Case	Relative rate constants for transfer, propagation and termination	Description	Effect on R_p	Effect on $\bar{P}_n$
1	$k_p \gg k_{tr}\ k_i' \cong k_p$	Normal chain transfer	None	Decrease
2	$k_p \ll k_{tr}\ k_i' \cong k_p$	Telomerisation	None	Large decrease
3	$k_p \gg k_{tr}\ k_i' < k_p$	Retardation	Decrease	Decrease
4	$k_p \ll k_{tr}\ k_i' < k_p$	Degradative chain transfer	Large decrease	Large decrease

One can simply derive that the following holds for chain transfer:

$$\bar{P}_n = \frac{1}{1/\bar{P}_{n'} + C_T[T]/[M]} \tag{2.31}$$

where $C_T = k_{tr,T}/k_p$ is the rate coefficient for chain transfer to T and P_n' is the degree of polymerisation in the absence of chain transfer, and

$$R_p = \left\{k_p[M^\bullet] + k_i'[T^\bullet]\right\}[M] \tag{2.32}$$

Chain transfer always leads to a decrease in $\bar{P}$. However, the influence on R_p depends on the relative magnitude of the re-initiation rate constant k_i' compared with k_p and on the ratio $[M^\bullet]/[T^\bullet]$ which is determined by C_T.

C_T also determines the relative rate of consumption of the transfer agent as compared to the monomer. Only in the case of $C_T = 1$ does the ratio [T]/[M] not change with conversion and therefore P_n does not change with conversion in a transfer dominated regime. In Table 2.2 an overview is given of the different effects of transfer on rate and molecular weight.

In the case of transfer reactions to monomer Equation 2.31 simplifies to Equation 2.31a and the average degree of polymerisation is independent of conversion in a transfer dominated regime!

$$\bar{P}_n = \frac{1}{1/\bar{P}_{n'} + C_T} \tag{2.31a}$$

2.3.2 Radical Transfer Reactions to Polymer

Chain transfer to polymer leads to branched polymer chains and thus greatly affects the physical and mechanical properties of a polymer, such as the ability to crystallize. Transfer to polymer does not necessarily lead to a decrease in molecular mass, but the chain length distribution becomes broader. If we consider transfer from a growing polymer chain to another dead chain, this is *intermolecular transfer to polymer*. If the radical attacks a proton in the same chain we call this *intramolecular transfer to polymer*, also called backbiting.

The *number* of side-chains can be expressed in terms of branching density BD = number of side-chains per reacted monomer. The rate of formation of chain radicals generated by transfer = the rate of formation of side-chains, which is:

$$\frac{d[M^\bullet]_{tr,p}}{dt} = k_{tr,p}[M^\bullet]\{[M]_0 - [M]\} \tag{2.33}$$

introducing conversion x,

$$\text{with } \frac{dx}{dt} = \frac{d}{dt}\frac{[M_0 - [M]}{[M]_0} = -\frac{1}{[M]_0}\frac{d[M]}{dt} = \frac{1}{[M]_0}k_p[M^\bullet][M] \tag{2.34}$$

$$\text{becomes } \frac{d[M^\bullet]_{tr,p}}{dx} = C_p[M]_0\frac{x}{1-x} \tag{2.35}$$

Integration between 0 and x and dividing by $[M]_0 - [M] = x[M]_0$ gives:

$$BD = -C_p\{1 + (1/x)\ln(1-x)\} \tag{2.36}$$

Independent of C_p, the extent of branching greatly increases with increasing conversion. To minimize the number of side-chains the conversion must be limited. Dilution of the system with solvent does not help according to Equation 2.36.

Studies for acrylates contradicted Equation 2.36. In a study on the chain transfer to polymer in the free radical solution polymerisation of n-butyl acrylate (Ahmed, Heathly and Lovell, 1998) showed that besides the expected high branch densities in PBA at high conversion (ρ = 2–3% at 95% conversion), also at low conversion high branch densities were found (3.8% at 8% conversion!). This striking result was explained by intramolecular chain transfer to polymer (back-biting). Moreover, for dilute solutions, as the initial monomer concentration decreases, the probability of chain transfer to polymer (and hence the mole percent of branches) increases (Table 2.3). This is because in the dilute solutions the overall polymer repeat unit concentration is too low for overlap of different polymer coils and intramolecular chain transfer to polymer dominates. Under these conditions, the local polymer repeat unit concentration within the isolated propagating chains is defined by the chain statistics and so is approximately constant, whereas the monomer is distributed uniformly throughout the solution. Thus, for dilute solutions, as $[M]_0$ decreases, the probability of chain transfer to polymer increases. Despite the fact that intramolecular chain transfer to polymer does, in principle, not affect the molecular mass distribution (if the radical after chain transfer has the same reactivity as the primary radical) still an effect can be expected on the GPC analysis. As the number of branches in the chain increases, the molecule is packed more densely. This means that the intrinsic viscosity of the branched chain as compared to its linear analogue decreases and, therefore, also the apparent molecular mass. In Table 2.3 branching densities as a function of concentration and conversion are shown.

Several effects on the molecular mass distribution and the apparent k_p value can be anticipated in the case of transfer to polymer processes:

1. *Intramolecular chain transfer to polymer:*
   Backbiting is rendering a branched chain that will appear at a lower apparent molecular mass in SEC (species 2a and 2b)

   The slow re-initiation of the backbone derived radical can affect the kinetics (Chiefari *et al.*, 1999)

Table 2.3 *Branching densities for butyl acrylate polymerisation. Reprinted under the terms of the STM agreement from [Ahmed, Heathly and Lovell, 1998] Copyright (1998) American Chemical Society.*

$[M]_0$ (mol kg^{-1})	Conversion (%)	Branches (mol%)
1.56	35	1.6
1.56	38	1.7
1.56	49	1.8
1.56	63	2.0
1.56	82	2.5
0.78	7	2.3
0.78	7	2.5
0.55	7	3.3
0.39	9	4.2
0.23	8	5.3
0.78	28	2.7
0.78	23	2.7
0.55	27	3.5
0.39	26	4.5
0.23	25	5.9
3.90	98	1.9
3.12	97	2.1
2.34	95	2.3
0.78	88	3.9
0.78	88	3.8

The transfer to polymer process can lead to β-scission (Plessis *et al.*, 2000) giving two shorter chains (species 3 and 4)

2. *In addition for intermolecular chain transfer to polymer:*
   A short chain is produced because it is formed before normal termination would occur.
   A dead chain is re-initiated and will grow longer than without the transfer process.
   Most of the effects will render smaller apparent k_p values at higher temperatures giving a downwards deviation from the Arrhenius plot, as indeed is observed for the acrylates. Actually the β-scission chemistry is claimed to be a useful route towards the production of macromonomers (Chiefari *et al.*, 1999) (species 3 in Figure 2.3).

The EPR experiments of Yamada on methyl acrylate also support this reaction scheme. He was able to show the occurrence of the mid-chain radical and also the occurrence of β-scission (Tanaka *et al.*, 2000; Azukizawa *et al.*, 2000) in the temperature range 40–85 °C.

2.4 Radical Copolymerisation

2.4.1 Derivation of the Copolymerisation Equation

We limit the discussion to copolymerisations where two monomers are built in unbranched polymers via the free radical mechanism. If the penultimate unit in the radical chain end

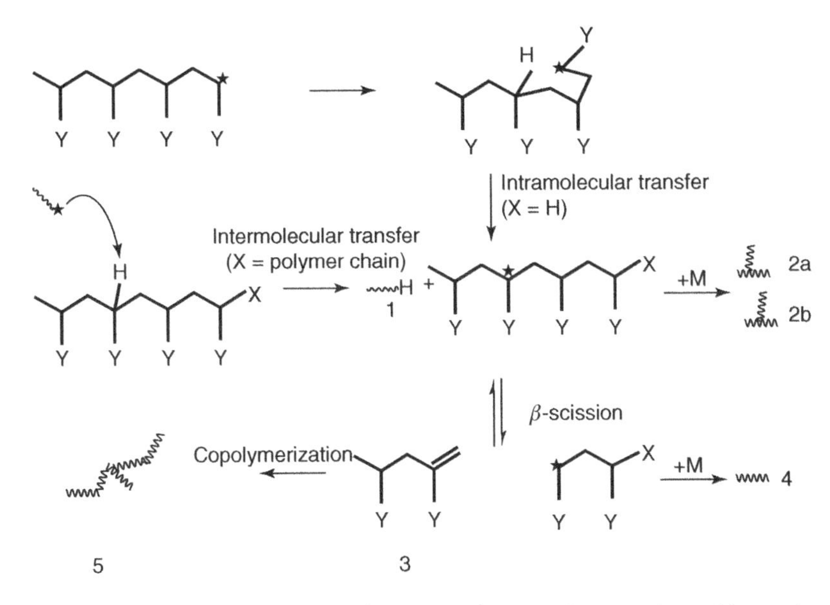

Figure 2.3 *Reactions that can occur during a polymerisation reaction with acrylates. Y is an acrylate ester side group (or the acid itself). Intramolecular transfer to polymer leads to a branched chain (2a). Intermolecular transfer to polymer results in a shorter chain (species 1) and re-initiates a dead polymer chain and results in branched chain 2b. The backbone radical can undergo β-scission that results in two shorter chains (species 3 and 4). The unsaturated species 3 can undergo copolymerisation resulting in species 5. Adapted with permission from [van Herk, 2001] Copyright (2001) Wiley-VCH.*

does not influence the reactivity of the radical and Flory's principle for constant reactivity is valid, only four propagation steps are involved in the copolymerisation reaction:

$$\begin{aligned} &-M_1^\bullet + M_1 \xrightarrow{k_{11}} -M_1^\bullet \\ &-M_1^\bullet + M_2 \xrightarrow{k_{12}} -M_2^\bullet \end{aligned} \qquad r_1 = \frac{k_{11}}{k_{12}} \tag{2.37}$$

$$\begin{aligned} &-M_2^\bullet + M_2 \xrightarrow{k_{22}} -M_2^\bullet \\ &-M_2^\bullet + M_1 \xrightarrow{k_{21}} -M_2^\bullet \end{aligned} \qquad r_2 = \frac{k_{22}}{k_{21}} \tag{2.38}$$

The reactivity ratios represent the preference of the chain end radical $-M_1^\bullet$ for addition of monomer 1 to monomer 2 (r_1) and of chain end radical $-M_2^\bullet$ for addition of monomer 2 to monomer 1 (r_2).

We can define the probability that a certain chain end radical reacts with a certain monomer:

$$p_{11} = \frac{k_{11}[M_1]}{k_{11}[M_1] + k_{12}[M_2]}, \qquad p_{12} = 1 - p_{11} \tag{2.39}$$

$$p_{22} = \frac{k_{22}[M_2]}{k_{22}[M_2] + k_{21}[M_1]}, \qquad p_{21} = 1 - p_{22} \tag{2.40}$$

The mole fraction of blocks M_1 with length i (mole fraction relative to all other blocks M_1) is:

$$x_i = p_{11}^{i-1}(1 - p_{11}) \tag{2.41}$$

so that the average block length is:

$$\bar{i}_n = \sum x_i \quad i = \frac{1}{1 - p_{11}} = r_1 \frac{[M_1]}{[M_2]} + 1 \tag{2.42}$$

Similarly for the M_2 blocks:

$$\bar{j}_n = \frac{1}{1 - p_{22}} = r_2 \frac{[M_2]}{[M_1]} + 1 \tag{2.43}$$

If $\bar{P}_n \rightarrow \infty$, or more precisely formulated, if there are many –M1M2– and –M2M1– transitions per copolymer chain, there are as many M_1 blocks as there are M_2 blocks. Consequently, at any time the following holds for the composition of the instantaneously formed copolymer:

$$\frac{d[M_1]}{d[M_2]} = \frac{\bar{i}_n}{\bar{j}_n} = \frac{r_1[M_1]/[M_2] + 1}{r_2[M_2]/[M_1] + 1} \tag{2.44}$$

This is the *copolymerisation equation*, which is often expressed as the fraction of free monomer $f_1 = [M_1]/\{[M_1] + [M_2]\}$ and the fraction of monomer built into the instantaneously formed copolymer $F_1 = d[M_1]/\{d[M_1] + d[M_2]\}$:

$$F_1 = \frac{r_1 f_1^2 + f_1 f_2}{r_1 f_1^2 + 2 f_1 f_2 + r_2 f_2^2} \tag{2.45}$$

For comparison of the various types of copolymer one can use the average block length of a 1:1 copolymer. For $\bar{j}_n = \bar{i}_n$ it holds that $[M_1]/[M_2] = (r_2/r_1)^{1/2}$, so that

$$\bar{i}_n = \bar{j}_n = 1 + (r_1 r_2)^{1/2} \tag{2.46}$$

Some general characteristics of copolymerisation:

1. k_i and k_t do not occur in the copolymerisation equation that is, the copolymer composition is independent of the overall reaction rate and the initiator concentration;
2. The reactivity ratios are generally independent of the type of initiator, inhibitor, retarder, chain transfer agent, but are dependent on temperature and (high) pressure. Occasionally they depend on the type of solvent;
3. In general, $f_l \neq F_1$, so that both the composition of the monomer feed mixture and the copolymer composition change as the conversion increases. This phenomenon is referred to as "*composition drift*". As the conversion increases the distribution of the chain composition broadens. This can lead to heterogeneous copolymers.

This can have a significant influence on the processability and mechanical behaviour of the products.

2.4.2 Types of Copolymers

Depending on the value of the reactivity ratios we can distinguish between a number of cases:

1. *Ideal copolymerisation* occurs if $r_1r_2 = 1$, that is, each radical end has the same preference for one of the monomers. In this case $d[M_1]/d[M_2] = r_1[M_1]/[M_2]$, and for a 1 : 1 copolymer $\bar{j}_n = \bar{i}_n = 2$.

   The monomers succeed each other in very short blocks of length 1–3.

   A special case is where $r_1 = r_2 = 1$ or $k_{11} = k_{12}$ and $k_{22} = k_{21}$.

   The reactivity of both radicals will in general be different, in other words $k_{11} \neq k_{21}$! In this case $f = F$ always holds, independently of conversion. The copolymerisation of (for instance) styrene (M_1) and *p*-methoxystyrene (M_2) is almost an ideal copolymerisation with $r_1 = 1.16$, $r_2 = 0.82$ and $r_1r_2 = 0.95$ (60°C). Ideal copolymers will not crystallise in general because of the absence of a periodical succession of monomer units and the absence of stereoregularity (atactic). Some typical F versus f curves are shown in Figure 2.4.
2. *Alternating copolymerisation* occurs if $r_1 = r_2 = 0$, that is, each radical reacts exclusively with the other monomer, not with its own. In this case $d[M_1]/d[M_2] = 1$ and for each monomer feed composition $\bar{j}_n = \bar{i}_n = 1$. An example is the copolymerisation of vinyl acetate (M_1) with maleic anhydride (M_2), for which $r_1 = 0.055$, $r_2 = 0.003$ and $r_1r_2 = 0.000\,17$.

   Alternating copolymers can crystallise only if the copolymers are purely alternating and if the product is stereoregular or if the monomers are symmetrical, like isobutene. Some typical F versus f curves are shown in Figure 2.5.

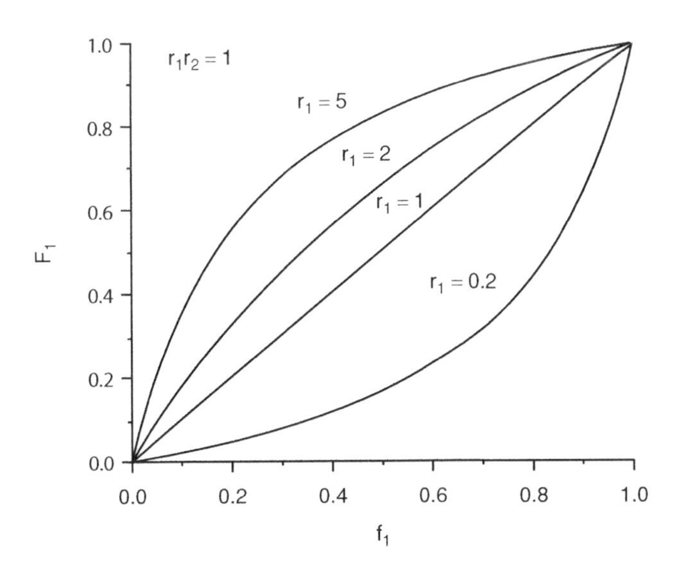

Figure 2.4 *Ideal copolymerisations.*

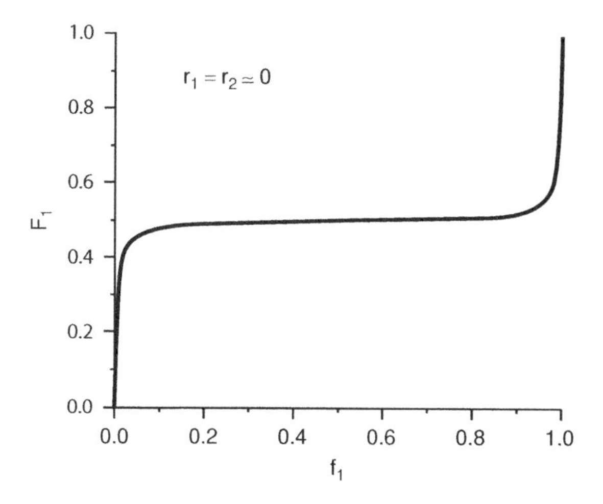

Figure 2.5 *Alternating copolymerisation.*

3. *Non-ideal copolymerisation*
   a. $0 < r_1 r_2 < 1$, this applies to most cases. The copolymerisation behaviour is in between alternating and ideal. The product $r_1 r_2$ is a measure of the tendency for alternating behaviour.

   An interesting case occurs if $r_1 < 1$ and $r_2 < 1$. The F–f curve intersects the diagonal and we speak of *azeotropic copolymerisation*. In the intersection the copolymer composition does not change as conversion proceeds, since $f_1 = F_1 = (1 - r_2)/(2 - r_1 - r_2)$. However, this is an unstable equilibrium. In general there will be no crystallisation. Some typical F versus f curves are shown in Figure 2.6.

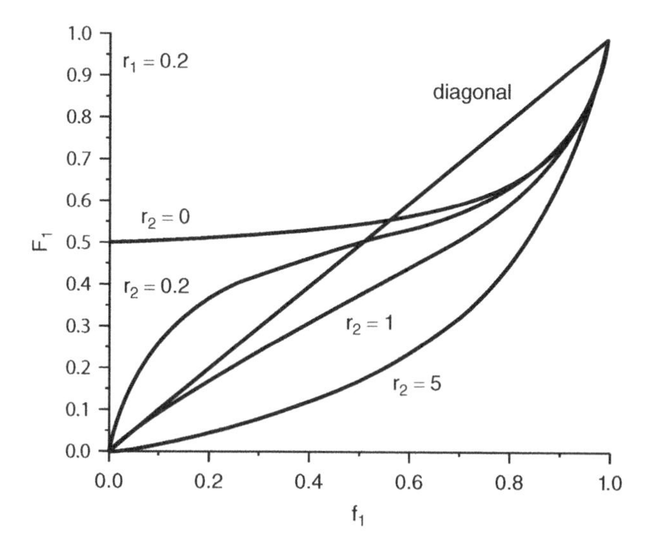

Figure 2.6 *Several copolymerisation f–F curves.*

b. $r_1r_2 > 1$ with $r_1 < 1$ and $r_2 > 1$. For a 1:1 copolymer $\bar{j}_n = \bar{i}_n > 2$. There is a certain tendency to form larger blocks because of the preference of M_2 for homopropagation. In general there is no crystallisation.
c. $r_1r_2 > 1$ with $r_1 > 1$ and $r_2 > 1$. This would lead to block copolymerisation and in the extreme case of $r_1 \gg 1$ and $r_2 \gg 1$ even to simultaneous homopolymerisations. This behaviour is very rare though. So if block copolymers are required, specific techniques must be employed.

Homogeneous – heterogeneous copolymers
Homogeneous copolymers are formed only if one of the following conditions is satisfied:

- low conversion
- constant monomer feed composition (f_1 = constant)
- azeotropic copolymerisation behaviour ($f_1 = F_1 = f_{az}$)
- $r_1 = r_2 = 1$ ($f_1 = F_1$)
- $r_1 = r_2 = 0$ ($F_1 = 1/2$)
- *Homogeneous copolymers* generally have the physical properties of a new polymer, such as the glass transition temperature.

If a batch polymerisation is performed to high conversion under non-azeotropic conditions, then the monomer feed composition changes and the composition of the copolymer (composition drift) also changes. It is then possible that the copolymer formed initially is no longer miscible in the copolymer formed in a later part of the reaction. Phase separation then occurs and one speaks of heterogeneous copolymers. Such a *heterogeneous copolymer* has the properties of the different polymers it is composed of (e.g., it may have two glass transition temperatures). Examining the glass transition temperature of the copolymer, the plot of specific volume against temperature shows a sharp bend and the curve of the modulus E of the material against temperature shows a strong decrease (Figure 2.7).

2.4.3 Polymerisation Rates in Copolymerisations

In copolymerisation, the average propagation rate coefficient $<k_p>$ is given by the general equation:

$$<k_p> = \frac{\bar{r}_1 f_1^2 + 2 f_1 f_2 + \bar{r}_2 f_2^2}{(\bar{r}_1 f_1/\bar{k}_{11}) + (\bar{r}_2 f_2/\bar{k}_{22})} \qquad (2.47)$$

where in the terminal model $\bar{r}_1 = r_1, \bar{r}_2 = r_2, \bar{k}_{11} = k_{11}, \bar{k}_{22} = k_{22}$

$$\text{with } r_1 = \frac{k_{11}}{k_{12}} \quad \text{and} \quad r_2 = \frac{k_{22}}{k_{21}}$$

In the case that also the penultimate unit influences the reactivity, the penultimate model applies.

There can be eight different propagation steps distinguished in the penultimate unit model.

The rate coefficients are defined as follows (example):

$$-M_1M_2^{\bullet} + M_1 \xrightarrow{k_{121}} -M_2M_1^{\bullet} \qquad (2.48)$$

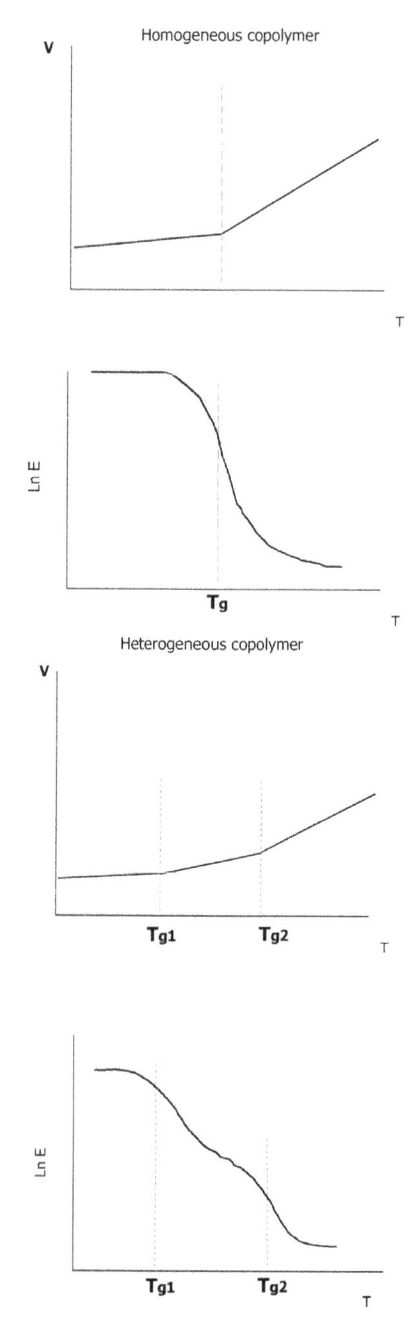

Figure 2.7 *Specific volume (V) and elasticity modulus (E) for homogeneous copolymers and heterogeneous copolymers.*

We can further define $\bar{k}_{11}$ *and* $\bar{k}_{22}$ as:

$$\bar{k}_{11} = \frac{k_{111}(f_1 r_1 + f_2)}{f_1 r_1 + f_2/s_1} \quad \text{and} \quad \bar{k}_{22} = \frac{k_{222}(f_2 r_2 + f_1)}{f_2 r_2 + f_1/s_2}$$

$$\bar{r}_1 = \frac{r_1'(f_1 r_1 + f_2)}{f_1 r_1' + f_2} \quad \text{and} \quad \bar{r}_2 = \frac{r_2'(f_2 r_2 + f_1)}{f_2 r_2' + f_1} \tag{2.49}$$

$$r_1 = \frac{k_{111}}{k_{112}},\ r_2 = \frac{k_{222}}{k_{221}},\ r_1' = \frac{k_{211}}{k_{212}},\ r_2' = \frac{k_{122}}{k_{121}},\ s_1 = \frac{k_{211}}{k_{111}},\ s_2 = \frac{k_{122}}{k_{222}}$$

In general the compositional data of a copolymerisation can be well described by the ultimate model. The k_p data in copolymerisations, for example obtained by the pulsed laser method (van Herk, 2000), are best described by the penultimate unit model. The description of the compositional data is less model sensitive.

2.5 Controlled Radical Polymerisation

In free radical polymerisation the growth time of a polymer chain is around 1 s. In other polymerisation mechanisms, like step reactions or living anionic polymerisations, the growth time of a chain is equal to the total reaction time. During this longer time period one can change the composition of the monomer feed and produce block copolymers or other interesting chain compositions. The mechanism of radical polymerisation can make use of a wealth of possible monomers, much more than for example in ionic polymerisations. Therefore, introducing a living character in the radical polymerisation is very interesting. Since 1980 there has been a tremendous increase in research in this field, which has led to several new polymerisation mechanisms.

Ideally, bimolecular termination should be eliminated from the reaction. The only way to do this is to have a very low concentration of radicals, which in turn would mean a very low rate of polymerisation. In many circumstances this would be too slow for use in industrial processes, and a compromise between molar mass control and the rate of the reaction should be made.

The active radical is turned into a dormant species. This dormant species is reactivated every once in a while. In total the time that the radical is in an active form is the same as for a free radical polymerisation (seconds), but because the active/dormant ratio is low the overall reaction times are large. Because the active/dormant ratio is low, the concentration of "free" radicals is low. Recalling Equation 2.5

$$R_t = 2k_t[\mathrm{M}^\bullet]^2$$

the rate of termination drops dramatically because this rate is second order in the radical concentration. The rate of polymerisation will also drop, but because this rate is first order in the radical concentration still reasonable rates can be obtained. While the radical is in its active form a few monomeric units add to the chain and then the chain is converted to the dormant species again. Of course some termination can still occur in the active period but in general this is negligible.

The number average molar mass, M_n, and the molar mass distribution (MMD) are controlled by interplay of kinetic parameters, which will be described in more detail in

Chapter 5. In principle, there are two ways of terminating a growing chain: bimolecular termination and transfer. These two modes of termination have led to the two main categories of living radical polymerisation; (i) reversible termination and (ii) reversible chain transfer.

Reversible termination requires the deactivation of active polymeric radicals through termination reactions to form dormant polymer chains, and activation of dormant polymer chains to form active chains. On the other hand, reversible chain transfer requires active chains to undergo transfer reactions with the dormant chains, and thus the reversible chain transfer end-group is transferred from a dormant to an active species. A narrow MMD is observed when this exchange reaction is fast. In all these techniques, the MMD can be controlled such that the polydispersity index (PDI) is below 1.1.

A further discussion of the living radical polymerisation techniques and their application in heterogeneous polymerisation techniques can be found in Chapter 5.

3

Emulsion Polymerisation

A.M. van Herk[1] and R.G. Gilbert[2]
[1] Institute of Chemical and Engineering Sciences, Jurong Island, Singapore and Department of Chemical Engineering & Chemistry, Eindhoven University of Technology, The Netherlands
[2] Centre for Nutrition & Food Science, University of Queensland, Australia, and Tongji School of Pharmacy, Huazhong University of Science and Technology, China

3.1 Introduction

Polymerisations may be categorised by both the polymerisation mechanism (e.g., radical polymerisation, anionic polymerisation, etc.), and by the polymerisation technique (e.g., solution polymerisation, emulsion polymerisation, etc.). A third factor is how the reactor is operated: in batch mode, or by adding monomers during the process (semi-continuous), or by continuous operation. Mechanism, technique and process strategies (mode of operation) all have an influence on the rates of polymerisation and the characteristics of the formed polymer. In this chapter we will focus on the special characteristics that can be distinguished in an emulsion polymerisation related to rate, development of molar mass and chemical composition. In Chapter 4 the effects of the process strategy will be discussed.

A radical polymerisation can be carried out with a range of polymerisation techniques. Those with only a single phase present in the system are *bulk* and *solution* polymerisations, involving the monomer, a solvent, if present, and the initiator. By definition, the formed polymer in a bulk or solution polymerisation remains soluble (either in the monomer or the solvent). A *precipitation* polymerisation is one in which the system starts as a bulk or solution polymerisation, but the polymer precipitates from the continuous phase to form polymer particles which are not swollen with monomer. A precipitation polymerisation where the polymer particles swell with monomer is called a *dispersion* polymerisation: arising from polymerisation in the continuous phase, the formed polymer particles are

Chemistry and Technology of Emulsion Polymerisation, Second Edition. Edited by A.M. van Herk.

an additional locus of polymerisation, and the particles in these systems are colloidally stabilized. Precipitation polymerisation is often performed in an aqueous medium. Dispersion polymerisation is usually performed in organic solvents that are poor solvents for the formed polymer (supercritical or liquid carbon dioxide may also be used as a continuous medium for dispersion polymerisation).

An *emulsion* polymerisation comprises water, an initiator (usually water-soluble), a water-insoluble monomer and a colloidal stabiliser, which may be added or may be formed *in situ*. The main locus of polymerisation is within the monomer-swollen latex particles which are either formed at the start of polymerisation, or may be added initially (in which case one has *seeded* emulsion polymerisation). The term 'emulsion polymerisation' is a misnomer (arising for historical reasons: the process was originally developed with the aim of polymerising emulsion droplets, although in fact this does not occur). The starting emulsion is not thermodynamically stable, although the final product is colloidally and thermodynamically stable. Several variants of conventional emulsion polymerisation are also sometimes used. An *inverse* emulsion polymerisation is where the continuous phase is organic in combination with an aqueous discrete phase containing a water-soluble monomer (e.g., acrylamide). Two variants of emulsion polymerisation are *micro-* and *mini-emulsion* polymerisations. In a micro-emulsion polymerisation, conditions are chosen so that the monomer droplets are so small (typical particle radius 10–30 nm) that they become the locus of polymerisation. A co-surfactant (e.g., hexanol) is usually used to obtain such small droplets. The starting micro-emulsion is thermodynamically stable (Guo *et al.*, 1992). A mini-emulsion polymerisation is one where the starting mini-emulsion comprises droplets with diameters in the range 50–1000 nm, more typically 50–100 nm. These mini-emulsions are thermodynamically unstable, but kinetically metastable, with lifetimes as long as months. They are stabilized against diffusion degradation (Ostwald ripening) by a hydrophobe: a compound that is insoluble in the continuous phase (Tang *et al.*, 1992). Mini-emulsion polymerizations have the advantage that transport of the monomer through the aqueous phase is not needed (the locus of polymerization is the mini-emulsion droplets!), and therefore it is suitable for polymerization of very hydrophobic monomers and also for encapsulation (Landfester, 2009). In both mini- and micro-emulsions, the polymerisation locus is within the micro- or mini-emulsion droplets.A *suspension* polymerisation is one which starts with a conventional emulsion, and in which the polymerisation is entirely within the (large) monomer droplets. The initiator is oil-soluble and a stabilising agent is used which does not form micelles.

3.2 General Aspects of Emulsion Polymerisation

An *ab initio* emulsion polymerisation involves the emulsification of one or more monomers in a continuous aqueous phase and stabilisation of the droplets by a surfactant. In a seeded emulsion polymerisation, one starts instead with a preformed seed latex. Usually, a water-soluble initiator is used to start the free-radical polymerisation. The locus of polymerisation is within submicron polymer particles (either formed during the process or added at the start), which are swollen with monomer during the polymerisation process, and dispersed in the aqueous phase. The final product is a latex comprising a colloidal dispersion of polymer particles in water. *Ab initio* emulsion polymerisation differs from suspension, mini- and micro-emulsion polymerisations in that the particles form as a separate phase during the polymerisation process. The particle size is much smaller than those formed in a suspension polymerisation, and also much smaller than the original monomer droplets.

It is essential to be aware that the polymer colloids which are the result of an emulsion polymerisation contain *many* polymer chains in each particle (despite the not uncommon misconception that there is only one chain per particle). Two observations make this apparent. First, the size of a typical polymer colloid, $\sim 10^2$ nm, is very much greater than the volume that could be occupied by a single polymer chain of the molecular weight ($\sim 10^6$) typical of that found in emulsion polymerisations. Second, when one considers that most particles have at least one growing radical in them over a significant fraction of the many hours that are required to complete a typical emulsion polymerisation, and that the upper bound for the growth time of a single chain is $k_{tr}[M]_p$ (where k_{tr} is the rate coefficient for transfer to monomer, and $[M]_p$ is the concentration of monomer in the particles), then one sees that the lifetime of a typical single chain in a styrene system (where at 50 °C one has $k_{tr} \sim 10^{-2}$ M^{-1} s^{-1} (Tobolsky and Offenbach, 1955) and $[M]_p \sim 6$ M (Hawkett, Napper, and Gilbert, 1980))can be no more than $\sim 10^1$ s, orders of magnitude less than the time during which the latex particle is polymerising.

The fact that particles in an emulsion polymerisation are small, much smaller than those in a (conventional) emulsion, indicates that polymerisation does not occur in the monomer droplets. If a surfactant is used in the system, above the critical micelle concentration, then micelles form. A micelle is an aggregate of $\sim 10^2$ surfactant molecules, usually spherically shaped with the dimension of a few nanometers. If present, micelles are the locus of the commencement of polymerisation, because they are much more numerous than the monomer droplets, and thus much more likely to capture aqueous-phase radicals generated from the initiator: *micellar nucleation*. Consistent with this, an increase in surfactant concentration results in an increase in the number of formed particles. If there is no added surfactant, or the system is below the critical micelle concentration, a latex can still form, stabilized by entities formed from the initiator. Particle formation is by the collapse (coil-to-globule transition) of aqueous-phase oligomers to form particles by *homogeneous nucleation*.

The emulsion polymerisation process is often used for the (co-)polymerisation of monomers, such as vinyl acetate, ethylene, styrene, acrylonitrile, acrylates and methacrylates. Conjugated dienes, such as butadiene and isoprene, are also polymerised on a large industrial scale with this method. One of the advantages of emulsion polymerisation is the excellent heat exchange due to the low viscosity of the continuous phase during the whole reaction. Examples of applications are paints, coatings (including paper coatings), adhesives, finishes and floor polishes (see Chapters 10 and 11). Emulsion polymerisation is frequently used to create core–shell particles, which have a layer structure. Core–shell products are in use in the coatings industry, in photographic and printing materials and in the production of high impact materials (a core of rubbery polymer and a shell of a glassy engineering plastic). In recent years, considerable interest has arisen in the preparation of block copolymers in emulsion polymerisation through the use of controlled radical polymerisation mechanisms. The formation of block copolymers within a latex particle can lead to interesting new morphologies and can lead to new latex applications. Chapter 5 is devoted to this new field. There are many texts on applications and structure–property relations of latexes (Lovell and El-Aasser, 1996; Blackley, 1997; Warson and Finch, 2001; Urban and Takamura, 2002a).

Emulsion polymerisation kinetics have important differences from solution and bulk polymerisations. These differences can lead to many advantages: for example, an increase in molar mass can be achieved without reducing the rate of polymerisation. Emulsion polymerisation is known for its relatively high rates of polymerisation and high molar masses compared to other process strategies. A disadvantage of emulsion polymerisation is the presence of surfactant and other additives, which may result in deleterious properties under some circumstances. At the same time, the presence of these additives gives the process considerable flexibility.

3.3 Basic Principles of Emulsion Polymerisation

The contemporary physical picture of emulsion polymerisation is based on the qualitative picture of Harkins (1947) and the quantitative treatment of Smith and Ewart (1948), with more recent contributions by Ugelstad and Hansen (1976), Gardon (1977), Gilbert and Napper (1974), Blackley (1975), Gilbert (1995) and Thickett and Gilbert (2007) (see also Chapter 1).

The main components of an emulsion polymerisation are the monomer(s), the dispersing medium (usually water), surfactant (either added or formed *in situ*) and initiator.

It is important to realize that the same basic free-radical polymerisation mechanisms operate in solution, bulk and emulsion polymerisations. The kinetic relations in Chapter 2 are therefore valid in an emulsion polymerisation. The differences commence with the actual concentrations of the various species at the locus of polymerisation. Instead of using the overall concentration of the monomer and radicals in the reactor, based on the volume of the reactor content, the concentration inside the latex particles must be used in the appropriate rate equations. A latex particle can be regarded as a nano-reactor. In another important difference to solution and bulk polymerisations, the radicals and monomers have to cross the interface between the particle and aqueous phases. The number of latex particles is, therefore, often an important factor in the overall rate of polymerisation.

During the progress of the polymerisation, three distinct intervals can be observed. Interval 1 is the initial stage where particle formation takes place. Several mechanisms of particle nucleation have been proposed, which will be discussed in Section 3.4.

Interval 2 is characterised by a constant number of particles (the polymerisation locus) and the presence of monomer droplets. The monomer-swollen particles grow and the monomer concentration within these particles is kept constant by monomer diffusing through the water phase from the monomer droplets. The beginning of Interval 2 is when particle formation stops, which – if micelles are present initially – is usually the conversion where the surfactant concentration drops below its critical micelle concentration. Interval 3 begins with the disappearance of monomer droplets, after which the monomer concentrations in both the particle and water phases decrease continuously.

The latex particles in all three Intervals are swollen with monomer, and thus the particle size in an emulsion polymerisation at a particular conversion will be the *swollen* value, that is, that due to monomer plus polymer. If one takes a sample out at a particular conversion and measures its size, for example, by electron microscopy or by light scattering, then this will be the *unswollen* volume, because the measuring process usually (but not invariably) involves disappearance of the monomer from the particle. The relation between swollen and unswollen size is simply given by mass conservation, found from the relative amounts of monomer and polymer. Specifically, assuming ideal mixing, the ratio of the swollen to unswollen diameters is given by $\{d_M/(d_M - [M]_p M_0)\}^{1/3}$. Here d_M is the density of monomer and M_0 is the molecular weight of the monomer.

3.4 Particle Nucleation

In a typical *ab initio* emulsion polymerisation, the starting emulsion is opaque. After initiator is added, particle formation commences (sometimes the suspension then has a bluish sheen for a few minutes, if the newly formed particles are sufficiently small to give rise to Mie scattering of incident light). As the polymerisation proceeds, the dispersion turns milky white. The number of particles increases in Interval 1, as does the rate of polymerisation. Interval 2 commences when particle formation is finished, whereafter the particle number is constant and frequently (although not invariably) the polymerisation rate is also constant.

Our discussion of particle nucleation will be couched in terms of the particle number density (number concentration in the aqueous phase), N_p. Particle number and particle size are trivially related by mass conservation for a system containing a given amount of monomer:

$$N_p = \frac{\text{mass of monomer}}{4/3\pi r d_p} \tag{3.1}$$

where r is the (unswollen) particle radius and d_p the density of the polymer.

Particle nucleation in emulsion polymerisation is a complex process and has been the subject of many investigations over the years. Rather than give a historical development, we shall cite a few milestones in modern understanding. The original development of emulsion polymerisation started with the premise that it was possible to polymerise in emulsion droplets. The realisation that micelles were "stung" to produce particles was put forward by Harkins (1947), and elegantly quantified by Smith and Ewart (1948), who also enunciated surfactant adsorption onto particles as the reason particle formation ceases. However, in the absence of added surfactant, it was pointed out (Priest, 1952; Fitch and Tsai, 1971) that *homogeneous nucleation* can occur, where monomer in the water phase propagates until the resulting species precipitates (undergoes a coil-to-globule transition).

A summary of the modern understanding of particle formation is now presented, with a sketch in Figure 3.1. First, consider the species present during Interval 1: monomer droplets, micelles (if there is sufficient added surfactant) and latex particles. The orders of magnitude of the number concentrations and diameters of these species are given in Table 3.1. Consider a system with a monomer which is sparingly water-soluble (e.g., styrene) and persulfate initiator, which thermally dissociates to sulfate radicals ($SO_4^{-\bullet}$) in the aqueous phase. These radicals are very hydrophilic and are extremely unlikely to enter any of the three organic phases just given. However, these sulfate radicals will propagate with the monomer in the water phase, albeit at a relatively slow rate because of the low monomer concentration in the water. Taking styrene as an example at 50 °C, the aqueous-phase monomer solubility $[M]_p$ is ~4 mM and the propagation rate coefficient of styrene with a polystyrene radical is ~2.4×10^2 $M^{-1}s^{-1}$ (Buback *et al.*, 1995), and thus a new monomer unit will be added to a sulfate radical in approximately 1 s (this is very much a lower bound, because the propagation rate coefficient of a sulfate radical is very much faster than this, but subsequent propagation steps will have rate coefficients closer to the styrene/polystyrene radical value).

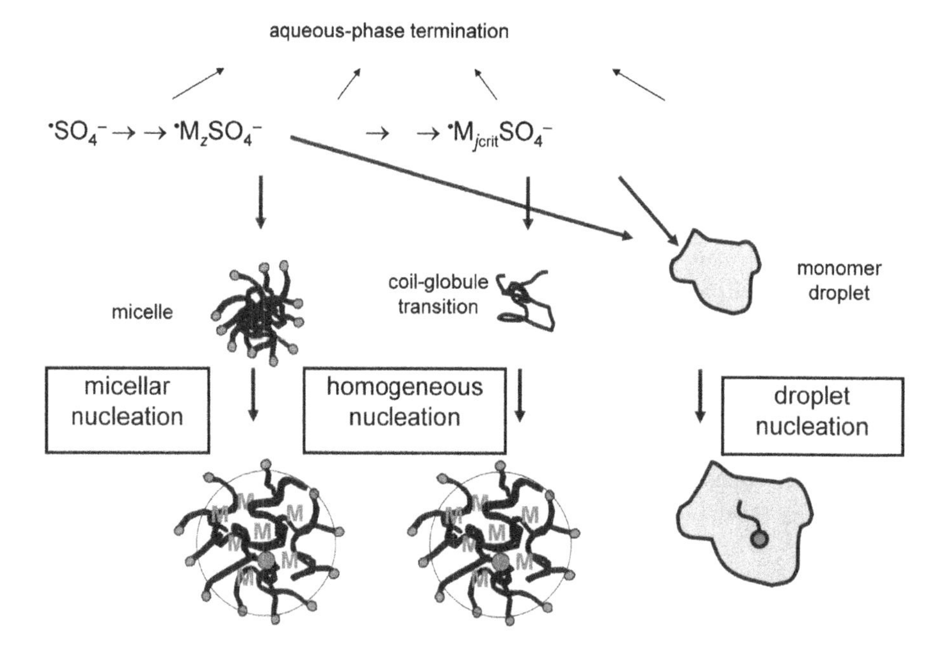

Figure 3.1 The most important events involved in particle formation. M denotes a monomer species.

After just a few additions, a species is formed with degree of polymerisation z (a z-mer) which will be surface active: for styrene this will be with $z \sim 3$, that is, $^{\bullet}M_3SO_4^-$ (Maxwell *et al.*, 1991).

Nucleation when micelles are present: If micelles are present, a surface-active z-mer will enter a micelle, and once its radical end is in the monomer-rich interior of the monomer-swollen micelle, propagation will be much more rapid. This rapid propagation will result in a latex particle. That is, latex particles form as a *new phase*.

It is instructive at this point to consider the fate of a newly-formed z-mer in the aqueous phase. It could (i) enter a micelle, (ii) enter a pre-existing particle, (iii) enter a monomer droplet, (iv) propagate further, or (v) undergo aqueous-phase termination. Now, the entry

Table 3.1 Indicative number densities and sizes of monomer droplets, micelles and latex particles.

	Monomer droplets	Micelles	Latex particles
Number density (ml^{-1})	10^7	10^{18}	10^{15}
Diameter (nm)	10^5	10^1	10^2
Rate of entry of z-meric radical (s^{-1})	10^1	10^8	10^6
Rate of aqueous-phase propagation of z-meric radical (s^{-1})	10^0		

rate coefficient of a surface-active species into any of these species is diffusion-controlled (Morrison *et al.*, 1992). This might seem surprising, because particles and micelles are covered with charged surfactant; however, modified DLVO theory indicates (Barouch, Matijevic and Wright, 1985) that electrostatic repulsion between the charged z-mer and these species is unimportant, because the electrical double layer is highly curved for a tiny species such as a charged radical. That is, this rate of radical entry is given by $k_e = 4\pi\, D N\, r$, where D is the diffusion coefficient of the radical species and N the number concentration of the latex particles, micelles or emulsion droplets. Choosing a diffusion coefficient of 1.5×10^{-5} cm^2 s^{-1} for a z-meric radical, the rates of entry of a z-meric radical into a particle, a droplet and a micelle are shown in Table 3.1. It is apparent that entry into micelles is favoured by about two orders of magnitude compared to entry into a pre-existing particle, and both are favoured by many orders of magnitude compared to entry into a droplet. The rate of further propagation is given by k_p $[M]_{aq}$, and the value of this rate for styrene is also shown in Table 3.1. The rate of aqueous-phase termination will be considered at a later stage in this discussion; suffice it to say here that termination of z-mers is usually unimportant if micelles are present.

Table 3.1 shows that propagation is orders of magnitude less likely than entry into micelles and pre-existing particles. Propagation is also slightly less likely than entry into droplets, but because the quantities in Table 3.1 are representative rather than always applicable, no significance can be placed on this difference.

The representative quantities of Table 3.1 immediately show that particle formation by entry into droplets (the third mechanism in Figure 3.1) is insignificant. There are three points of note here. (i) This droplet nucleation mechanism was the original rationale for the development of emulsion polymerisation, thus providing an interesting example of a valuable process being developed on incorrect reasoning. (ii) Droplet nucleation *is* the mechanism in mini-emulsion polymerisations, where in the ideal case each droplet becomes a particle (in these systems, the droplet size is much smaller, and the droplet number very much greater, than for conventional emulsion polymerisations). (iii) There are certain emulsion polymerisation systems where droplet nucleation *is* significant, for example, systems such as chlorobutadiene emulsion polymerisation (neoprene), where there may be a very large contribution to radical formation in droplets by adventitious species such as peroxides (Christie *et al.*, 2001), and certain RAFT-controlled emulsion polymerisations (Prescott *et al.*, 2002a, 2002b).

The preceding discussion shows that, while micelles are present, the predominant fate of new z-meric radicals in the water phase will be to enter micelles and thus form new particles. These particles are colloidally stabilized by adsorbed surfactant. As new particles are formed, and pre-existing ones grow by propagation, the increase in surface area of the particles will result in progressively more and more surfactant being adsorbed onto the particle surface, and the aqueous-phase surfactant is replenished from that in the micelles. Eventually, the concentration of free surfactant falls below the cmc, and micelles disappear.

Once micelles have disappeared, there can be no further particle formation by this micellar-nucleation mechanism, and a newly formed z-mer can have one of four remaining fates: entry into pre-existing particles, further propagation, entry into droplets, and aqueous-phase termination. Usually, although not invariably, the number of particles that have been formed at this stage is such that the rate of entry of z-mers into these pre-formed particles is much greater than that of the other fates (although it is essential to be aware that aqueous-phase termination is significant for lower degrees of polymerization, that is, there can be attrition of radicals by termination on the way to gaining surface activity). Particle formation in a system which initially contained surfactant above the cmc therefore ceases approximately when the surface area of pre-existing particles is sufficient to reduce the concentration of surfactant below the cmc. Typical surfactant concentrations are such that micelles are not exhausted particularly quickly, and thus the particle number in a system initially above the cmc is relatively high.

The pioneering work of Smith and Ewart (1948) introduced some, but not all, of these basic notions. The most important mechanism, which was unknown at that time, was that entry was by z-mers rather than by sulfate radicals directly entering particles. Smith and Ewart set out an elegant mathematical development leading to the particle number being given by:

$$N_p = \frac{1}{3}(5a_e)^{3/5}\left(\frac{k_d}{\pi^{1/2}K}\right)^{2/5}[\mathrm{I}]^{2/5}[\mathrm{S}]^{3/5} \tag{3.2}$$

where [I] and [S] are initiator and surfactant concentrations, respectively, a_e is the saturation value of the area per adsorbed surfactant molecule, k_d is the initiator dissociation rate coefficient, and $K = k_p M_0\,[\mathrm{M}]_p/(N_{Av} d_p)$, where N_{Av} is the Avogadro constant. This work was carried out in the absence of modern mechanistic knowledge, and thus this equation is only rarely quantitatively in accord with experiment (despite many claims in the early literature). This lack of agreement with experiment includes the dependence on [I] and [S]: the predicted exponents of 2/5 and 3/5 may indeed be seen, but only over a limited range of conditions. Nevertheless, Equation 3.2 correctly implies the increase in particle number with both initiator concentration (and radical flux) and surfactant concentration.

Homogeneous nucleation In the absence of added surfactant, the fates of a z-mer are the same as in a surfactant-containing system after the exhaustion of surfactant by adsorption: further propagation, entry into a pre-existing particle, and aqueous-phase termination. At the commencement of polymerisation, there are no particles present (contrasting with a micellar system when they form at the very beginning) and one of the main fates of a new z-mer is to undergo further propagation until it reaches a critical degree of polymerisation j_{crit}, which of course is greater than z. At this degree of polymerisation, the oligomers become "insoluble", or more precisely undergo a coil-to-globule transition. The resulting collapsed chain is hydrophobic, and thus is swollen by monomer. Hence, the radical end is in a monomer-rich environment, whereupon rapid propagation ensues and a particle forms. In the absence of added surfactant, charged end-groups from the initiator (e.g., $-SO_4^{-\bullet}$) provide colloidal stability. This mechanism is *homogeneous nucleation*. All new radicals that escape aqueous-phase termination will form particles. After a sufficient number of particles are formed, capture of z-mers by pre-existing particles becomes competitive with homogeneous nucleation (propagational growth to a j_{crit}-mer) and as there are no micelles present to capture z-mers (which would lead to new particle formation), particle formation finishes early. Thus the number of particles is much smaller, and the ultimate size of particles much larger, than in the presence of surfactant. Early publications on homogeneous nucleation are Goodall, Wilkinson and Hearn (1977), Fitch and Tsai (1971) and El-Aasser and Fitch (1987).

It is important to be aware that growth to j_{crit}-mers is always possible, even in a micellar-nucleated system after the exhaustion of surfactant. If conditions are such that this growth is significant compared to the alternative fates of entry into pre-existing particles and aqueous-phase termination, then new particles can form. This can occur, for example, in a seeded system wherein the particles are large and hence N_p is small. Recall that the rate of entry of a z-mer goes as particle size, but (for a given amount of polymer) the particle number is inversely proportional to the *cube* of particle size. This can lead to *secondary nucleation*: a new crop of particles formed in a seeded system. Means of quantifying this are now available (Morrison and Gilbert, 1995; Coen *et al.*, 1998).

Many quantitative models are now available for particle number which take into account the many events involved (Smith and Ewart, 1948; Priest, 1952; Fitch and Tsai, 1971; Ugelstad and Hansen, 1976; Hansen and Ugelstad, 1978; Hansen and Ugelstad, 1979a, 1979b, 1979c; Fitch *et al.*, 1984; Dougherty, 1986a, 1986b; Feeney, Napper and Gilbert, 1987; Richards, Congalidis and Gilbert, 1989; Giannetti, 1990; Schlüter, 1990, 1993; Hansen, 1992a, 1992b; Fontenot and Schork, 1992; Casey, Morrison and Gilbert, 1993; Tauer and Kühn, 1995; Coen and Gilbert, 1997; Coen *et al.*, 1998; Herrera-Ordonez and Olayo, 2000; Herrera-Ordonez and Olayo, 2001; Tauer, 2001; Gao and Penlidis, 2002; Coen *et al.*, 2004), including models which take into account some or all of the mechanisms omitted from Figure 3.1, such as intra-particle kinetics and coagulation. As exemplified by Coen *et al.* (Coen and Gilbert, 1997; Coen *et al.*, 2004), the more detailed models seem to be able to predict particle numbers from first principles which are in quite acceptable agreement with experiment for simple systems.

3.5 Particle Growth

Particle growth occurs throughout Intervals 1, 2 and 3. The kinetics are mainly controlled by the distribution and exchange of radicals over the various phases and cannot be oversimplified. Models are numerous but well described in various reviews (Ugelstad and Hansen, 1976; Hansen and Ugelstad, 1982; Gilbert and Napper, 1983, Gilbert, 1995). The basic rate equation for homogeneous batch free radical polymerisation is:

$$R_p = -\frac{d[M]}{dt} = k_p[M][R^\bullet] \tag{3.3}$$

where R_p is the rate of polymerisation per unit volume, [M] the monomer concentration, and $[R^\bullet]$ the radical concentration. This is modified in emulsion polymerisation to take account of the fact that the locus of polymerisation is within the latex particles, and one obtains:

$$R_p = k_p[M]_p\bar{n}\frac{N_p}{N_{Av}V_s} \tag{3.4}$$

where $\bar{n}$ is the average number of radicals per particle, and V_s is the (monomer-)swollen volume of a particle.

N_p changes throughout Interval 1. In Interval 2, the presence of monomer droplets keeps $[M]_p$ approximately constant, and N_p is also constant. In Interval 3, N_p remains constant while $[M]_p$ decreases, as determined by simple mass conservation from the behaviour of R_p.

The value of k_p can be determined by pulsed-laser polymerisation (PLP) (van Herk 2000; Olaj and Bitai, 1987; Coote, Zammit and Davis, 1996). A list of reliable values of k_p determined by this method is given in the Polymer Handbook and a number of papers by a IUPAC Working Party (e.g., Asua *et al.*, 2004). The number of particles, N_p, is determined from the particle size (Chapter 9) and conversion, through Equation 3.1.

The rate also requires a knowledge of $\bar{n}$. The classic work of Smith and Ewart (1948) again set out the ground-rules, but there have been significant advances since then. More recent mechanistic knowledge, such as the realisation that termination is diffusion-controlled (Benson and North, 1962) and hence depends on the degrees of polymerisation of both terminating chains, is the principal cause for the revised means of expressing the rates of polymerisation as now presented.

3.5.1 The Zero-One and Pseudo-Bulk Dichotomy

In free radical polymerisations, rates are controlled by the processes of initiation, propagation, transfer and combination. Although these same processes operate in an emulsion polymerisation, the kinetics in an emulsion polymerisation particle are in general different from those in bulk or solution polymerisations because of the confinement of radicals within particles. Thus, for example, systems where particles containing at most one radical are such that a radical in one particle is isolated from those in other particles, and thus the rate of termination (but not the rate coefficient) will be different from a bulk system with the same radical concentration. For this reason, one has to treat particles containing one radical as being quite distinct from those with two, and so on: that is, there will be different rate equations describing the kinetics of particles containing zero, one, two, ... radicals. Moreover, emulsion polymerisations exhibit events which have no counterpart in bulk or solution polymerisations: *phase transfer* processes, where radicals move between the particle and the water phases. There are two types of these: radical entry and radical exit.

It is now well established (e.g., (Maxwell *et al.*, 1991; van Berkel, Russell and Gilbert, 2003)) that entry arises when a z-meric radical irreversibly attaches itself to a particle. There is also strong evidence (e.g., (Morrison *et al.*, 1994)) that exit occurs by transfer within a particle to a monomer, producing a radical which is slightly soluble (e.g., it will be recalled that the water solubility of styrene monomer is ~ 4 mM, and it is reasonable to suppose that a styrene radical will have a similar solubility). This new monomeric radical can diffuse away from the parent particle (Nomura *et al.*, 1970).

A major cause of complexity in describing the kinetics of emulsion polymerisations is the dependence of the termination rate coefficient on the lengths of the two chains. In the most general case, this means that in a particle containing n growing chains, one must know the distribution of degrees of polymerisation of each of these chains. The result is an infinite hierarchy of variables and rate equations which cannot even be written in closed form, much less solved, except by Monte Carlo means, which cannot be used for routine interpretation of experimental data. Moreover, such equations would have a huge number of rate parameters whose values are uncertain, again thwarting a meaningful interplay between theory and experiment.

The way around this impasse has been to create a division of emulsion polymerisation kinetics into two simple but widely applicable cases: *zero-one* and *pseudo-bulk* kinetics.

Zero-one kinetics apply to a system in which the entry of a radical into a particle that contains a growing radical results in termination before significant propagation has occurred. Thus in a zero-one system, radical termination within a particle is not rate-determining. This type of behaviour commonly occurs for small particles; the size of particles for which this limiting behaviour is applicable depends on the monomer type and other polymerisation conditions. A zero-one system shows 'compartmentalisation behaviour', where the rate and molar mass distribution may be strongly influenced by radicals being isolated from each other within the latex particles. The value of $\bar{n}$ for a zero-one system can never exceed 1, and when the system is in a pseudo-steady state $\bar{n}$ can never exceed 1/2. (Historically, the zero-one classification incorporates both Cases 1 and 2 in the original Smith and Ewart paper; Smith–Ewart Case 2, where $\bar{n}$ is exactly $^1/_2$, is rarely met in practice).

A system obeying pseudo-bulk behaviour is one wherein the kinetics are such that the rate equations are the same as those for polymerisation in the bulk. In these systems, $\bar{n}$ can take any value in a pseudo-bulk system. Common cases are (i) when the value of $\bar{n}$ is so high that each particle effectively behaves as a micro-reactor, and (ii) when the value of $\bar{n}$ is low, exit is very rapid and the exited radical rapidly re-enters another particle and may grow to a significant degree of polymerisation before any termination event. (This case is *not* the same as Smith and Ewart's Case 3 kinetics, because these were applicable only to systems with $\bar{n}$ significantly above 1/2).

As will be seen, this simple dichotomy permits data to be interpreted in a way that avoids having so many fitting parameters that no meaningful mechanistic information can be obtained.

A straightforward method has been developed (Maeder and Gilbert, 1998; Prescott, 2003) for determining into which of the two categories a given system falls. This method calculates the probability as a function of chain length that termination will not occur after a z-meric radical has entered a latex particle containing a single growing chain and commenced propagation. This treatment is based on the diffusion-controlled model for termination (Russell, Gilbert and Napper, 1992; Strauch *et al.*, 2003). For common systems, the values of parameters needed for this calculation are available in the literature or can be estimated from analogous systems. This is illustrated in Figure 3.2, showing this probability for styrene and for butyl acrylate systems at 50 °C with 1 mM persulfate initiator and a particle number of 1.6×10^{17} dm^{-3}. It is seen that, for styrene, z-meric radicals entering into particles with 50 nm swollen radius are terminated at very low degrees of polymerisation, while the probability of not terminating starts to become significant beyond 70 nm unswollen radius. Hence for these conditions, styrene follows zero-one kinetics for swollen particle radii less than ~70 nm. However, for butyl acrylate, which has a very high propagation rate coefficient (Lyons *et al.*, 1996; Asua *et al.*, 2004), the system is only marginally zero-one for extremely small particles, and follows pseudo-bulk kinetics for particles with unswollen radii greater than ~30 nm.

3.5.2 Zero-One Kinetics

In the zero-one limit, particles can only contain one or no free radicals. The spirit of the original Smith–Ewart formulation is followed by defining N_0 and N_1 as the number of particles containing zero and one radical, respectively. These are normalized so that $N_0 + N_1 = 1$, and thus for a zero-one system, $\bar{n} = N_1$. The kinetic equations describing

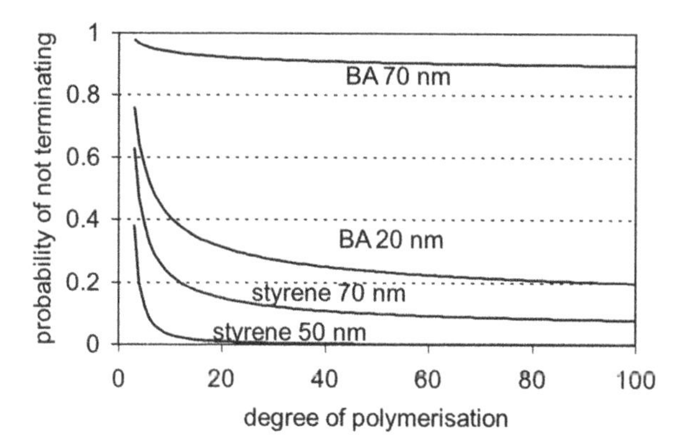

Figure 3.2 *Calculated probabilities for not terminating as a function of degree of polymerisation of an entering z-meric radical for styrene and butyl acrylate (z = 2 for both) for various swollen particle radii as indicated, for 1 mM persulfate initiator at 50 °C,* $N_p = 1.6 \times 10^{17}$ *dm*$^{-3}$.

the evolution of the radical population take into account changes due to radical entry (with entry frequency ρ, giving the number of entering radicals per particle per unit time; this terminology follows the IUPAC recommendation (Slomkowski *et al.*, 2011)), and exit (with exit frequency k, giving the number of radicals lost per particle per unit time by radicals going from the particle to the aqueous phase). These kinetic equations are:

$$\mathrm{d}N_0/\mathrm{d}t = -\rho N_0 + (\rho + k)N_1; \quad \mathrm{d}N_1/\mathrm{d}t = \rho N_0 - (\rho + k)N_1 \tag{3.5}$$

These equations are derived by noting that exit turns a particle containing one radical into a particle containing no radicals, as does entry, because entry causes instantaneous termination if the particle already contains a growing radical. Entry into a particle containing no radicals creates a particle containing a single growing radical.

The next step in developing these equations is to note that entry can be both by radicals which are derived directly from initiator, that is, z-mers such as $^{\bullet}M_3SO_4^-$, and by radicals which have exited from another particle. Recalling that exit is held to arise from transfer to monomer leading to a relatively soluble monomeric radical, it is realized that this exiting radical is chemically quite distinct from a z-mer. Because both z-meric (initiator-derived) and exited radicals can enter, it is necessary to take exit into account when considering entry. Now, consideration of the aqueous-phase kinetics of the various radical species (Morrison *et al.*, 1994) shows that the fate of an exited radical is overwhelmingly to enter another particle rather than undergo aqueous-phase termination (except for a few special systems such as vinyl acetate (De Bruyn, Gilbert and Ballard, 1996)). Since the rate of exit is $k\bar{n}$, and each exit leads to re-entry, and since $\bar{n} = N_1$ in a zero-one system, Equation 3.5 becomes:

$$\frac{\mathrm{d}\bar{n}}{\mathrm{d}t} = \rho_{\mathrm{init}}(1 - 2\bar{n}) - 2k\bar{n}^2 \tag{3.6}$$

where ρ_{init} is the entry rate coefficient for radicals *derived directly from the initiator*.

It is apparent why the zero-one limiting case is useful to interpret and predict data: it contains only two rate parameters, ρ and k. These can be obtained unambiguously by a combination of initiating using γ-radiolysis and then following the rate after removal from the radiation source (these data are very sensitive to radical loss processes, especially the exit rate coefficient k), and steady-state rate data in a system with chemical initiator, which is sensitive to both ρ and k. Moreover, the transfer model for exit makes specific predictions as to the dependence of the value of k on, for example, particle size (Ugelstad and Hansen, 1976). Specifically, this model predicts that k should vary inversely with particle area, and that the actual value of k can be predicted a priori from rate parameters measured by quite different means, such as the transfer constant to monomer. This size dependence, and the quantitative accord between model and experiment, have been verified for the exit frequency (Morrison *et al.*, 1994).

This type of data also yields the frequency for entry, ρ_{init}. The dependence of this on initiator concentration for systems containing ionic surfactant can be quantitatively fitted by the "Maxwell–Morrison" model for exclusive entry by z-mers (Maxwell *et al.*, 1991; van Berkel, Russell and Gilbert, 2003); the situation is more complex but still well understood for systems containing polymeric surfactant (Thickett, Morrison and Gilbert, 2008). Moreover, such experiments have also verified the prediction of this model that ρ_{init} should be independent of both particle size (in two systems with the same initiator and particle concentrations but two different particle sizes) (Coen, Lyons and Gilbert, 1996) and the charge on both initiator and particle (van Berkel, Russell and Gilbert, 2003).

These mechanistic deductions, which have led to considerable physical understanding of the kinetics of particle growth, illustrate the usefulness of the zero-one limiting case.

It is noted that a limiting value of zero-one kinetics occurs when the radical flux is so high that the steady-state value of $\bar{n}$ takes the value of exactly $1/2$. This arises because if there is a high radical flux, then entry into a particle containing a growing radical is the dominant chain-stopping event ("instantaneous" termination). Each entry event then either creates a radical if there were none in the particle, or destroys radical activity if the particle did contain a growing chain; it is then apparent that, on average, half the particles will contain a single growing radical and half will contain none, that is, $\bar{n} = 1/2$. This elegant limiting case was deduced by Smith and Ewart (1948). It is characterized by a polymerisation rate that is independent of initiator concentration. However, such cases are very rare: the only examples showing $\bar{n} = 0.5$ that are deemed reliable are for certain styrene systems (e.g., (Hawkett, Napper, and Gilbert, 1980)). There were many erroneous reports in the early literature that it was *always* the case that $\bar{n} = 0.5$, based on misinterpretation of the original Smith–Ewart paper. The advent of reliable values of k_{p} through PLP (Olaj and Bitai, 1987; Buback *et al.*, 1992, 1995) meant that $\bar{n}$ could be determined directly from observed polymerisation rates using Equation 3.4, and showed that the often-presumed "universality" of $\bar{n} = 1/2$ was science fiction rather than fact.

3.5.3 Pseudo-Bulk Kinetics

In this limit, termination within the particle is rate-determining. An important implication of termination being diffusion-controlled (and hence chain-length dependent) is that, in conventional free-radical polymerisation, termination events are dominated by termination between a mobile short chain (formed in an emulsion polymerisation by entry of a z-mer, by re-entry of an exited radical, or by transfer to monomer) and a long, relatively immobile one. This is known as "short-long termination".

The compartmentalisation, which is a major effect in zero-one systems, is by definition absent in pseudo-bulk systems because either (i $\bar{n}$ is sufficiently large as to render insignificant the isolation of radicals from each other which is so important in zero-one systems, or (ii) there is a very rapid exchange of radicals between particles. The latter circumstance can occur in systems with low $\bar{n}$ (which is why the pseudo-bulk classification is qualitatively different from Smith–Ewart Case 3). This rapid exchange is typical of the emulsion polymerisations of butyl acrylate and methyl methacrylate. It arises when the product of the transfer constant and the water solubility of the monomer is sufficiently high that there is an adequate rate of radicals exiting the particles, *and* that either these exited radicals re-enter and re-exit a number of times (so that the boundaries between particles are not "visible"), or they propagate so quickly on re-entry that rapid short-long termination is diminished.

Under any of these circumstances, a pseudo-bulk emulsion polymerisation follows the same kinetics as the equivalent bulk system:

$$\frac{d\bar{n}}{dt} = \rho - 2\frac{< k_t >}{N_{AV} V_S}\bar{n}^2 \tag{3.7}$$

Here the *average* termination rate coefficient k_t is the average of the chain-length dependent termination rate coefficients $k_t^{i,j}$ over the distribution of radicals of each chain length, with concentrations R_i (Russell, Gilbert and Napper, 1992):

$$< k_t >= \frac{\Sigma_\iota \Sigma_\varphi \kappa_\tau^{\iota\varphi} P_\iota P_\varphi}{(\Sigma_\iota P_\iota)^2} \tag{3.8}$$

It is seen that $<k_t>$ depends on radical concentrations and, hence, on the concentration of initiator. This dependence is not strong, but is significant, and means, for example, that the value of $<k_t>$ changes if the radical source is switched off (which is one of the reasons why the traditional rotating-sector method for determining k_p can give false results). However, simulations of the type performed by Russell *et al.* (Russell, Gilbert and Napper, 1993) show that $<k_t>$ quickly goes to its low radical flux value after the source of radicals is removed. It is essential to be aware that the value of $<k_t>$ from a relaxation experiment will be slightly but significantly different from that in the same system with a constant concentration of chemical initiator.

What is important about Equation 3.7 is that, just as with Equation 3.6, it is an equation with only two parameters, which can thus be used for unambiguous data interpretation and prediction, without a plethora of adjustable parameters. An example of this is the extraction of $<k_t>$ from gamma relaxation rate data in the emulsion polymerisation of styrene (Clay *et al.*, 1998). The data so obtained are in accord (within experimental scatter) of $<k_t>$ values inferred from treatment of the molecular weight distributions, and also from a priori theory. This data reduction method has also been performed recently for a corresponding methyl methacrylate emulsion polymerisation (van Berkel, Russell and Gilbert, 2005). The information gained from these data is particularly useful: it supports the supposition that termination is indeed controlled by short-long events. Moreover, for the MMA system, the data show that radical loss is predominantly caused by the rapid diffusion of short radicals generated by transfer to monomer (i.e., the rate coefficient for termination is a function of those for transfer and primary radical termination). Such mechanistic information is clearly useful for the interpretation and design of emulsion polymerisation systems in both academia and industry.

3.5.4 Systems between Zero-One and Pseudo-Bulk

Systems whose kinetics do not fall unambiguously into the zero-one or pseudo-bulk categories pose a problem for routine interpretation and prediction, let alone for obtaining useful mechanistic information such as that discussed in the preceding section. One can always use Monte Carlo modelling (Tobita, 1995), but the enormous amount of computer time this requires, and the plethora of unknown parameters, precludes its use for obtaining mechanistic information from experiment.

A recent breakthrough in this area (Prescott, Ballard and Gilbert, 2005b) has solved this problem for conventional emulsion polymerisation systems. In essence, one uses the "zero-one-two" model (Lichti, Gilbert and Napper, 1980), where instead of assuming instantaneous termination if a radical enters a particle containing a growing chain, one allows for a finite rate of termination in this case, but assumes instantaneous termination if a radical enters a particle containing *two* growing chains. Based on the "distinguished-particle" approach (Lichti, Gilbert and Napper, 1980, 1982), an expression is derived for k_t which takes chain length dependence into account while allowing for compartmentalisation effects. This formulation is readily implemented with minimal computational resources. However, the approximations used to derive this result are unfortunately inapplicable to the important case of controlled radical polymerisation in emulsion, where most growing chains are long, and where compartmentalisation effects are important (Prescott, 2003); in this case, there is no alternative at present to the Monte Carlo treatment (Prescott, 2003).

These various kinetic models can be used to predict and interpret the molar mass distributions in emulsion polymerisations, for example, (Clay and Gilbert, 1995; Clay *et al.*, 1998; van Berkel, Russell and Gilbert, 2005). In many systems, for example, (van Berkel, Russell and Gilbert, 2005), the molecular weight distribution (MWD) is dominated by transfer to monomer, even when there is extensive termination. The latter result, which is at first surprising, is because of short-long termination: transfer to monomer results in a highly mobile short radical which often terminates with a relatively immobile long one before it has undergone significant propagation. Contrary to what is sometimes believed, transfer to monomer frequently dominates the MWD in a zero-one system, despite the fact that entry results in instantaneous termination and therefore one might intuitively expect that entry would be the dominant chain-stopping event. However, examination of the typical values for entry rate coefficients, for example, (van Berkel, Russell and Gilbert, 2003), shows that the time between entry events is much longer than that between transfer events. In systems with high $\bar{n}$ (which will always be pseudo-bulk), the dominant chain-stopping event is often termination between two chains of at least moderate degree of polymerisation, for example, (Clay *et al.*, 1998). In this case, the molecular weights (or $\bar{M}_n$ to be precise) are significantly less than the transfer limit.

Other publications on the prediction of molecular weight distributions are Tobita, Takada and Nomura (1994) and Giannetti, Storti and Morbidelli (1988).

3.6 Ingredients in Recipes

This section provides an overview of the major ingredients in emulsion polymerisation. A laboratory scale recipe for an emulsion polymerisation contains monomer, water, initiator, surfactant, and sometimes a buffer and/or chain transfer agents. Commercial emulsion polymerisation recipes are usually much more complicated, with 20 or more ingredients. The complexity of components, and the sensitivity of the system kinetics, mean that small changes in recipe or reaction conditions often result in unacceptable changes in the quality of the product formed.

3.6.1 Monomers

Particles in an emulsion polymerisation comprise largely monomers with a limited solubility in water. The most common monomers are styrene, butadiene, vinyl acetate, acrylates and methacrylates, acrylic acid and vinyl chloride. Besides monomers that make up a large part of the latex, other monomers are often added in smaller quantities and have specific functions, like stabilisation ((meth)acrylic acid) and reactivity in crosslinking (epoxy-group containing monomers, amine- or hydroxy functional groups etc.). These are often denoted "functional monomers".

3.6.2 Initiators

The most commonly used laboratory water-soluble initiators are potassium, sodium and ammonium persulfates. Next in line are the water-soluble azo-compounds, especially those with an ionic group, such as 2,2′-azobis(2-methylpropionamidine) dihydrochloride, or V-50. Another important group are the peroxides (benzoyl peroxide, cumene hydroperoxide).

In cases where the polymerisation should be performed at lower temperatures (less than 50 °C), a redox system can be used. Lower polymerisation temperature gives the advantage of lowering chain branching and crosslinking in the synthesis of rubbers. Usually the redox couple reacts quickly to produce radicals, and thus one or both components must be fed during the course of the emulsion polymerisation process. For this reason, redox initiators are very useful for safety in commercial emulsion polymerisations because, in the case of a threatened thermal runaway (uncontrolled exotherm), the reaction can be quickly slowed by switching off the initiator feed. A typical example of a redox system is *tert*-butyl hydroperoxide and sodium metabisulfite. While reductants such as Fe^{2+} can be used, these tend to produce discolouration and also may induce coagulation of the latex particles.

There are also other methods to create free radicals, such as γ-radiolysis, light in combination with photoinitiators, and electrons from high energy electron beams. One of the advantages is that these alternative methods can produce pulses of radicals and in this way an influence on the growth time of the polymer can be obtained. On the laboratory scale these methods are used to obtain kinetic parameters (Gilbert, 1995; van Herk, 2000).

3.6.3 Surfactants

A surfactant (surface active agent), also referred to as emulsifier, soap or stabiliser, is a molecule having both hydrophilic and hydrophobic segments. The general name for this group is amphipathic, indicating the molecules' tendency to arrange themselves at oil/water interfaces. In emulsion polymerisation surfactants serve three important purposes: stabilisation of the monomer droplets, generation of micelles, and stabilisation of the growing polymer particles leading to a stable end product.

Surfactants are mostly classified according to the hydrophilic group:

- Anionic surfactants, where the hydrophilic part is an anion
- Cationic surfactants, where the hydrophilic part is a cation
- Amphoteric surfactants, where the properties of the hydrophilic function depend on the pH,
- Non-ionic surfactants, where the hydrophilic part is a non-ionic component, for instance polyols, sugar derivatives or chains of ethylene oxide.

Other types of surfactants are the polymeric (steric) stabilizers such as partially hydrolysed polyvinyl acetate. Also oligomeric species formed *in situ*, when $SO_4^{-\bullet}$ radicals react with some monomer units in the aqueous phase, will have surface active properties, and can even form a colloidally stable latex. Electrosteric stabilisers combine steric and electrostatic functionalities: for example, inclusion of acrylic acid in a recipe results in chains with blocks comprised largely of poly(acrylic acid) which, in the aqueous phase, then pick up enough hydrophobic monomer to enter the particle and continue polymerisation in the particle interior. The hydrophilic component remains in the aqueous phase and provides colloidal stability both sterically and, under the appropriate conditions of pH, electrostatically. This mode of stabilisation is very common in surface coatings, because it gives excellent freeze–thaw stability.

Common anionic surfactants include sodium dodecyl sulfate (SDS) and the Aerosol series (sodium dialkyl sulfosuccinates), such as Aerosol OT (AOT, sodium di(2-ethylhexyl)sulfosuccinate) and Aerosol MA (AMA, sodium dihexyl sulfosuccinate). These particular surfactants tend to result in monodisperse latexes.

3.6.4 Other Ingredients

Electrolytes. Electrolytes are added for several reasons. For example, they can control the pH (buffers), which prevents hydrolysis of the surfactant and maintains the efficiency of the initiator. The addition of electrolytes can lead to more monodisperse particles but also to coagulation.

Chain Transfer Agents. Emulsion polymerisation may result in an impractically high molecular mass polymer. To reduce the molar mass, chain transfer agents (CTA), usually mercaptans, are frequently used. The mercaptan is introduced into the reactor together with the monomer phase. The consumption of the mercaptan taking place in the loci should be properly kept in balance with monomer consumption.

Sequestering agents. In redox systems, adventitious metal ions can catalyse radical formation in an uncontrolled way. So-called sequestering agents (for example EDTA) are added to prevent this.

3.7 Emulsion Copolymerisation

3.7.1 Monomer Partitioning in Emulsion Polymerisation

Because an emulsion polymerisation comprises several phases (water, monomer droplets, particles) it is important to know what the local concentrations are at the locus of polymerisation (usually the particles).

The monomer concentration in the latex particles directly determines the rate of polymerisation while the monomer ratio in the latex particles determines the chemical composition of the copolymer formed. The monomer concentrations can be accessed both experimentally and through models. These models can be useful in the design of polymerisation reactors, process control and product characteristics, such as molar mass and chemical composition distributions of the copolymers formed. In this section a thermodynamic model based on the Flory–Huggins theory (Flory, 1953) of polymer solutions will be discussed and applied to experimental results on the partitioning of monomer(s) over the different phases present during an emulsion (co)polymerisation.

At equilibrium the partial molar free energies of the monomer will have the same value in each of the phases present, that is, the monomer swollen colloid (micelles, vesicles and/or latex particles), the monomer droplets and the aqueous phase (Morton, Kaizerman and Altier, 1954; Gardon, 1968):

$$\Delta G_{\mathrm{l}} = \Delta G_{\mathrm{d}} = \Delta G_{\mathrm{a}} \tag{3.9}$$

where ΔG_{l}, ΔG_{d} and ΔG_{a} are the partial molar free energies of the latex phase, monomer droplets, and the aqueous phase, respectively. Utilising the appropriate equations for the partial molar free energy of the colloidal and aqueous phase (see for derivation e.g., Maxwell *et al.* (1992a)), Equation 3.10 is obtained. This is known as the Vanzo equation (Vanzo, Marchessault and Stannett, 1965), which describes the partitioning of monomer between the aqueous phase and the latex particles in the absence of monomer droplets.

$$\ln(1 - v_{\mathrm{p}}) + v_{\mathrm{p}}\left[1 - \frac{1}{\bar{X}_{\mathrm{n}}}\right] + \chi v_{\mathrm{p}}^{2} + \frac{2V_{\mathrm{m}}\gamma v_{\mathrm{p}}^{1/3}}{r_{0}RT} = \ln\left[\frac{[\mathrm{M}]_{\mathrm{a}}}{[\mathrm{M}]_{\mathrm{a,sat}}}\right] \tag{3.10}$$

Here v_{p} is the volume fraction of polymer (related to the conversion), X_{n} is the number average degree of polymerisation of the polymer, χ is the Flory–Huggins interaction parameter between the monomer and the polymer, R is the gas constant and T the temperature. V_m is the molar volume of the monomer, γ is the particle–water interfacial tension and r_{o} is the radius of the unswollen micelles, vesicles and/or latex particles. $[\mathrm{M}]_{\mathrm{a}}$ is the concentration of monomer in the aqueous phase, and $[\mathrm{M}]_{\mathrm{a,sat}}$ the saturation concentration of monomer in the aqueous phase. Figure 3.3 shows the contributions of the different terms of Equation 3.10 to the Vanzo equation.

For a more detailed discussion see also Section 4.2 in Chapter 4 and Figure 4.5.

The partitioning of monomer between the aqueous phase and latex particles, below and at saturation, can be predicted by Equation 3.10. However, this requires that both the Flory–Huggins interaction parameter and the interfacial tension be known. These parameters may be polymer volume fraction dependent (see Maxwell *et al.* (1992a, 1992b) for prediction of monomer partitioning). Equations similar to Equation 3.20 can be derived for the swelling of micelles and vesicles with one of more monomers, and of homopolymer, co- and terpolymer latex particles with two or more monomers at and below saturation (Noel, Maxwell and German, 1993; Schoonbrood *et al.*, 1994). However, one must be careful with the use of the Vanzo equation, because data fitting can under some circumstances lead to negative values of χ (Kukulj and Gilbert, 1997), which is of course unphysical.

Equation 3.10 can describe the swelling behaviour of a latex particle with one monomer below and at saturation. In the case of saturation swelling with two monomers, substituting the appropriate expression for the partial molar free energy of the different phases into

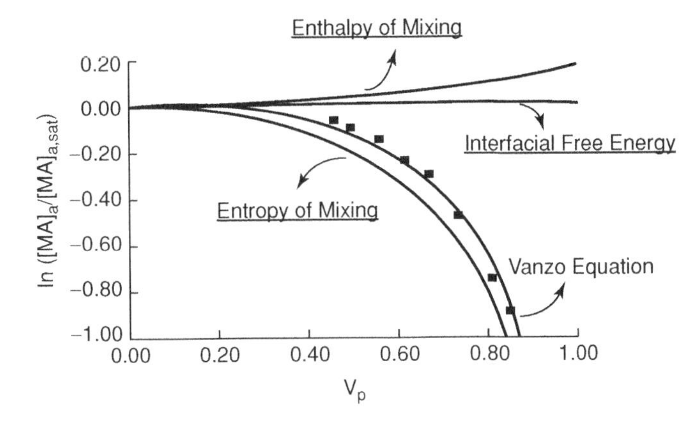

Figure 3.3 *Comparison of theoretical predictions and experimental measurements of methyl acrylate partitioning at 45 °C for a poly(methyl acrylate) seed latex with an unswollen radius of 91 nm (closed squares). Theoretical predictions: different terms of Equation 3.10 are depicted, for the Vanzo equation $\chi = 0.2$ and $\gamma = 45$ mN m^{-1} were taken from the literature (Maxwell* et al.*, 1992a).*

Equation 3.10 leads to Equation 3.11 (for an exact derivation see e.g., Maxwell *et al.* (1992a, 1992b) and Noel, Maxwell and German (1993)) which is quite similar to Equation 3.10, except that it is assumed that the molecular weight is sufficiently high that the term in X_n can be ignored.

$$\begin{aligned} &\ln v_{p,i} + (1 - m_{ij}) * v_{p,j} + v_p + \chi_{ij} * v_{p,j}^2 + \chi_{ip} * v_p^2 \\ &+ v_{p,j} * v_p * (\chi_{ij} + \chi_{ip} - \chi_{jp} * m_{ij}) + \frac{2 * V_{m,i} * \gamma * v_p^{1/3}}{r_0 * RT} \\ &= \ln v_{d,i} + (1 - m_{ij}) * v_{d,j} + \chi_{ij} * v_{d,j}^2 = \ln\left[\frac{[M_i]_a}{[M_i]_{a,sat}}\right] \end{aligned} \tag{3.11}$$

Here $v_{p,i}$ and $v_{p,j}$ are the volume fractions of monomers i and j in the latex particles, respectively, $\chi_{i,j}$ is the Flory–Huggins interaction parameter between monomer i and j, while $\chi_{i,p}$ and $\chi_{j,p}$ are the Flory–Huggins interaction parameters between monomer i and j and the polymer, respectively; m_{ij} is the ratio of the molar volume of monomer i over monomer j, and $v_{d,i}$ and $v_{d,j}$ represent the volume fraction of monomer i and j, respectively, in the monomer droplets. It can be shown from Equation 3.11 that, at saturation swelling, the mole fraction of monomer i in the monomer droplets ($f_{i,d}$) equal is to the mole fraction of monomer i in the latex particles ($f_{i,p}$). This holds also for monomer j. This is shown in Equation 3.12 (Maxwell *et al.*, 1992b):

$$f_{i,p} = f_{i,d} \quad \text{and} \quad f_{j,p} = f_{j,d} \tag{3.12}$$

where $f_{j,d}$ and $f_{j,p}$ are the mole fractions of monomer j in the monomer droplets and in the latex particles, respectively. Making the assumption that the total monomer concentration in the latex particles is equal to the sum of the concentrations of the individual monomers, together with Equation 3.12, the concentration of monomer i in the latex particles can

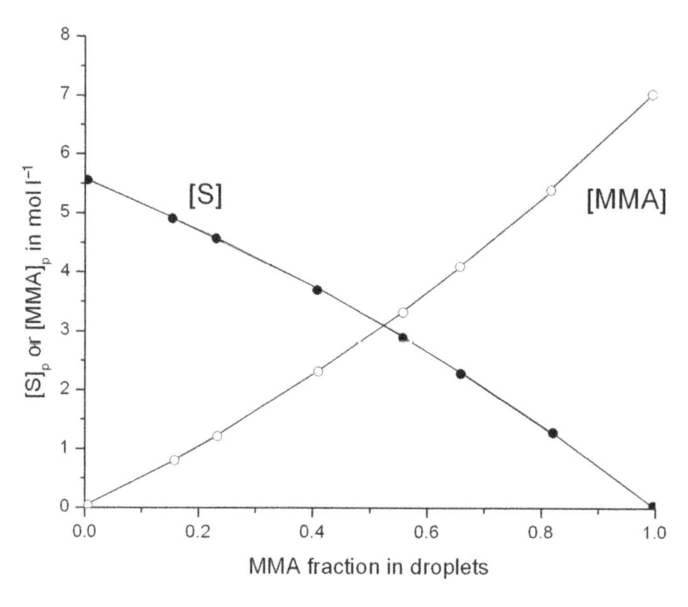

Figure 3.4 *Experimentally determined monomer concentrations in the latex particles as a function of the fraction of methyl methacrylate in the droplet phase (symbols) compared with theoretical predictions according to Equation 3.13 (full lines). Methyl methacrylate concentrations (open circles), styrene concentrations (closed squares) in poly(styrene-co-methyl methacrylate)(25/75).*

be predicted from the individual saturation concentration of monomer i and j in the latex particles, that is, $[M_{i,sat}]$ and $[M_{j,sat}]$, respectively. For a given seed latex, the concentration of monomer i in the latex particles ($[M_i]$) is related to the mole fraction of monomer i in the monomer droplets ($f_{i,d}$) and given by the following equation (Maxwell *et al.*, 1992a, 1992b):

$$[M_i] = f_{i,d}\left(\left(\left[M_{i,sat}\right] - \left[M_{j,sat}\right]\right) f_{i,d} + \left[M_{j,sat}\right]\right) \tag{3.13}$$

A similar expression can be deduced for monomer j. Figure 3.4 represents the partitioning of styrene and methyl methacrylate between monomer droplets and latex particles consisting of poly(styrene-*co*-methyl methacrylate) and polybutadiene-*graft*-poly(styrene-*co*-methyl methacrylate), respectively (Aerdts, Boei and German, 1993). As is shown in Figures 3.4 and 3.5, the experimental data can be described by this model, that is, by Equation 3.13. Figure 3.5 shows the mole fraction of methyl methacrylate in the latex particles as a function of the mole fraction of methyl methacrylate in the monomer droplets. Here again, the model can describe the experimental results extremely well.

The above discussed thermodynamic model for describing, explaining and predicting monomer partitioning during an emulsion polymerisation has also been successfully applied to the swelling of phospholipid bilayers by an organic solvent (Maxwell and Kurja, 1995).

In conclusion, it can be said that the partitioning behaviour of monomers between the different phases present during an emulsion polymerisation can be described and predicted using a simple thermodynamic model derived from the classical Flory–Huggins theory

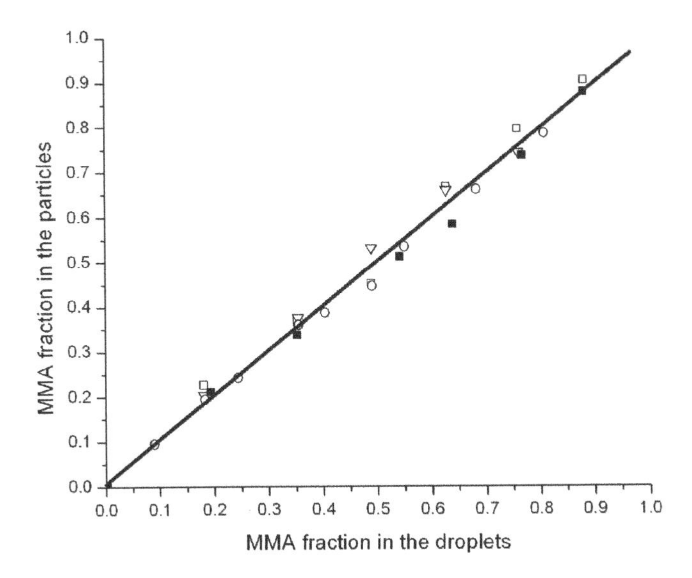

Figure 3.5 *Experimentally determined fractions of methyl methacrylate (MMA) in the droplet phase as a function of the fraction of methyl methacrylate in different latex particles. MMA and styrene in PB (open circles), SMMA-free (open squares), SMMA-graft (open triangles) from polybutadiene-graft-poly(styrene-co-methyl methacrylate) latex particles, while the closed squares represent a poly(styrene-co-methyl methacrylate) latex swollen with styrene and MMA. The solid line gives the theoretical prediction according to Equation 3.13.*

for polymer solutions. In general, therefore, the monomer concentrations at the locus of polymerisation are relatively easily accessible. However, this is not the case for more water-soluble monomers (acrylic acid etc.). For these monomers suitable models are not readily available and one has to rely on the experimental data.

3.7.2 Composition Drift in Emulsion Co- and Terpolymerisation

A special aspect of (emulsion) copolymerisation compared to (emulsion) homopolymerisation is the occurrence of composition drift. In combination with the instantaneous heterogeneity (statistical broadening around the average chemical composition), this phenomenon is responsible for the chemical heterogeneity of the copolymers formed. Composition drift is a consequence of the difference between instantaneous copolymer composition and overall monomer feed composition. This difference is determined by: (i) the reactivity ratios of the monomers (kinetics), and (ii) the monomer ratio in the main loci of polymerisation (viz., latex particles) which can differ from the overall monomer ratio of the feed (as added according to the recipe), which in turn is caused by monomer partitioning. In many cases the monomer ratio in the polymer particles equals the monomer ratio in the monomer droplets; the water solubility of the monomers is then the main factor that has an effect on the monomer ratio in the polymer particles.

In principle, when one compares solution or bulk copolymerisation to emulsion copolymerisation, two situations can be distinguished: (i) If the more reactive co-monomer is the less water-soluble one, then there will be a stronger composition drift as the amount of water increases in the recipe (e.g., styrene-methyl acrylate (Schoonbrood *et al.*, 1995a)); (ii) If the more reactive co-monomer is the more water-soluble one, then a smaller composition drift can occur as the amount of water increases (e.g., indene-methyl acrylate, methyl acrylate-vinyl 2,2-dimethylpropanaoate (Noel *et al.*, 1994)). In the latter cases the composition drift may even be reversed at very high water contents.

In order to be able to describe and control an emulsion copolymerisation, both the reactivity ratios and monomer partitioning have to be known.

Batch processes are known to give two-peaked distributions of copolymer composition when a strong composition drift occurs during the course of the (emulsion) copolymerisation. Moreover, in emulsion copolymerisation the degree of bimodality appears to depend on the monomer/water ratio (Van Doremaele, 1990; Van Doremaele, van Herk and German (1990) Schoonbrood *et al.*, 1995a). Semi-continuous processes (i.e., addition of monomer during polymerisation) can be used to prepare more homogeneous copolymers. Dynamic mechanical spectroscopy or differential scanning calorimetry and transmission electron microscopy combined with preferential staining techniques have been used to determine the possible occurrence of phase separation due to double-peaked chemical composition distributions (CCDs). It has been shown that the compositional heterogeneity of the copolymer has a dramatic effect on the mechanical properties (Schoonbrood *et al.*, 1995a).

Ternary emulsion copolymerisation. In the fundamental investigations described in literature dealing with emulsion copolymerisation most attention has been given to binary copolymerisation, that is, polymerisation of two monomers. Far less attention has been paid to ternary emulsion copolymerisation (three monomers), that is, terpolymerisation. Emulsion terpolymerisation investigations, mostly dealing with properties and applications, have been published mainly in the patent literature.

It is obvious that the typical aspects that distinguish emulsion copolymerisation from homopolymerisation, for example, monomer partitioning, dependence of kinetics on the local monomer concentration ratio and so on, become much more complex when three monomers are involved, not to mention the complications in terpolymer analysis.

However, since it can easily be understood that using three monomers gives the possibility to obtain an even larger variety and refinement of copolymer properties, there is considerable research on emulsion terpolymerisation, although it is to be expected that there will be little or no fundamental mechanistic differences between binary and ternary emulsion copolymerisation systems.

Schoonbrood studied the emulsion terpolymerisation of styrene, methyl methacrylate and methyl acrylate (Schoonbrood et. al., 1996a, 1996b) and for the first time also determined the propagation rate coefficients for this ternary system by means of pulsed laser polymerisation (Schoonbrood *et al.*, 1995b). He also determined and predicted the microstructure, in terms of CCD, of these terpolymers (Schoonbrood et al., 1996a, 1996b).

3.7.3 Process Strategies in Emulsion Copolymerisation

The emulsion polymerisation strategy, that is, the kind of process, can have a considerable effect on molecular structure and particle morphology. The intrinsic factors as well as the process conditions determine the colloidal aspects of the copolymer latex (particle diameter, surface charge density, colloidal stability, etc.), the characteristics of the polymeric material in the particles and the structure of the particles (copolymer composition as a function of particle radius, etc.). In turn, these factors determine the properties of the latex and the copolymer product.

The ultimate goal of most of the investigations on emulsion copolymerisation is to be able to control the process in such a way as to produce a copolymer product (latex or coagulate) with the desired properties. For this purpose the semi-continuous (sometimes called semi-batch) emulsion copolymerisation process is widely used in industry. The main advantages of this process as compared with conventional emulsion batch processes include a convenient control of emulsion polymerisation rate in relation to heat removal, and control of the chemical composition of the copolymer and particle morphology. These are important features in the preparation of speciality or high performance polymer latexes.

Semi-continuous emulsion copolymerisation processes can be performed by applying various monomer addition strategies.

Constant addition strategy. The most widely investigated and described procedure is the addition of a given mixture of the monomers (sometimes pre-emulsified monomers) at a constant rate.

For instance, this procedure is followed in many papers dealing with the semi-continuous emulsion copolymerisation of vinyl acetate and butyl acrylate (e.g., El-Aasser *et al.*, 1983). Two main situations can be distinguished with respect to the monomer addition rate: (i) Flooded conditions: the addition rate is higher than the polymerisation rate. (ii) Starved conditions: the monomers are added at a rate lower than the maximum attainable polymerisation rate (if more monomer were to be present). The latter process (starved conditions) is often applied in the preparation of homogeneous copolymers/latex particles. In this case after some time during the reaction, because of the low addition rates, a steady state is attained in which the polymerisation rate of each monomer is equal to its addition rate and a copolymer is made with a chemical composition identical to that of the monomer feed. Sometimes semi-continuous processes with a variable feed rate (power feed) are used to obtain latex particles with a core–shell morphology (Bassett, 1983).

Controlled Composition Reactors. Intelligent monomer addition strategies in copolymerisations rely on the monitoring of monomer conversions. In copolymerisation, control of the copolymer composition can also be obtained when applying monomer addition profiles. These monomer addition profiles can either be based on the direct translation of on-line measurements to monomer addition steps (controlled composition reactor) or the profiles can be predicted by emulsion copolymerisation models on a conversion basis. The required conversion–time relation is then obtained by on-line measurements. On-line methods of determining monomer conversion are, as outlined above, important for controlling the emulsion (co)polymerisation process. Developments in on-line Raman spectroscopy are very promising (McCaffery and Durant, 2002)(Van den Brink *et al.*, 2001). Reviews have appeared on on-line sensors for polymerisation reactors (Schork 1990; German *et al.*, 1997).

Optimal Addition Profile. Arzamendi and Asua (1989) developed a so-called optimal monomer addition strategy. By using this method they demonstrated that within a relatively short period of time homogeneous vinyl acetate (VAc)-methyl acrylate (MA) emulsion copolymers can be prepared in spite of the large difference in the reactivity ratios. The reactor was initially charged with all of the less reactive monomer (viz., VAc) plus the amount of the more reactive monomer (viz., MA) needed to initially form a copolymer of the desired composition. Subsequently, the more reactive monomer (MA) was added at a computed (time variable) flow rate (optimal addition profile) in such a way as to ensure the formation of a homogeneous copolymer.

Van Doremaele (1990) applied a more pragmatic approach: a method which can be applied without actually calculating $\bar{n}(t)$ or $\bar{n}(f_p)$ and may, therefore, be more generally applicable. This method was applied to the emulsion copolymerisation of styrene (S) and MA. The batch emulsion copolymerisation of S and MA is known to often produce highly heterogeneous copolymers (styrene being the more reactive and less water-soluble monomer).

Rather than a large difference between the reactivity ratios (VAc-MA), here the large difference between the water solubilities of S and MA is the main problem. As stated, the time-evolution of $\bar{n}$ was not actually calculated but it was set equal to 0.5 as a first estimation. It would be highly fortuitous if the first estimated addition profile based on $\bar{n} = 0.5$ would be optimal, because the average number of radicals will generally deviate from this first estimation. Nevertheless, a first addition profile was calculated, presuming this value of $\bar{n}$. Separately, the correlation between the amount of styrene to be added and the conversion was calculated from thermodynamic equilibrium data that would lead to the desired copolymer composition. Combining the results, that is, the conversion–time curve from the experiment carried out with this addition profile and the correlation between the amount of styrene to be added and conversion, a new addition profile was calculated. In the case of the S-MA system the iteration converged rapidly, only four iteration steps being required in S-MA emulsion copolymerisation to arrive at indistinguishable monomer addition rate profiles.

In order to evaluate the results, Van Doremaele analysed the copolymers formed by means of high performance liquid chromatography (HPLC) providing detailed microstructural information (viz., chemical composition distribution, CCD) of the copolymers.

In Figure 3.6 the CCDs are depicted for three high conversion S-MA copolymers having the same average chemical composition but prepared by different processes. The one prepared by the conventional batch process exhibits bimodality, has two glass transition temperatures and has a minimum film formation temperature of 17 °C. Both the one prepared in a semi-batch process under starved conditions (32 h) and that obtained in a semi-continuous process while applying the optimal monomer addition strategy (5 h) are homogeneous with respect to chemical composition and have a minimum film formation temperature of 27 °C. In general, the reactions based on the optimal addition rate profile proceed more rapidly than those based on constant addition rate strategies.

With the further development of on-line Raman spectroscopy the controlled composition reactor seems to become realistic. In Figure 3.7 the CCD of a copolymerisation of butyl acrylate and veova 9 is depicted, for both a batch and a Raman-controlled reaction.

3.8 Particle Morphologies

Composite latex particles are usually prepared by seeded emulsion polymerisation. In the first stage well-defined particles are prepared, while in the second stage another monomer is polymerised in the presence of these well-defined particles. Multistage emulsion polymerisation produces structures such as core–shell, "inverted" core–shell and phase separated structures, such as sandwich structures, hemispheres, "raspberry-like" and void particles.

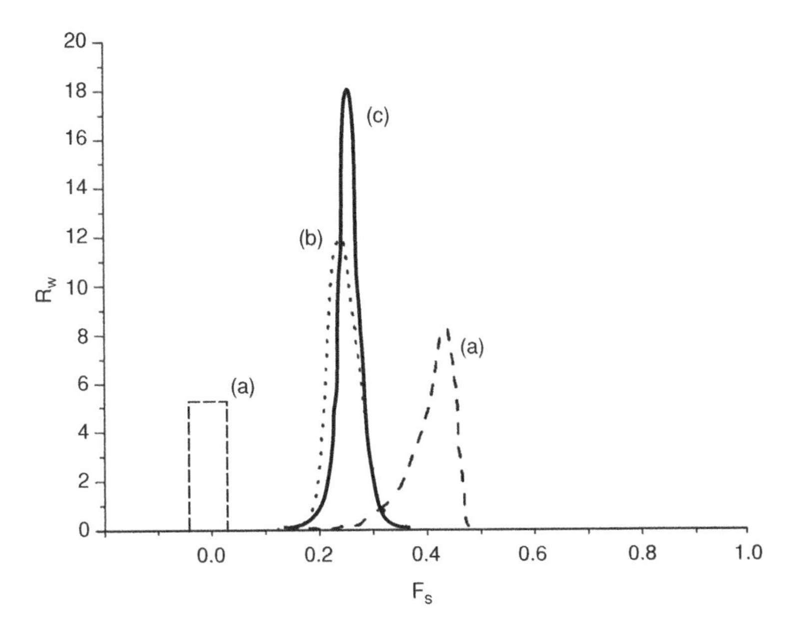

Figure 3.6 *CCDs, experimentally determined with HPLC, of three styrene/methyl acrylate emulsion copolymers, all with* $F_s = 0.25$ *and* $(M/W)_0 = 0.2$ *(g/g). (a) semi-continuous, starved conditions (32 h), (b) semi-continuous, optimal addition profile (5 h), (c) conventional batch process (3 h). Reprinted with permission from [van Doremaele, 1990] Copyright (1990) G.H.J. Doremaele.*

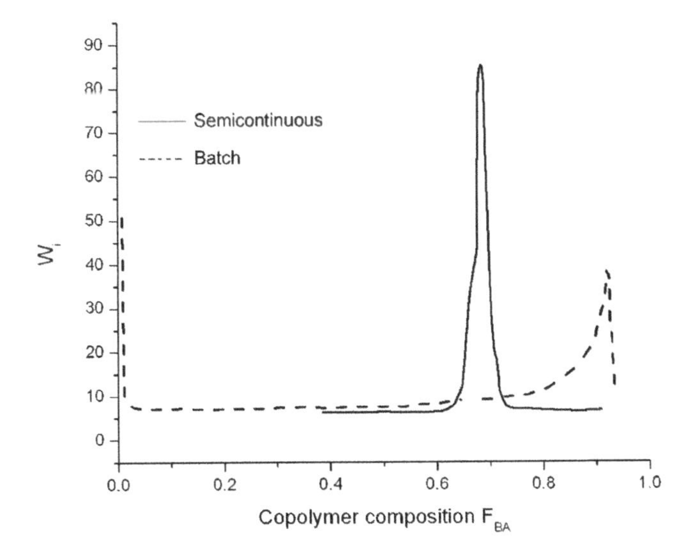

Figure 3.7 *Butyl acrylate - Veova 9 control of copolymer composition distribution by on-line Raman spectroscopy.*

Control of the composite latex particle morphology is important for many latex applications, such as adhesives and coatings (Urban and Takamura, 2002a) and impact modification and toughening of polymer matrices (Lovell, 1995). These structures have a major influence on the properties. The particle morphology can be affected by many of the polymerisation parameters and conditions, for example, the water solubility of the monomers; type, amount and addition mode of other ingredients such as surfactant, initiator, chain transfer, or crosslinking agents; degree of compatibility of the polymers; viscosity of the polymerisation loci (swelling of the core particle and the molar mass of polymer); the degree of grafting of the second stage polymer onto the core particle; the polarity of the polymers; the interfacial tension at the polymer–polymer and polymer–water interphases; the degree of crosslinking; the methods of monomer addition; and the polymerisation temperature.

Modelling the particle morphology is extremely complex and no broadly applicable approach is available yet. There is a thermodynamic approach where the interfacial tensions between the two polymers and between each of the polymers and water are the determining factors. Calculations of the latex particle morphology on the basis of minimisation of the interfacial energy change have been reported by Sundberg *et al.* (1990) and Chen, Dimonie and El-Aasser (1991a, 1991b and 1992a, 1992b). The morphology also may be determined by kinetic processes, as described by Chern and Poehlein (1987, 1990a, 1990b) and Mills, Gilbert and Napper (1990). The interfacial tension seems to be one of the main parameters controlling particle morphology in composite latexes. Depending on the type of initiator, the surface polarity can be different and, therefore, also the particle surface polarity, rather than the polymer bulk hydrophilicity, could be the controlling parameter in determining which phase will be inside or outside in composite particles. The kinetic parameters (such as the viscosity at the polymerisation loci, molar mass of the polymers, and mode of addition of the second stage monomer) influence the rate of formation of a certain morphology that is basically determined by the interfacial tensions (Chen *et al.*, 1993). Asua *et al.* (González-Ortiz and Asua, 1996a, 1996b) developed a mathematical model for the development of particle morphology in emulsion polymerisation based on the migration of clusters. The clusters are formed if the newly formed polymer chain is incompatible with the polymer existing at the site where it is formed, thus inducing phase separation. The equilibrium morphology is reached when the polymer chains diffuse into the clusters and the clusters migrate in order to minimize the Gibbs free energy.

The motion of the clusters is due to the balance between the van der Waals forces and the viscous forces.

For an overview of developments of morphologies we refer to Chapter 6.

Particle morphologies with more than three phases have been studied by Sundberg and Sundberg (1993).

The characterisation of the particle morphologies is very important in order to control seeded emulsion polymerisations and is described in Chapter 9.

3.8.1 Core–Shell Morphologies

It is not obvious that seeded emulsion polymerisation readily leads to core–shell morphologies, as discussed in Section 3.7. The design of the core–shell particles is dictated by the desired properties and applications. The properties that core–shell latex particles exhibit depends on a number of parameters, such as the polymer or copolymer type, the molar mass, the amount of grafted material between the core and the shell, the particle size and particle size distribution, the relative proportion of the core to the shell material, the glass transition temperature T_g of the polymer in the core and in the shell. Three main types of core–shell composite particles can be distinguished, viz. composite particles with organic cores, with inorganic cores and those with an 'empty' core, the so-called hollow particles. These three types will be discussed in the following sections.

3.8.1.1 Organic Cores

The core of organic composite latex particles can be varied along with the desired properties. The most important parameters of the polymer in the core are the T_g, the molar mass, the crosslink density, and the type of (co)polymer. Composite latex particles used for impact modification consist of a rubbery core and a glassy shell that is miscible or can react with the matrix. Examples of such types of polymers are the very important acrylonitrile-butadiene-styrene (ABS) composite polymer and also the methyl methacrylate- butadiene-styrene transparent composite polymer for the impact modification PVC. For coating applications the latex particles often consist of a glassy core and a low T_g shell that facilitates the film-forming properties (Vandezande and Rudin, 1994).

3.8.1.2 Encapsulation of Inorganic Particles

Application of Microencapsulated Particles. The micro-encapsulation of pigment and filler particles is an important area of research, both in academia and in industrial laboratories (van Herk and Landfester, 2010). Many activities in the past decade have been aimed at obtaining inorganic powders coated with an organic polymer layer. Such systems are expected to exhibit properties other than the sum of the properties of the individual components. In general, several benefits from this encapsulation step can be expected when the particles are incorporated in a polymeric matrix (e.g., plastics or emulsion paints):

- Better particle dispersion in the polymeric matrix
- Improved mechanical properties
- Improved effectiveness in light scattering in a paint film
- Protection of the filler or pigment from outside influences
- Protection of the matrix polymer from interaction with the pigment
- Improved barrier properties of a paint film.

The applications of these encapsulated particles relate to the above-mentioned benefits and can be found in filled plastics, paints, inks, paper coatings and so on (van Herk and German, 1999).

Very important applications of encapsulated pigment and filler particles are in emulsion paints. One of the more expensive components of water-borne paints is the white pigment, usually titanium dioxide. The pigment is added to obtain hiding power. The hiding power or opacity depends on the occurrence of light absorption and light scattering. For pigments with a high refractive index, like titanium dioxide, light scattering forms the main contribution to the hiding power. The light scattering effectiveness of the pigment particles depends on their particle size and on the interparticle distance. Agglomerates of pigment, already present in the wet paint film or formed by flocculation during the drying process, will reduce the scattering effectiveness of the dispersed pigment particles. By encapsulating the pigment particles, it is expected that the chance of flocculation is reduced and that the dispersion in the final paint film is improved. It has been suggested that the layer thickness could be optimised to obtain optimum spacing between titanium dioxide particles to achieve maximum light scattering (Templeton-Knight, 1990).

In encapsulating the pigment particle an important adverse effect of the pigment could be influenced, that is the generation of radicals under the influence of UV light. These radicals can lead to degradation of the matrix polymer and thus to reduced durability. With the proper choice of the polymer layer the durability might also be improved. Other advantages are improved block resistance, less dirt pick-up, better adhesion and improved chemical resistance.

For the above-mentioned reasons most commercial pigments already involve inorganic and/or organic surface modifications. An additional benefit can be brought about by the formation of multiple layers of polymer on inorganic particles where additional, for example rubber toughening, effects can be introduced. Other applications of encapsulated pigments can be found in inks, paper coatings and electro-photographic toners.

When the inorganic particles are magnetically responsive this opens pathways to special applications, like coupling of enzymes and antibodies to the surface of the magnetic particles after which drug targeting becomes possible (see Chapter 11). These particles can also be used in biochemical separations (Arshady, 1993).

Encapsulation of inorganic particles can also be achieved by using controlled radical polymerization, for example in the approach developed by Hawkett (Nguyen *et al.*, 2008) RAFT is used to produce random copolymer oligomers that are charged. These oligomers are adsorbed on the surface of the oppositely charged inorganic particles and, subsequently, fresh monomer is added to encapsulate the inorganic particles. With this method and other methods clay platelets were also encapsulated (Mballa Mballa *et al.*, 2012).

In conclusion it can be said that the smaller the inorganic particles , the more efficiently do encapsulation reactions proceed. This effect can easily be related to Figure 3.1 where capturing the formed oligomers by an available surface prevents the formation of new particles. To encapsulate particles larger than 500 nm other approaches than emulsion polymerisation should be applied, for example heterocoagulation with *in situ* formed (unstable) latex particles (van Herk and German, 1999).

3.8.1.3 Hollow Particles

The preparation of hollow particles through emulsion polymerisation is very interesting, for instance in the light of using these particles as drug carriers in controlled drug release. Other applications are in surface coatings and as opacifiers.

In principle, there are two routes to obtain hollow particles through emulsion polymerisation; one possibility is preparing particles that, after isolation, undergo a further treatment to render them hollow, the other route is designing the synthesis in such a manner that hollow particles are obtained directly. The first route starts with the preparation of core–shell particles. The core then can either be removed by dissolving it in an appropriate solvent or it could shrink more strongly than the shell upon drying or treatment with acid or base. The second route can be based on various vesicle polymerisations strategies.

Core–shell emulsion polymerisation. Vanderhoff, Park and El-Aasser (1991) prepared particles consisting of a core of a copolymer of methacrylic acid and methyl methacrylate and a shell of crosslinked material. After neutralisation with NH_4OH, the core material collapses and the particles contain voids of between 130 and 760 nm. A similar approach was applied by Okubo and Ichikawa (1994a) where the particles were produced by an emulsion-free terpolymerisation of styrene, butyl acrylate and methacrylic acid. The effect of pH, temperature and time of acid treatment on the multi-hollow structure formed were investigated.

A somewhat different approach is where an organic solvent is used to extract core material. In one example (Okubo, Ichikawa and Fujimura, 1991; Okubo and Nakagawa, 1994b), large polystyrene seed particles, produced by dispersion polymerisation, are used as a seed in a second stage polymerisation where a shell of polystyrene-divinylbenzene was polymerised around the core. The core material was then extracted with toluene under reflux. Depending on the divinylbenzene content, particles could be obtained with structures ranging from one void to a fine multi-voided structure.

Vesicle polymerisation. There are several options to achieve polymerisation in or of vesicles (Paleos, 1990):

- Polymerise the surfactant molecules when these contain polymerisable groups
- Polymerise the counterions of the surfactants
- Polymerise monomer that is contained in the bilayer.

The most flexible route is the one where the bilayer is swollen with the monomers of choice. In that case the glass transition temperature, permeability, layer thickness and degradability of the polymer layer can be varied more easily than in the other approaches.

Jung *et al.* (2000) described the polymerisation in vesicles, leading to different types of morphologies including hollow particles. Also here the use of the RAFT approach has turned out to be successful to produce hollow latex particles through a vesicle templating reaction (Ali, Heuts and van Herk, 2011).

3.8.1.4 Reactive Latexes

In the development of water-borne coatings, a main area of current research activities is the crosslinking of the polymer film. Traditionally, solvent-based coatings yield a crosslinked film after the drying process, whereas water-borne coatings result in a thermoplastic polymer film. The result of this is that, for example, solvent resistance of solvent-based coatings is superior to that of water-borne coatings.

It is well known that the process of cohesive strength development in a water-borne polymeric coating consists of three main mechanisms:

- Molecular interdiffusion of polymer chains from one particle into another
- Interfacial crosslinking
- Residual crosslinking.

This process of cohesive strength development is the final stage in the complex process of film formation. The two preceding stages are the evaporation of water and the coalescence of the latex particles. These two stages have been investigated extensively and a few different models have been proposed to describe these physical processes. In this chapter, we will mainly focus on the third stage of film formation, that is, the cohesive strength development.

The first process of interest in the cohesive strength development is the interdiffusion of polymer chains. It is well known that the diffusion of polymer chains in a polymer matrix is strongly dependent on the molar mass of the chains. In terms of development of the cohesive strength, two opposing effects can be recognized, as follows.

Polymer with a relatively low molar mass ensures facile diffusion of chains from one particle into the other, after coalescence of the particles in the film formation process. However, the effect of this interdiffusion on the strength development is not very large.

Polymer with a higher molar mass is hindered in its diffusion to a larger extent. The contribution of this diffusion process to the development of the cohesive strength is much larger than in the case of low molar mass polymer.

The method of crosslinking determines to some extent the requirements with respect to polymer–polymer interdiffusion. Below, two examples will be given. One consists of a polymer that is to be crosslinked by a low molar mass crosslink agent. The other consists of two different polymers containing complementary reactive groups. Terms like interfacial crosslinking followed by residual crosslinking apply to the former of these examples, but hardly to the latter.

Crosslinking of polymers by low molar mass crosslink agents. The most elementary form of a crosslinking water-borne coating is where the emulsion polymer contains functional groups that are crosslinked in a reaction with a low molar mass crosslink agent. The crosslinking agent will generally be added to the latex immediately prior to application on the substrate. This type of system is referred to as a two-component coating, for obvious reasons. In general the crosslink agent will reside in the aqueous phase. Diffusion of the crosslinking agent into the polymer particles is crucial in order to obtain a homogeneously crosslinked film. One of the concerns here is that, upon coalescence of the particles, a relatively high concentration of crosslinking agent is present on the interface between the particles. This may result in a densely crosslinked film at the interface, which greatly reduces mobility of polymer chains across the interface, and may result in inhomogeneous crosslinking. In the previously indicated scheme, the third step, that is, residual crosslinking, is hindered to some extent. One solution to this problem is the homogeneous distribution of crosslink agent throughout the polymer phase. In the regular systems this will result in crosslinking of the latex particles before film formation. These crosslinked particles will not be able to undergo film formation, hence an inferior quality of the coating will be achieved. However, when the crosslinking reaction is intrinsically slow, but when its rate can be enhanced by some catalyst, this problem may be solved.

One of the common functional monomers to induce crosslinking, is 2-hydroxyethyl methacrylate (HEMA). This is a highly water-soluble monomer compared to typical co-monomers applied in waterborne coatings (butyl (meth)acrylate, 2-ethylhexyl (meth)acrylate). This large difference in water solubility results in strong variations of the co-monomer ratio between the different phases of the polymerisation mixture.

3.8.1.5 Transparent Latexes

There is a special interest in transparent core–shell latexes, being latexes with small particle sizes. Applications can be in transparent gels used for intra-ocular eye lenses, for example

(Pusch and van Herk, 2005). Referring to Equation 3.2, there are several ways to obtain small particles through creating a high number of particles: for example, increasing the surfactant concentration, the initiator concentration or both. Reducing the volume growth rate (K) in Equation 3.2 is also an option. This can be achieved by reducing the swellability of the latex particles with the aid of crosslinking monomers (microgels, see for example Pusch and van Herk, 2005), reducing the monomer concentration in the particles. Another way to reduce the monomer concentration during the nucleation stage is basically not stirring the water–monomer mixture, making the monomer transport strongly diffusion controlled (which is not the case when the monomers are well emulsified). This approach has been studied by Sajjadi (Sajjadi and Jahanzad, 2006; Chen and Sajjadi, 2009).

In the case of a transparent latex photoinitiation becomes possible. Transparent latexes produced by photopolymerization were also reported (Hu, Zhang and Yang, 2009).

Finally, it is noted that the successful achievement of RAFT-controlled emulsion polymerisation (Ferguson *et al.*, 2002a) has opened the way to achieve all of the above morphologies in a novel way. Thus it is possible to make core–shell particles wherein the same polymer chain is both core and shell (Leswin *et al.*, 2009), for example by polymerising first a water-soluble monomer (e.g., acrylic acid) to about degree of polymerisation 5, then controlled feed of a hydrophobic monomer (e.g., butyl acrylate) to about degree of polymerisation 20, so that the resulting diblocks self-assemble into particles wherein each chain can continue polymerising under RAFT control. The BA feed can then be continued to any desired degree of polymerisation, and this can then be followed by another monomer (e.g., styrene) to form a triblock. The final result is a latex particle containing only poly(acrylic acid)-*b*-poly(butyl acrylate)-*b*-polystyrene chains, with the BA portions in the shell and the styrene portions in the core. The same strategy can be used to make a wide variety of different compositions and morphologies. These polymer colloids have novel properties which cannot be achieved by conventional emulsion polymerisation strategies.

4

Emulsion Copolymerisation, Process Strategies

Jose Ramon Leiza[1] and Jan Meuldijk[2]
[1]POLYMAT, University of the Basque Country UPV/EHU, Spain
[2]Department of Chemical Engineering & Chemistry, Eindhoven University of Technology, The Netherlands

4.1 Introduction

In many cases latex products are composed of more than one monomer. In copolymerisation two or more monomers are incorporated into the polymer chains. The copolymer chains are produced by simultaneous polymerisation of two or more monomers in emulsion. Emulsion copolymerisation allows the production of materials with properties which cannot be obtained by latex products consisting of one monomer, that is, homopolymer latexes, or by blending homopolymers. The properties of the materials required are usually dictated by the market. Nowadays most of the materials properties are achieved by combination of more than two monomers in the copolymer product. Typical industrial emulsion polymerisation formulations are mixtures of monomers giving hard polymers, and monomers leading to soft polymers. Styrene and methyl methacrylate are examples of monomers giving hard polymers, that is, polymers with a high glass transition temperature, T_g. Soft polymers, that is, polymers with a low T_g, are, for example, formed from n-butyl acrylate. The industrial emulsion polymerisation formulations also contain small amounts of functional monomers, such as acrylic and methacrylic acid, to impart improved or special characteristics to the latex product. Note that the colloidal stability of the latex product can be seriously improved by acrylic and methacrylic acid. Furthermore, some applications may demand the addition of other specialty monomers which make the kinetics of the copolymerisation even more complex.

Chemistry and Technology of Emulsion Polymerisation, Second Edition. Edited by A.M. van Herk.

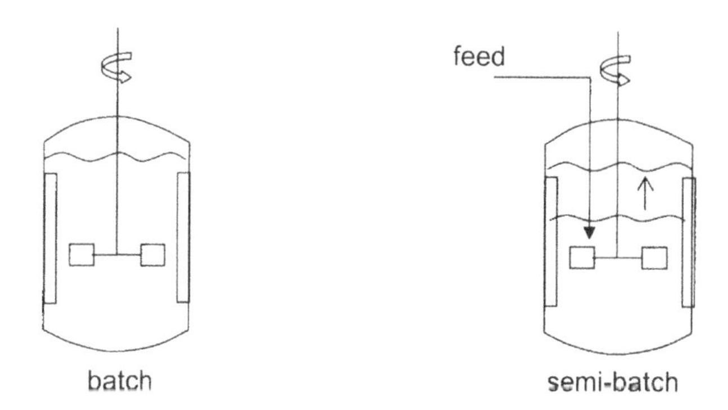

Figure 4.1 *Schematic representation of batch and semi-batch operation.*

This chapter focuses on key features to understand the emulsion copolymerisation kinetics and on the influence of the mode of operation on the copolymer composition of the final latex products. The focus is on batch and semi-batch or semi-continuous operation, see Figure 4.1. Only the free radical emulsion copolymerisation of two monomers is considered but the concepts can be directly applied for formulations containing more than two monomers. The reacting monomers usually having different reactivities and polymerise simultaneously. The reactivities and the individual concentrations of the monomers at the locus of polymerisation, that is, the particle phase, govern the ratio of the monomers incorporated into the polymer chains at a certain time.

Note that when a certain monomer homopolymerises first in some kind of controlled radical polymerisation and the second one is supplied after complete conversion of the first one, and so on, block copolymers will be obtained. Controlled radical polymerisation and block copolymer production are discussed in detail in Chapters 2 and 5.

Characteristic features of emulsion copolymerisation appear when discussing the copolymerisation process in a perfectly mixed batch reactor. In a batch process no materials enter or leave the reactor during the polymerisation, see Figure 4.1. For monomer i the following mass balance should be obeyed:

$$\frac{dN_i}{dt} = -R_{p,i} \cdot V_w \text{ or } N_i^0 \cdot \frac{d\bar{x}_i}{dt} = R_{p,i} \cdot V_w \quad \text{with} \quad N_i = N_i^0 \cdot (1 - \bar{x}_i) \tag{4.1}$$

where N_i, N_i^0, V_w and $\bar{x}_i$ stand for the amount of monomer i at a certain moment of time, the amount of monomer at the start of the polymerisation, the volume of the water phase in the reactor and the conversion of monomer i, respectively. $R_{p,i}$ is the rate of polymerisation of monomer i in the particles, expressed in units: $\text{mol}_i\ \text{m}_w^{-3}\ \text{s}^{-1}$.

In Equation 4.1 only polymerisation in the particle phase is taken into account. Aqueous phase polymerisation is ignored, which is a proper approximation for sparsely water soluble monomers. In a batch process the conversion rate of the more reactive monomer is larger than the conversion rate(s) of the other monomer(s). So theratio of the monomers in the copolymer chains changes with time. Initially a relatively large amount of the more reactive monomer is incorporated, that is, the momentary fraction of the more reactive monomer incorporated into the copolymer is relatively high. Note that the momentary or instantaneous fraction of a monomer in a copolymer stands for the fraction of that monomer at a certain

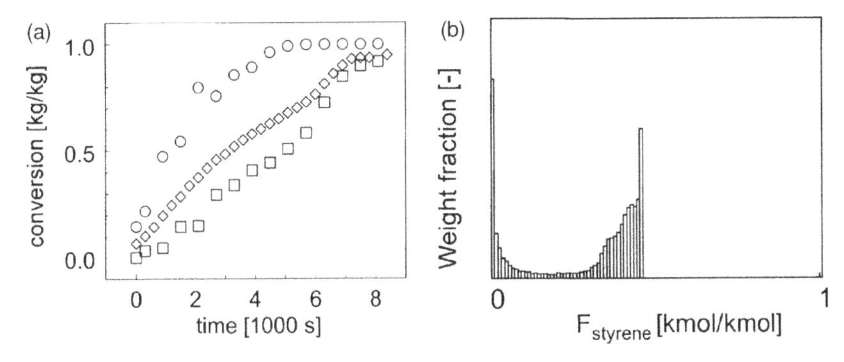

Figure 4.2 *Typical example of the conversion time history (a) and chemical composition distribution (b) of a batch emulsion copolymerisation of styrene and methyl acrylate. Initial overall molar ratio of styrene and methyl acrylate is 0.33. Styrene is the more reactive monomer. ◇: overall monomer conversion; ○: partial conversion of styrene; □: partial conversion of methyl acrylate. Note the fast homopolymerisation of methyl acrylate when the styrene has vanished.*

moment of time. During the course of the batch process the momentary fraction of the more reactive monomer incorporated into the copolymer gradually changes to lower values when the polymerisation reaction proceeds. This phenomenon is referred to as composition drift. As a consequence, a batch process usually results in a product with a rather broad composition distribution, see Figure 4.2 for the emulsion copolymerisation of styrene and methyl acrylate, where styrene is the more reactive monomer. Note that styrene vanishes at about 60% overall monomer conversion. The copolymer composition distribution as well as the cumulative copolymer composition of a latex product can be controlled by controlling the monomer composition in the particle phase during the course of the process, for example by feeding monomers to the reaction mixture in a semi-batch process, see Figure 4.1.

The operational strategy determines the concentrations, $[M_i]_p$, or the mole fractions, f_i, of the individual monomers i in the particle phase, see Section 4.3.

The monomer concentrations of the polymerising monomers in the particle phase, together with copolymerisation kinetics, govern the rates by which the monomers are incorporated into the polymer chains. The key quantities in describing the performance of copolymerisation processes are the instantaneous copolymer composition, y_i, and the cumulative copolymer composition, $y_{i,c}$. The instantaneous copolymer composition and the cumulative copolymer composition are defined as:

$$y_i = \frac{R_{p,i}}{\sum_{i=1}^{m} R_{p,i}} \quad \text{and} \quad y_{i,c} = \frac{1}{X_T} \cdot \int_0^{X_T} y_i \cdot dX_T = \frac{N_{p,i}}{\sum_{i=1}^{m} N_{p,i}} \tag{4.2}$$

where m, $R_{p,i}$, X_T and $N_{p,i}$ stand for the number of monomers involved in the polymerisation, the rate of polymerisation of monomer i, the overall monomer conversion and the number of moles of monomer i polymerised, respectively. The cumulative copolymer composition can be measured by spectroscopic techniques, for example, IR and NMR. The rate of polymerisation of monomer i depends on the concentration of monomer i in the particles:

$$R_{p,i} = \left(\sum_{j=1}^{m} k_{p,ji} \cdot P_j \right) \cdot [M_i]_p \cdot \frac{\bar{n}}{N_A} \cdot N_p \text{ mol}_i \text{ m}_w^{-3} \text{ s}^{-1} \quad (4.3)$$

where $\bar{n}$, N_p and P_j stand for the average number of radicals per particle, the number of particles per unit volume of the water phase and the fraction of radical chains in the particles bearing a radical end group of monomer j. The propagation rate coefficient for reaction of monomer i with a growing chain ending with monomer j is denoted as $k_{p,ji}$. For copolymerisations with two monomers $\bar{n}$ as well as the fractions ending with monomer 1 and 2 can be calculated according to Nomura, Kubo and Fujita (1983), Forcada and Asua (1985), Storti *et al.* (1989):

$$P_1 = \frac{k_{p,21} \cdot [M_1]_p}{k_{p,21} \cdot [M_1]_p + k_{p,12} \cdot [M_2]_p} \quad \text{and} \quad P_2 = \frac{k_{p,12} \cdot [M_2]_p}{k_{p,21} \cdot [M_1]_p + k_{p,12} \cdot [M_2]_p} \quad (4.4)$$

It should be realised that: $P_1 + P_2 = 1$.

According to the terminal model for copolymerisation of two monomers (Mayo and Lewis, 1944; Alfrey and Goldfinger, 1944) the instantaneous copolymer composition y can be related to the monomer fractions in the locus of polymerisation:

$$y_1 = \frac{r_1 \cdot f_1^2 + f_1 \cdot f_2}{r_1 \cdot f_1^2 + 2 \cdot f_1 \cdot f_2 + r_2 \cdot f_2^2} \quad \text{and} \quad y_2 = \frac{r_2 \cdot f_2^2 + f_1 \cdot f_2}{r_1 \cdot f_1^2 + 2 \cdot f_1 \cdot f_2 + r_2 \cdot f_2^2} \quad (4.5)$$

In Equation 4.5 r_1 and r_2 stand for the reactivity ratios, defined as:

$$r_1 = \frac{k_{p,11}}{k_{p,12}} \quad \text{and} \quad r_2 = \frac{k_{p,22}}{k_{p,21}} \quad (4.6)$$

Note that $f_1 + f_2 = 1$ and $y_1 + y_2 = 1$.

$$f_1 = \frac{[M_1]_p}{[M_1]_p + [M_2]_p} \quad \text{and} \quad f_2 = \frac{[M_2]_p}{[M_1]_p + [M_2]_p} \quad (4.7)$$

Figure 4.3 gives a graphical representation of Equation 4.5 for the monomer pairs methyl methacrylate (MMA)/n-butyl acrylate (BA) and styrene (S)/methyl acrylate (MA). Plots as shown in Figure 4.3 allow estimations about the sensitivity of a copolymerisation recipe for composition drift.

Local concentrations of the monomers in the polymerisation loci, that is, the polymer particles at a certain moment of time, determine the composition, y_i, of instantaneously formed copolymer. The momentary monomer concentrations in the particles follow from the monomer partitioning between the different phases involved, that is, the water phase, the droplet phase and the particle phase. The time evolution of the concentrations of the monomers in the polymer particles are governed by the kinetics of the copolymerisation, see Chapter 2, as well as by the strategy used to feed the monomers to the reactor. Indeed, to control the copolymer composition produced during emulsion copolymerisation processes monomer feeding strategies are implemented. This chapter addresses monomer partitioning, see Section 4.2, and operational strategies, see Section 4.3. In Section 4.3 different alternatives are presented to produce copolymers with a desired composition in semi-continuous operation.

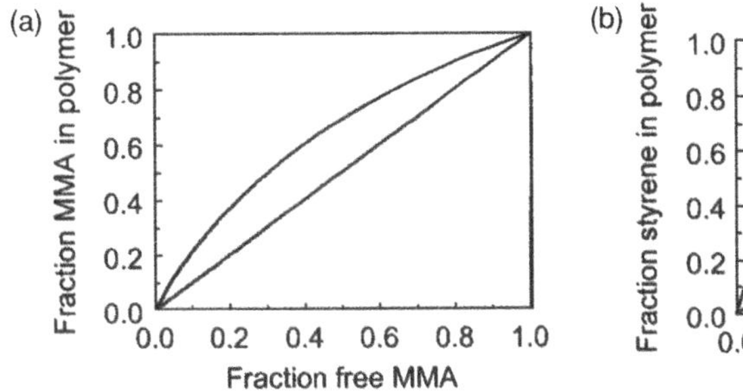

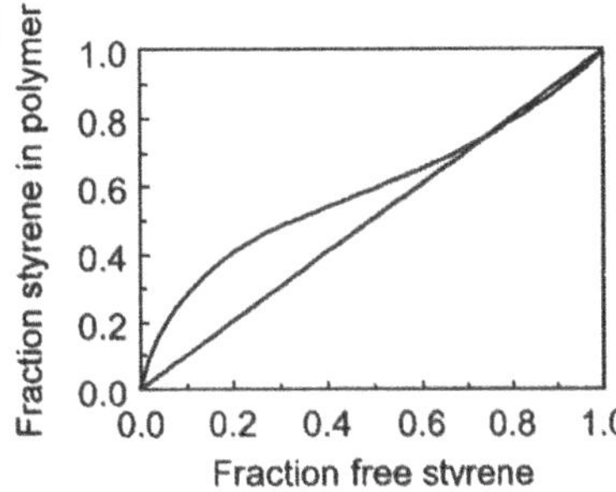

Figure 4.3 *Mol fraction of the more reactive monomer incorporated into the copolymer as a function of the mol fraction of that monomer in the locus of polymerization. Monomer pairs: (a) methyl methacrylate–n-butyl acrylate* ($r_{MMA} = 2.24$, $r_{BA} = 0.414$)*; (b) styrene–methyl acrylate* ($r_S = 0.73$, $r_{MA} = 0.19$)*.*

4.2 Monomer Partitioning

During the Intervals I and II of a batch emulsion polymerisation, monomers are divided, that is, partition, over the monomer droplets, the aqueous phase and the polymer particles. The monomer that is consumed by polymerisation in the polymer particles is replaced by monomer that is transferred from the monomer droplets through the aqueous phase into the particle phase. In Interval III, there are no droplets and the monomer is mostly located in the polymer particles. In semi-batch processes, monomers are continuously fed into the reactor, usually under starved conditions, namely, at high instantaneous conversions for example, polymer/monomer ratios close to 90/10 on a weight basis. Under these circumstances, only the newly fed monomer droplets are present in the reactor and the life-time of these droplets is short because the monomers are transferred through the aqueous phase to the polymer particles where they are consumed by polymerisation.

The concentration of monomer in the polymer particles depends on the relative time constants for mass transfer and polymerisation. Except for poorly emulsified highly water-insoluble monomers, the time constant for mass transfer is negligible with respect to the time constant for polymerisation. Hence the concentrations of the monomers in the different phases are given by the thermodynamic equilibrium:

$$[\text{Monomer}]_{\text{droplets}} \rightleftarrows [\text{Monomer}]_{\text{particles}} \rightleftarrows [\text{Monomer}]_{\text{water}}$$

In this section the equilibrium of monomer over the different phases is addressed to allow calculation of the rate of polymerisation, see Equation 4.3 and the instantaneous and cumulative copolymer composition, see Equation 4.2.

4.2.1 Slightly and Partially Water Miscible Monomers

Prediction of the instantaneous chemical composition as well as the rate of polymerisation during an emulsion copolymerisation of monomers 1 and 2 requires knowledge of the concentrations of the reacting monomers in the particle phase. Monomers 1 and 2 are only sparsely or moderately water soluble. The amounts of the monomers 1 and 2 in the droplet phase, the water phase and the particle phase are related by mass balances for the monomers. Equation 4.8 gives the mass balance for monomer 1:

$$N_{1,\text{tot}} = N_{1,\text{w}} + N_{1,\text{d}} + N_{1,\text{p}} \tag{4.8}$$

$N_{1,\text{tot}}$, $N_{1,\text{w}}$ $N_{1,\text{d}}$ and $N_{1,\text{p}}$ stand for the total amount of monomer 1 in the reaction mixture, the amount of monomer 1 in the water phase, the droplet phase and the particle phase, respectively. N is in mol. Equation 4.8 can be rewritten in terms of the concentration in the aqueous phase, $[\text{M}_1]_\text{w}$, the volume of the water phase, V_w, the concentration in the droplet phase, $[\text{M}_1]_\text{d}$, the volume of the droplet phase, V_d, the concentration in the particle phase, $[\text{M}_1]_\text{p}$, and the volume of the particle phase, V_p:

$$N_{1,\text{tot}} = [\text{M}_1]_\text{w} \cdot V_\text{w} + [\text{M}_1]_\text{d} \cdot V_\text{d} + [\text{M}_1]_\text{p} \cdot V_\text{p} \tag{4.9}$$

Concentrations are expressed in mol m^{-3} of the phase involved, that is, the water phase, the droplet phase and the particle phase. Using the volume fraction of polymer in the particles, $\phi_{\text{pol,p}}$, the polymer volume in the particle phase, $V_{\text{pol,p}}$, and the volume fraction of monomer 1 in the particles, $\phi_{1,\text{p}}$, as well as in the droplets, $\phi_{1,\text{d}}$, Equation 4.9 can be rearranged into:

$$N_{1,\text{tot}} = [\text{M}_1]_\text{w} \cdot V_\text{w} + \frac{\varphi_{1,\text{d}} \cdot V_\text{d}}{v_1} + \frac{\varphi_{1,\text{p}} \cdot \left(V_{\text{pol,p}}/\varphi_{\text{pol,p}}\right)}{v_1} \tag{4.10}$$

v_1 stands for the molar volume of pure monomer 1. Note that molar volume changes of the monomers due to mixing with monomer 2 and/or the polymer have been neglected on going from Equations 4.9 to 4.10. For the mass balance of monomer 2 over the three phases equations analogous to 4.8–4.10 can be derived. Note that in the absence of monomer droplets, for example, Interval III for a batch emulsion polymerisation process, Equation 4.10 reduces to Equation 4.11:

$$N_{1,\text{tot}} = [\text{M}_1]_\text{w} \cdot V_\text{w} + \frac{\varphi_{1,\text{p}} \cdot \left(V_{\text{pol,p}}/\varphi_{\text{pol,p}}\right)}{v_1} \tag{4.11}$$

4.2.1.1 *Saturation Swelling*

For sparsely and moderately soluble monomers the following relation holds for the volume fractions of 1 and 2 in the droplet and the particle phase (Maxwell *et al.*, 1992a):

$$\frac{\varphi_{1,\text{d}}}{\varphi_{2,\text{d}}} = \frac{\varphi_{1,\text{p}}}{\varphi_{2,\text{p}}} \tag{4.12}$$

By negligible volume changes due to mutual mixing of the monomers 1 and 2 in the droplet phase and by mixing the monomers 1 and 2 with the polymer phase in the particles, Equation 4.12 can be rewritten in concentrations or mol fractions:

$$\frac{[\text{M}_1]_\text{d}}{[\text{M}_2]_\text{d}} = \frac{[\text{M}_1]_\text{p}}{[\text{M}_2]_\text{p}} \quad \text{or} \quad \frac{f_{1,\text{d}}}{f_{2,\text{d}}} = \frac{f_{1,\text{p}}}{f_{2,\text{p}}} \tag{4.13}$$

where f stands for the mol fraction of the particular monomer related to the total amount of monomer in either the droplet phase or the particle phase.

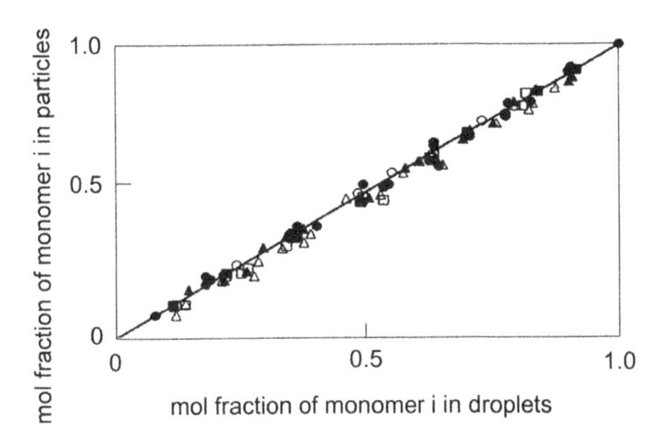

Figure 4.4 *Experimentally determined monomer fractions in latex particles as a function of the monomer fraction in the monomer droplets. ○: methyl acrylate–vinyl acetate in a poly-(MA-VAc) copolymer latex. Δ: methyl acrylate–styrene, □: n-butyl acrylate–styrene, ▲: methyl acrylate–n-butyl acrylate, ■: methyl acrylate–methyl methacrylate and •: methyl methacrylate–styrene on several (co)polymer seeds. The solid line represents the prediction by Equation 4.13. Reprinted from [Verdurmen-Noël, E.F.J. 1994] Copyright (1994) E.F.J. Verdurmen-Noël.*

Experimental data clearly demonstrate Equation 4.13, see Figure 4.4.

Assuming that the monomers in the droplet phase form an ideal liquid mixture, the equilibrium water phase concentrations of monomer 1 and monomer 2, $[M_1]^*_{w,sat}$ and $[M_2]^*_{w,sat}$ can simply be related to the water phase concentrations of the monomers 1 and 2 in equilibrium with pure monomer 1 and 2:

$$[M_1]^*_{w,sat} = f_{1,d} \cdot [M_1]_{w,sat} \quad \text{and} \quad [M_2]^*_{w,sat} = f_{2,d} \cdot [M_2]_{w,sat} \tag{4.14}$$

In Equation 4.14 $[M_1]_{w,sat}$ and $[M_2]_{w,sat}$ stand for the water phase concentrations of the monomers 1 and 2 in equilibrium with the pure monomers 1 and 2, respectively.

It should be noticed that Equation 4.14 for liquid–liquid equilibria has its analogy with Raoults law for vapour–liquid equilibria.

Saturation swelling data for latex particles with pure monomer 1 or pure monomer 2 allow calculation of the concentrations of monomers 1 and 2 in the particles from the mol fractions in the droplets, see Figure 4.4. Having access to phase equilibria, as given in Figure 4.4, and to reactivity ratios allows estimation of the cumulative copolymer composition as a function of conversion in seeded emulsion polymerisation during the stage of saturation swelling, that is, in the presence of monomer droplets.

4.2.1.2 Partial Swelling

In the absence of monomer droplets the monomer concentrations in the particles are directly related to the concentrations of monomers 1 and 2 in the aqueous phase. Equation 4.15, known as the Vanzo equation (Vanzo, Marchessault and Stannett, 1965) describes the partitioning of monomer between the aqueous phase and the latex particles in the absence of monomer droplets for homopolymerisation of monomer 1:

$$\ln\left(\frac{[M_1]_w}{[M_1]_{w,sat}}\right) = \ln(1-\varphi_{pol,p}) + \varphi_{pol,p}\cdot\left(1-\frac{1}{\bar{M}_n}\right) + \chi\cdot\varphi_{pol,p}^2 + \frac{4\cdot v_1\cdot\gamma\cdot\varphi_{pol,p}^{1/3}}{d_0\cdot R\cdot T} \tag{4.15}$$

where $\bar{M}_n$, χ, γ, R, T and d_0 stand for the number average molecular weight of the polymer, the Flory–Huggins interaction parameter, the particle–water interfacial tension, the gas constant, absolute temperature and the unswollen particle diameter. Equation 4.15 is based on the calculation of the excess Gibbs energy for polymer solutions as described by Flory (1953). The expression: $\ln(1-\varphi_{pol,p}) + \varphi_{pol,p}\cdot\left(1-\frac{1}{\bar{M}_n}\right)$ is proportional to the conformational contribution to the excess entropy of mixing on mixing monomer 1 with the polymer of monomer 1. For emulsion polymerisation $\bar{M}_n$ is so large that:

$$\ln(1-\varphi_{pol,p}) + \varphi_{pol,p}\cdot\left(1-\frac{1}{\bar{M}_n}\right) \approx \ln(1-\varphi_{pol,p}) + \varphi_{pol,p} \tag{4.16}$$

The term: $\chi\cdot\varphi_{pol,p}^2$ is related to specific energetic interactions between monomer 1 and the polymer of 1. For monomer/polymer systems with no or weak specific interactions, for example, styrene, butadiene and n-butyl acrylate the specific interactions are usually very small and the contribution of the term $\chi\cdot\varphi_{pol,p}^2$ to the excess Gibbs energy of mixing is then small as compared to the contribution of the terms: $\ln(1-\varphi_{pol,p}) + \varphi_{pol,p}$. The term: $\frac{4\cdot v_1\cdot\gamma\cdot\varphi_{pol,p}^{1/3}}{d_0\cdot R\cdot T}$ refers to the contribution of the particle–water surface tension to the partial molar Gibbs energy of the monomer 1 in the particle phase. The contribution of the surface tension to monomer partitioning is, in most cases, negligible as compared to the contribution of the entropy of mixing. Figure 4.5 shows the contribution of the individual terms of the right-hand side of Equation 4.15 to its total value for the partitioning of methyl acrylate over a poly methyl acrylate latex and water.

The partitioning of the monomers 1 and 2 over latex particles consisting of a copolymer of the monomers 1 and 2 and the water phase, is described by Equation 4.17, which is a somewhat extended form of Equation 4.15 (Maxwell *et al.*, 1992b):

$$\ln\left(\frac{[M_1]_w}{[M_1]_{w,sat}}\right) = \ln\varphi_{1,p} + \varphi_{pol,p}\cdot\left(1-\frac{1}{\bar{M}_n}\right) + \left(1-\frac{v_1}{v_2}\right)\cdot\varphi_{2,p} + \chi_{1,2}\cdot\varphi_{2,p}^2 + \chi_{1,pol}\cdot\varphi_{pol,p}^2 + \varphi_{2,p}\cdot\varphi_{pol,p}\cdot\left(\chi_{1,2}+\chi_{1,pol}-\chi_{2,pol}\cdot\frac{v_1}{v_2}\right) + \frac{4\cdot v_1\cdot\gamma\cdot\varphi_{pol,p}^{1/3}}{d_0\cdot R\cdot T} \tag{4.17}$$

In Equation 4.17 $\chi_{1,2}$, $\chi_{1,pol}$ and $\chi_{2,pol}$ stand for the Flory–Huggins interaction parameters between the monomers 1 and 2, between monomer 1 and the polymer, and between monomer 2 and the polymer, respectively. Note that $[M_1]_{w,sat}$ is the concentration of monomer 1 in water in equilibrium with *pure* monomer 1.

For the monomer pair styrene (1)- n-butyl acrylate (2), $\chi_{1,2}$, $\approx$ 0.20–0.25 and $\frac{v_1}{v_2}\approx 0.8$, This means, together with the considerations on going from Equations 4.15 to 4.16, that

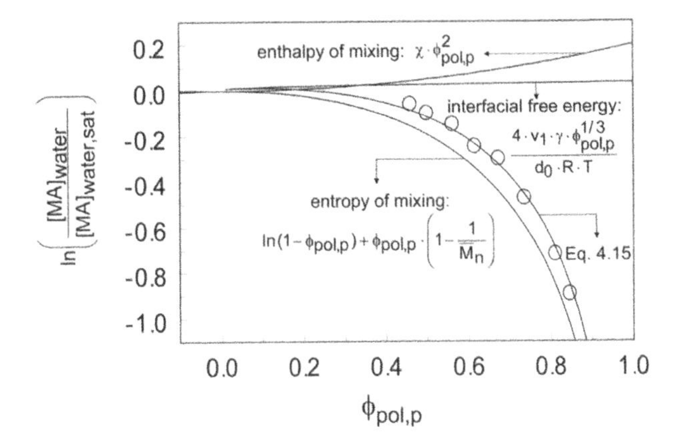

Figure 4.5 *Partitioning of methyl acrylate over the aqueous phase and a poly-methyl acrylate seed latex with an unswollen diameter (d_0) of 192 nm. ○: experimental data; solid line calculation results using Equation 4.15 with $\gamma = 45 \times 10^{-3}$ N m^{-1} and $\chi = 0.2$ (Maxwell* et al.*, 1992b). Other lines represent the contribution of the individual terms in Equation 4.15.*

omitting all terms in which a Flory–Huggins parameter appears, only leads to deviations in the monomer composition in the particles of not more than 10%. Equation 4.17 then reduces to:

$$\ln\left(\frac{[\mathrm{M}_1]_\mathrm{w}}{[\mathrm{M}_1]_\mathrm{w,sat}}\right) = \ln\varphi_{1,\mathrm{p}} + \varphi_{\mathrm{pol,p}} \tag{4.18}$$

for monomer 1. The relation between the concentration of monomer 2 in the water phase and the volume fraction of monomer 2 in the particles is given by Equation 4.19, using identical considerations as for monomer 1.

$$\ln\left(\frac{[\mathrm{M}_2]_\mathrm{w}}{[\mathrm{M}_2]_\mathrm{w,sat}}\right) = \ln\varphi_{2,\mathrm{p}} + \varphi_{\mathrm{pol,p}} \tag{4.19}$$

Note that:

$$\varphi_{1,\mathrm{p}} + \varphi_{2,\mathrm{p}} + \varphi_{\mathrm{pol,p}} = 1 \tag{4.20}$$

If the overall concentrations of the monomers 1 and 2, as well as the polymer volume, $V_{\mathrm{pol,p}}$, and the volume of the water phase, V_w, in the reaction mixture are known, the use of the Equations 4.18–4.20, combined with Equation 4.10 and its corresponding relation for monomer 2, allows calculation of $[\mathrm{M}_1]_\mathrm{w}$, $[\mathrm{M}_2]_\mathrm{w}$, $\varphi_{1,\mathrm{p}}$ and $\varphi_{2,\mathrm{p}}$. Note that the water phase concentrations of the monomers 1 and 2 in equilibrium with the pure monomers 1 and 2, respectively $[\mathrm{M}_1]_\mathrm{w,sat}$ and $[\mathrm{M}_2]_\mathrm{w,sat}$, should be known for this calculation (Gardon, 1968).

In many practical cases, instead of using the complete equilibrium equations that led to Equations 4.18 and 4.19, the description of the equilibrium between the particle and the aqueous phases can be simplified if partition coefficients are applied:

$$K^i_{\mathrm{a,b}} = \frac{[\mathrm{M}_i]_\mathrm{a}}{[\mathrm{M}_i]_\mathrm{b}} \tag{4.21}$$

where a and b stand for the phase involved. Thus, the computation of the concentration of monomers in the polymer particles reduces to solving the material balances of Equation 4.10 or Equation 4.11 and the equilibrium Equations 4.21. Gugliotta *et al.* (1995a), Gugliotta, Arzamendi and Asua (1995b), Gugliotta *et al.* (1995c) have shown that in high solids content recipes (>50 wt%) the difference between using partition coefficients or the Vanzo equation to account for the thermodynamic equilibrium is negligible. In addition the solution of Equation 4.10 or Equation 4.11 is easier and computationally faster when using partition coefficients.

4.2.2 Consequences of Monomer Partitioning for the Copolymer Composition

The instantaneous copolymer composition depends on the mol fractions of the monomers involved in the polymerisation at the locus of polymerisation. The monomer to water ratio in the recipe may have considerable consequences for the composition of the monomer droplets and, therefore, also for the free monomer composition in the particle phase. Figure 4.6 demonstrates the effect of the monomer to water ratio on the composition of the monomer phase in a methyl acrylate(MA)–styrene(S)/water emulsion. Because the water solubility of MA is more than two orders of magnitude larger than the solubility of styrene in water a decreasing monomer to water ratio at constant overall MA–styrene molar ratio leads to a considerably lower mol fraction of MA in the monomer droplet phase. In the presence of particles the MA–styrene molar ratio will also decrease on decreasing the monomer to water ratio and, therefore, also the copolymer composition, see Figure 4.7. Figure 4.7 demonstrates that the instantaneous ratio of the monomers incorporated into the copolymer

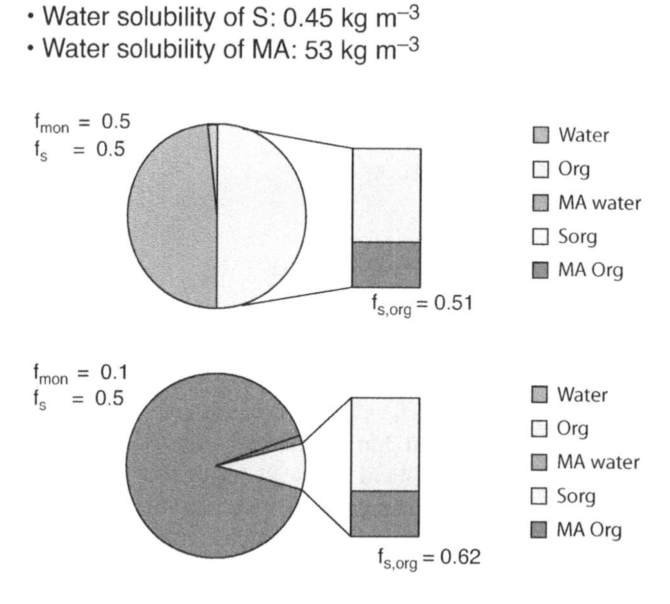

Figure 4.6 *Influence of the monomer weight fraction in monomer in water emulsions on the styrene (S) and methyl acrylate (MA) partitioning for the system S–MA.*

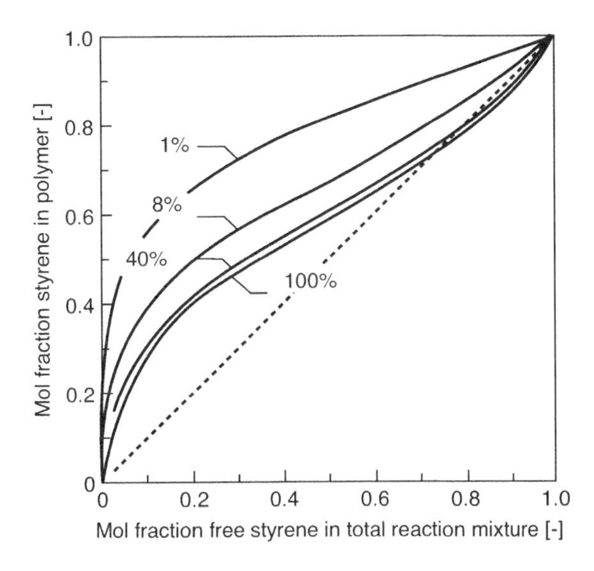

Figure 4.7 *Calculated mol fraction of styrene in the copolymer for the emulsion copolymerisation of styrene and methyl acrylate as a function of the mol fraction styrene in the total reaction mixture for some monomer weight fractions in the reaction mixture. Calculations with the ultimate approach, Equation 4.5, monomer 1 is styrene, monomer 2 is methyl acrylate,* $r_S = 0.73$ *and* $r_{MA} = 0.19$*. Temperature 50 °C, solubility data from Schoonbrood* et al.*, 1995a.*

for emulsion copolymerisation of a sparsely water soluble monomer and a moderately or even water miscible monomer depends strongly on the volume ratio of the phases, see for example, Van Doremaele *et al.* (1992), Verdurmen-Noël (1994), Noël *et al.* (1996).

When both monomers are sparsely water soluble, for example, the monomer pairs styrene–butadiene and styrene–n-butyl acrylate, the effect of the monomer to water ratio on the monomer composition in the droplet phase as well as in the particle phase is negligible. So for sparsely water soluble monomer pairs the monomer to water ratio has hardly any influence on the instantaneous and cumulative copolymer composition.

As a conclusion it can be stated that for monomer pairs of which one monomer is sparsely water soluble and one is moderately water soluble or even completely miscible with water, the concentration ratio of the monomers in the particle phase depends strongly on the volume ratio of the phases involved. However, the effect of the monomer to water ratio, M/W, is only important for too small values of M/W. For industrial recipes $M/W \geq 1$ and the influence of M/W on the copolymer composition will be negligible.

4.2.2.1 One of the Reacting Monomers Completely Water Miscible

If a completely water miscible monomer is involved in the copolymerisation, e.g acrylic acid or 2-hydroxyethyl methacrylate (HEMA) simple relations between volume fractions and concentrations in the droplet and the particle phase, see Equations 4.12 and 4.13, are no longer valid. Obviously, the mass balances describing the distribution of the monomers over the different phases, as given in Equations 4.8–4.11, are still applicable. However, more experimental data are necessary as compared to systems of sparsely and moderately soluble monomers. Specific intermolecular interactions between the monomer molecules and between the monomer molecules and the polymer chains, such as hydrogen bridges, have a large influence on the partitioning behaviour. Partitioning of acrylic type acids depends strongly on the pH of the aqueous phase (Slawinski *et al.*, 2000)

4.3 Process Strategies

Batch, semi-continuous and continuous reactors are used in emulsion polymerisation. Typically, these reactors are stirred tank reactors and the most common operation mode is the semi-continuous one because of its versatility, as will be shown in this section. Because of their large heat transfer area/reactor volume ratio continuous operation in tubular reactors is an attractive alternative, but it is not often used in emulsion polymerisation, principally due to the high risk of phase segregation, fouling and pipe clogging. Loop reactors (Abad *et al.*, 1994; Abad, De La Cal and Asua, 1997), that is, tubular reactors with a high recirculation ratio, and pulsed reactors (Paquet and Ray, 1994; Meuldijk and German, 1999) have been used, but the main drawback of these tubular reactors is a strong demand for proper emulsification, see for example, Meuldijk *et al.* (2003) and for recipes with high colloidal stability to prevent shear-induced coagulation and fouling.

In this section batch and semi-batch operation modes to produce copolymers in emulsion polymerisation will be discussed.

4.3.1 Batch Operation

A batch reactor is a closed system, that is, no materials enter or leave the reactor during the polymerisation reaction, in which the time is the only independent variable. The batch operation can be used for some small-scale production of homopolymers from monomers with a relatively small heat of polymerisation. However, the drawbacks associated with this type of operation limit its industrial use. These drawbacks are:

- The control of the polymer properties is impracticable.
- Low productivity considering the charging, discharging and cleaning times.
- Because all of the monomer is initially charged in the reactor, the heat generation rate during the reaction is high and the control of the reactor temperature is very difficult.
- Batch to batch variations due to irreproducible particle nucleation may jeopardize product consistency. In order to avoid this problem, seeded emulsion polymerisation may be employed.

Batch reactors are commonly employed in research laboratories because of their simplicity and low cost of operation.

The composition of the copolymers produced in batch reactors will be dictated by the reactivity ratios of the monomers, r_i, see Equation 4.6, as well as by the mol fractions of the monomers, f_i in the polymer particles, see Equation 4.7. The instantaneous composition can then be predicted by the terminal model, see Equation 4.5. Most of the common monomers employed in emulsion polymerisation recipes present different reactivities, and a consequence of this is the compositional drift (non-constant copolymer composition) produced in batch operation. The composition drift can be easily calculated by computing the instantaneous, see Equation 4.2, and cumulative copolymer compositions.

In order to calculate the rate of polymerisation, the monomer concentration in the particles, $[M_i]_p$, the average number of radicals per particle, $\bar{n}$, and the number of particles, N_p, as well as the reactivity ratios should be available. The calculation of $[M_i]_p$ has been described in Section 4.2. The calculation of $\bar{n}$ and N_p can be found in Chapters 2 and 3. Note that to calculate the time evolution of the instantaneous copolymer composition, the time history of the variables $[M_i]_p$,$\bar{n}$ and N_p must also be known. The unreacted or free monomer present in the reactor can be computed from the general macroscopic material balance for a perfectly mixed stirred tank reactor by omitting inlet and outlet streams:

$$\frac{dN_i}{dt} = -R^*_{p,i} \cdot V_w + F_{i,in} - F_{i,out} \tag{4.22}$$

where N_i is the number of moles of monomer i in the reactor (mol). $F_{i,in}$ and $F_{i,out}$ are the inlet and outlet molar flow rates of monomer i (mol s^{-1}), respectively. $R^*_{p,i}$ and V_w, respectively stand for the rate of polymerisation of monomer i and the volume of the water phase in the reactor. Note that $R^*_{p,i} = R_{p,i}+$ contribution of the water phase polymerisation of monomer i. $R_{p,i}$ follows from Equation 4.3.

The cumulative composition is the average composition of the copolymer formed up to a given time:

$$y_{1,c} = \frac{N_{p,1}}{\sum N_{p,i}} \tag{4.23}$$

where $N_{p,i}$ stands for the number of moles of monomer i polymerised. For sparsely water soluble monomers, $N_{p,i}$ can be calculated from Equation 4.1 for all polymerising monomers i. An example of the copolymer composition produced in a batch reactor is shown in Figures 4.8 and 4.9 for a MMA/BA comonomer system. A seeded batch emulsion copolymerisation is simulated with a particle concentration N_p of 2.6×10^{20} particles per m^3 water phase and an initial molar ratio of MMA and BA equal to one. The reactivity ratios r_{MMA} and r_{BA} used in the simulation are 2.24 and 0.414, respectively (Vicente, 2001). Partition coefficients, see Equation 4.21 were used to account for monomer partitioning (Vicente, 2001).

Figure 4.8 clearly shows the evolution of the instantaneous and the cumulative copolymer composition for this comonomer system. The copolymer formed in the initial stages of the polymerisation is rich in MMA: the instantaneous composition referred to BA is 0.3. However, when the polymerisation proceeds more and more BA is incorporated into the copolymer chains. For instance, for reaction times longer than 35 min, when overall conversion and the MMA conversion become larger than 60 and 80 %, respectively, the copolymer chains produced are richer in BA than in MMA units. The cumulative copolymer composition corresponds to 50 mol% BA and 50 mol% MMA at the beginning of the reaction because the seed used this composition. Obviously the drift in the cumulative composition is less pronounced and when all the monomer is consumed the composition is 50 mol% BA and 50 mol% MMA. It is worth pointing out here that experimentally one can

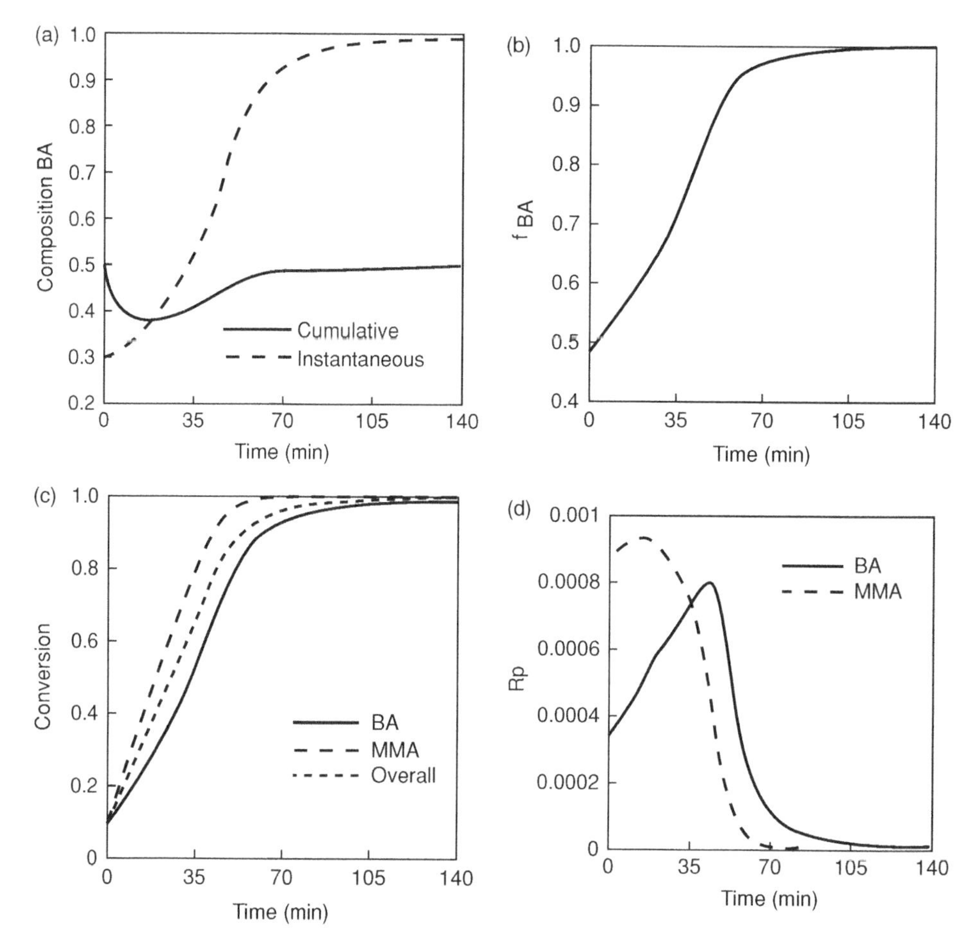

Figure 4.8 *Simulated data for the seeded batch emulsion copolymerisation of MMA and BA, initial molar ratio of BA and MMA is 1: (a) Instantaneous and cumulative copolymer composition; (b) ratio of the concentration of monomer in the polymer particles referred to BA,* $f_{BA} = \frac{[BA]_p}{[BA]_p + [MMA]_p}$; *(c) partial and overall conversions; (d) rates of polymerisation.*

measure the cumulative copolymer composition by combining the results of instrumental analysis techniques, for example, IR, NMR and, gas chromatography with material balances. However, the instantaneous composition cannot be determined in this way. Figure 4.8b displays the time evolution of the molar ratio of the monomer concentrations in the polymer particles. The ratio, as referred to BA, increases as polymerisation proceeds because MMA is consumed faster than BA and eventually is depleted, that is, $f_{BA} = 1$. The Figures 4.8a and b indicate that production of a copolymer with a constant instantaneous composition of 50 mol% BA and 50 mol% MMA requires a ratio of the concentrations of BA and MMA in the polymer particles of approximately 0.72. Figure 4.8c shows the overall and partial monomer conversions. Note that the MMA is consumed faster than BA and that after 60 min of reaction, corresponding to an overall conversion of

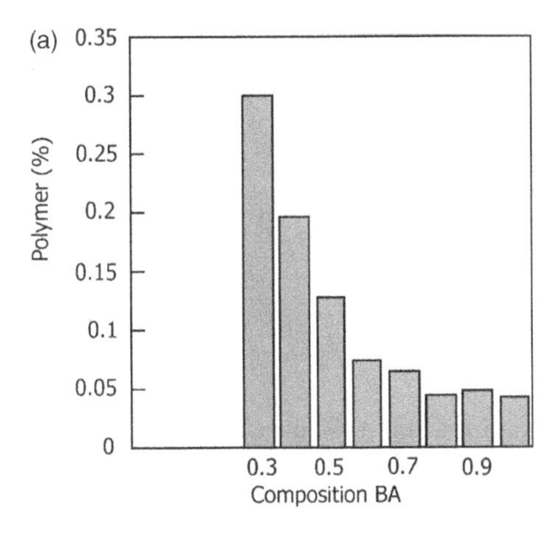

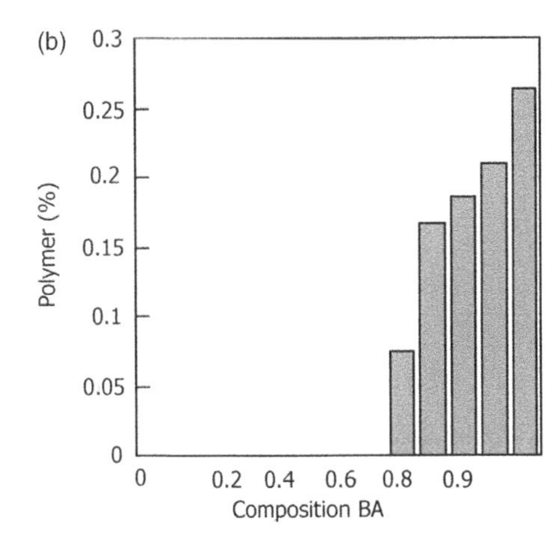

Figure 4.9 *Copolymer composition distribution of the BA/MMA latex produced in seeded batch emulsion copolymerisation: (a) initial molar ratio of BA and MMA is 1; (b) initial molar ratio of BA and MMA is 9.*

90%, a homopolymer of BA is being formed at a very low rate of polymerisation, see Figure 4.8d.

The copolymer composition distribution can be calculated from the data collected in Figure 4.8 for the instantaneous composition and the time evolution of the overall conversion. Figure 4.9 displays the copolymer composition distribution for batch seeded emulsion copolymerisations of BA and MMA simulated with two initial monomer compositions: one with 50 mol% BA and 50 mol% MMA and one with 90 mol% BA and 10 mol% MMA.

Figure 4.9 also shows that the copolymer is heterogeneous for both feed compositions; namely copolymer chains with compositions richer and poorer than the feed composition are produced in a batch operation.

4.3.2 Semi-Batch Operation

In semi-batch operation mode, some fraction of reactants, that is, the initial charge, is charged into the reactor, and the rest of the formulation is continuously fed over some period of time. Most commercial products are manufactured by semi-batch operated reactors. The main characteristic of this type of operation is the great flexibility. By varying the composition and amount of the initial charge, as well as the composition and flow rates of the feeds, both the temperature of the reaction mixture and the polymer quality may be controlled. A wide range of products is accessible using semi-batch operation. It allows the tailoring of any polymer property including copolymer composition, molecular weight distribution, polymer architecture, particle morphology (see Chapter 6) and particle size distribution (Leiza and Asua, 1997). Examples of composition control are shown below. In addition, a large portfolio of products can be produced in one single reactor. The main drawback of the semi-batch operation mode is the relatively low productivity, which is compensated by using larger reactors.

The macroscopic material and energy balances for semi-batchwise operated perfectly mixed stirred tank reactors are given by the Equations 4.24 and 4.25, respectively:

$$\frac{dN_i}{dt} = F_{i,\text{in}} - R_{\text{p},i} \cdot V_{\text{w}} \tag{4.24}$$

$$m_{\text{r}} \cdot C_{p,\text{r}} \cdot \frac{dT_{\text{r}}}{dt} = \sum_{i=1}^{m} R_{\text{p},i} \cdot (-\Delta H_{\text{r},i}) \cdot V_{\text{w}} + \dot{m}_{\text{feed}} \cdot C_{p,\text{feed}} \cdot (T_{\text{feed}} - T_{\text{r}}) + U \cdot A \cdot (T_{\text{jacket}} - T_{\text{r}}) \tag{4.25}$$

In Equation 4.25 m stands for the number of polymerising monomers. $C_{p,\text{r}}$, $C_{p,\text{feed}}$, m_{r}, $\dot{m}_{\text{feed}}$, $-\Delta H_{\text{r},i}$, T_{r}, T_{feed}, T_{jacket}, U and A stand for the heat capacity of the polymerising mixture in the reactor in units J kg^{-1} K^{-1}, the heat capacity of the feed in units J kg^{-1} K^{-1}), the mass of the reaction mixture at time t in kg, the mass flow rate of the feed in units kg s^{-1}, the heat of polymerisation in units J mol$_i^{-1}$, the temperature of the reaction mixture at time t in K, the temperature of the feed in K, the temperature of the liquid in the jacket in K, the overall heat transfer coefficient in units W m^{-2} K^{-1}) and the heat transfer area in m^2, respectively. In Equation 4.25 the terms$\sum_{i=1}^{m} R_{\text{p},i} \cdot (-\Delta H_{\text{r},i}) \cdot V_{\text{w}}$, $\dot{m}_{\text{feed}} \cdot C_{p,\text{feed}} \cdot (T_{\text{feed}} - T_{\text{r}})$ and $U \cdot A \cdot (T_{\text{jacket}} - T_{\text{r}})$ represent the heat production by polymerisation, the heat consumption rate by increasing the temperature of the feed from the feed temperature to the temperature of the reaction mixture and the rate of heat transfer rate from the reaction mixture to the cooling liquid in the jacket, respectively. Every term is expressed in W.

4.3.2.1 *Operation and Feeding Alternatives*

In general the initial charge contains a seed, which is used principally to avoid the lack of reproducibility of the nucleation stages when the seed is produced *in situ*. In addition it should be remarked that nucleation is very scale sensitive (Meuldijk *et al.*, 2003). Besides the seed latex the initial charge in a semi-batch process contains a fraction of the total amount of water to be used, surfactant and initiator. Under some circumstances a certain amount of the monomer(s) can also be present. The rest of the formulation ingredients are fed to the reaction mixture at a constant flow rate. Note that in many cases time-dependent feed rates are used. These feed rate profiles can be calculated by using empirical knowledge of the process or by application of optimization techniques based on mathematical models.

4.3.2.1.1 Effect of the Seed in the Initial Charge. The amount of polymer in the seed usually represents less than 5% by weight of the total polymer in the final latex product and, hence, the properties of the seed polymer are negligible with respect to the copolymer formed afterwards. However, the amount of polymer and the particle size of the seed are important. Thus, for a given amount of seed polymer the lower the particle size, the higher the number of particles and, as a consequence, the specific particle surface area per unit volume of the water phase will be larger. As a result of this higher specific particle surface area secondary nucleation is less likely to occur during the monomer addition period. If secondary nucleation is avoided during the process, latex products with narrow particle size distributions can be obtained.

4.3.2.1.2 Effect of the Amount of Emulsifier and Its Distribution. The higher the total emulsifier concentration in the recipe the higher the total number of polymer particles and the lower the particle size achieved during the polymerisation. If the emulsifier is fully loaded in the initial charge there is a risk of (limited) coagulation of the particles in the reaction mixture. As a result of particle growth by simultaneous polymerisation and monomer absorption the fractional surface coverage of the particles with emulsifier decreases if no emulsifier is fed to the reaction mixture. Coagulation will occur as a result of a decrease in the fractional surface coverage of the particles with emulsifier below its critical value for colloidal stability (Kemmere *et al.*, 1998). By charging all emulsifier at the beginning of the reaction there will also be a risk of macroscopic coagulum formation during the process. It is strongly recommended to split the surfactant between the initial charge and the feeding. As a result of this splitting of the total amount of surfactant, there will be surfactant available to avoid the fractional surface coverage of the particles with emulsifier falling below its critical value for colloidal stability during the monomer addition period.

When latex products with a bimodal PSD are desired, secondary nucleation can be induced by mid-course pulse-wise emulsifier additions to the reaction mixture.

4.3.2.1.3 Effect of the Amount of Initiator and Its Distribution. The effect of the amount of initiator on the product properties varies from system to system. In general the addition of initiator is not a good control variable. It should, however, be realised that the radical production rate by initiator decomposition has an influence on the rate of formation of oligomer radicals in the aqueous phase and so on the entry rate, that is, the entry frequency, of surface active oligomer radicals into the particles. The time that a polymer chain in a particle can grow before bimolecular termination occurs is directly related to the entry rate. So the molecular weight distribution is influenced by the entry rate and by that on the radical production rate which is directly related to the initiator concentration. Initiator feeding might also have an influence on the development of the particle size distribution. Furthermore, care should be taken with the initiator concentration in the feed stream into the reactor. Relatively high local initiator concentrations may lead to coagulation of the latex because the colloidal stability limit can be exceeded. In any case, fast mixing of the feed stream with the reactor contents, that is, fast mesomixing, is a prerequisite for colloidal stability. Fast mesomixing can be achieved by feeding close to the impeller tip.

4.3.2.1.4 Feeding of Monomer as a Monomer in Water Emulsion or as Neat Monomer. The part of the monomer not initially charged into the reactor is added at a constant flow rate or with a predefined flow rate profile. The monomer addition can be done by feeding neat monomer or by feeding monomer pre-emulsified in water. Experience shows that the result of adding neat or pre-emulsified monomer leads to different polymerisation rates and copolymer properties in terms of molecular weights (Poehlein, 1997; Zubitur and Asua, 2001). This is basically due to monomer transport from the monomer phase to the reacting latex particles. If neat monomer is fed to the reaction mixture the resistance against monomer transport from the monomer phase via the aqueous phase to the reacting particles may be much larger than in the case of addition of pre-emulsified monomer. The reason for this difference is the specific mass transfer area per unit volume of the aqueous phase which is, on average, considerably smaller when neat monomer is fed than when pre-emulsified monomer is fed. For pre-emulsified monomer feed the polymerisation and its outcome is usually governed by intrinsic polymerisation kinetics. In the case of neat monomer feeding the rate of polymerisation and the outcome of the process may be for a large part governed by mass transfer limitation. The differences between the outcome of the polymerisation on feeding neat monomer or pre-emulsified monomer is more pronounced when sparsely water-soluble monomers are used and/or when the polymerisation is carried out on a larger scale. Also, when long chain mercaptans, for example, *tert*-docedyl mercaptane or *n*-dodecyl mercaptane, are employed as chain transfer agents in the formulation, their efficiency in controlling the molecular weights is different for neat and pre-emulsified monomer addition. For pre-emulsified monomer addition the apparent reactivity of the chain transfer agent is larger than for neat monomer addition. As a consequence molecular weight control for pre-emulsified monomer addition is better than for neat monomer addition (Zubitur and Asua, 2001; Mendoza *et al.*, 2000.

4.3.3 Control Opportunities

As pointed out above the great advantage of semi-batch operation is the possibility of controlling both reactor temperature and polymer quality by means of manipulating the feed flow rates of the different reagents that is, monomer(s), chain transfer agent and emulsifier. In this section special emphasis is made on the control of the copolymer composition by manipulating co-monomer feed flow rates.

4.3.3.1 Temperature Control

When emulsion copolymerisation is scaled-up to industrial size reactors, that is, reaction volumes larger than 10 m^3, temperature control becomes an issue. A larger reaction volume is always accompanied by a smaller reactor wall heat transfer area at the reactor wall per unit volume of the reaction mixture. As a consequence the heat transfer rate per unit volume of the reaction mixture through the reactor wall to the cooling fluid flowing through the jacket decreases when scaling-up, see Equation 4.26.

$$\dot{Q}_{\text{generated}} = \sum_{i=1}^{m} R_{\text{p},i} \cdot (-\Delta_{\text{r}} H_i); \ \dot{Q}_{\text{transfer}} = U \cdot \frac{A}{V_{\text{w}}} \cdot (T_{\text{r}} - T_{\text{jacket}}) \quad (4.26)$$

In Equation 4.26 $\dot{Q}_{\text{generated}}$ and $\dot{Q}_{\text{transfer}}$ stand for the heat generation due to polymerisation per unit volume of the water phase in units W m_{w}^{-3} and the heat transfer rate per unit volume of the aqueous phase in units W m_{w}^{-3}, respectively. Note that, as a result of the dimension of the rate of polymerisation of monomer *i*, $R_{\text{p},i}$, being mol m_{w}^{-3} s^{-1}, the water volume in the reaction mixture has been used instead of the volume of the reaction mixture.

A way to control the rate of heat generation is by controlling the monomer concentration in the polymer particles, where polymerisation takes place. This can be done by manipulating the feed flow rate of the monomer. The lower the monomer feed flow rate the lower the concentration of monomer in the polymer particles and the lower the heat generation rate. Furthermore, under certain low monomer flow rates a pseudo-steady-state is achieved in semi-batch operation. Under this operational condition the polymerisation rate equals the feed flow rate of the monomer and, hence, the monomer concentration in the reactor remains constant. These conditions are usually known as *starved conditions* and are

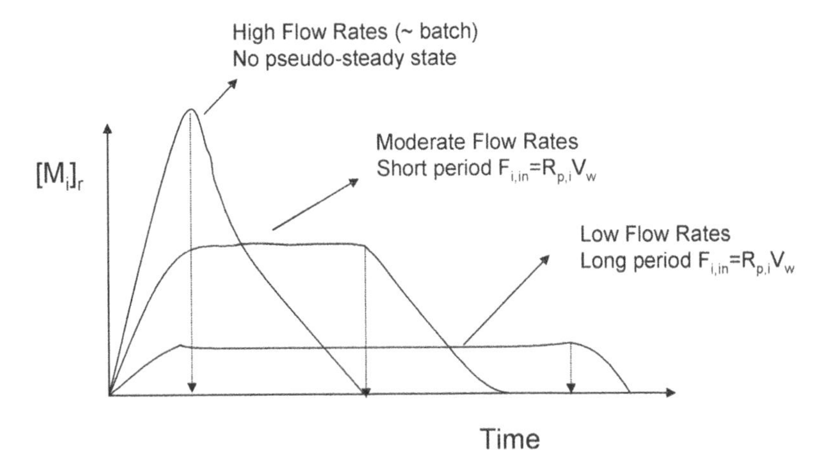

Figure 4.10 *Effect of the monomer feed flow rate on the concentration of monomer in the reactor.*

illustrated in Figure 4.10. For these pseudo-stationary conditions $\dot{Q}_{\text{generated}} < \dot{Q}_{\text{transfer}}$ and the temperature of the reaction mixture can be controlled at the desired value by manipulating the inlet jacket temperature. Figure 4.10 also demonstrates that at high monomer flow rates, that is, nearly batch operation, such a pseudo-steady-state is not achieved. Hence, the monomer concentration goes up when the monomer feed is on, that is, $F_{i,\text{in}} > R_{\text{p},i} \cdot V_{\text{w}}$. Once the feed is finished, the monomer concentration decreases, as in a batch process. The heat generated by polymerisation presents a similar profile and may exceed the maximum heat removal capacity, $U \cdot A \cdot (T_{\text{r}} - T_{\text{jacket}}^{\text{min}})$, of the reactor. Note that $T_{\text{jacket}}^{\text{min}}$ is the lowest possible temperature of the fluid in the jacket for the system chosen. In the case of exceeding the maximum heat removal capacity of the reactor, that is, $\dot{Q}_{\text{generated}} > \dot{Q}_{\text{transfer}}$, the reactor temperature increases, which eventually may lead to a thermal runaway. For the lower flow rates the pseudo-steady-state is achieved. The lower the flow rate the longer the time period where the monomer concentration in the reactor remains constant, that is, $F_{i,\text{in}} = R_{\text{p},i} \cdot V_{\text{w}}$. Therefore, the lower the feed flow rate, the lower the heat generation rate and the easier the temperature control, though the longer the process time and the lower the productivity.

4.3.3.2 Copolymer Composition Control

To produce latex with a given copolymer composition the ratio of the monomer concentrations in the polymer particles must be kept at the value that ensures the production of the desired composition. This co-monomer ratio can be easily determined for the terminal model for copolymerisation using Equation 4.5. Writing the monomer mol fraction f_1 in the particles as the dependent variable and the desired copolymer composition y_1 as the independent variable, Equation 4.27 is obtained:

$$f_1 = \frac{(2 \cdot y_1 \cdot (1 - r_2) - 1) + \{[2 \cdot y_1 \cdot (1 - r_2) - 1] - 4 \cdot [y_1 \cdot (r_1 + r_2 - 2) - (r_1 - 1)] \cdot r_2 \cdot y_1\}^{0.5}}{2 \cdot [y_1 \cdot (r_1 + r_2 - 2) - (r_1 - 1)]} \tag{4.27}$$

To avoid the composition drift that occurs in batch reactors when monomers of different reactivities are copolymerised, the monomer mol fraction in the polymer particles, f_1 (or f_2) must be maintained at the desired value during the whole polymerisation. If a constant (homogeneous) copolymer composition is sought f_1 must be maintained at the value calculated from Equation 4.27, which is only a function of the desired copolymer composition y_1 and the reactivity ratios of the monomers, r_1 and r_2. Note that this mol fraction f_1 is the mol fraction in the polymer particles and not in the reactor. Thus the required mol fraction in the reactor can be calculated by accounting for the partitioning of the monomers in the different phases present at each time in the reactor, see Equation 4.9. For high solids content recipes, that is, high monomer to water ratios in the formulation, and sparsely water soluble monomers there is basically no significant difference between f_1 in the polymer particles and in the reactor; namely all the free monomer is within the polymer particles (Gugliotta *et al.*, 1995a; Gugliotta, Arzamendi and Asua, 1995b; Gugliotta *et al.*, 1995c).

In what follows several monomer feeding strategies to control copolymer composition in semi-batch operation will be presented, namely feeding strategies to maintain f_1 at the value required to produce the desired copolymer composition distribution.

4.3.3.3 Copolymer Composition (Distribution) Control

Copolymer composition influences strongly the end-use properties of the copolymers. For instance, the T_g of the copolymers is defined by the co-monomer composition in the polymer chains, but if the chain composition is not homogeneous, that is, if the copolymer is produced under compositional drift conditions, the copolymer may exhibit more than one T_g; this being deleterious for the final application. The adhesive and mechanical properties have been reported as properties that are significantly affected by the copolymer composition distribution (Laureau *et al.*, 2001). Since the copolymer composition is defined during the polymerisation, it is necessary to control this property during latex production. Several strategies to control the copolymer composition distribution will be discussed in the following.

4.3.3.3.1 Starved Monomer Addition. As described earlier, if the monomer feed flow rates are slow enough, a pseudo-steady-state might be achieved. During this pseudo-steady-state the rate of polymerisation equals the flow rate of the monomer, see Figure 4.10. When this condition is obeyed, the instantaneous composition of the copolymer produced is given by the ratio of the flow rates of the monomer in the feed stream, or by the ratio of the monomer in the feeding tank if a unique monomer stream is used for the addition:

$$y_1 = \frac{R_{p,1}}{R_{p,1} + R_{p,2}} \approx \frac{F_{m,1}}{F_{m,1} + F_{m,2}} \tag{4.28}$$

Therefore, production of a latex product with a given desired copolymer composition under starved conditions only requires adjustment of the flow rates of the monomers at the desired ratio by keeping the overall flow rate slow enough to reach the pseudo-steady-state conditions. It is worth stressing that Equation 4.28 is only obeyed during the plateau region, as sketched in Figure 4.10. Outside this region, at the beginning of the monomer addition and once the feeding of the monomer is completed, the copolymer produced has a different copolymer composition. This means that to ensure a reasonable homogeneity

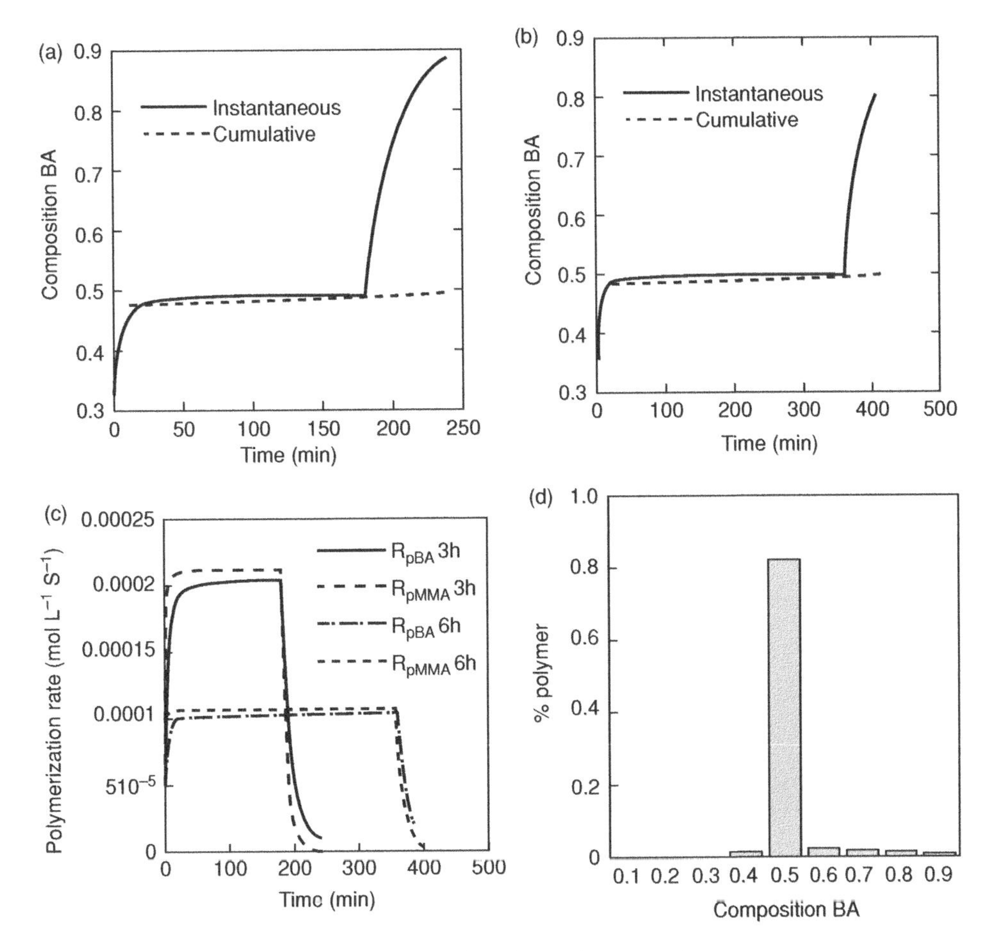

Figure 4.11 *Simulated data for the starved feed semi-batch emulsion copolymerisation of BA and MMA. Initial molar ratio of BA and MMA is 1. Instantaneous and cumulative copolymer composition for (a) 3 and (b) 6 h monomer addition, respectively; polymerization rates (c); CCD for 3 h addition (d).*

of the copolymer the plateau region should be long enough. The drawback of this process is that productivity is too low, because of the very long lasting monomer addition periods. However, most of the industrial plants produce copolymer latexes by implementing the starved monomer feed strategy because of its simplicity.

Figure 4.11 shows simulated data for two seeded semi-batch emulsion copolymerisations of BA and MMA carried out with feeding times of 3 and 6 h, respectively. Figures 4.11a and b present the cumulative and instantaneous copolymer compositions. The results clearly demonstrate that the steady-state is achieved in both cases. However, for the addition period of 6 h the fraction of copolymer with a composition deviating from the desired composition of 0.5 is smaller than for the addition period of 3 h. Furthermore, the cumulative composition is closer to 0.5 for the 6 h addition period than for the 3 h addition. In comparison with the batch process the composition drift is almost negligible, as displayed in Figure 4.11d which shows a very narrow CCD centred at 0.5.

4.3.3.3.2 Power Feed Addition. The power feed addition strategy is also conducted under monomer starved conditions but aiming at producing a copolymer with varying composition, see Basset and Hoy (1981b). In this case a heterogeneous copolymer composition distribution, that is, a copolymer product with a predefined composition distribution, is sought. This is obviously achieved by varying the ratio of the monomer flow rates into the reactor continuously with time.

The simplest way to achieve this is by feeding to the reactor a co-monomer mixture from one (or more) reservoir(s), as shown in the two cases in Figure 4.12. For instance, in the case of two monomers, two reservoirs 1 and 2 are used, see Figure 4.12a. Initially reservoir 1 is filled with monomer A and reservoir 2 with monomer B. During polymerisation the content of reservoir 2 is continuously pumped to the reactor and the content of reservoir 1 is

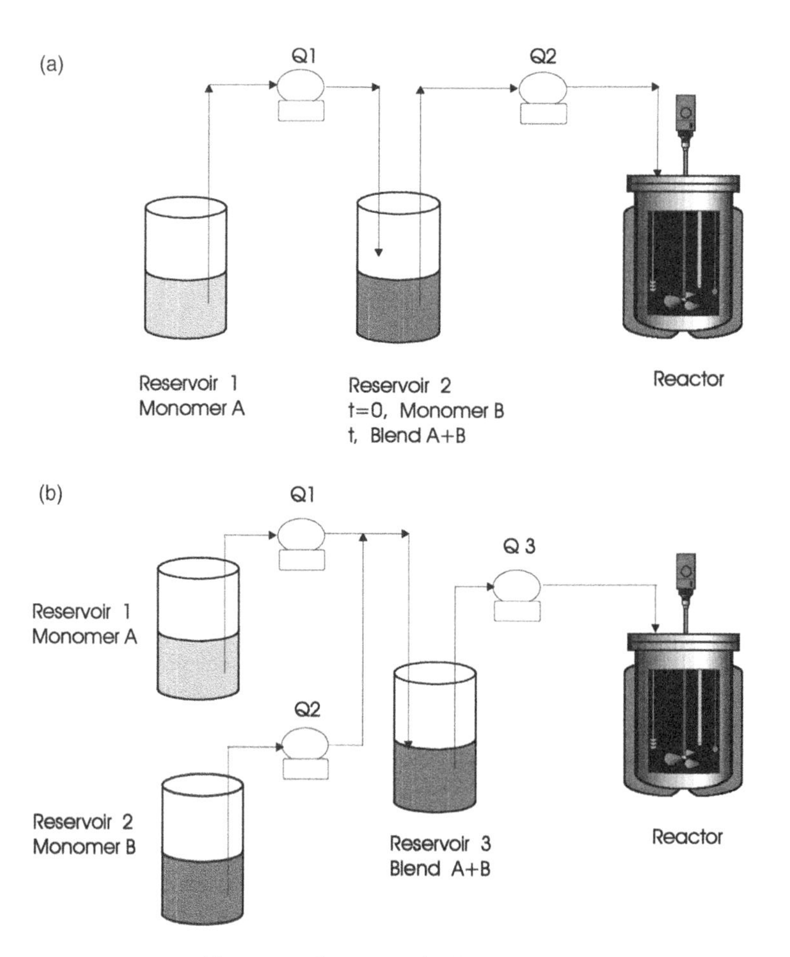

Figure 4.12 *Two possible set-ups for power-feed control of copolymer composition.*

continuously pumped to reservoir 2. Thus by varying the flow rates of pumping reservoir 1 to 2 and reservoir 2 to the reactor, different profiles of the copolymer composition distribution can be produced. The configuration of Figure 4.12b is more versatile because it allows the production of composition profiles that cannot be produced with the top configuration. However, note that to have a proper control of the composition, starved conditions, that is, polymerisation rate ~ overall feed flow rate, must be obeyed. Outside these starving conditions the composition will differ from the ratio achieved at each time corresponding with the monomer compositions in reservoir 2 (Figure 4.12a) or 3 (Figure 4.12b).

4.3.3.4 Optimal Control Strategy Based on a Mechanistic Model: Open-Loop Control

The starved monomer addition is a composition control strategy that does not require any knowledge of the kinetics of the process and is, therefore, easy to implement. However, starved feed monomer addition is not very demanding from a productivity point of view. So there is a lot of room for improvement.

Optimal control strategies for copolymer composition in emulsion polymerisation have attempted to cover the low productivity drawback of the monomer starved addition strategies. The objective of the optimal monomer feeding strategies is obviously to maintain the copolymer composition at the desired value but at the same time these optimal feeding strategies try to increase productivity by reduction of the process time. In kinetic terms operation with an optimal feeding strategy means that the process should be run at the maximum achievable polymerisation rate. Strictly speaking, optimising the monomer feed profile requires solution of an optimisation problem of a nonlinear dynamic process under certain constraints (Arzamendi and Asua, 1989; Arzamendi and Asua, 1991; Canu *et al.*, 1994; Canegallo *et al.*, 1994; de la Cal *et al.*, 1995; Echeverria, De la Cal and Asua, 1995). The objective function to be minimized is usually the process time and constraints have to be included to account for the required instantaneous copolymer composition and for large scale reactors the maximum heat removal rate from the reaction mixture. Other constraints are related to the equipment used in the plant. Obviously to solve this optimisation problem, a mathematical model of the process is required. Furthermore, plant–model mismatch must be reduced as much as possible to improve the experimental performance of the optimal solution.

Nevertheless, the optimisation problem is significantly reduced if heuristic knowledge of the emulsion copolymerisation process is used. To keep everything clear the emulsion copolymerisation of BA and MMA is considered. For the emulsion copolymerisation of BA and MMA the batch and semi-batch operation under starved monomer addition has already been discussed in terms of the copolymer composition distribution of the latex product. The emulsion copolymerisation of BA and MMA is considered with the following characteristics:

- Seeded copolymerisation of monodisperse particles with composition y.
- Constant number of polymer particles during the process.
- Aqueous phase polymerisation negligible.
- Time constant for monomer mass transfer is much lower than the time constant for polymerisation. So the resistance against mass transfer between the different phases is negligible and the polymerisation obeys intrinsic kinetics. Therefore, the distribution of monomer between the phases is controlled by a thermodynamic equilibrium.

The optimum strategy to produce a BA/MMA copolymer with a homogeneous, that is, a constant instantaneous composition y during the whole course of the process, is described in the following part (Arzamendi and Asua, 1989, 1990, 1991). The reactor is initially charged with all the less reactive monomer (butyl acrylate, BA) together with the amount of the more reactive monomer (methyl methacrylate, MMA) needed to initially form a copolymer with the desired composition. Then the remaining MMA is added to the reaction mixture at a flow rate that ensures the formation of a copolymer of the desired composition y. This optimal feeding policy involves the calculation of the amount of MMA to be initially charged into the reactor, as well as its time-dependent MMA addition flow rate.

4.3.3.4.1 Calculation of the Initial Amount of the Most Reactive Monomer. The amount of MMA in the initial charge can be calculated by using Equation 4.27 together with equilibrium equations and the overall material balances, see Section 4.2. Equation 4.27 provides the mol fraction of the BA or MMA in the polymer particles that ensures the formation of BA/MMA copolymer of the desired composition y. The equilibrium of the monomer in the different phases can be accounted for as described in Section 4.2. For simplicity partition coefficients, see Equations 4.29–4.32, are considered here:

$$K_{\mathrm{p,w}}^{\mathrm{MMA}} = \frac{[\mathrm{MMA}]_{\mathrm{p}}}{[\mathrm{MMA}]_{\mathrm{w}}} = \frac{\varphi_{\mathrm{MMA,p}}}{\varphi_{\mathrm{MMA,w}}} \tag{4.29}$$

$$K_{\mathrm{p,w}}^{\mathrm{BA}} = \frac{[\mathrm{BA}]_{\mathrm{p}}}{[\mathrm{BA}]_{\mathrm{w}}} = \frac{\varphi_{\mathrm{BA,p}}}{\varphi_{\mathrm{BA,w}}} \tag{4.30}$$

$$K_{\mathrm{d,w}}^{\mathrm{MMA}} = \frac{[\mathrm{MMA}]_{\mathrm{d}}}{[\mathrm{MMA}]_{\mathrm{w}}} = \frac{\varphi_{\mathrm{MMA,d}}}{\varphi_{\mathrm{MMA,w}}} \tag{4.31}$$

$$K_{\mathrm{d,w}}^{\mathrm{BA}} = \frac{[\mathrm{BA}]_{\mathrm{p}}}{[\mathrm{BA}]_{\mathrm{w}}} = \frac{\varphi_{\mathrm{BA,d}}}{\varphi_{\mathrm{BA,w}}} \tag{4.32}$$

In Equations 4.29–4.32 $K_{a,b}^{i}$ stands for the partition coefficient of monomer i between phases a and b, see Equation 4.21. In the Equations 4.29–4.32, p represents the polymer particles; d the monomer droplets and w the aqueous phase.

Together with the equilibrium equations, material balances are used, see Equations 4.33–4.39:

$$\varphi_{\mathrm{MMA,p}} + \varphi_{\mathrm{BA,p}} + \varphi_{\mathrm{pol,p}} = 1 \tag{4.33}$$

Note that Equation 4.33 is equivalent to Equation 4.20.

$$\varphi_{\mathrm{MMA,w}} + \varphi_{\mathrm{BA,w}} + \varphi_{\mathrm{w,w}} = 1 \tag{4.34}$$

$$\varphi_{\mathrm{MMA,d}} + \varphi_{\mathrm{BA,d}} = 1 \tag{4.35}$$

$$V_{\mathrm{MMA}} = V_{\mathrm{p}} \cdot \varphi_{\mathrm{MMA,p}} + V_{\mathrm{w}} \cdot \varphi_{\mathrm{MMA,w}} + V_{\mathrm{d}} \cdot \varphi_{\mathrm{MMA,d}} \tag{4.36}$$

$$V_{\mathrm{BA}} = V_{\mathrm{p}} \cdot \varphi_{\mathrm{BA,p}} + V_{\mathrm{w}} \cdot \varphi_{\mathrm{BA,w}} + V_{d} \cdot \varphi_{\mathrm{BA,d}} \tag{4.37}$$

$$V_{\mathrm{pol}} = V_{\mathrm{p}} \cdot \varphi_{\mathrm{pol,p}} \tag{4.38}$$

$$W = V_{\mathrm{w}} \cdot \varphi_{\mathrm{w,w}} \tag{4.39}$$

$\varphi_{i,\mathrm{a}}$ is the volume fraction of monomer i in phase a; $\phi_{\mathrm{w,w}}$ is the volume fraction of water in the water phase, V_i is the volume of monomer i in the reactor, V_{w} is the volume of the aqueous phase; W is the volume of water, and V_{pol} the volume of polymer. All volumes are in units m^3. In Equations 4.33 and 4.35 it is assumed that water is not present in either monomer droplets or polymer particles. Since the volume of the less reactive monomer BA is known from the recipe as well as the volumes of the polymer (seed) and water, the simultaneous solution of Equations 4.27 and 4.29–4.39 allows calculation of the amount of the most reactive monomer MMA in the initial charge.

4.3.3.4.2 Calculation of the Flow Rate of MMA During the Process. The amount of MMA that has to be present in the reactor at any time to produce a copolymer of constant composition y depends on the amounts of both free monomer BA and copolymer. These amounts vary during the polymerisation and hence the system of Equations 4.27 and 4.29–4.39 has to be coupled to the material balances of the butyl acrylate:

$$\frac{\mathrm{d}N_{\mathrm{BA}}}{\mathrm{d}t} = -(k_{\mathrm{p,BABA}} \cdot P_{\mathrm{BA}} + k_{\mathrm{p,MMABA}} \cdot P_{\mathrm{MMA}}) \cdot [BA]_{\mathrm{p}} \cdot \frac{\bar{n} \cdot N_{\mathrm{p}}}{N_{\mathrm{A}}} \tag{4.40}$$

Equation 4.40 can be solved provided that the time evolution of both the average number of radicals per particle, $\bar{n}$, and the number of polymer particles, N_{p}, are available. The amount of MMA incorporated into the copolymer, $N_{\mathrm{p,MMA}}$, at any time can be calculated from the desired instantaneous composition if the amount of BA incorporated into the copolymer is known, see Equation 4.41:

$$N_{\mathrm{p,MMA}} = \frac{(N_{\mathrm{BA}}^0 - N_{\mathrm{BA}})}{y} \tag{4.41}$$

where N_{BA}^0 is the number of moles of BA in the initial charge. The total amount of MMA added to the reactor, $N_{\mathrm{MMA}}^{\mathrm{T}}$, at any time is given by the amount of free monomer MMA, N_{MMA}, plus the amount that was already polymerized with BA, $N_{\mathrm{p,MMA}}$, see Equation 4.42.

$$N_{\mathrm{MMA}}^{\mathrm{T}} = N_{\mathrm{MMA}} + N_{\mathrm{p,MMA}} \tag{4.42}$$

The molar feed rate of MMA to be added to the reactor to ensure the formation of the desired copolymer composition is calculated as:

$$F_{\mathrm{MMA,in}} = \frac{\mathrm{d}N_{\mathrm{MMA}}^{\mathrm{T}}}{\mathrm{d}t} \tag{4.43}$$

In order to calculate the optimal feed rate profiles from Equation 4.43 the values of the parameters involved in Equation 4.27 and Equations 4.29–4.40 are required. These parameters include the partition coefficients of the monomers, the propagation rate constants, the reactivity ratios, the initiator decomposition rate constants and all the other parameters affecting the control of the average number of radicals per particle; namely all the parameters affecting the rates of entry and exit of radical species into or out of the polymer particles.

Figure 4.13a shows the optimal flow rate profile of MMA calculated for a formulation used in the simulations of the batch and starved semi-batch emulsion polymerisation processes of BA and MMA described earlier in this chapter. The objective was to produce a homogeneous copolymer containing 50 mol% MMA and 50 mol% BA, that is, the

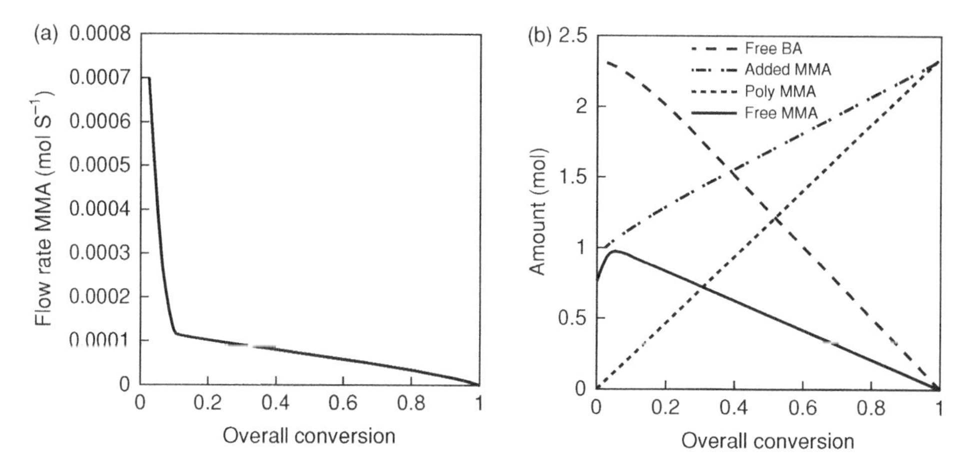

Figure 4.13 *Optimal flow rate of MMA (a) and optimum profiles of the free MMA and BA as well as the poly MMA as a function of conversion (b).*

instantaneous composition y is 0.5 during the whole polymerisation process. The process time should be as low as possible. The flow rate of MMA is high at the beginning of the process because of the high polymerisation rate and then decreases gradually to keep the concentrations of MMA and BA in the polymer particles at the desired value. Figure 4.13b presents the free amounts of MMA and BA as a function of conversion; the values at zero conversion indicate the initial charge in the reactor where the entire BA is charged and only a fraction of MMA. Figure 4.13b also shows the amount of MMA incorporated into the copolymer, see Equation 4.41, as well as the total amount of MMA added at each value of the conversion, see Equation 4.42.

Although significant work has been devoted to the determination of the entry and exit rates, see Chapter 2, the accurate prediction of $\bar{n}$ is still uncertain and hence the optimal feed rates calculated by using the approach just described might fail. In other words by tracking the trajectories calculated in the approach depicted above there is a risk of obtaining some compositional drift. This has been shown in the literature (Arzamendi and Asua, 1990; Arzamendi, Leiza and Asua, 1991; van Doremaele *et al.*, 1992; Leiza, Arzamendi and Asua, 1993a; Leiza *et al.*, 1993b; Gugliotta *et al.*, 1995a; Gugliotta, Arzamendi and Asua, 1995b; Gugliotta *et al.*, 1995c; Schoonbrood *et al.*, 1995b). The solution adopted to reduce the composition drift was to iterate the approach presented by reducing model-plant mismatch by estimating $\bar{n} \cdot N_p$ from the experiments carried out in the iteration procedure. Usually two or three iteration experiments were necessary to completely suppress the composition drift and to produce a copolymer with a homogeneous composition.

The optimal strategy just described was developed for small reactors, reaction volumes were between 1 and 2 dm^3. In these small reactors the heat removal rates per unit volume of the reaction mixture are high. However, the heat removal rate per unit volume decreases on scaling-up from laboratory reactors to large scale industrial reactors, which have limited capacity for heat removal. When a semi-batch emulsion copolymerisation is carried out following the optimal monomer addition strategy, a large amount of heat is generated at the beginning of the process because of the large concentrations of monomers in the initial charge. If the rate of heat generation exceeds the heat removal rate of the reactor, the temperature of the reaction mixture increases and, hence, the rate of polymerisation will also increase. As a consequence, the optimal monomer addition profile that was calculated at constant temperature will not lead to a copolymer of the desired composition. Therefore, for large scale industrial reactors the optimal strategy has to be modified. An approach that takes the limited heat removal rates of the reactors into account comprises an initial charge of the reactor with only a fraction of the less reactive monomer (BA) plus some amount of the more reactive monomer (MMA) (Arzamendi and Asua, 1991). The amounts of BA and MMA are chosen in such a way that the mol fractions of the free monomers in the polymer particles, that is, f_{MMA} and f_{BA}, still obey Equation 4.27. However, the absolute concentrations of the reacting monomers in the particles are kept at a level for which the rate of polymerisation, see Equations 4.3 and 4.40, will adopt such a value that the heat generated will never exceed the safe maximum heat removal rate of the reactor. Then both monomers should be added at a time-dependent flow rate that ensures the formation of the desired copolymer and without violating the additional constraint of the maximum heat removal rate.

4.3.3.5 Optimal Control Strategy Based on a Mechanistic Model: Closed-Loop Control

Even if an iterated monomer addition profile, as described in the previous section, is used, deviations from the desired copolymer composition may be found in a latex production plant. These deviations may be found if disturbances show up which are not considered in the model. A typical example is unexpected inhibition due to changes in the inhibitor content in the monomer feedstock.

This drawback, typical for any open-loop control strategy, can be circumvented by using closed-loop control strategies. A closed-loop strategy requires measurement of the property to be controlled or at least a measurement from which the property of interest, that is, in this case the copolymer composition, can be inferred. This measurement will allow manipulation of the flow rates of the monomers to force the copolymer composition to the target value. Several closed-loop strategies to control copolymer composition in emulsion copolymerisation systems have been reported in the literature (Dimitratos *et al.*, 1989; Leiza, Arzamendi and Asua, 1993a; Leiza *et al.*, 1993b; Urretabizkaia, Leiza and Asua, 1994; Saenz de Buruaga *et al.*, 1996, 1997a, 1997b; Saenz De Buruaga, Leiza and Asua, 2000; van den Brink *et al.*, 2001). Basically these closed-loop strategies differ in the design of the controller and the on-line sensor used to measure or to infer the copolymer composition. In this section a closed-loop control strategy that uses on-line measurements of the monomer conversion based on reaction calorimetry is briefly described.

The closed-loop strategy based on reaction calorimetry is shown in Figure 4.14.

During the polymerisation the heat released by the polymerisation reaction can be determined online from temperature measurements, see these references for details about reaction calorimetry (MacGregor, 1986; Moritz, 1989; Bonvin, Valliere and Rippin, 1989; Schuler and Schmidt, 1992). Calorimetric measurements can be used to infer the free amount of monomers in the copolymerisation and also the overall conversion (Urretabizkaia *et al.*, 1993; Gugliotta *et al.*, 1995a; Gugliotta, Arzamendi and Asua, 1995b; Gugliotta *et al.*, 1995c; Hammouri, McKenna and Othman, 1999; Saenz De Buruaga, Leiza and Asua, 2000) by means of an observer/estimator. Basically the estimator solves the monomer material balance differential equations using the heat of reaction as the input variable instead

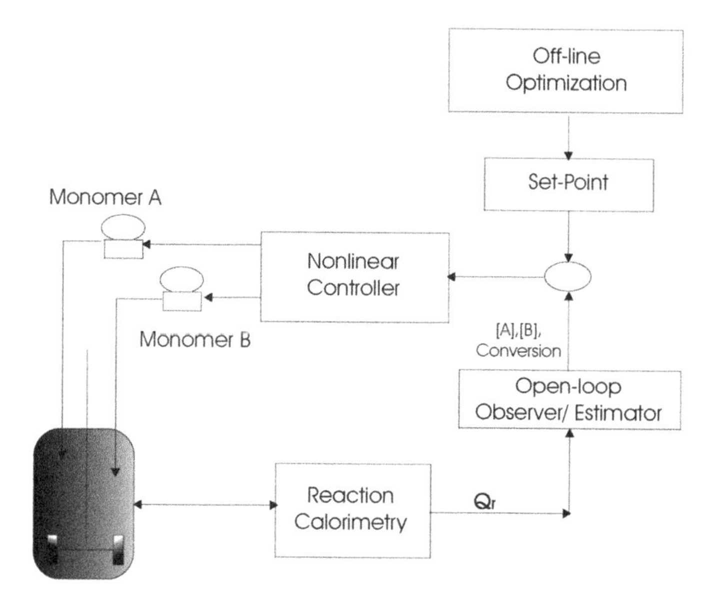

Figure 4.14 *Closed-loop control strategy to control copolymer composition based on reaction calorimetry.*

of the theoretical polymerisation rate. Therefore, the estimator uses the enthalpies of polymerisation of the monomers and the reactivity ratios as parameters. This information is compared with the optimal trajectories of the free monomer in the reactor calculated on a conversion basis. The nonlinear controller uses these data to calculate the flow rates of the monomers to be added to the reaction mixture to track the optimal trajectories of the monomers. In this section it will only be explained how those master curves are calculated. Details on the estimation of the heat of reaction from temperature measurements, the estimation of the free monomer from heat of reaction and the design of the nonlinear controllers can be found elsewhere (Saenz de Buruaga *et al.*, 1996, 1997a, 1997b; Saenz De Buruaga, Leiza and Asua, 2000).

Ignoring the effect of the monomer concentration on the number of polymer particles and on the number of free radicals per particle, the minimum process time to obtain a homogeneous copolymer is obtained by minimizing the following objective function:

$$J = \int_0^{X_\mathrm{T}} \frac{\mathrm{d}X_\mathrm{T}}{\sum_{i=1}^{2} R_{\mathrm{p}i}} \tag{4.44}$$

where X_T is the overall molar monomer conversion. The following constraints must be considered:

1. The copolymer must have the desired composition

$$f_1 = \frac{[\mathrm{M}_1]}{[\mathrm{M}_1] + [\mathrm{M}_2]} = f(X_\mathrm{T}) \tag{4.45}$$

In the case of homogeneous composition, $f(X_T)$ is a constant, but it can be any function, depending on the copolymer composition distribution (de la Cal *et al.*, 1995).

2. There is a limit to the maximum swelling of the particles by monomer:

$$[M_i]_p \leq [M_i]_{pref} \tag{4.46}$$

There are several reasons for imposing such a limit $[M_i]_{pref}$. It is a way of controlling the heat generation rate and also the monomer to polymer ratio in the particles, which might affect the MMD. Also if the limit is set below the saturation swelling of the particles, monomer droplet formation is avoided, which will imply a loss of control.

3. Monomer cannot be removed from the reactor;

$$\frac{\partial M_i}{\partial X_T} \geq 0 \tag{4.47}$$

4. There is a finite limit to the amount of monomer that can be added to the reactor:

$$M_{it} + M_{ipol} - M_{iT} \leq 0 \tag{4.48}$$

Where M_{it} is the free monomer in the reactor, M_{ipol} the amount of monomer *i* in polymer and M_{iT} the total amount of monomer *i* in the recipe.

The only data needed to solve this optimization are the monomer reactivity ratios and their partition coefficients. The monomer feed profiles can be calculated a priori, and hence the closed-loop control strategy requires only that these profiles are followed, for which the overall conversion must be measured online. It is worth noting that the optimum trajectories are independent of the polymerisation rate and that they can be seen as master curves for each co-monomer pair system for the operation conditions considered.

Figure 4.15a shows the master curves necessary to produce a homogeneous BA/MMA copolymer containing 50 mol% BA and 50 mol% MMA at a temperature of 70 °C. The

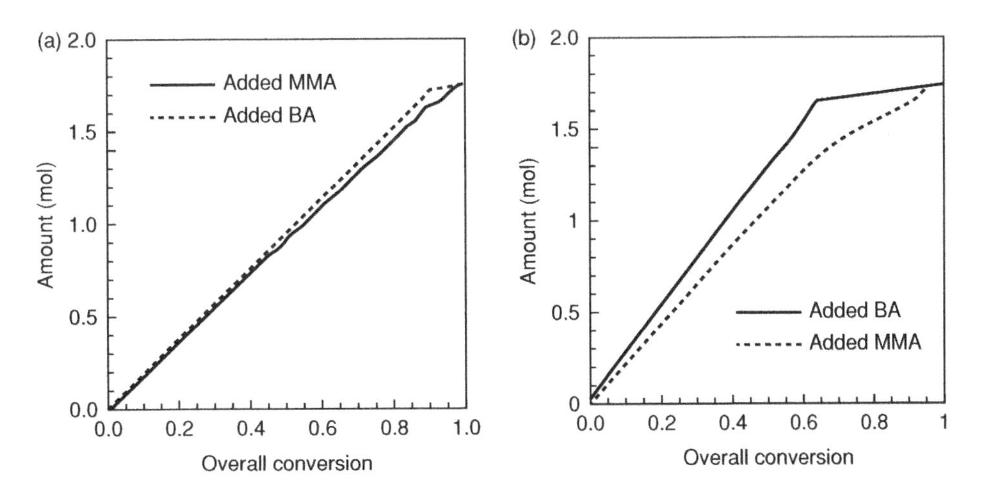

Figure 4.15 *Master curves of the total amount of MMA and BA that must be added to the reactor as a function of the conversion for two limits in the maximum swelling: (a)* $[BA]_p = 0.25$ *kmol m*$^{-3}$*; (b) no limit.*

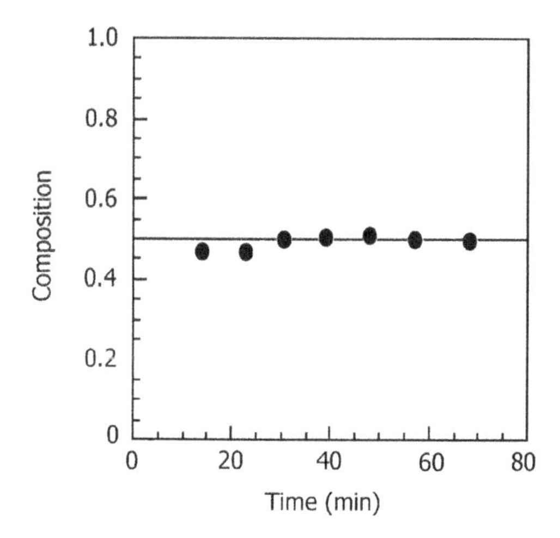

Figure 4.16 *Cumulative copolymer composition referred to BA for BA/MMA controlled copolymerisation tracking the master curves of Figure 4.15a.*

master curves are calculated for two different monomer swelling values (constraint 2). The master curves show the amounts of monomer BA and MMA that must be added to the reactor as a function of conversion. The values at zero conversion represent the monomer that must be initially present in the reactor. The plot also shows that the addition of the less reactive monomer must end at a lower conversion than the more reactive monomer. This difference in the optimal monomer addition profiles calculated for case (a) (a limit is imposed in the free amount of BA in the particles) and case (b) (no limit is imposed) show that the complete addition of the monomers can be achieved at lower overall conversion (in less process time) when no limit is imposed on the concentration of monomer in the polymer particles.

Figure 4.16 shows the experimental results obtained when the trajectory of Figure 4.15 was tracked online by measuring the conversion from the heat of reaction.

5

Living Radical Polymerisation in Emulsion and Miniemulsion

Bernadette Charleux,[1] *Michael J. Monteiro,*[2] *and Hans Heuts*[3]

[1]*Chemistry, Catalysis, Polymers & Processes, Université de Lyon, France*

[2]*Australian Institute for Bioengineering and Nanotechnology, The University of Queensland, Australia*

[3]*Department of Chemical Engineering & Chemistry, Eindhoven University of Technology, The Netherlands*

5.1 Introduction

Polymers with well-designed architectures prepared by living/controlled radical polymerisation (LRP) have invoked the interest of both academia and industry (Matyjaszewski *et al.*, 2000a, 2000b, 2000c; Matyjaszewski, 2000d; Braunecker and Matyjaszewski, 2007). Two main polymerisation mechanisms are currently used to control molar mass distribution, composition and structure. The first is based on reversible termination, comprising nitroxide-mediated polymerisation (Solomon, Rizzardo and Cacioli, 1985; Georges *et al.*, 1994; Hawker, Bosman and Harth, 2001; Nicolas *et al.*, 2012) (NMP) and atom transfer radical polymerisation catalyzed by a transition metal complex (Wang and Matyjaszewki, 1995a, 1995b) (ATRP) or more generally metal-mediated polymerisation including also single-electron transfer LRP (Rosen and Percec, 2009). The second relies on a reversible chain transfer reaction based either on the direct exchange of a terminal functional group (iodine atom (Goto, Ohno and Fukuda, 1998) or organotellurium group (TeRP, Yamago, Lida and Yoshida, 2002; Kwak *et al.*, 2006), or on an addition-fragmentation step (Le *et al.*, 1998; Charmot *et al.*, 2000; Barner-Kowollik, 2008) (RAFT). Many different architectures can be prepared in bulk or solution, and the applications for such architectures are slowly being realised. The challenge is to prepare these architectures in an environmentally friendly medium, water. The main LRP techniques have been applied under heterogeneous

Chemistry and Technology of Emulsion Polymerisation, Second Edition. Edited by A.M. van Herk.

conditions in suspension, dispersion, *ab initio* emulsion, seeded emulsion and miniemulsion using water as the reaction medium (see the following review articles: Qiu, Charleux and Matyjaszewski, 2001; Cunningham, 2002d, 2008; Save, Guillaneuf and Gilbert, 2006; McLeary and Klumperman, 2006; Charleux and Nicolas, 2007; Oh *et al.*, 2008b; Zetterlund, Kagawa and Okubo, 2008; Charleux, D'Agosto and Delaittre, 2010; Monteiro and Cunningham, 2012; Charleux *et al.*, 2012). The purpose of this chapter is to survey the literature on the use of LRP in dispersed media, provide some mechanistic understanding and describe the advantages and limitations of the various techniques.

The major advantage of water over bulk monomer or organic solvents is the environmentally friendly nature of the reaction medium. Water-borne processes are cheap, can be used for a broad range of monomers and a wide range of experimental conditions, the heat transfer is highly efficient, high conversions with low monomer residuals can be reached, there are no organic volatile compounds, and one can obtain high polymer solids ($\sim$50% wt) in a low viscosity environment. Another advantage in emulsion copolymerisation is that the morphology of the particle can be controlled. Coupling the advantages of LRP (such as the preparation of well-defined polymer architectures) with the above advantages of carrying out the polymerisation in water will produce a new class of specialty polymer materials for use in the coatings industry, materials and biomedical applications.

Although LRP techniques are relatively well understood in bulk or solution, in heterogeneous polymerisations the already complex kinetics are further complicated by partitioning of the activating species in the various environments, the rate of transportation of these species and larger dormant ones to the reaction locus, aqueous phase reactions, choice of surfactant, and control of the particle size distribution (PSD). However, the major kinetic advantage of classical free-radical emulsion polymerisations is compartmentalisation of the propagating radicals in either particles or droplets, which diminishes the amount of bimolecular termination, and consequently enhances the rate of polymerisation with better control of the molar mass distribution (MMD) compared to both bulk and solution polymerisations. It is thus important to utilize these advantages for LRP, especially when scaling-up to an industrial process.

5.2 Living Radical Polymerisation

5.2.1 General/Features of a Controlled/Living Radical Polymerisation

This section will describe the general features of the various LRP techniques in bulk and solution, which have already been extensively reviewed (Matyjaszewski and Davis, 2002; Matyjaszewski, Gnanou and Leibler, 2007). It will provide the reader with the mechanistic pathways of each technique and the kinetic parameters used to control the rate of polymerisation, the molar mass and molar mass distribution. The major difference between a 'living' radical polymerisation and a conventional radical polymerisation is the growth of polymer chains over the extent of the polymerisation, which generally lasts from a few minutes to many hours. In conventional radical polymerisation, chains are formed continuously throughout the polymerisation, and each chain grows on a very short time scale (in the range from 10^{-2} s to a few s). Conversely, in a 'living' radical polymerisation, all or most chains are formed after a small percentage of conversion, and grow slowly and continuously throughout the polymerisation (this may take minutes or even hours).

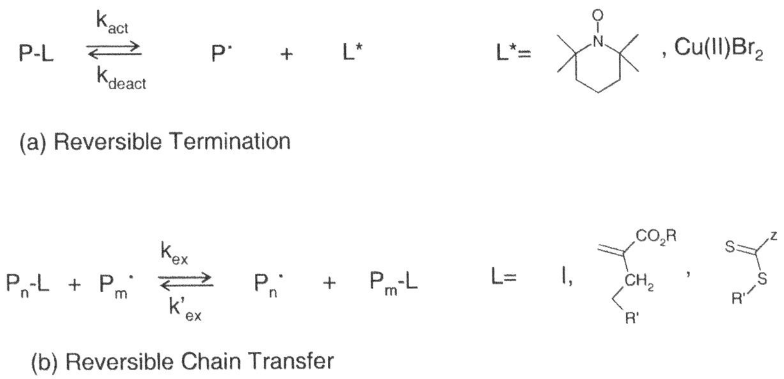

$$\text{P-L} \underset{k_{deact}}{\overset{k_{act}}{\rightleftharpoons}} \text{P}^{\cdot} + \text{L}^{*} \qquad \text{L}^{*} = \text{(TEMPO)}, \text{Cu(II)Br}_2$$

(a) Reversible Termination

$$\text{P}_n\text{-L} + \text{P}_m^{\cdot} \underset{k'_{ex}}{\overset{k_{ex}}{\rightleftharpoons}} \text{P}_n^{\cdot} + \text{P}_m\text{-L} \qquad \text{L} = \text{I}, \ \text{CH}_2\text{=C(CO}_2\text{R)CH}_2\text{R'}, \ \text{S=C(Z)SR'}$$

(b) Reversible Chain Transfer

Figure 5.1 *Two general mechanisms for LRP.*

5.2.1.1 *Control of Molar Mass and Molar Mass Distribution*

Control of the molar mass and molar mass distribution using LRP has important implications for the mechanical properties of the final polymer, making these techniques highly attractive for industrial applications (Monteiro and Cunningham, 2012).

The number average molar mass, M_n, and MMD are controlled by interplay of kinetic parameters, which will be described in more detail below. In general, all the LRP techniques fall into two main categories (see Figure 5.1): (i) reversible termination (e.g. NMP (Solomon, Rizzardo and Cacioli, 1985), ATRP (Kato *et al.*, 1995; Wang and Matyjaszewki, 1995a, 1995b) and Iniferter (Otsu, 2000)) and (ii) reversible chain transfer (e.g. degenerative chain transfer, RAFT (Le *et al.*, 1998), TeRP (Yamago, Lida and Yoshida, 2002; Kwak *et al.*, 2006). Active species are defined as species (polymeric radicals) that can add monomer in an elementary reaction step, and dormant species are defined as non-active chains, which nonetheless have the capacity to form active species during the polymerisation. The M_n can be controlled by varying the concentration ratio of monomer to control agent, the greater this ratio the higher the value of M_n.

Reversible termination requires the deactivation of active polymeric radicals through termination reactions to form dormant polymer chains, and activation of dormant polymer chains to form active chains with, for example, heat, light, or a redox reaction. The width of the MMD (as quantified by the polydispersity index, PDI) is controlled by the number of activation–deactivation cycles. On the other hand, reversible chain transfer requires active chains to undergo transfer reactions with the dormant chains, and thus the reversible chain transfer end-group is transferred from dormant to active species. A narrow MMD is observed when this exchange reaction is fast. In all these techniques, the MMD can be controlled such that the PDI is below 1.5.

The nemesis of LRP is the unwanted bimolecular termination reaction of active chains to form 'dead' polymer that can no longer participate in the polymerisation process, and which consequently broadens the MMD. Ideally, bimolecular termination should be eliminated from the reaction. The only way to do this is to have a very low concentration of active species, which in turn means a very slow rate of polymerisation. In many circumstances this would be too slow for use in industrial processes, and a compromise between MMD control

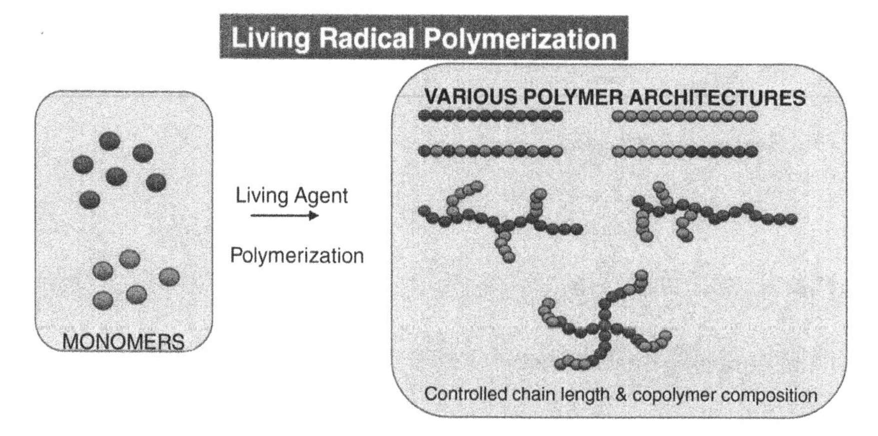

Figure 5.2 *Various architectures that can be prepared with LRP.*

and the speed of the reaction must then be made. Importantly, for the polymerisation to be considered under good control the amount of 'dead' polymer formed through termination or other side reactions should be negligible in comparison to the amount of dormant species. It should be noted that control of a living radical polymerisation does not mean that the PDI of the final polymer be less than 1.1. In many cases, polymerisations using less active controlling agents are still considered as living polymerisations even though the PDI is greater than 1.1. The real test of a successful LRP is the retention of the chain-end functionality.

5.2.1.2 Synthesis of Block Copolymers and More Complex Architectures

The advance of LRP to the field of polymer chemistry has allowed not only precise control of the MMD but also the synthesis of complex polymer architectures (Figure 5.2) that were previously absent from the polymer chemist's toolbox. A drawback when synthesising more complex architectures is that termination reactions play a more critical role in preparing well-defined structures. In the synthesis of star polymers (Figure 5.2), for example, termination reactions lead primarily to star–star coupling, with the number of arms on the star influencing the weight fraction of the star–star by-product. For example, if we compare the synthesis of 3- and 6-arm stars, where the termination rate and the M_n's are equal for both reactions, the weight fraction of star–star coupling for the 6-arm will be approximately double that of the 3-arm one.

5.2.2 Reversible Termination

One way to control free-radical polymerisation is the so-called reversible termination technique. It employs a chemically stable deactivator molecule to react with propagating carbon-centred radicals to form covalent bonds, that is, a dormant species (Matyjaszewski, 1998; Matyjaszewski *et al.*, 2000a, 2000b, 2000c; Matyjaszewski, 2000d, 2003; Matyjaszewski and Davis, 2002; Braunecker and Matyjaszewski, 2007; Matyjaszewski, Gnanou and Leibler, 2007). The chain-end of the dormant species is activated into the free-radical form (active state) either thermally or via a redox process. Only in the active state can the polymer chains grow by monomer addition, and undergo reactions typical of free-radical polymerisations, such as combination, disproportionation and chain transfer. In the dormant state, the polymer chains cannot react with monomer and are additionally protected against side reactions. In this way, all polymer chains undergo a large number of activation–deactivation cycles throughout the polymerisation, which ensures that they all experience the same probability of growth. Usually, the activation–deactivation equilibrium reaction favours the inactive state, and the concentration of active species is orders of magnitude smaller than the overall concentration of polymer chains. However, the active species concentration should not be much lower than in conventional free-radical polymerisation to obtain rates of polymerisation commensurate with industrial processes.

Polymer chains produced by LRP can be classified into three major categories: dormant, active (free radical) and dead (killed by any kind of irreversible termination reaction). The first two classes correspond to the so-called living chains and the degree of livingness of a polymer can be quantified by the proportion of those living chains as shown below.

$$\text{Proportion of living chains in a classical free-radical polymerisation} = \frac{[\text{radicals}]}{[\text{dead chains}] + [\text{radicals}]} \tag{5.1}$$

$$\text{Proportion of living chains in a controlled system} = \frac{[\text{radicals}] + [\text{dormant chains}]}{[\text{dead chains}] + [\text{radicals}] + [\text{dormant chains}]} \tag{5.2}$$

In a classical free-radical polymerisation, the concentration of active radicals is very small compared to the concentration of dead polymer chains, which make up most of the polymer. In contrast, in a living radical polymerisation the concentration of dead chains is only a small percentage of the concentration of dormant chains. Usually, the concentration of dormant chains is in the range 10^{-3}–10^{-2} mol L^{-1}, and the concentration of radicals in the range 10^{-9}–10^{-7} mol L^{-1}.

Practically, the polymerisation can be carried out in two different ways (Figure 5.3). The first method is to simply add the deactivator into the polymerisation medium containing

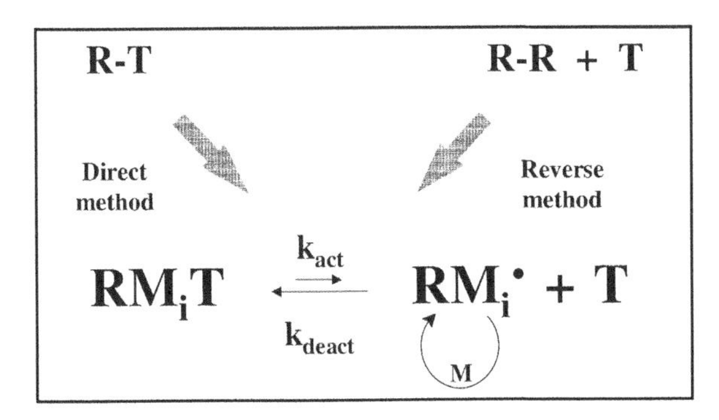

Figure 5.3 *General scheme of a living radical polymerisation by reversible termination. T is the reversible termination agent.*

monomer and radical initiator and then adapt the experimental conditions so that the trapped radical can be reactivated. The second method consists of using a pre-made dormant species of small size (considered as a unimolecular initiator) to start the polymerisation without the need for a radical initiator. In both cases, initiation must be fast, so that all growing chains are created within a short time span. It should, however, be noted that the second method leads to a better control over molar mass as the initiator efficiency is usually close to 1, which might not always be the case when a classical radical initiator is used.

If the criteria of fast activation–deactivation together with fast and efficient initiation are fulfilled, the number average degree of polymerisation (x_n) should increase linearly with monomer conversion according to:

$$x_n = \frac{[M]_0}{[I]_0}x \tag{5.3}$$

with $[M]_0$ the initial monomer concentration, $[I]_0$ the initial concentration of initiator and x the monomer conversion. The polymer chains in a LRP should conform to a Poisson distribution, with the polydispersity index given by:

$$\text{PDI} = \frac{x_w}{x_n} \xrightarrow[x_n \to \infty]{} 1 + \frac{x_n}{(x_n + 1)^2} \cong 1 + \frac{1}{x_n} \tag{5.4}$$

Typically, the polydispersity index is low, PDI < 1.5, which cannot be reached using conventional free-radical polymerisation. Ideally, the PDI should continuously decrease during the course of the polymerisation reaction (Litvinenko and Müller, 1997). In the case of slow initiation, x_n is larger than the theoretical value, but reaches it when all the initiator molecules are consumed. This leads to a broadening of the molar mass distribution. As most of the polymer consists of dormant chains with a small amount of dead macromolecules, the polymer can then be re-used as a macroinitiator (in the appropriate experimental conditions) to add the same monomer or another one, leading to the formation of a block copolymer.

The polymerisation kinetics are not regulated by the steady-state approximation like in classical free-radical polymerisation, but depend on the activation–deactivation equilibrium (Fukuda, Goto and Ohno, 2000; Goto and Fukuda, 2004). The concentration of propagating radical is proportional to the concentration of dormant chains and to $K = k_{act}/k_{deact}$ (Figure 5.3), the equilibrium constant. Additionally, it is inversely proportional to the deactivator concentration, which is regulated by the so-called persistent radical effect (PRE) (Fischer, 2001); because the propagating radicals can terminate with each other while the deactivator cannot, a continuous increase in the concentration of deactivator is generally observed. This trend favours the reversible deactivation process and results in the self-regulation of the concentration of the active radicals in the presence of a deactivator (Figure 5.4).

5.2.2.1 *Nitroxide-Mediated Living Radical Polymerisation (NMP)*

Nitroxides represent a very important class of radical deactivators (Bertin, Destarac and Boutevin, 1998; Hawker, Bosman and Harth, 2001; Hawker, 2002; Nicolas *et al.*, 2012). They are stable radicals able to terminate with carbon-centred radicals at near diffusion controlled rates. The trapping reaction forms an alkoxyamine (Figure 5.5) that is very stable at low temperatures, and could be considered an irreversible termination step. However, at elevated temperature, the C–O bond may undergo homolytic cleavage, producing back

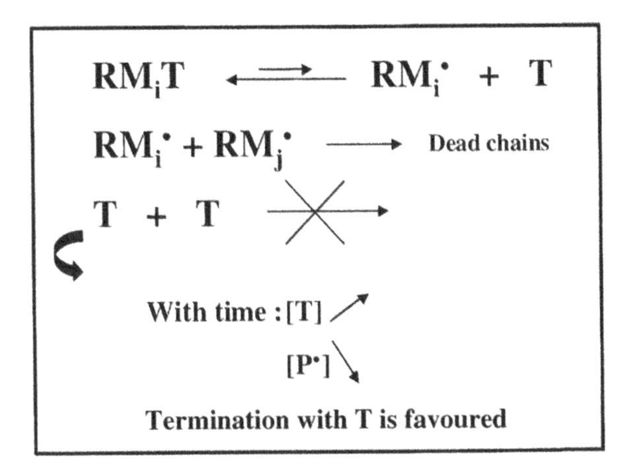

Figure 5.4 Simplified scheme of the persistent radical effect.

the propagating radical and nitroxide. This equilibrium between propagating radical and inactive alkoxyamine is the key step in nitroxide-mediated LRP; activation is thus purely a thermal process. The polymerisation can be started either from a bicomponent initiating system, that is, a conventional radical initiator in conjunction with free nitroxide, or a monocomponent initiating system, that is, using a preformed alkoxyamine instead of the radical initiator. Generally, polymerisations are performed in bulk at temperatures above 100 °C, depending on the selected nitroxide, and are slow reactions.

A wide variety of nitroxides have been synthesised and some have been used as mediators in free-radical polymerisations (Bertin, Destarac and Boutevin, 1998; Hawker, Bosman and Harth, 2001; Hawker, 2002; Matyjaszewski, 1998; Matyjaszewski *et al.*, 2000a, 2000b, 2000c; Matyjaszewski, 2000d, 2003; Nicolas *et al.*, 2012) (Figure 5.6). TEMPO (2,2,6,6 tetramethylpiperidine-l-oxyl, **N1** in Figure 5.6) was the first to be tested (Georges *et al.*, 1993) and remained for a long time the most widely used and studied nitroxide for LRP. TEMPO-mediated LRP was successfully performed for styrene and its derivatives, leading to the synthesis of well-defined block copolymers and star-shaped structures. The application of this method to other monomers appeared to be less straightforward. The poor results that were initially obtained by LRP of acrylic ester monomers could be overcome by controlling the concentration of free nitroxide in the system (Georges, Lukkarila and Szkurhan, 2004). Until recently, however, no living radical polymerisation could be obtained in the case of methacrylic esters owing to the preferred TEMPO-induced β-hydrogen elimination from the propagating radicals leading to the formation of ω-unsaturated dead chains (Burguière *et al.*, 1999). Other cyclic nitroxides were also proposed, but their behaviour

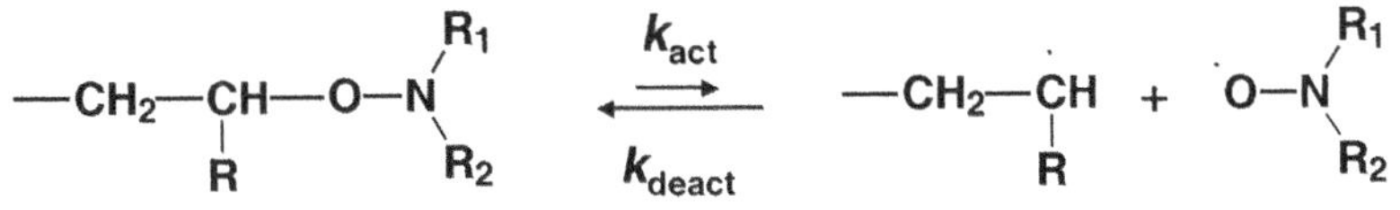

$$-CH_2-\underset{R}{CH}-O-N\begin{matrix}R_1\\R_2\end{matrix} \underset{k_{deact}}{\overset{k_{act}}{\rightleftharpoons}} -CH_2-\underset{R}{\dot{C}H} + \dot{O}-N\begin{matrix}R_1\\R_2\end{matrix}$$

Figure 5.5 Activation–deactivation in nitroxide-mediated living radical polymerisation.

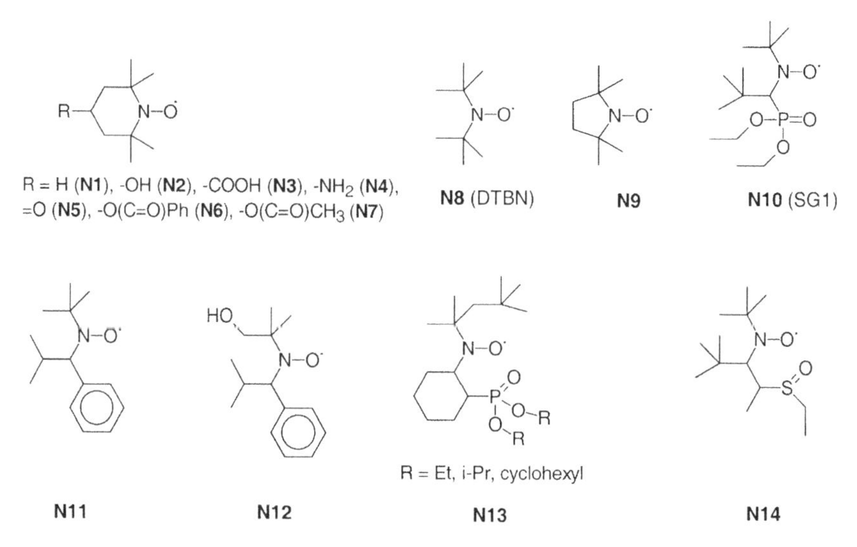

Figure 5.6 *Examples of nitroxides used as mediators in living radical polymerisation.*

was not very different from that of TEMPO (Matyjaszewski, 1998; Matyjaszewski *et al.*, 2000a, 2000b, 2000c; Matyjaszewski, 2000d, 2003).

A new class of acyclic nitroxides was more recently reported (**N10–N14** in Figure 5.6) (Nicolas *et al.*, 2012). First, because of a larger activation–deactivation equilibrium constant, faster rates of polymerisation than with TEMPO were observed for styrene polymerisation. In addition, these nitroxides were shown to be well suited for the living radical polymerisation of several other monomers, including acrylates, acrylamides, acrylic acid, acrylonitrile, maleic anhydride and isoprene (Hawker, Bosman and Harth, 2001; Hawker, 2002; Nicolas *et al.*, 2012). Like with TEMPO, the controlled homopolymerisation of methacrylate monomers remained unsuccessful (McHale, Aldabbagh and Zetterlund, 2007; Dire *et al.*, 2008) but the difficulty was overcome by copolymerisation with a very low percentage of styrene or acrylonitrile. The method led to well-defined alkoxyamines with penultimate unit effect reducing significantly the dissociation temperature below 90 °C (Charleux, Nicolas and Guerret, 2005; Nicolas *et al.*, 2006a; Nicolas, Brusseau and Charleux, 2010).

This feature opened the way to the synthesis of complex copolymer architectures using nitroxide-mediated polymerisation. Alkoxyamines that incorporate these nitroxides were synthesised by various chemistry procedures, as described by Hawker, Bosman and Harth, 2001 and Nicolas *et al.* (2012); examples are shown in Figure 5.7.

5.2.2.2 *Atom Transfer Radical Polymerisation (ATRP) and Related Techniques*

The second family of techniques based on a reversible termination mechanism is a catalyzed reversible redox process, named atom transfer radical polymerisation (ATRP) (Kamigaito, Ando and Sawamoto, 2001; Matyjaszewski and Xia, 2001, 2002; Matyjaszewski, 2012). Control over the polymerisation is realised based on the same principle as in nitroxide-mediated polymerisation, that is, an activation–deactivation equilibrium where both

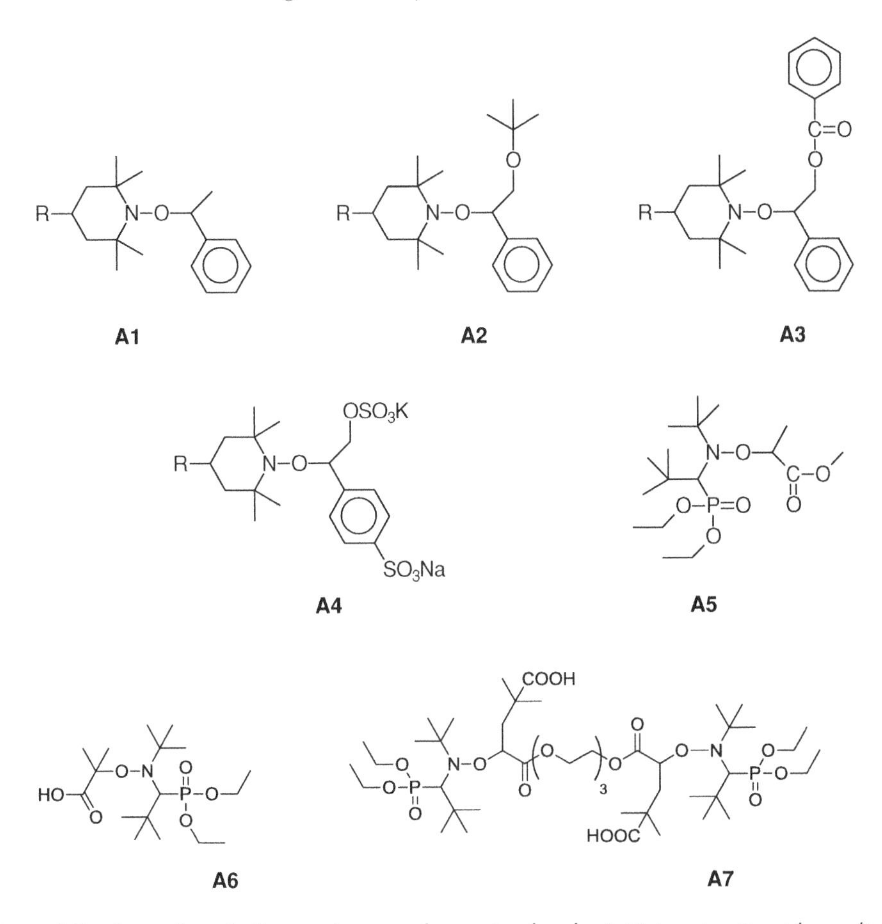

Figure 5.7 Examples of alkoxyamines used as unimolecular initiators in nitroxide-mediated living radical polymerisation.

activation and deactivation involve an atom transfer reaction (Figure 5.8). The method is based on the reversible transfer of a halogen atom between a dormant alkyl halide and a transition metal catalyst using redox chemistry. The alkyl halide forms a growing radical and the transition metal is oxidised via an inner sphere electron transfer process. In the first reaction, the role of the activator is often played by a copper (I) species complexed by two nitrogen-containing ligands (see Figure 5.9 for a few examples of relevance to emulsion polymerisation) and the role of the deactivator by the corresponding copper (II) species. Other transition metal complexes (Ru, Fe, Ni, ...) and other ligands were also

$$-CH_2-\underset{R}{CH}-X + Mt^{(+n)}X_n / \text{Ligand} \underset{k_{deact}}{\overset{k_{act}}{\rightleftharpoons}} -CH_2-\underset{R}{\dot{C}H} + Mt^{(n+1)}X_{n+1} / \text{Ligand}$$

Figure 5.8 Activation–deactivation equilibrium in ATRP.

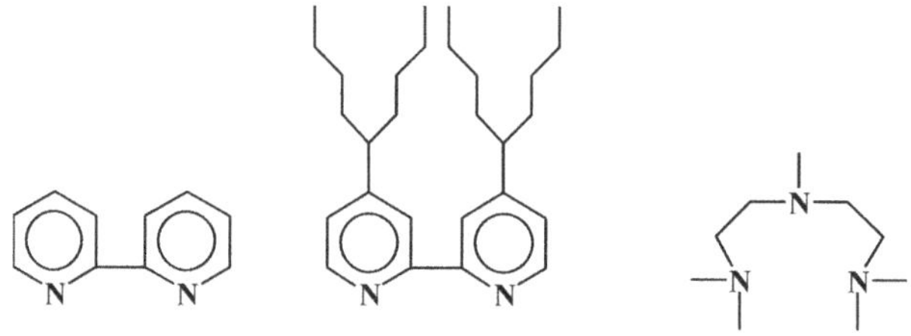

Bpy dNbpy PMDETA

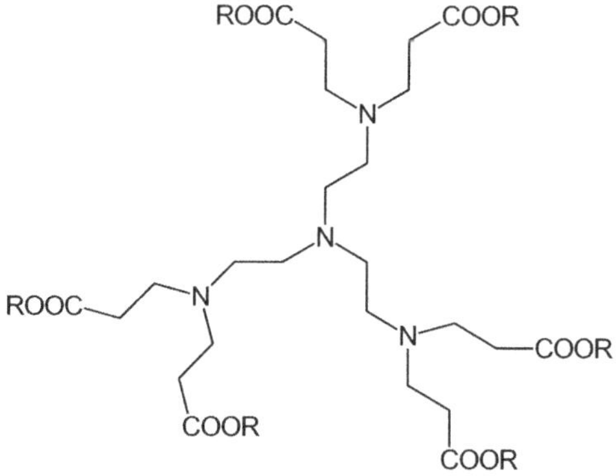

R = lauryl (LA_6TREN), 2-ethylhexyl (EHA_6TREN) or butyl (BA_6TREN)

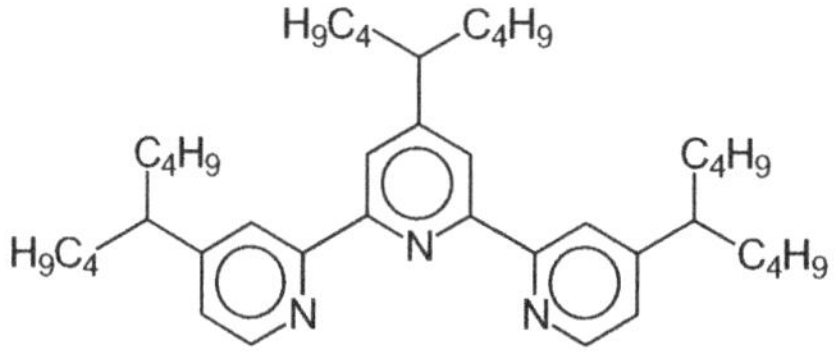

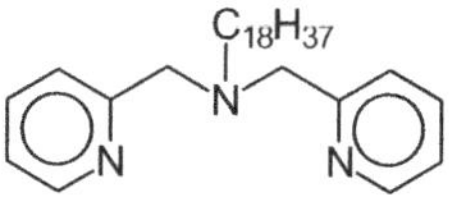

substituted terpyridine (tNtpy) picolyamine (BPMODA)

Figure 5.9 Examples of ligands used in ATRP in conjunction with copper catalyst.

used. Since the Cu system has been most widely used, we focus here on the Cu-mediated ATRP. When the polymerisation is started with an alkyl halide and a transition metal in the lower oxidation state (Cu(I)), the system behaviour can be compared with an alkoxyamine-initiated nitroxide-mediated polymerisation. The polymerisation can also be conducted in an alternative way, that is, starting the polymerisation with a conventional radical initiator and a metal complex in the higher oxidation state (i.e. the deactivator, Cu(II)). This process, which is called reverse ATRP, has the advantage that Cu(II) is insensitive to oxidation, but the disadvantage that in the case of block copolymer synthesis one always produces a small amount of homopolymer of the second block (as is the case in RAFT). Furthermore, since in this process the "ATRP-initiator" is formed *in situ*, stoichiometric amounts of $Cu(II)X_2$ are required. To reduce the amount of Cu and "secondary" polymer in the system, the so-called simultaneous reverse and normal initiation (SRNI) was developed by Matyjaszewski's group (Gromada and Matyjaszewski, 2001; Braunecker and Matyjaszewski, 2007; Matyjaszewski, 2012). Now an alkyl halide initiator is used in such a concentration that it defines the target molar mass, while simultaneously, a classical radical initiator is added (0.1–0.2 equivalent with respect to the alkyl halide) in conjunction with a highly active Cu(II) complex (also 0.1–0.2 equivalent). The Cu(I) complex activator is generated from the added Cu(II), upon deactivation of the radicals produced by the dissociation of the initiator.

Although SRNI allows a reduction in the overall Cu concentration, it still suffers from the formation of a small fraction of chains initiated by the conventional initiator. Soon after the invention of SRNI it was realised that this problem could be circumvented by using reducing agents, other than free radicals (from conventional initiators), which are not capable of initiating polymerisation. Examples of such reducing agents are Cu(0) or ascorbic acid (AsAc). This process was termed activators generated by electron transfer (AGET), (Min, Jakubowski and Matyjaszewski, 2006a; Matyjaszewski, 2012) and to date this method has yielded the most promising results in emulsion polymerisation (Monteiro and Cunningham, 2012).

A further reduction in Cu concentration has been made possible by the development from AGET to activators regenerated by electron transfer (ARGET) ATRP (Jakubowski, Min and Matyjaszewski, 2006; Matyjaszewski, 2012) and, in a different way, to initiators for continuous activator regeneration (ICAR) (Matyjaszewski *et al.*, 2006; Matyjaszewski, 2012). The reduction in overall Cu concentration is possible because the rate of polymerisation only depends on the ratio of activator (Cu(I)) and deactivator (Cu(II)) concentrations and not on the actual concentrations. An unlimited reduction in concentration, however, is not possible as the MMD will become broader upon decreasing the Cu(II) concentration; the practical limit is in the ppm range. Since the unavoidable radical–radical termination reactions would lead to an unacceptable build-up of Cu(II) at these low overall Cu(II) concentrations, the catalyst requires continuous regeneration by reducing agents (ARGET) or radicals (ICAR). Both methods, for which the mechanisms are shown schematically in Figure 5.10, also have the potential benefit of starting with the oxygen-insensitive Cu(II) complex

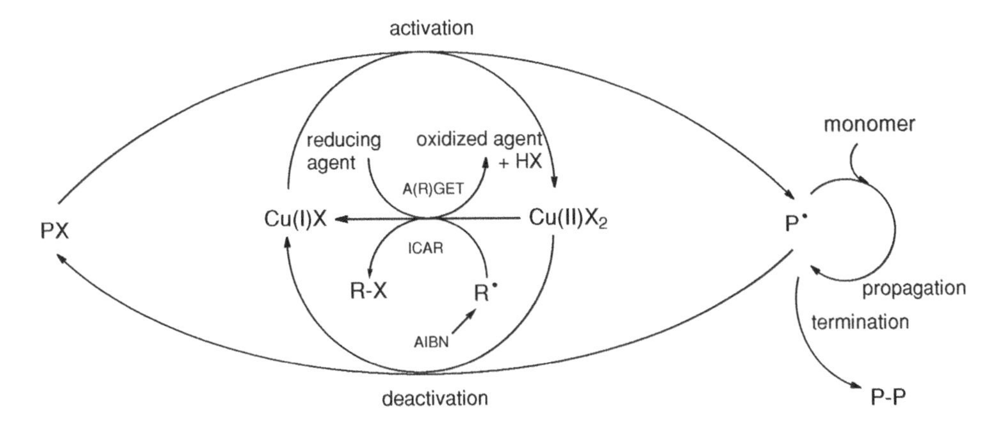

Figure 5.10 *Simplified reaction mechanism of A(R)GET and ICAR ATRP.*

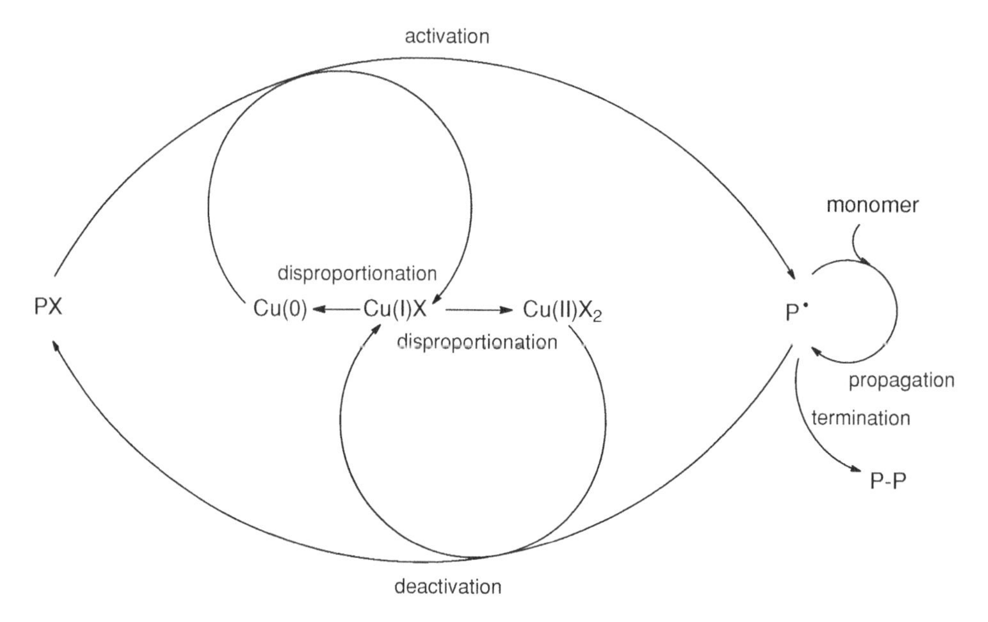

Figure 5.11 *Simplified reaction mechanism of SET-LRP.*

(Matyjaszewski, 2012). Since these methods will always result in some impurities (e.g. excess reducing agents, homopolymer chains, oxidised reducing agents) Matyjaszewski's group also developed an electrochemical version of ATRP, where the Cu(I)/Cu(II) equilibrium can be regulated via the electrode potential (Magenau *et al.*, 2011).

An interesting development in the field has also been the discovery of so-called single electron transfer - living radical polymerisation (SET-LRP) by Percec's group (Percec *et al.*, 2002, 2006). This method, which applies to polymerisation of polar monomers in polar solvents, relies on the disporportionation reaction of Cu(I) complexes in polar solvents (2 Cu(I) $\leftrightarrow$ Cu(0) + Cu(II)). Now, Cu(0), in the presence of an (ATRP-type) ligand and polar solvent, acts as the activator (and not as a reducing agent as in the case of A(R)GET ATRP) and Cu(II) still acts as the deactivator (see Figure 5.11). As this method relies on the presence of polar solvents, SET-LRP has not yet been used in (mini)emulsion polymerisation systems (where the polymerisation takes place inside the particles).

To finish this section, it should be noted that the concentration of propagating radicals in ATRP is generally larger than that in nitroxide-mediated polymerisation, and the equilibrium constant can be adjusted by changing the type of initiator, of transition metal and/or of ligand along with their respective concentrations. The large equilibrium constant indicates that faster polymerisation rates can be achieved and/or lower polymerisation temperatures can be chosen. In addition, a variety of monomers can be polymerised using ATRP, ranging from styrenes, (meth)acrylates, acrylonitrile, (meth)acrylamides, methacrylic acid and vinylpyridine.

5.2.3 Reversible Chain Transfer

The mechanism for reversible chain transfer relies on an exchange reaction between the dormant and active species (Figure 5.1b). The reaction mixture consists of a specific reversible chain transfer agent (RCTA), monomer (solvent is optional) and initiator. The initiation step to produce active species is identical to conventional free-radical polymerisation (e.g. thermal decomposition of an initiator). The concentration of initiator decomposed produces the same concentration of dead polymer through bimolecular radical–radical termination. It is recommended that to produce a well-controlled molar mass distribution, the initiator concentration should be kept significantly lower than the initial concentration of RCTA, resulting in a greater proportion of dormant to dead species. Once again, a compromise should be reached between MMD control and the speed of the reaction.

The reactivity of RCTA towards the active species has a great influence on the M_n and MMD evolution with monomer conversion (x), and is controlled by the value of C_{ex} ($=k_{ex}/k_p$); where k_p is the rate constant for propagation. Analytical equations for the evolution of x_n and PDI with x are given as follows (Müller *et al.*, 1995).

$$x_n = \frac{\gamma_0\, x}{1-(1-\alpha)(1-x)^{\beta}} \quad \text{or} \quad M_n = \frac{\gamma_0\, x}{1-(1-\alpha)(1-x)^{\beta}} M_o \tag{5.5}$$

$$\text{PDI} = \frac{1}{\gamma_0 x} + \frac{1}{x}\left[2 + \frac{\beta-1}{\alpha-\beta}(2-x)\right] - \frac{2\alpha(1-\alpha)}{(\beta-\alpha)x}\left[1-(1-x)^{1+\beta/\alpha}\right] \tag{5.6}$$

where M_0 is the monomer molar mass, $\gamma_0 = [M]_0/[RCTA]_0$, x is the fractional conversion, $\alpha = [P^\bullet]/[RCTA]$ (with $P^\bullet$, the concentration of propagating radicals) and $\beta = C_{ex}$.

These equations can be used for reversible chain transfer where termination, transfer to monomer and all other side reactions are negligible. Figure 5.12 shows the evolution of M_n and PDI with conversion by varying C_{ex} over a range of values. At $C_{ex} = 1$, M_n reaches its maximum value early in the polymerisation and remains constant at that value until all the monomer is consumed (i.e. at $x = 1$). Similarly, the PDI at early conversion reaches 2 and remains constant to $x = 1$, which is similar to what is found for the addition of a conventional chain transfer agent with a chain transfer constant of 1. The evolution of M_n and PDI with conversion starts to resemble that of an 'ideal' living polymerisation with an increase in C_{ex}. The results show that for a C_{ex} value greater than 10, M_n increases linearly with x, and there is no change in the M_n evolution as C_{ex} becomes larger. In contrast, the PDI is more sensitive to the value of C_{ex}, and the greater the value of C_{ex}, the lower the PDI at early conversion. The simulations clearly show that the reversible chain transfer technique allows polymers with controlled MMD to be made. This, coupled with the ability to make complex architectures, offers a wide range of new materials that can be synthesized.

5.2.3.1 *Degenerative Transfer*

The simplest form of reversible chain transfer is degenerative transfer that involves the transfer of a halide from a dormant to an active species (Figure 5.1b). The iodine-mediated polymerisation (L = iodine in Figure 5.1b) is a well-known example (Gaynor, Wang and Matyjaszewski, 1995; Gaynor, Qiu and Matyjaszewski, 1998). It has been found that the C_{ex} equals 3.6 for iodine-mediated polymerisation of styrene initiated with AIBN or benzoyl peroxide when the RCTA is an iodine-terminated polystyrene oligomer (Goto, Ohno and Fukuda, 1998). The M_n and PDI versus conversion plots are similar to the profiles given in Figure 5.12 for a C_{ex} equal to 3, and suggest that the theory provides a good approximation to experiment. The data also suggest that the exchange reaction is not fast enough to produce polymer with a narrow MMD.

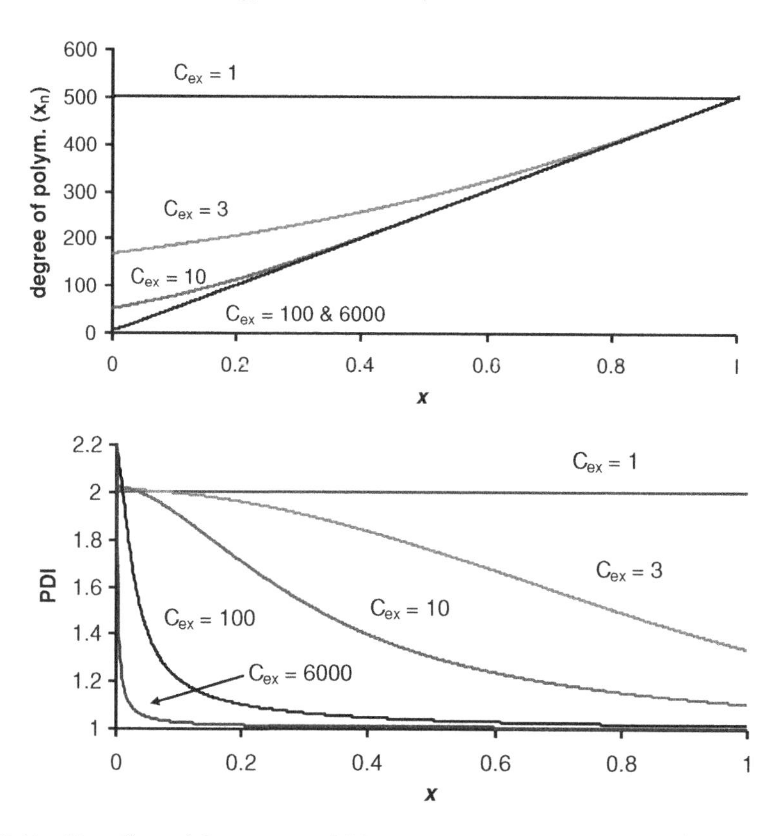

Figure 5.12 *The effect of C_{ex} on x_n and PDI versus conversion in a reversible chain transfer LRP.*

5.2.3.2 Reversible Addition–Fragmentation Chain Transfer (RAFT)

A more complicated reversible chain transfer mechanism involves the reversible addition–fragmentation chain transfer (RAFT) reactions. In this case, the mechanism is identical to degenerative chain transfer but the exchange reaction goes from reactants to products through an intermediate radical species (Figure 5.13). These reactions are kinetically similar to degenerative chain transfer only if the intermediate radical does not react with other radicals or monomer in the system and remains at a sufficiently low concentration level. If this is the case then the equations above can be used to describe as a first approximation the evolution of M_n and PDI with conversion (Monteiro, 2005a, 2005b).

The general structure of RAFT agents is A = C(Z)-Y-X (Figure 5.13) where the Z group activates radical addition to the double bond, and the X group is the leaving group after fragmentation. Reversibility of the transfer process in ensured when A = Y. The end-groups of the final polymer will be X- at one end and A = C(Z)-Y- at the other, resulting in the synthetic control of the end-group structures (i.e. telechelic polymers), which is similar to all other LRP techniques. The types of monomer and A, Y, Z groups on the RAFT agent play an important role in dictating the value of C_{ex}. For C_{ex} less than 1, RAFT agents kinetically

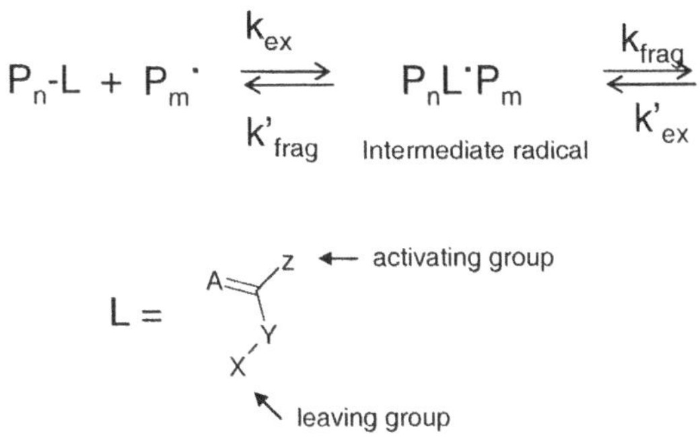

$$P_n\text{-}L + P_m^{\bullet} \underset{k'_{frag}}{\overset{k_{ex}}{\rightleftharpoons}} P_nL^{\bullet}P_m \underset{k'_{ex}}{\overset{k_{frag}}{\rightleftharpoons}} P_n^{\bullet} + P_m\text{-}L$$

Intermediate radical

L = A=C(Z)–Y–X

Z ← activating group

X ← leaving group

Figure 5.13 *General equilibrium mechanism of reversible addition–fragmentation chain transfer (RAFT).*

act as conventional chain transfer agents, but are capable of making block copolymers or more complex architectures. The most reactive of the RAFT agents are those where Z is a phenyl group and A (=Y) is a sulfur atom. The A sulfur atom highly activates the double bond toward polymeric radical addition, and the phenyl group activates it even further by providing resonance stabilization to the intermediate radical formed. The leaving group has a marked effect on the C_{ex} value, and suggests that polymeric groups are by far superior in this role. There are now some excellent reviews on RAFT, and the reader is recommended to the following publications that described the history and progression of RAFT since its invention (Moad, Rizzardo and Thang, 2005, 2006, 2009).

5.3 Nitroxide-Mediated Polymerisation in Emulsion and Miniemulsion

5.3.1 Introduction

Several processes were initially selected to investigate nitroxide-mediated polymerisation in aqueous dispersed systems, such as suspension, dispersion, seeded emulsion, batch emulsion and miniemulsion polymerisations (Qiu, Charleux and Matyjaszewski, 2001; Cunningham, 2002d, 2008; Charleux and Nicolas, 2007; Zetterlund, Kagawa and Okubo, 2008; Monteiro and Cunningham, 2012; Charleux *et al.*, 2012). In this chapter we shall mainly focus on emulsion and miniemulsion polymerisations (see the quoted review articles for more information on the other processes). The latter can be considered as a derived method leading to the same type of final latex (see Chapter 3). According to the selected system, that is, emulsion or miniemulsion, the type of initiator might differ significantly. Indeed, emulsion polymerisation requires the use of a water-soluble initiator, whereas miniemulsion can be carried out with either a water-soluble or an oil-soluble initiator. In all cases, however, as its main role is to interfere with the polymerisation kinetics, the nitroxide should be sufficiently soluble in monomer, as well as in monomer/polymer mixtures.

Two main nitroxide families have been examined: TEMPO (**N1** in Figure 5.6) and derivatives (**N2** to **N7**) and SG1 (**N10**). With TEMPO, most of the results were related to styrene homopolymerisation and only a few articles reported the homopolymerisation of n-butyl acrylate (Georges, Lukkarila and Szkurhan, 2004) in addition to its copolymerisation with styrene. With SG1, the homo-, random and block copolymerisations were investigated for styrene, n-butyl acrylate and methyl methacrylate monomers. Initially, most authors turned their attention towards miniemulsion polymerisation, in which nucleation and latex stability were rather easy to control, and only recently was emulsion polymerisation thoroughly studied.

5.3.2 Control of Molar Mass and Molar Mass Distribution

5.3.2.1 Homopolymerization of Styrene in Miniemulsion

5.3.2.1.1 Monomer-Soluble Radical Initiator in Conjunction with Free Nitroxide. In a miniemulsion process, the monomer droplets behave as independent microreactors, which allows the use of the same reagents as in the bulk. In that way, benzoyl peroxide (BPO) was applied as an initiator in the miniemulsion polymerisation of styrene at 125 °C, in conjunction with free TEMPO (Prodpan *et al.*, 2000) according to the recipe initially reported by the Georges group in its pioneering work (Georges *et al.*, 1993). The miniemulsions were stabilized against coalescence by a classical anionic surfactant stable at high temperature, and against Ostwald ripening by the addition of hexadecane. A good control over molar mass and molar mass distribution was achieved (M_n followed the theoretical line and PDI was in the 1.4–1.7 range) but the polymerisation rate was somehow slower than in the corresponding bulk system. The produced latexes exhibited 20% solids with good stability, but the mean diameters were larger and the particle size distribution broader than in analogous non-living miniemulsion polymerisations. Similar systems of styrene miniemulsion polymerisation received careful attention from Cunningham *et al.* (Cunningham *et al.*, 2002a, 2002b); in particular, their purpose was to assess the effect of the water-solubility of the initiator and the nitroxide on the quality of control. Their results confirmed that the choice of a proper nitroxide with optimal hydrophilicity is actually crucial in miniemulsion.

When the acyclic SG1 nitroxide was selected instead of TEMPO, the miniemulsion polymerisation of styrene could be performed at 90 °C in the presence of AIBN (azobisisobutyronitrile) as a monomer-soluble radical initiator (Lansalot *et al.*, 2000). However, even after 24 h the monomer conversion was incomplete owing to the persistent radical effect leading to the release of a high free nitroxide concentration. With the slow thermal autoinitiation of styrene at 90 °C and the good stability of SG1 in the studied medium, no consumption of the nitroxide excess could be achieved under such experimental conditions. Consequently, although the polymers exhibited good molar mass characteristics, the polymerisation was too slow to be really viable.

None of the observed results were actually fundamentally different from those found in bulk polymerisation. The main reasons are the choice of similar reagents together with the possibility for all of the polymerisation events to take place in the same locus, that is, the monomer droplet/polymer particle behaving as an individual microreactor.

5.3.2.1.2 Water-Soluble Radical Initiator in Conjunction with Free Nitroxide. When using a water-soluble initiator in conjunction with a monomer-soluble nitroxide, the polymerisation starts in the aqueous phase and should be transported towards the monomer droplets in the early stage of the polymerisation, via entry of the oligoradicals or the oligomeric alkoxyamines generated in the aqueous phase. This might affect the polymerisation kinetics, the initiator efficiency and, hence, the control over molar mass and molar mass distribution.

When BPO was replaced by potassium persulfate, a water-soluble radical initiator, and the polymerisation was carried out at 135 °C with a $[TEMPO]_0/[K_2S_2O_8]_0 = 2.9$ initial molar ratio, the miniemulsion polymerisation of styrene was particularly fast as conversion reached 87% within 6 h (McLeod *et al.*, 2000). This feature was actually assigned to the partition of TEMPO between the organic and the aqueous phase that would lead to a substantial decrease in the deactivator concentration in the polymerisation locus. Although molar masses and molar mass distributions were correctly controlled, further investigations of chain extension allowed the authors to demonstrate that quite a large proportion of chains was actually dead (McLeod *et al.*, 2000). It can thus be supposed that, besides partitioning of the nitroxide, side reactions of degradation occurred, leading to both a faster polymerisation and an enhanced rate of irreversible termination. When TEMPO was replaced by the more water-soluble hydroxy-TEMPO (Cunningham *et al.*, 2002b), the kinetics were indeed not affected whereas the M_n versus conversion profile was. This main difference was assigned to different initiator efficiencies. The kinetics in the aqueous phase, together with the partition coefficient of the nitroxide and of the oligomeric alkoxyamines, were indeed shown to have a large effect on the outcome of the polymerisation (Cunningham *et al.*, 2002c; Cunningham *et al.*, 2002b).

When SG1 was used as a deactivator in bicomponent initiating systems in conjunction with potassium persulfate/sodium metabisulfite as a redox initiator, the polymerisation rate was strongly enhanced with respect to the same polymerisations initiated with AIBN at 90 °C (Lansalot *et al.*, 2000; Farcet *et al.*, 2000). The optimal $[SG1]_0/[K_2S_2O_8]_0$ initial molar ratio was 1.2 (best compromise between fast rate and good quality of control). In such conditions, after an induction period of less than 1 h (corresponding to the *in situ* formation of alkoxyamines), the conversion progressed quite rapidly to reach more than 90% within 8 h. At this stage, polymerisation was restricted to the dispersed organic phase of the system and the kinetics regulated by the activation–deactivation equilibrium. In all cases molar masses followed the predicted values, with polydispersity indexes in the 1.5–2.0 range. An important parameter that strongly affected the conversion rate was pH. When insufficiently buffered, the water phase became acidic and the decrease in pH was accompanied by an increased polymerisation rate along with low initiator efficiency (larger M_n than theoretically predicted). These features were assigned to side reactions between SG1 and the two components of the initiating system leading to a decrease in the concentration of alkoxyamines produced *in situ*. At neutral pH, these side reactions were unimportant, which ensured a better control over the polymer characteristics. The persulfate/metabisulfite initiator concentration was also varied to target different molar masses, but the pH had to be adjusted simultaneously to reach the goal and avoid the above-mentioned features.

5.3.2.1.3 Monomer-Soluble Alkoxyamine Initiator. As shown above, the use of a bicomponent initiating system with a conventional radical initiator and free nitroxide is a very simple way to achieve LRP in a miniemulsion system. From a practical viewpoint, very few parameters have to be changed with respect to a classical polymerisation. Nevertheless, such system suffers from low and uncontrolled initiator efficiency. For this main reason, the use of a preformed alkoxyamine initiator (i.e. monocomponent initiating system) is an excellent alternative. With TEMPO, the alkoxyamines used were usually TEMPO-capped polystyrenes synthesized in a preliminary bulk polymerisation (Pan *et al.*, 2001; Keoshkerian, McLeod and Georges, 2001a). Surprisingly, the results were not as good as expected and seemed to depend strongly upon the experimental conditions. In the first example (Pan *et al.*, 2001), the polystyrene macroinitiator was isolated and purified prior to its use in miniemulsion. Differently, in the other case (Keoshkerian, McLeod and Georges, 2001a), the polymerisation of styrene in the presence of BPO and TEMPO was started in bulk, then stopped at 5% conversion and finally this organic phase was subjected to emulsification in the presence of water and surfactant without any purification step. Pan *et al.* (2001) observed the occurrence of thermal autoinitiation of styrene leading to molar masses slightly lower than expected with polydispersity indexes increasing with monomer conversion. They also observed, for the same reason, and similarly to bulk systems, that the polymerisation rate was independent of the initiator concentration (Fukuda *et al.*, 1996). Additionally, the polymerisation rate was independent of the surfactant concentration (Pan *et al.*, 2002) and, hence, of the particle number because the compartmentalisation effect did not apply in those systems exhibiting large particles (Charleux, 2000; Zetterlund, 2011). Improved results were obtained more recently using TEMPO-capped polystyrene macroinitiators in the miniemulsion polymerisation of styrene, achieving high conversions in short reaction time, with good control over chain growth (Lin, Cunningham and Keoshkerian, 2004; Cunningham *et al.*, 2006; Alam, Zetterlund and Okubo, 2008).

With SG1 as a mediator, the preformed alkoxyamine used in miniemulsion was an oil-soluble low molar mass compound with the 1-(methoxycarbonyl)eth-1-yl initiating radical (the so-called MONAMS **A5**). This type of well-defined initiator allowed good control over the initiation step, the concentration of living chains and the concentration of free nitroxide. Both molar mass and rate of polymerisation were similar to those obtained when the reaction was carried out in bulk (Farcet, Burguière and Charleux, 2003). The target M_n could be larger than in the previously presented bicomponent initiating system with persulfate and metabisulfite and the polydispersity indexes were systematically lower. Recent advances in the SG1-mediated miniemulsion polymerisation of styrene were achieved via the *in situ* formation of the surfactant at the oil/water interface leading to stable miniemulsions without a homogenisation device (oleic acid solubilised in the organic phase, which was mixed with an alkaline water-phase) (Guo, Liu and Zetterlund, 2010).

5.3.2.1.4 Water-Soluble Alkoxyamine Initiator. A new SG1-based alkoxyamine bearing a methacrylic acid initiating group (BlocBuilder **A6**, Figure 5.7) allowed the miniemulsion polymerisation of styrene to be initiated for the first time by a water-soluble alkoxyamine initiator (in alkaline conditions) (Nicolas *et al.*, 2004a; Charleux and Nicolas, 2007). The initiating efficiency was rather low but was increased to 100% by the addition of a small amount of methyl acrylate, which enhanced the propagation rate in the water-phase before entry of the low molar mass macroalkoxyamines in the monomer droplets. In these conditions, a good control over M_n and PDI was observed, and the final latexes were highly stable.

5.3.2.2 Homopolymerisation of n-Butyl Acrylate

Keoshkerian, Szkurhan and Georges (2001b) used two different monomer-soluble alkoxyamines to initiate the miniemulsion polymerisation of n-butyl acrylate at 135 °C:

a polystyrene oligomer ($x_n = 11$) terminated with the nitroxide **N12** well suited for acrylate living polymerisation and the TEMPO-based low molar mass alkoxyamine **A3**. In the former case, very few experimental data were actually given but it was concluded that good control was reached. In the presence of TEMPO, experimental conditions had to be adjusted to eliminate the excess nitroxide (via addition of ascorbic acid), and hence allow the polymerisation to take place. The effect of ascorbic acid was thoroughly examined in a more complete study of the same system in bulk and miniemulsion polymerisations (Georges, Lukkarila and Szkurhan, 2004). Although better results were obtained with the former initiating system, the authors still preferred the second alkoxyamine because of the commercial availability of the nitroxide.

Using the alkoxyamine initiator **A5** and the nitroxide SG1 as a mediator, living radical homopolymerisation of n-butyl acrylate was successfully performed in batch aqueous miniemulsion at 112 °C under 3 bar pressure (Farcet, Charleux and Pirri, 2001a; Farcet, Nicolas and Charleux, 2002). For some experiments, the alkoxyamine was used alone, while in other cases a small fraction of free SG1 was added to regulate the polymerisation rate and to decrease the extent of macroradical self-termination, resulting in a reduction of the polydispersity index. Different molar masses were targeted and in all cases, M_n increased linearly with monomer conversion, matching the predicted value and the molar mass distribution was narrow. For example, M_n up to 50 000 g mol^{-1} could be obtained with polydispersity indexes ranging between 1.2 and 1.4. The performed miniemulsion polymerisations led to stable latexes with 20 to 45 wt% solids. However, like in the previous examples of styrene miniemulsion polymerisations, the average particle diameters of the final latexes were rather large and the particle size distribution was broad and often multimodal.

With the BlocBuilder alkoxyamine **A6** used in alkaline conditions in which it is soluble in the water phase, the miniemulsion polymerisation was better controlled than in the case of styrene, with high initiator efficiency at 112 °C, leading to a very good control of the polymerisation and stable latexes, with diameters in the 200 300 nm range (Nicolas *et al.*, 2004a).

5.3.2.3 Emulsion Polymerisation in the Presence of Surfactants

In order to avoid the complexity of the nucleation step in *ab initio* systems, the very first study of nitroxide-mediated emulsion polymerisation (Bon *et al.*, 1997) was carried out in a seeded second stage emulsion polymerisation in the absence of monomer droplets, that is to say in the typical conditions of Interval III. The alkoxyamine **A2** was selected as an oil-soluble initiator. The quality of control was, however, not as good as in the bulk. Indeed, the experimental M_n values remained below the theoretical ones and the polydispersity index increased from 1.41 to 1.54 throughout the polymerisation. The low values of M_n and the observed broadening of the molar mass distribution on the low molar mass side were assigned to the occurrence of thermal self-initiation of styrene, creating new chains. Nevertheless, this work perfectly demonstrated the robustness of the nitroxide-mediated polymerisation chemistry and, hence, opened the way to the future works in the domain.

The requirement for a water-soluble initiator incited other groups to start the polymerisation with a bicomponent initiating system using a water-soluble radical initiator in conjunction with TEMPO or a derivative (Marestin *et al.*, 1998; Cao *et al.*, 2001). For instance, Marestin *et al.* (1998) studied the *ab initio* emulsion polymerisation of styrene at 130 °C in the presence of a water-soluble initiator and with several nitroxides from the TEMPO family. It appeared that only the amino-TEMPO (**N4**) gave satisfactory results, from both the colloidal and the macromolecular viewpoints. Broad particle size distribution was, however, always observed as the consequence of thermal self-initiation of styrene in the monomer droplets. When the bicomponent initiating system was replaced by the negatively charged water-soluble alkoxyamine **A4**, a stable latex could be obtained without any added surfactant, providing that polymerisation was carried out at low solids content (<2 wt%). The addition of surfactant considerably improved the stability, allowing the formation of a stable latex with 10 wt% solids, but where the particle size distribution was bimodal. Polymerisation was faster than with the bicomponent initiating system and the formed polymer had $M_n = 42\,000$ g mol^{-1} with PDI = 1.7. This polydispersity index could be reduced down to 1.2 upon the addition of 0.3 equivalent of free amino-TEMPO. However, this was at the expense of the polymerisation rate, which showed a significant drop. In a similar work, Cao *et al.* (2001) also assessed the effect of hydrophilicity of TEMPO-based nitroxides for batch emulsion polymerisations of styrene conducted at 120 °C in the presence of a water-soluble initiator. Among the studied nitroxides, only the acetoxy derivative (**N7**) achieved living polymerisation. In contrast to the previous study, stable latexes with small particle size (below 100 nm) and good regularity were obtained.

Ab initio emulsion polymerisations of styrene were also conducted at 90 °C, using the stable acyclic phosphonylated nitroxide radical SG1 (**N10**) as a mediator together with a water-soluble redox initiator (Lansalot *et al.*, 2000). A long induction period was observed, assigned to the formation of water-soluble oligomeric alkoxyamines before nucleation. In this system, the molar mass of the polymer increased with conversion following the theoretical line, but the molar mass distribution was broad (PDI between 2.0 and 2.5). Rather small particles were obtained (average diameter was 120 nm) with a broad particle size distribution. It was also found that a small percentage of coagulum usually formed.

Here again, a significant breakthrough was witnessed with the use of BlocBuilder alkoxyamine (**A6**, Figure 5.7) in a two-step emulsion polymerisation (Nicolas *et al.*, 2004b). The first step consisted in a microemulsion-like system with the polymerisation of a low amount of n-butyl acrylate, sufficient to swell the surfactant micelles and to avoid the presence of large monomer droplets. After polymerisation, well-defined, SG1-capped poly(n-butyl acrylate) oligomers were obtained within small particles. Then, a second load of monomer (either styrene or n-butyl acrylate) was added in "one shot", resulting in chain extension and increase in the latex solids content. The quality of control was very good and the final latexes were stable but the particle size distributions were rather broad with average diameters in the 260–660 nm range, depending on the initial concentration of surfactant. When using a water-soluble, difunctional alkoxyamine (DIAMA, **A7** in Figure 5.7) under similar experimental conditions, the particle size distribution narrowed significantly and the average diameters were lower (Nicolas *et al.*, 2005). This was assigned to the double negative charge of the neutralized DIAMA that improved colloidal stability. Changing the process into a semi-batch addition of the second load of monomer allowed the overall polymerisation time to be drastically reduced at no expense to the quality of control (Nicolas, Charleux and Magnet, 2006b).

5.3.3 Synthesis of Block and Random or Gradient Copolymers via (Mini)Emulsion Polymerisation

5.3.3.1 *TEMPO-Mediated Miniemulsion Polymerisation*

The very first synthesis of a polystyrene-*b*-poly(n-butyl acrylate) block copolymer produced in an aqueous dispersed system was reported by Keoshkerian *et al.* (Keoshkerian, McLeod and Georges, 2001a). The method used a short TEMPO-capped polystyrene macroinitiator previously synthesised in bulk. Tortosa, Smith and Cunningham (2001) then described a different way to reach the same goal: first, styrene miniemulsion polymerisation was carried out using $K_2S_2O_8$ or BPO/TEMPO or hydroxy-TEMPO as a bicomponent initiating system; then, the formed nitroxide-terminated polystyrene was purified from the residual styrene and free nitroxide by precipitation, dissolved in n-butyl acrylate, emulsified in water and the second miniemulsion step was started by heating. Thetructural quality of the block copolymer was assessed. Nevertheless, the method is only viable on the laboratory scale and was essentially used for mechanistic understanding.

5.3.3.2 *SG1-Mediated Miniemulsion Polymerisation*

SG1-mediated random copolymerisation of styrene and n-butyl acrylate was performed in miniemulsion using the same experimental conditions as for n-butyl acrylate homopolymerisations (Farcet, Charleux and Pirri, 2001a). In addition to the good control over molar mass and molar mass distribution, a narrow composition distribution was observed and the chains exhibited a gradient composition. This feature was demonstrated by liquid adsorption chromatography, which is an analytical technique that gives information on the copolymer composition distribution (conditions to get separation according to the composition independently of the molar mass were established). Results confirmed that the composition distribution of the final living copolymers was much narrower than exhibited by analogous non-controlled copolymers.

Diblock copolymers were also synthesized in miniemulsion polymerisation, using a sequential addition of the monomers, n-butyl acrylate first, then styrene (Farcet, Charleux and Pirri, 2001b). The linear increase in M_n after styrene addition, the complete shift of the size exclusion chromatography traces and the decrease in the polydispersity indexes clearly indicated that chain extension from the first poly(n-butyl acrylate) living segment was effective. Furthermore, liquid adsorption chromatography did not show any detectable poly(n-butyl acrylate) homopolymer, signifying an efficient re-initiation by the first block.

5.3.3.3 *SG1-Mediated Emulsion Polymerisation*

The two-step emulsion polymerisation process using BlocBuilder or DIAMA water-soluble initiators and presented above (Nicolas *et al.*, 2004b, 2005; Nicolas, Charleux and Magnet, 2006b) was also applied to the synthesis of diblock, and triblock copolymer latexes, based on polystyrene and poly(n-butyl acrylate) by simply adding styrene after the synthesis of the poly(n-butyl acrylate) block. The molar mass of the block copolymers was well controlled up to high M_n values, while keeping low PDIs and the obtained latexes were stable with solids as high as 26 wt.%. The block copolymers, moreover, exhibited phase separation properties within the particles, leading to onion-like morphologies (Nicolas *et al.*, 2007).

The copolymerisation of n-butyl methacrylate mediated by SG1 with a small amount of styrene was reported in surfactant-free emulsion polymerisation using the BlocBuilder alkoxyamine as an initiator (Thomson *et al.*, 2010). The presence of styrene was needed to ensure a good quality of control for the employed methacrylate monomer (Charleux, Nicolas and Guerret, 2005). The polymerisations conducted in the absence of surfactant or below the critical micelle concentration allowed monomodal PSD to be obtained. However, in the former case, poor initiator efficiency and slow polymerisation were observed and were improved by the addition of a small amount of methyl acrylate in the first stage.

5.3.4 Surfactant-Free Emulsion Polymerisation Using the Polymerisation-Induced Self-Assembly Technique

The most recent developments are related to the successful implementation of surfactant-free, *ab initio,* batch emulsion polymerisation processes using water-soluble macroalkoxyamines as initiators. The method is called *polymerisation-induced self-assembly* as it allows amphiphilic block copolymers to be formed *in situ* and to self-assemble simultaneously to the growth of the hydrophobic block (Charleux *et al.*, 2012). The first system used was based on SG1-capped poly(sodium acrylate) macroalkoxyamines for the polymerisation of styrene, n-butyl acrylate (Delaittre *et al.*, 2005, 2006) and 4-vinylpyridine (Delaittre *et al.*, 2009). With the first two monomers, well-defined spherical particles were obtained, with diameters below 100 nm and very narrow PSD, in stable latexes exhibiting 20 to 40 wt% solids. With 4-vinylpyridine, cylindrical micelles and vesicles were formed, showing the possibility for this process to target morphologies usually found in self-assembled preformed amphiphilic block copolymers (Zhang and Eisenberg, 1995; Discher *et al.*, 2007).

The drawback of the poly(sodium acrylate) macroalkoxyamines was their low initiating efficiency, which was overcome by using poly(sodium methacrylate)-based macroalkoxyamines. The latter contained approximately 10 mol% of styrene (Dire *et al.*, 2007) or 4-styrene sulfonate (Brusseau *et al.*, 2010) to enhance the control over methacrylic acid by the nitroxide SG1 and to allow the formation of stable alkoxyamines able to dissociate at temperatures below 100 °C. Their use in the emulsion polymerisation of methyl methacrylate with a low percentage of styrene led to a highly efficient re-initiation and to the *in situ* formation of well-defined amphiphilic block copolymers, self-assembled into small, spherical particles (Dire *et al.*, 2009; Brusseau *et al.*, 2011). By changing the molar mass of the hydrophobic block, the system led, moreover, to various morphologies, such as fibers and vesicles (Groison *et al.*, 2012), confirming the possibility to tune the particle morphology in the same way as for self-assembled block copolymers, with the advantages of high concentrations and short elaboration times.

5.4 ATRP in Emulsion and Miniemulsion

5.4.1 Introduction

Because ATRP requires the use of an activator and a deactivator to control both the polymerisation kinetics and the chain growth, its application to heterogeneous systems seems less straightforward than for nitroxide-mediated polymerisation. Nevertheless, numerous studies have been carried out via ATRP, using mostly copper complexes as catalyst, and there are now general strategies available for carrying out ATRP in emulsion, with those based on AGET-ATRP (Min, Gao and Matyjaszewski, 2005b; Min, Jakubowski and Matyjaszewski, 2006a) probably being the most robust and the method of choice at the moment (Monteiro and Cunningham, 2012).

Before getting into detailed discussions of the studies that have been carried out, we will address a few general points and make a few generalizations (on which there will always be some exceptions). A very important criterion to fulfil should be the high solubility in the monomer phase of both the Cu(I) and the Cu(II) complexes. In general, the Cu(II) complexes will be more water-soluble, and preferential partitioning to the water phase by Cu(II) would lead to a decrease in deactivator inside the particles, which in turn leads to an uncontrolled polymerisation (Kagawa *et al.*, 2007a; Min *et al.*, 2009; Monteiro and Cunningham, 2012). Much of the initial research has pointed out the crucial role of the ligand in ATRP aqueous dispersed systems (Gaynor, Qiu and Matyjaszewski, 1998; Qiu *et al.*, 1999b; Chambard, De Man and Klumperman, 2000; Matyjaszewski *et al.*, 2000a; Peng *et al.*, 2003), and in general it can be concluded that indeed only the ligands with a sufficiently high hydrophobicity are efficient (see Figure 5.9). Additionally, selection of a suitable surfactant is a nontrivial task; indeed, in general, only nonionic and cationic surfactants that do not interact with the catalyst lead to living polymerisations and stable latexes (Jousset *et al.*, 2001; Simms and Cunningham, 2006). For the nonionic surfactants a rule of thumb seems to be that HLB (hydrophilic/lipophilic balance) values of about 15 are required, but this is only a first indication as the molecular structure has also been shown to be very important (Jousset *et al.*, 2001). Both Brij 98 and Tween 80 (polyoxyethylene(20) oleyl ether and polyoxyethylene sorbitan monooleate, respectively, see Figure 5.14) are frequently used as nonionic surfactants and yield stable latexes, but often relatively high amounts (more than 5 wt%, and sometimes even as much as 20 wt%, w.r.t. monomer) are required. Furthermore, one always needs to be aware of the fact that when increasing the temperature, the stabilization properties of these oligo(ethylene glycol)-based surfactants diminishes (Zetterlund, Kagawa and Okubo, 2008). The most successful cationic surfactant to date has been CTAB (cetyltrimethylammonium bromide, see Figure 5.14). In the following sections, a more detailed overview of the studies to date is given and these are divided according to the polymerisation technique: direct ATRP, reverse ATRP and SRNI/AGET ATRP.

5.4.2 Direct ATRP

5.4.2.1 Control of Molar Mass and Molar Mass Distribution

The very first reports in the domain used an emulsion polymerisation approach, but the alkyl halides that were selected as ATRP initiators were not sufficiently water soluble, leading to significant droplet nucleation. Consequently, the systems were closer to (micro)suspension polymerisations, resulting in large micrometric particles with broad particle size distributions (Makino, Tokunaga and Hogen-Esch, 1998; Jousset *et al.*, 2001). Even though the colloidal characteristics were far from perfect, these aqueous dispersed ATRP systems provided examples of well controlled homopolymers, such as polyacrylates, polymethacrylates and polystyrene, demonstrating the viability of the method at temperatures ranging

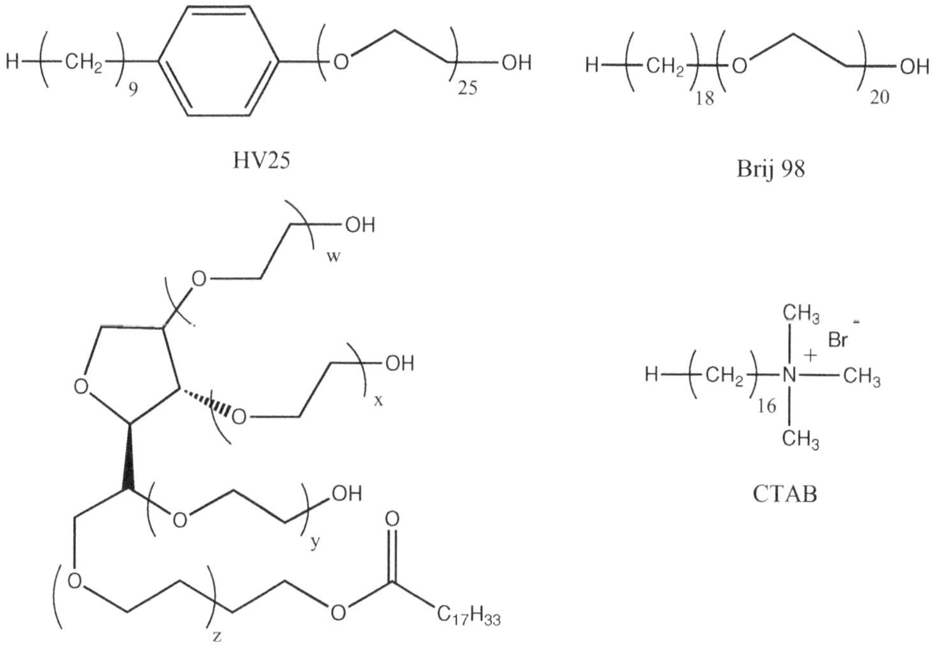

HV25

Brij 98

CTAB

w + x + y + z = 20
Tween 80

Figure 5.14 Commonly used nonionic and cationic surfactants.

from 60 to 90 °C (Gaynor, Qiu and Matyjaszewski, 1998; Qiu, Gaynor and Matyjaszewski, 1999a; Chambard, De Man and Klumperman, 2000; Wan and Ying, 2000). More recent studies by Eslami and Zhu (Eslami and Zhu, 2005, 2006) on the direct ATRP in emulsion of 2-ethyl hexyl methacrylate (EHMA) in which again several different surfactants and polymerisation conditions were tested, basically confirmed the earlier results. Using ethyl 2-bromoisobutyrate (EBiB) as the initiator and the CuBr/dNbpy catalyst system the best results were obtained using either Tween 80 or Brij 98 as surfactants at a temperature of 30 °C. Large particle sizes of around 700 nm were obtained (no mention about the width of the PSD), but replacing 5% of the CuBr by $CuBr_2$ resulted in a reduction in particle size to about 300 nm and a narrower MMD. A further increase in $CuBr_2$ led to severe coagulation.

In order to avoid problems with the transport of reagents through the aqueous phase, mini- and microemulsion approaches may be more appropriate and both techniques have been successfully applied using direct ATRP. One of the first examples in miniemulsion was the homopolymerisation of n-butyl methacrylate (*n*BMA) at 70 °C with EBiB as the initiator. A nonionic surfactant in large amount (13.5 wt% with respect to monomer) was, however, required to stabilize the 300 nm latex particles (Matyjaszewski *et al.*, 2000b). An early report on direct ATRP of methyl methacrylate (MMA) in microemulsion by Matyjaszewski's group using a CuBr/BPMODA catalyst system, EBiB initiator and Brij98 as surfactant was not successful and a bimodal MMD was obtained (Min and Matyjaszewski,

2005a). Later work by Okubo's group (Kagawa *et al.*, 2007b) using a different stabilization system was successful. Here *i*BMA (isobutyl methacrylate) was polymerized using a CuBr/dNbpy catalyst system, EBiB initiator and mixed surfactant system consisting of *n*-tetradecyltrimethylammonium bromide (TTAB) and a polyoxyethylene nonyl phenyl ether (Emulgen 911 - E911 with HLB $\approx$ 14 - and Emulgen 931 - E931 with HLB $\approx$ 17). The best results were obtained using TTAB/E911 with high conversions, low PDI ($\sim$1.3), and stable particles with diameters around 13 nm.

Crosslinked nanoparticles (diameters $\approx$40–140 nm) and block copolymers of polystyrene have also been prepared via simultaneous Click reactions and direct ATRP in emulsion (Xu *et al.*, 2009). In these studies, alkyne-containing initiators, CuBr/PMDETA as the catalyst system and Tween 20 as the surfactant (contains a laureate group instead of an oleate group as in Tween 80) were used. In the case of the block copolymer synthesis a mono-alkyne initiator was used and *p*-xylylene diazide as the coupling agent, and for the crosslinked nanoparticles, a di-alkyne initiator and 4-vinylbenzyl azide as cross-linkable co-monomer.

Finally, the "nanoprecipitation" work as a type of seeded emulsion polymerisation by Georges and coworkers is worth mentioning in this section (Chan-Seng and Georges, 2006; Chan-Seng *et al.*, 2008). In this work, a low molar mass polymer was synthesized in solution via ATRP, thus yielding a macroinitiator. This macroinitiator was then dissolved together with CuBr/BPMODA (1.5 : 1) in a small amount of acetone which was subsequently poured into an aqueous solution containing Brij 98 ($\sim$5 wt% w.r.t. monomer) as a surfacant. Upon evaporation of the acetone, the low molar mass polymer precipitates into nanosized seed particles. These were subsequently swollen with monomer and polymerized with a good molar mass control yielding colloidally stable latex particles in the 200–300 nm range, but with a slightly broad distribution. Polystyrene-Br (Chan-Seng and Georges, 2006) and poly(butyl acrylate)-Br (Chan-Seng *et al.*, 2008) have been chain extended this way in direct emulsion polymerisation.

5.4.2.2 *Synthesis of Statistical and Block Copolymers*

Statistical copolymers were prepared via direct ATRP in aqueous dispersed systems, in a similar manner to the synthesis of the corresponding homopolymers (Matyjaszewski *et al.*, 2000c). The synthesis of block copolymers, however, was shown to be more complicated than for bulk or solution systems. The alkyl halide end-group might not be very stable in the presence of water and/or a too low deactivator concentration in the organic phase might result in an increase in the termination rate. Regardless of the way this arises it would result in incomplete block copolymer formation. In addition, the cross-over reaction might be less efficient than in homogeneous systems (Qiu, Charleux and Matyjaszewski, 2001). Successful block copolymerisations were then accomplished in two steps by using a poly(n-butyl acrylate) macroinitiator prepared in bulk to initiate the polymerisation of styrene in the aqueous dispersed medium. Clean block copolymers were obtained with varying end groups and ligands (Matyjaszewski *et al.*, 2000c).

Eslami and Zhu (Eslami and Zhu, 2006) reported on the synthesis of PMMA-*b*-PEHMA-*b*-PMMA triblock copolymers in emulsion via direct ATRP. First, an emulsion polymerisation of EHMA was carried out using a bifunctional initiator (1,4-butylene di(2-bromoisobutyrate), a CuBr (95%)/$CuBr_2$ (5%)/dNbpy catalyst system and Tween 80 surfactant. The obtained latex was then used as a seed for the polymerisation of MMA. A stable latex with an average diameter of around 300 nm (and relatively broad distribution) was obtained with well-controlled molar masses (PDI ~ 1.2).

An interesting example of block copolymer formation is that reported by Okubo and coworkers (Kagawa *et al.*, 2005) who performed a direct ATRP of *i*BMA in miniemulsion using a CuBr/dNbpy catalyst system, EBiB as initiator and 6–10 wt% w.r.t. monomer of Tween 80 surfactant. This latex was mixed with a styrene emulsion and heated to result in block copolymers in particles that showed a very interesting layered "onion-like" morphology. Later studies by the same group (but now using AGET ATRP, see below) showed that to obtain this "onion" morphology of the particles, a very good blocking efficiency is required (Kitayama *et al.*, 2009).

Finally, Xu and coworkers reported two relevant studies on the preparation of stable nanoparticle dispersions with particle diameters in the range 30–80 nm. In the first study (Xu *et al.*, 2008), they described the microwave-assisted synthesis of stable PEG-*b*-PS nanoparticles in the range 30–60 nm (~10% solids in water) starting from a PEG-Cl macroinitiator, styrene monomer, CuCl/bpy catalyst and 5% Tween 20 surfactant. The second study (Liu *et al.*, 2009) used a polypropyleneimine dendrimer-derived initiator for the emulsion polymerisation of styrene using CuCl/bpy as the catalyst system and Tween 20 as the surfactant.

5.4.3 Reverse ATRP

In contrast to nitroxide-mediated polymerisation, direct ATRP might be advantageously replaced by reverse ATRP in aqueous dispersed systems for several reasons. First, the Cu(II) deactivator is tolerant to oxygen, which facilitates the preparation of the emulsion, in particular the miniemulsion systems that require homogenisation prior to reaction. Second, numerous water-soluble radical initiators are available, which significantly reduces the probability of droplet nucleation (for similar reasons as in conventional emulsion polymerisation). The fundamental requirements for a successful polymerisation remain the same as those previously identified for direct ATRP with the use of sufficiently hydrophobic ligand along with a nonionic or a cationic surfactant. In contrast to direct ATRP, however, reverse ATRP usually displayed an induction period, depending on the temperature and the ratio of the deactivator to the initiator. It was assigned to the time required for deactivation of the radicals by Cu(II) before reaching the atom transfer equilibrium (Qiu, Gaynor and Matyjaszewski, 1999a; Qiu *et al.*, 2000; Simms and Cunningham, 2006, 2007, 2008).

5.4.3.1 Control of Molar Mass and Molar Mass Distribution

Most studies were initially focused on the polymerisation of *n*BMA, using Cu(II) in conjunction with dialkylbipyridine ligand (Figure 5.9) as the deactivator, and Brij 98 as the surfactant (Qiu, Gaynor and Matyjaszewski, 1999a; Qiu *et al.*, 2000). In an emulsion process, azo initiators (2,2′-azobis(2-methylpropionamidine) dihydrochloride,V-50; 2,2′-azobis[2-(2-dimidazolin-2-yl) propane] dihydrochloride, VA-044) led to a better control than persulfate and the most effective temperature was 70 to 90 °C. Well controlled polymerisations were usually achieved, exhibiting a linear increase in molar mass with monomer conversion and low polydispersity index, but the initiator efficiency remained invariably low (below 50%), as the consequence of extensive termination reactions in the water-phase.

From the colloidal viewpoint, the latexes were very stable, with particle diameter ranging from 150 to 300 nm, depending on the experimental conditions.

Successful miniemulsion polymerisations were also achieved with the same monomer and the same Cu(II)/dialkylbipyridine complex, in the additional presence of hexadecane; high shear of the initial system was provided by ultrasonication. A monomer-soluble initiator (azobisisobutyronitrile, AIBN) and a water-soluble one (V-50) were employed (Matyjaszewski *et al.*, 2000b). The latter, however, led to better controlled molar masses, with higher initiator efficiency. Stable latexes were achieved with particle diameter of 300 nm, but the amount of surfactant required was very large (i.e. 13.5 wt% based on monomer). In a more recent study (Li and Matyjaszewski, 2003a, 2003b), the conditions to improve both the colloidal characteristics and the control over molar masses were reported. Indeed, latexes with doubled solids content (>20 wt%), and 1/6 the amount of the nonionic surfactant (Brij98, 2.3 wt% based on monomer) with respect to the previous study were prepared in miniemulsion at 70 °C. The water-soluble radical initiator was VA-044. The key to success was the replacement of dialkylbipyridine by nitrogen-based tetradentate ligands, such as EHA_6-TREN, as shown in Figure 5.9. The catalytic activity of their copper complex is much higher than that of the copper/dialkylbipyridine. This allowed the authors to decrease the polymerisation temperature to 70 °C, which was in favour of a better colloidal stability. In addition, the catalyst hydrophobicity could be easily adjusted by modification of the substituents on the nitrogen atoms (Figure 5.9). With a lauryl ester in the substituent structure, the ligand was very hydrophobic and, hence, essentially located in the monomer droplets. As a consequence, the amount of surfactant had to be reduced to 2.3 wt% based on monomer to avoid the presence of micelles, which favoured the coexistence of a double nucleation mechanism: droplet nucleation led to living polymer, whereas micellar nucleation led to non-controlled polymer formed in the absence of catalyst, via an emulsion process. When lauryl was replaced by 2-ethylhexyl, the polymerisation was fully controlled whatever the surfactant concentration, as a result of a better mobility of the catalyst in the system. In all cases, with large surfactant concentration, the particle size distribution was bimodal owing to the two nucleation mechanisms. Molar masses ranging from 30 000 to 100 000 g mol^{-1} were reported, with PDI between 1.5 and 1.7. More recent studies, again on the miniemulsion polymerisation of *n*BMA, by Simms and Cunningham, using as the catalyst system $CuBr_2$/EHA_6-TREN, showed good molar mass control and yielded stable latexes (~15 wt% solids) both with CTAB (0.5–2.5 wt% w.r.t. monomer (Simms and Cunningham, 2006)) and Brij 98 (10 wt% w.r.t. monomer (Simms and Cunningham, 2007)) surfactants. In both these studies VA-044 was used as the water-soluble azo initiator, but in one of the studies the redox initiator H_2O_2/ascorbic acid (AsAc) was also used (Simms and Cunningham, 2007). In this study, very high molar masses (~10^6 g mol^{-1}) with very narrow MMDs were obtained and the authors ascribed this to some specific, as yet unclear, reactions of the initiator system. A subsequent study on the same system (Simms and Cunningham, 2008) shows the compartmentalization effect, which effectively reduces both the rate of the polymerisation and the polydispersity index.

Reverse ATRP of MMA using $CuBr_2$/BPMODA/Brij 98/V-50 was also carried out in microemulsion by Min and Matyjaszewski (Min and Matyjaszewski, 2005a) in the same study as mentioned above in the section on direct ATRP. Although the results were clearly better than those for the direct ATRP, still a high PDI (~1.6) was obtained for the polymers. Only when carried out via (AGET-ATRP, see following section) was a low PDI achieved.

5.4.3.2 Synthesis of Block Copolymers

The latexes described above (Li and Matyjaszewski, 2003b) were subjected to chain extension by feeding a new load of n-butyl methacrylate and surfactant after 98.3% monomer conversion from the first stage. Such results that demonstrated the livingness of the first block can be considered as the first step towards the synthesis of block copolymers via ATRP in an aqueous dispersed system.

5.4.4 Next Generation ATRP Techniques: SRNI and AGET

5.4.4.1 Control of Molar Mass and Molar Mass Distribution

Both the SNRI and AGET techniques are particularly well suited for miniemulsion polymerisation since they use a Cu(II) complex initially, which is less sensitive to air than the corresponding Cu(I) complex. In a typical SNRI recipe (Li and Matyjaszewski, 2003a; Li *et al.*, 2004a), an oil-soluble alkyl halide (such as methyl 2-bromopropionate or ethyl 2-bromoisobutyrate) was used as an initiator. It was initially dissolved in the monomer droplets in the presence of hexadecane, 0.2 equivalent of Cu(II)/ligand complex and 0.125 equivalents of the radical initiator when it was oil-soluble. The water phase contained the Brij98 nonionic surfactant. After homogenisation, the polymerisation was started by heating the miniemulsion at 80 °C. When a water-soluble radical initiator was selected, it was added in the medium after homogenisation. For this method, the type of ligand is crucial as it should be highly hydrophobic and able to form a very active copper complex. In this work a substituted terpyridine and a picolyl amine (BPMODA) were selected (Figure 5.9). Homopolymerisations of n-butyl methacrylate, n-butyl acrylate and styrene were carried out with such experimental conditions. The expected criteria of a controlled system were fulfilled with good control of molar mass and low polydispersity indexes (typically 1.2–1.4), in particular when an oil-soluble initiator was used.

Although SNRI already performs really well, it still suffers from the fact that there will be some chain formed from the conventional initiator-derived radicals. Hence, after the quick development of AGET-ATRP, which does not use initiator-derived radicals to reduce Cu(II) to Cu(I), but a non-radical forming reducing agent, such as ascorbic acid, this procedure soon after became the method of choice (Min, Gao and Matyjaszewski, 2005b). In AGET-ATRP the Cu is generallly added as $CuBr_2$ using BPMODA or dNbpy as the ligand, and using a simple initiator, such as EBiB. With the use of the appropriate surfactants as mentioned before, stable latexes containing polymers with controlled molar mass are obtained. The first report of AGET-ATRP in aqueous dispersion was reported for the miniemulsion polymerisation of n-butyl acrylate (*n*BA) (Min, Gao and Matyjaszewski, 2005b). In this study a solution was made of $CuBr_2$/BPMODA in the *n*BA containing about 3 wt% hexadecane. This solution was then added to an aqueous Brij 98 solution after which it was sonicated and AsAc was added to start the reaction (at 80 °C). Good colloidal stability and molar mass control were reported.

Matyjaszewski and coworkers also extended AGET-ATRP to microemulsion polymerisation and in a study of the system MMA/$CuBr_2$/BPMODA/Brij 98/EBIB and AsAc or V-50 which was carried out using direct ATRP, reverse ATRP (both discussed above) and AGET-ATRP; the latter clearly outperformed the former two (Min and Matyjaszewski,

2005a). This success in microemulsion polymerisation led to the development of a very useful polymerisation technique that approximates an *ab initio* AGET-ATRP emulsion polymerisation. This technique, first illustrated for the AGET-ATRP of *n*BA, (Min, Gao and Matyjaszewski, 2006b) basically consists of first preparing a microlatex seed by creating a microemulsion from a large amount of surfactant, the $CuBr_2$ catalyst, ligand, initiator and a small amount of the monomer and starting the polymerisation by the addition of ascorbic acid. To this microlatex seed, the remainder of the monomer is subsequently fed to yield the final latex.

5.4.4.2 Complex Architectures

A considerable advantage of SNRI ATRP over simple reverse ATRP concerns the possibility to synthesize complex (co)polymer architectures, while maintaining the advantages of a reverse ATRP system. It results from the possible use of alkyl halide initiator with functionality larger than 1. This is illustrated by the synthesis of three-arm star polystyrene and poly(n-butyl acrylate) in a miniemulsion process, using a trifunctional alkyl halide initiator (Li, Min and Matyjaszewski, 2004b). In miniemulsion, star-block copolymers were also prepared using multifunctional macroinitiators (Li *et al.*, 2004a) and spontaneous gradient copolymers by copolymerisation of *n*BA with *n*BMA (Min, Li and Matyjaszewski, 2005c).

It will also come as no surprise that AGET-ATRP has been used to produce more complex polymer architectures; even in the first reported AGET-ATRP study in miniemulsion (Min, Gao and Matyjaszewski, 2005b), the authors report the successful synthesis of PMA-*b*-PS block copolymers and 3-arm star block copolymers of poly(MA-*b*-S)$_3$ starting from a trifunctional initiator. Forced gradient copolymers of the following pairs *n*BA/*t*BA, *n*BA/S and *n*BMA/MMA in AGET-ATRP miniemulsion polymerisation were also reported by the same authors (Min, Oh and Matyjaszewski, 2007). In these studies the AGET-ATRP miniemulsion polymerisation was carried out in a similar way to that described before, starting with the first monomer and feeding in the second monomer. The block copolymer work by Okubo's group (Kitayama *et al.*, 2009) leading to onion-like layered particle morphologies has already been mentioned. Here, an AGET-ATRP in miniemulsion of *n*BMA was carried out using a system with $CuBr_2$/dNbpy catalyst, Brij98 surfactant and AsAc reducing agent. This seed was subsequently swollen with styrene and polymerized in the presence and in the absence of EBiB initiator. In the absence of EBiB a good blocking efficiency was observed, leading to onion-like layered morphologies, but in the presence of EBiB a significant amount of PS homopolymer was deliberately formed (thus the process had a low blocking efficiency) and the particles displayed a disordered morphology.

As a final example of the application of AGET-ATRP in emulsion we will mention here the work by Matyjaszewski and coworkers on the preparation of core–shell particles with a crosslinked core (called "hairy particles" in the paper) (Min *et al.*, 2009). In this study the microemulsion seed latex technique was used to first prepare a crosslinked seed from MMA and EGDMA (ethyleneglycol dimethacrylate) crosslinker and an AGET-ATRP system consisting of $CuBr_2$/BPMODA/EBiB/Brij 98 which was started by AsAc. The resulting 30 nm crosslinked nanoparticles were subsequently grown by the addition of *n*BA, which basically forms linear poly(n-butyl acrylate) chains extending from the particles.

5.4.4.3 *Inverse Miniemulsion Polymerisation*

Matyjaszewski and coworkers also reported on the application of AGET-ATRP for the preparation of particles from water-soluble monomers via an inverse miniemulsion technique. The initial reports (Oh *et al.*, 2006a; Oh, Perineau and Matyjaszewski, 2006b) dealt with the preparation of stable poly[oligo(ethylene oxide) methacrylate] (POEOMA) particles in cyclohexane (10% solids contents). To this end the catalyst system $CuBr_2$/tris[(2-pyridyl)methyl]amine was dissolved in a small amount of water together with a PEG-derived initiator (PEG_{5000}-Br). This was dispersed in cyclohexane containing Span 80 surfactant (sorbitan monooleate, HLB = 4.3). Polymerisation was started by the addition of ascorbic acid and average particle sizes in the range of 120–150 nm were obtained. Similarly, nanogels were prepared in the presence of a crosslinker. For the polymerisation of hydroxyethyl methacrylate (HEMA) the same procedure was used, with the exception of the surfactant (Oh *et al.*, 2007). Neither Span 80, nor Brij 52 yielded a stable dispersion and the block copolymer surfactant poly(ethylene-*co*-butylene)-*b*-PEG needed to be used. In this way stable particles of PEG-*b*-PHEMA and PHEMA-*b*-PEG-*b*-PHEMA were prepared.

5.4.4.4 *Surfactant-Free Emulsion Polymerisation*

In order to reduce the amount of surfactant and to counteract possible surfactant migration upon film formation of the prepared latexes, in recent years several studies have been reported on the use of reactive surfactants in emulsion ATRP. These studies include the use of surface-active initiators (inisurfs) and monomers (surfmers). The work on inisurfs can be considered as an extension of the use of macroinitiators in nitroxide-mediated polymerisation and one of the first groups to publish on inisurfs in emulsion ATRP was that of Charleux. (Stoffelbach *et al.*, 2007, 2008) In a first attempt, the amphiphilic block copolymer initiator PEG_{111}-*b*-PS_{33}-Br was used in the AGET-ATRP miniemulsion polymerisation of *n*BMA, *n*BA and S. Although the polymerisation of *n*BMA was fast, its initiation was quite slow, leading to relatively high PDIs. This problem was solved by adding a small amount of S, which slowed down the polymerisation rate. Styrene and *n*BA both proceeded at a slow polymerisation rate. Stable particles were reported with average particle diameters in the range130–230 nm (the diameters increased with decreasing amount of inisurf, as expected), but the distributions were fairly broad. In a subsequent study (Li *et al.*, 2008) this latter problem was solved by the addition of a small molecule ATRP initiator, in this case EBiB. A better colloidal stability was observed and less surfactant could be used. In the same study, PEG_{44}-Br was also used as an initiator instead of the block copolymer, and also in this case good control and stable latexes were found.

In subsequent work by the Charleux group (Stoffelbach *et al.*, 2008), the authors used an initiating analogue of CTAB, that is, 11′-(*N*,*N*,*N*-trimethylammonium bromide) undecyl-2-bromo-2-methyl propionate (TABUB) (see Figure 5.15) in the miniemulsion polymerisation of MMA via AGET and ARGET ATRP mechanisms. First, a solution of the catalyst system $CuBr_2$/BPMODA was dissolved in the monomer containing 2.3 wt% of hexadecane. This solution was added to the water containing the inisurf and started with ascorbic acid. The polymerisations were fast, reached high conversions and showed a good molar mass control. Average particle sizes were in the range 150–250 nm, with slightly broad particle size distributions.

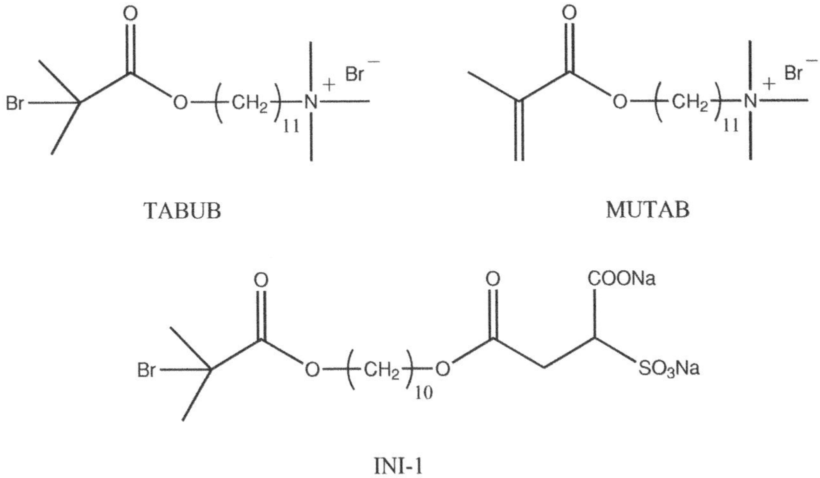

TABUB

MUTAB

INI-1

Figure 5.15 *Reactive surfactants in AGET-ATRP emulsion polymerisation.*

The same inisurf was later used by Li and Matyjaszewski (Li and Matyjaszewski, 2011) in the AGET-ATRP emulsion polymerisation of *n*BMA using the "microemulsion-seed technique" (Min, Gao and Matyjaszewski, 2006b). This approach was successful, but since very high surfactant concentrations are required in this technique, only low molar mass polymers can be produced. Therefore, they also used an analogous surfmer, 11′-(methacryloyloxy)-undecyl-(trimethyl ammonium bromide) (MUTAB) (Figure 5.15). First a solution was made of $CuBr_2$/BPMODA/EBiB in a small part of the *n*BMA monomer, which was slowly added to an aqueous MUTAB/NaCl solution. The polymerisation was started by addition of ascorbic acid, creating a microlatex which acted as a seed for the subsequent polymerisation of the remainder of the *n*BMA. Good molar mass control was achieved with all the MUTAB incorporated and the latexes all displayed a good stability with particle diameters ranging from 40 to 200 nm.

Finally, we would like to mention the use of an *anionic* inisurf in AGET-ATRP emulsion polymerisation. This inisurf (**INI-1** in Figure 5.15) was first used in the *ab initio* A(R)GET-ATRP emulsion polymerisation of MMA ($\sim$20% solids) catalyzed by $CuBr_2$/bpy/ascorbic acid at 80 °C (Cheng *et al.*, 2010). Good molar mass control was achieved, but (probably as expected) broad particle size distributions with averages in the range 250–650 nm were obtained. This inisurf was subsequently (in combination with an additional oligo(ethylene oxide)-based nonionic surfactant) used successfully to prepare fluoropolymer latexes with particle sizes in the range 240–290 nm (Shu *et al.*, 2011). The authors used BPMODA as a ligand this time and Matyjaszewski's microemulsion-seed technique (Min, Gao and Matyjaszewski, 2006b).

5.4.5 Some Concluding Remarks on ATRP in Emulsion

It is probably safe to conclude that currently the most efficient ATRP systems in aqueous dispersions are those based on AGET-ATRP, that is, those in which the Cu catalyst is added as the Cu(II) complex with very hydrophobic ligands, which is subsequently reduced *in situ* by, for example, ascorbic acid. Using appropriate nonionic or cationic surfactants, the polymerisations can then conveniently be carried out in miniemulsion or via a seeded polymerisation in which the seed was prepared by Matyjaszewski's microemulsion (Min, Gao and Matyjaszewski, 2006b) or George's nanoprecipitation (Chan-Seng and Georges, 2006) techniques. The use of multifunctional, oil-soluble initiators, then allows the synthesis of more complex polymer architectures.

5.5 Reversible Chain Transfer in Emulsion and Miniemulsion

Many industrial polymers are made through *ab initio* emulsion polymerisation, in which surfactant, monomer, water and initiator are mixed in a reactor, and polymerisation of the monomer at the required temperature is carried out. The drawback of this process is that the surfactant is not bound to the polymer chains and can leach out of the polymer film over time. Below, we will describe how macromolecular RAFT agents can be used as a reactive stabilizer to avoid this problem. *Ab initio* emulsion polymerisation is usually characterized by three intervals. Interval I is where the monomer swollen micelles are initiated to form particles. In this period, nucleation of particles dictates the number of particles for the rest of the polymerisation and hence has a large influence on the rate of polymerisation. Interval II is where micelles are no longer present, and particles grow with the monomer supplied from monomer droplet reservoirs. Interval III starts when these droplets are depleted, and the monomer concentration is no longer constant during this period and decreases with conversion.

It seems from the advantages of dispersion polymerisation that applying RAFT in an emulsion would be a major advance in many industrial polymers, especially in the preparation of novel polymers with controlled MMD, M_n, copolymer composition and particle morphology (Monteiro, 2010; Monteiro and Cunningham, 2012; Zetterlund, Kagawa and Okubo, 2008).

5.5.1 Low C_{ex} Reversible Chain Transfer Agents

An ideal *ab initio* polymerisation using low C_{ex} RCTAs (lower than a value of ~5) should, in principle, behave in a manner similar to that with a conventional chain transfer agent. However, the M_n and PDI should be under reversible chain transfer control, and allow the production of polymer structures such as blocks and stars. At the start of Interval I, the relative concentrations of monomer and RCTA in the droplets and swollen micelles should be equal to the starting concentrations. Upon commencement of polymerisation, the transfer of monomer and RCTA from droplets to particles is equal to their rate of consumption, as based on the 'twin films' theory. In the case where C_{ex} is equal to 1, the consumption rate of RCTA will be equal to that of monomer. This means that the ratio of monomer to RCTA will always be equal in the droplets and particles. The evolution of M_n and PDI with conversion for the three Intervals will therefore be identical to a bulk or solution polymerisation. The rate on the other hand will be affected by the exit of the leaving radical, X, after fragmentation. The effect of the leaving radical on the rate is dictated by a number of factors: the size and number of the particles, the partition coefficient of the leaving group in the monomer and water phases, the reactivity of the leaving group towards monomer and the dormant species, concentration of radicals in the water phase, and the average number of radical per particle (Pepels *et al.*, 2010; Jia and Monteiro, 2012).

Several studies reported the successful application of reversible chain transfer techniques in waterborne systems. All of these studies apply RCTA species with low C_{ex} constants to control the polymerisation. The alkyl iodides (degenerative transfer) used by several groups (Lansalot *et al.*, 1999; Butté, Storti and Morbidelli, 2000) have a transfer constant only slightly higher than unity. The *ab initio* emulsion polymerisation of styrene using $C_6F_{13}I$ was carried out at 70 °C (Lansalot *et al.*, 1999). It was found that the rate of polymerisation was not affected by the presence of $C_6F_{13}I$. However, the evolution of M_n with conversion was not in accord with the C_{ex} value. The authors postulated that, due to the hydrophobic character of $C_6F_{13}I$, its transfer from droplets to particles was slower than the rate of consumption of $C_6F_{13}I$ within the particles. To overcome the slow diffusion of $C_6F_{13}I$ to the particles, the authors carried out the reaction in miniemulsion (essentially Interval III kinetics), in which polymerisation takes place in the monomer droplets, excluding the need for $C_6F_{13}I$ transportation. The results gave the theoretically desired M_n and PDI (close to 1.5).

A similarly slow consumption of the compound can be expected for the RAFT agents, either because of a poor homolytic leaving group (Uzulina, Kanagasabapathy and Claverie, 2000) or a rather unactivated carbon–sulfur double bond (Monteiro *et al.*, 2000b). The emulsion polymerisation of xanthates, showed that the PDI and the M_n evolution with conversion could be predicted from theory for both styrene and n-butyl acrylate systems, suggesting that diffusion of the xanthates between the droplets and particles was sufficiently fast (Monteiro and de Barbeyrac, 2001b; Monteiro *et al.*, 2000b). The main observation was that the rate was retarded with increasing amount of xanthate. This was suggested to be a result of the enhanced exit rate due to the leaving group on the xanthate agent (Pepels *et al.*, 2010). It was also suggested that the increase in the exit rate led to a slight increase in the final particle number. Smulders, Gilbert and Monteiro (2003) studied the effect of entry and exit of such xanthate RAFT agents in seeded emulsion polymerisations of styrene. As expected due to the fragmentation of the leaving group from the xanthate, the enhanced exit led to retardation. But more surprisingly the entry rate was lowered, suggesting that xanthate RAFT agents were surface active. Zeta potential and conductivity experiments were carried out and showed that these RAFT agents were indeed pushing the surfactant sodium dodecyl sulfate (SDS) into the aqueous phase. For the xanthates this could be due to their ionic canonical forms.

The advantage of the low C_{ex} reversible chain transfer agents is that latexes with controlled particle size distributions using seeded polymerisations can be made. These latex particles in a second stage polymerisation can be further reacted with other monomers to make block copolymers with core–shell particle morphology or even latex particles in which the second block has functional groups (i.e. reactive latexes (Monteiro and de Barbeyrac, 2002)). This methodology has been used to produce nanoparticles that could sequester heavy metals at ppb levels from water solutions (Bell *et al.*, 2006).

5.5.2 High C_{ex} Reversible Chain Transfer Agents

The transition from reversible chain transfer agents with low activity to those with a high activity intuitively appears to be straightforward, but in practice turns out to be more complicated.

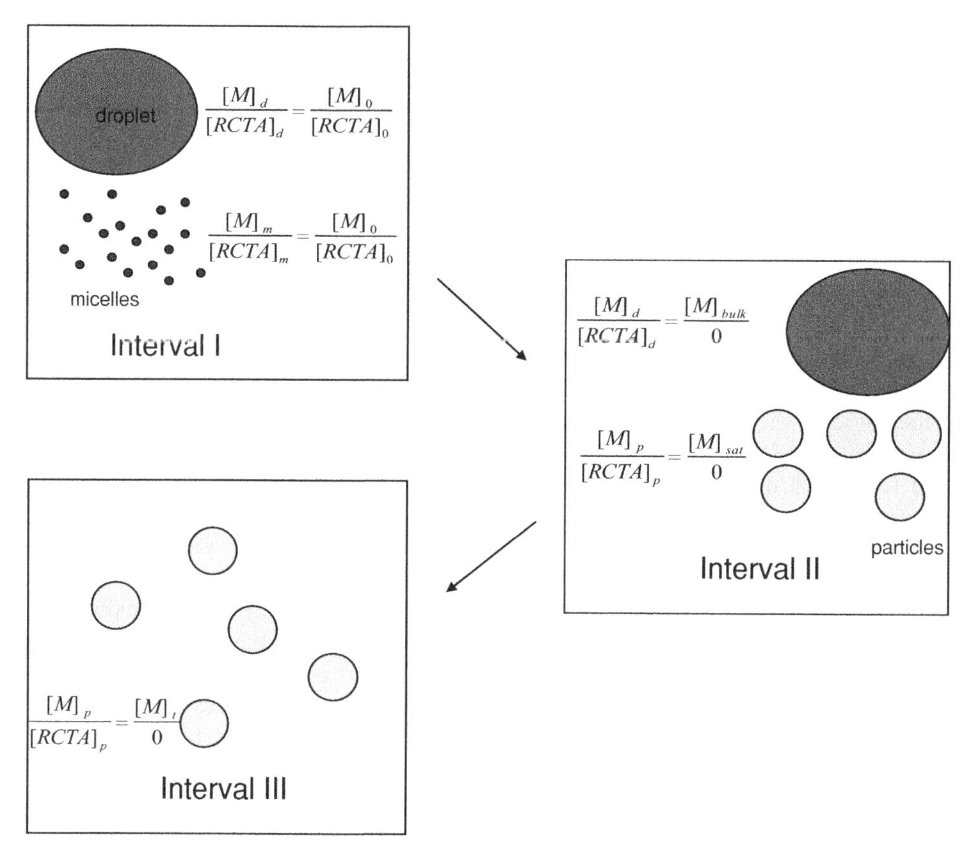

Figure 5.16 *Ideal mechanism for the use of RAFT in an* ab initio *emulsion polymerisation.*

5.5.2.1 Ab Initio *Emulsion Polymerisation*

An ideal *ab initio* emulsion polymerisation using the reversible chain transfer agents with a very high C_{ex} can be visualized, as shown in Figure 5.16. At the beginning of Interval I, the concentration ratio of monomer to RCTA is close to the bulk concentrations of both species. However, due to the high C_{ex}, the RCTA should, in principle, be consumed within the first few percent of monomer conversion. During this early period, exit from the particles will additionally be proportional to the concentration of the RCTA and the partitioning coefficient between the monomer and water phases of the leaving group X. The exited X radical can undergo several fates, it can terminate with other radicals in the aqueous phase, re-enter other micelles and particles to either terminate already growing radicals, react with dormant chains or initiate growth in new particles. Exit, in general, should result in a lowering of the rate of polymerisation (Jia and Monteiro, 2012).

After the initial stage (consumption of all reversible chain transfer agent), the polymer chains will consist exclusively of dormant polymeric chains located in the micelles or particles. The advantage of reversible chain transfer over reversible termination techniques is that the leaving group X becomes larger with conversion and has a low probability to exit the particles. In contrast, for the NMP and ATRP techniques, the partitioning of either the nitroxide or the copper deactivating species (Cu^{II}/ligand) into the water phase will occur throughout the polymerisation, irrespective of the chain length of the dormant species, which will alter the rate of polymerisation (Charleux, 2000; Zetterlund, Kagawa and Okubo, 2008). This suggests that for reversible chain transfer the polymer chains in the particle grow at a rate equivalent to that in the absence of RCTA, since exit and entry coefficients with or without RCTA should be identical at this stage. The polymer chains will grow in the particles in an environment with constant monomer concentration until Interval III is reached. During Intervals I and II there is a linear growth of M_n with conversion, but since the monomer concentration is assumed constant, Equation 5.6 cannot be used to determine the PDI index. During Interval III, the monomer concentration decreases with time and the M_n and PDI can be predicted with the simple bulk polymerisation equations.

In previous work, several highly reactive RAFT agents were applied in conventional emulsion polymerisations (SDS as surfactant), both seeded (Monteiro, Hodgson and de Brouwer, 2000a) and *ab initio* (Hodgson, 2000). While low activity xanthates could easily be used (Monteiro *et al.*, 2000b), high reactivity agents based on the dithiobenzoate group invariably led to colloid stability problems and formation of a conspicuous red layer (Monteiro, Hodgson and de Brouwer, 2000a). A large amount of the transfer agent would be lost in the form of an (oligomeric) coagulant, resulting in a much higher molar mass than expected for the emulsion material. It was also found that with increasing RAFT concentration the rate of polymerisation was severely retarded when a low concentration of initiator was used. The results showed that, in practice, the theoretical mechanism given above requires reconsideration.

The observed loss of control of the MMD and severe retardation in rate are directly related. It has been found that in bulk and solution polymerisations for the RAFT agent (cumyl dithiobenzoate, CDB) the rate was severely retarded. Many researchers have tried to explain this phenomenon through either termination reactions with the intermediate radical (Monteiro and de Brouwer, 2001a; Kwak *et al.*, 2002; Barner-Kowollik *et al.*, 2006) or slow fragmentation of the intermediate radical (Barner-Kowollik *et al.*, 2001, 2006). Therefore, retardation in the above emulsion polymerisation is possibly due to a combination of the chemical nature of the RAFT agent (CDB) and exit of the leaving group. However, the above conclusion alone does not explain all the data in the literature (see below), and in addition should not result in loss of control of the MMD. As indicated above, it was found that on transition from Interval II to III a red layer was formed. This could be a result of the severe retardation in rate, such that nucleation of the monomer droplets could be competing with particle or micellar nucleation. In conventional *ab initio* polymerisations, droplet nucleation is usually kinetically negligible since there are approximately one to two orders of magnitude more micelles or particles than droplets. Nucleation of droplets results in the formation of oligomeric species that cannot diffuse from droplets to particles (Prescott *et al.*, 2005a). At the end of Interval II, these oligomers (red in colour) will form a separate phase or layer.

Moad *et al.* (2000) overcame this problem by using a semi-batch process, where all the RAFT agent together with SDS and water-soluble initiator in the presence of a small amount of monomer was polymerized at 80 °C for 40 min, after which monomer was fed in slowly. The methodology allowed polymerisation in the absence of monomer droplets to give a system with good control over the MMD. Prescott *et al.* (2002b) provided strong support for this methodology by using an acetone transport technique to localize the RAFT agent inside seeded particles. Although they also observed good control of M_n with conversion, they observed both inhibition and retardation in rate even though they used a RAFT agent which showed no retardation in either bulk or solution.

The mechanism for retardation in emulsion polymerisation is non-trivial and depends on a number of competing factors (Monteiro, Hodgson and de Brouwer, 2000a; de Brouwer *et al.*, 2000; Prescott *et al.*, 2002a, 2002b). In the early stages of polymerisation, the high rate of termination due to short–short chain radical coupling will ensure that particles will be under zero-one conditions (Prescott, 2003). This should give a similar rate to the polymerisation without RAFT. The only mechanism for retardation would be exit of the leaving group after fragmentation from the RAFT agent. As the polymer chains increase in length, exit will no longer occur, and the kinetics would deviate from zero-one to pseudo-bulk (where more than one radical can reside in the particle). However, this would have the opposite effect of only increasing the rate of polymerisation. It was postulated that the only way for retardation to occur at chain lengths of the dormant species much greater than a z-mer is through 'frustrated entry' (Smulders, Gilbert and Monteiro, 2003). This postulate states that a z-mer formed in the aqueous phase will enter a particle and react with a dormant species of chain length, k (where $k > z$) to form a z-mer dormant species and a polymeric radical (k-mer). Since the C_{ex} is so high for this RAFT system, k-mer radicals will react many times with all the dormant species in the particle, and when it reacts back with a z-mer dormant agent the z-mer can exit the particle to terminate with other aqueous phase radicals or radicals in other particles. The cycle is only broken if monomer can add to the z-mer in the particle to reduce the probability of exit. It should be noted that a method to overcome this is to simply use a larger particle as the exit rate coefficient is proportional to the inverse of the radius squared.

Based on these new mechanistic insights miniemulsion appears to be the most logical technique to overcome many of the above problems. However, miniemulsions require greater amounts of surfactants and co-stabilizers than *ab initio* or seeded emulsion polymerisations, which will inevitably have an effect on the film properties.

Ferguson *et al.* (Ferguson, Russell and Gilbert, 2002b, Ferguson *et al.*, 2005) used a surfactant-free system to overcome many of these problems in RAFT-mediated polymerisations. This methodology has since been denoted as the *polymerisation-induced self-assembly* process (also shown previously in NMP). In the initial work by Ferguson *et al.* (2002a), they first polymerized acrylic acid (AA) in water mediated by an amphipatic RAFT agent, targeting approximately 5 monomer units. n-Butyl acrylate was then slowly fed into the reactor to form diblock copolymers, and when the hydrophobic poly(n-butyl acrylate) block grew to a sufficient length, the diblock self-assembled into polymer micelles stabilized by the poly(acrylic acid) hydrophilic blocks (see Figure 5.17). The particle size distribution was well controlled and the PDI of the polymer chains at the end of the polymerisation was close to 1.5, suggesting that the MMD was not well controlled. However, this methodology shows the advantages of using LRP in dispersed media to make latex particles without surfactant, and solves a long-standing problem in the coatings industry, in which surfactant migration to the surface of the coating results in deleterious coating properties.

Extensive effort was further placed on carrying out the polymerisations under batch conditions. The first system to be used with success was based on poly(ethylene oxide)

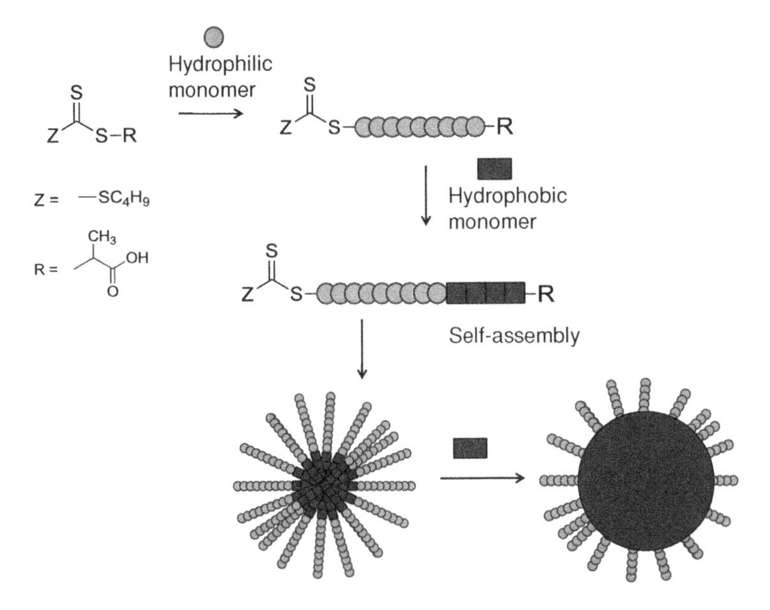

Figure 5.17 *General mechanism for the RAFT polymerisation-induced self-assembly process.*

(PEO) bearing a trithiocarbonate RAFT group. While the polymerisation of styrene was slow and not so well controlled, the polymerisation of n-butyl acrylate (Rieger *et al.*, 2008) and its copolymerisation with methyl methacrylate (Rieger *et al.*, 2009) were fast and quantitative, leading to the formation of self-stabilized, amphiphilic block copolymer particles, with well-defined chains of narrow MMD. Later, PEO was replaced by macroRAFT agents resulting from the direct RAFT polymerisation of hydrophilic monomers (for instance poly(*N*,*N*-dimethyl acrylamide), (Rieger *et al.*, 2010) polyacrylamide (Ji, Yan and Xie, 2008), poly(methacrylic acid) (Chaduc *et al.*, 2012), poly(acrylic acid-*co*-PEO acrylate) (Boissé *et al.*, 2010), poly(methacrylic acid-*co*-PEO methacrylate) (Zhang *et al.*, 2011a)). In the latter case the emulsion polymerisation was performed in a one-pot system, with the macroRAFT agent prepared via aqueous solution polymerisation in the first step. One further advantage of the *polymerisation-induced self-assembly* is the production of not only spherical particles, but also of a wide range of morphologies, including worm-like micelles, fibers, vesicles and others. The conditions that favor a given morphology have been studied, in particular for the poly(acrylic acid-*co*-PEO acrylate) (Boissé *et al.*, 2010, 2011) and the poly(methacrylic acid-*co*-PEO methacrylate) (Zhang *et al.*, 2011a, 2011b, 2012; Charleux *et al.*, 2012) macroRAFT agents.

Another method to overcome the kinetic problems is to carry out the polymerisations in nanoreactors (Urbani and Monteiro, 2009a, 2009b; Sebakhy, Kessel and Monteiro, 2010). In one case, a reactive nanoreactor consisting of diblock copolymer with a RAFT end-group, in which one block was hydrophilic and the other a thermoresponsive polymer (e.g. poly(*N*-isopropylacrylamide), PNIPAM), was heated above its lower critical solution temperature (LCST) to form polymer micelles (Urbani and Monteiro, 2009b). The polymerisation of styrene in these reactive nanoreactors gave M_n values close to theory with low PDIs

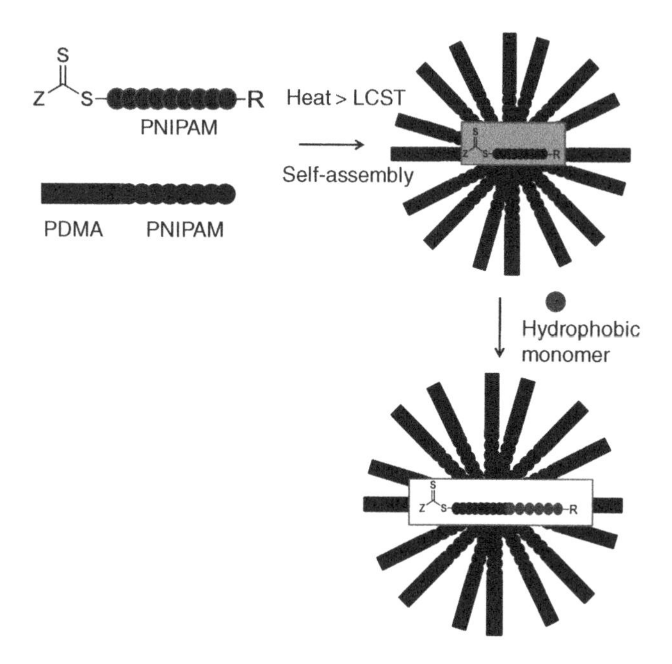

Figure 5.18 *General mechanism for the RAFT polymerisation using diblock nanoreactors.*

(<1.2). The size of the particles could be controlled by changing the amount of styrene, producing very narrow size distributions. In another variant on this thermoresponsive system (see Figure 5.18), a diblock copolymer consisting of a hydrophilic block and PNIPAM block was used as the nanoreactor (i.e. without a RAFT end-group). An oligomeric PNIPAM homopolymer with a RAFT end-group was then mixed with the diblock, monomer and initiator, and polymerized to produce both a narrow MWD and narrow particle size distribution (Urbani and Monteiro, 2009a; Sebakhy, Kessel and Monteiro, 2010). The advantage of this method was that particles were produced with a desired size and desired molar mass, allowing the decoupling of the size from the M_n. In a variant of this method, SDS was used with a PNIPAM macromolecular RAFT agent not only to control the polymerisation of styrene but also to produce worms, rods, vesicles and donuts when cooled below the LCST (Kessel, Urbani and Monteiro, 2011).

5.5.2.2 *Miniemulsion*

This section describes RAFT polymerisations carried out under miniemulsion conditions. The first detailed study of the use of SDS or for that matter any ionic stabilizer with a hydrophobe led to destabilization of the miniemulsion for polymerisations of styrene or ethyl hexyl methacrylate (EHMA) (Tsavalas *et al.*, 2001). However, a linear increase in M_n was found with conversion with polydispersities as high as 2. Conductivity measurements were also carried out to study the extent of surfactant migration into the aqueous phase. The results showed that SDS did migrate, but only after polymerisation was initiated. Lansalot, Davis and Heuts (2002) showed that they could find conditions where colloidal stability was observed when using SDS. However, the polydispersity found in their experiments ranged between 1.7 and 2, suggesting that these systems gave non-ideal RAFT polymerisation.

It was thought that destabilization could be similar to that observed for ATRP in the presence of SDS, therefore, ionic surfactants were substituted for a non-ionic surfactant (e.g. Brij 98) (de Brouwer *et al.*, 2000). This allowed well-defined polymer (PDI <1.2) to be prepared with no stability problems. However, retardation compared to polymerisation without a RAFT agent was found. This was stateded to be due to termination of the intermediate radical species, which lowered the propagating radical concentration considerably (Monteiro, de Brouwer, 2001a). Although the rate is significantly reduced by intermediate radical termination, it should have little or no effect on the MMD since the amount of RAFT dormant chains lost through intermediate termination is less than 5%. It was also found that a wide range of monomers could be polymerized with control and could be further used to prepare block copolymers with low polydispersities (<1.2) (de Brouwer *et al.*, 2000). One major advantage of the RAFT process is that acidic monomers can be used, which provide very efficient stability to the polymer latex particles. For example, using Brij 98 (non-ionic surfactant) a block copolymer of PEHMA-block-poly(methyl methacrylate-co-methacrylic acid) was prepared.

Luo, Tsavalas and Schork (2001) argued that the growing particles in a LRP would have a lower chemical potential than non-nucleated droplets due to the 'superswelling' effect of small oligomers. This would result in monomer transfer from high to low chemical potential, where monomer would swell the growing particles until equilibrium was reached. The authors suggested that by simply increasing the co-stabilizer level the problems found by using ionic stabilizers would be eliminated. Further work (McLeary *et al.*, 2004) by carrying out miniemulsions with higher levels of surfactant and co-stabilizer (hexadecane) tentatively supported Luo *et al.*'s postulate. The 'superswelling' effect accounts for many of the observations found not only in miniemulsions but also in *ab initio* emulsion polymerisations. Therefore, one must consider in surfactant-based emulsion polymerisations both the kinetic (e.g. exit) and thermodynamic (e.g. 'superswelling') effects, and the interplay between them.

5.6 Conclusion

Living radical polymerisation in aqueous dispersed systems such as emulsion and miniemulsion has been shown to be promising in the synthesis of new polymer particles. The three main techniques of control (NMP, ATRP and RAFT) are now able to produce well-defined polymers in an aqueous environment with a similar degree of livingness as in the bulk. Various levels of polymer design may now be reached, not only including the design of macromolecular architectures but also new particle morphologies. This will open a new methodology for the preparation of novel nanocomposite polymer colloids. All the research first showed that miniemulsion polymerisations gave the best results both from chemical and colloidal viewpoints. But now, optimal conditions have been found to apply living polymerisation in a classical *ab initio* emulsion process, close to those currently used in industrial applications. The development of surfactant-free systems based on *polymerisation-induced self-assembly* is now the way to prepare a variety of self-stabilized structures with various morphologies.

6
Particle Morphology

Yuri Reyes Mercado,[1] Elena Akhmastkaya,[2] Jose Ramon Leiza,[1] and Jose M. Asua[1]
[1] POLYMAT, University of the Basque Country UPV/EHU, Spain
[2] Basque Center for Applied Mathematics (BCAM), Spain

6.1 Introduction

Waterborne polymers are used in a wide range of applications, including synthetic rubber, paints, adhesives, additives in paper and textiles, leather treatment, impact modifiers for plastic matrices, additives for construction materials, cosmetics, flocculants, diagnostic tests and drug delivery (Asua, 1997; Lovell and El-Aasser, 1997; Urban and Takamura, 2002a). Although homogeneous particles meet the requirements of many of the applications, heterogeneous particles provide advantages in the more demanding cases. Thus, 2-phase soft–hard particles have been used for coatings, which combine a low minimum film-forming temperature and a high blocking resistance (Schuler *et al.*, 2000). Rubber-thermoplastic core–shell particles are useful to impart toughness to thermoplastic resins (Sato and Tateyama, 1988). Waterborne polymer–polymer hybrids (e.g., alkyd–acrylic (Tsavalas, Schork and Landfester, 2004; Goikoetxea *et al.*, 2009), polyurethane–acrylic (Li, Chiu and Don, 2007) and epoxy–acrylic (Kawahara *et al.*, 2001)) have been developed in an attempt to combine the positive properties of both polymers, avoiding their drawbacks. Structured latex particles are also used to overcome the limitations of some copolymerisation systems. One interesting example is the styrene (S)/vinyl acetate (VAc) system which consists of monomers having complementary properties. Although the system does not copolymerize ($r_S = 55$ and $r_{VAc} = 0.01$ (Odian, 2004), structured latex particles of the corresponding polymers can enhance the mechanical and resistance properties of the latex films due to the PS, while keeping the film-forming properties at room temperature of the PVAc.

Encapsulation of inorganic materials, such as TiO_2, within the polymer particles is a way to improve the dispersion of the inorganic material in coatings and, therefore, enhance its

Chemistry and Technology of Emulsion Polymerisation, Second Edition. Edited by A.M. van Herk.

hiding power (Caris, Kuijpers and van Herk, 1990; Erdem *et al.*, 2000a, 2000b, 2000c). Hollow particles can substitute pigments due to their special optical properties and also serve as catalyst supports. Other morphologies, such as nanocapsules, are widely used in drug delivery technologies (Anton, Benoit and Saulnier, 2008). In all cases, the properties depend strongly on particle morphology. Therefore, the control of the morphology of the latex particles is of primary importance from the industrial and scientific points of view.

This chapter discusses the morphology of latex particles obtained mainly by (mini)emulsion polymerisation. Some applications of these particles are described and the factors that influence the particle morphology are discussed. Mathematical models that describe and predict the particle morphology as a function of polymerisation variables are presented along with some experimental examples.

6.2 Synthesis of Structured Polymer Particles

Structured polymer particles can be synthesized by chemical and physical methods. The chemical methods involve polymerisation of monomers. The choice of a specific technique depends on the properties of interest for the particles. Structured particles composed of polymers formed by free radical polymerisation are mainly synthesized by emulsion polymerisation in a two-step process. In the first step, a monomer or mixture of monomers is polymerized in emulsion. The resulting latex is used as a seed in the second step, that is, a seeded emulsion polymerisation of a mixture of monomers that yields a polymer incompatible with the polymer forming the seed. Both stages are often semicontinuous processes in order to achieve a better control of temperature and polymer characteristics. In emulsion polymerisation, inorganic material can be placed at the surface of the polymer particles during the formation of the seed (Percy *et al.*, 2002; Colver, Colard and Bon, 2008; Schmid, Tonnar and Armes, 2008; Schmidt and Boodmeier, 2009), but this process cannot be used to incorporate the inorganic material within the polymer particles.

Miniemulsion polymerisation is used to synthesize structured particles that, in addition to the polymer obtained by free radical polymerisation from monomers showing a minimal water solubility, contain polymers produced by step growth polymerisation or by free radical polymerisation of highly water-insoluble monomers (Asua, 2002; Antonietti and Landfester, 2002). Miniemulsion polymerisation is also used to incorporate inorganic materials both within and at the surface of the particles (Hu, Chen and Wu, 2011).

Among the physical methods, heterocoagulation and solvent evaporation have been used to produce capricious particle morphologies.

6.2.1 Emulsion Polymerisation

In this process, the formation of the structured particle occurs during the second stage (seeded emulsion polymerisation). The formation of the morphology can be illustrated by considering the case of two-phase polymer–polymer waterborne particles. Let Polymer 1 be the polymer of the seed particles and Monomer 2 the mixture of monomers polymerised in the second stage. As Polymer 2, formed by polymerisation of Monomer 2, is incompatible with Polymer 1, the polymers phase separate, forming clusters of Polymer 2 within the Polymer 1 matrix. Polymerisation of Monomer 2 continues in both clusters and matrix and, hence, the clusters grow in size and new clusters are formed. The system is not at thermodynamic equilibrium because of the surface energy associated with the large Polymer 1–Polymer 2 interfacial area. In order to minimize the free energy of the system, the clusters migrate towards the equilibrium morphology. During the migration, the clusters may coagulate among them. The driving forces for the motion of the clusters are the van der Waals attraction–repulsion forces and the Brownian motion due to the thermal energy. The movement of the clusters is ruled by the balance between the driving forces and the resistance to flow resulting from the viscous drag. The viscous drag arises during polymerisation because the internal viscosity of the particle increases as the amount of monomer in the particle decreases. If the driving forces are able to overcome the resistance offered by the viscous drag, the system reaches the equilibrium morphology. Otherwise, the motion of the clusters stops and kinetically controlled metastable final particle morphologies are obtained. Therefore, the particle morphology depends on the interplay between thermodynamics and kinetics (Chen, Dimonie and El-Aasser, 1991a, 1991b, 1992; González-Ortiz and Asua, 1995, 1996a, 1996b; Sundberg and Durant, 2003). Obviously, consecutive addition of different monomer mixtures allows one to obtain multiphase polymer particles with different morphologies.

6.2.2 Miniemulsion Polymerisation

In this process, the water-insoluble material (such as polymer formed by step growth polymerisation, highly water-insoluble monomer, hydrophobic inorganic material) is dissolved or dispersed in a monomer mixture (which is polymerizable by free radical polymerisation) and a miniemulsion is formed by dispersing this solution/dispersion in water in the presence of emulsifiers. Static mixers (Ouzineb *et al.*, 2006), sonifiers (Landfester, Bechthold and Antonietti, 1999), rotor-stators (Lopez *et al.*, 2008) and high pressure homogenizers (Manea *et al.*, 2008) can be used to produce the miniemulsion. The high pressure homogenizers are the most promising for commercial applications (Goikoetxea *et al.*, 2011). The miniemulsion is stabilized against monomer diffusional degradation (Oswald ripening) by using co-stabilizers (low molecular weight, highly water-insoluble compounds). When the miniemulsion is polymerized, the development of the particle morphology is conceptually similar to that described above for the second stage of the emulsion polymerisation, although multiple phases may be involved.

Achieving narrow composition distribution among particles is challenging in miniemulsion polymerisation. On the one hand, thermodynamics determines that the large droplets are richer in monomer than the smaller ones. On the other hand, in batch processes very stable and small-sized monomer droplets are needed to achieve the nucleation of most of the droplets, avoiding the formation of particles by secondary nucleation. However, batch processes are not well suited to synthesize high solids content dispersions at large scale because of the high rate of heat generation during polymerisation. Semicontinuous miniemulsion polymerisation, that is, the feeding of a miniemulsion to the reactor, overcomes the problems related to heat removal, but only a fraction of the entering monomer droplets are nucleated (Rodriguez, Barandiaran and Asua, 2007). An alternative strategy is to prepare a miniemulsion with a fraction of the relatively water-soluble monomers and all the water-insoluble compounds to use as the initial charge in the reactor; polymerize it batchwise, and then feed semicontinuously the rest of the monomer mixture.

6.2.3 Physical Methods

Composite waterborne dispersed polymer particles can also be prepared by physical means such as heterocoagulation and solvent evaporation.

6.2.3.1 Heterocoagulation

In this method, structured particles are formed by coagulating particles of different nature (Teixeira and Bon, 2010). Thus, raspberry particles can be produced by mixing large particles bearing a charge with smaller particles of the opposite charge, keeping the proper size and population ratio. Additional layers can be formed by changing the sign of the particles in the next coagulation step (Caruso, Caruso and Möhwald, 1999a; Caruso *et al.*, 1999b). The colloidal stability of the system is an issue and usually low solids content dispersions can be obtained.

6.2.3.2 Solvent Evaporation

In this approach, the polymers that will constitute the particles are dissolved in a common good solvent. This organic phase is then dispersed in a continuous medium using a suitable emulsifying technique and, finally, the solvent is evaporated. Evaporation of the solvent leads to phase separation between the polymers. Equilibrium morphologies can be attained by slow evaporation of the solvent (Tanaka, Saito and Okubo, 2009).

A variation of this technique is the addition of water to a solution of block copolymers in THF, with the subsequent elimination of the organic solvent. In this way, polymer nanoparticles with capricious morphologies, such as lamella, tennis ball, and so on are obtained (Higuchi *et al.*, 2008).

6.3 Two-Phase Polymer–Polymer Structured Particles

The simplest, and most commonly found in practice, waterborne structured particles comprise two polymeric phases. Table 6.1 shows that even for this simple system a wide variety of morphologies (both thermodynamically and kinetically controlled) can be obtained. The thermodynamic equilibrium morphology is the one that has minimum interfacial energy. For a two-phase system, the equilibrium morphology is easily determined by minimizing the interfacial energy, calculated as the product of the interfacial areas (A_{ij}) and the interfacial tensions (γ_{ij}) (Torza and Mason, 1970; Chen, Dimonie and El-Aasser, 1991b; Winzor and Sundberg, 1992a, 1992b; González-Ortíz and Asua, 1995). Figure 6.1 presents the calculations conveniently summarized as a map using the ratios of the interfacial tensions as coordinates (González-Ortíz and Asua, 1995). It can be seen that for a two-phase system, the number of possible equilibrium morphologies is small: core–shell, inverted core–shell, hemispherical and separated particles. In this context, the inverted core–shell morphology refers to core–shell particles in which the core is formed by the polymer produced in the second stage of the synthesis. The necessary condition for inverted core–shell morphologies is that the second stage polymer is more hydrophobic than the first stage one. Figure 6.1 applies for non-crosslinked polymers. In the case of crosslinked polymers, the energy associated with expanding the network should be taken into account.

Table 6.1 *Illustrative examples of two-phase polymer–polymer particle morphologies.*

Morphology	Polymers	TEM Image	Reference
Core–Shell	PS(dark)/PVAc	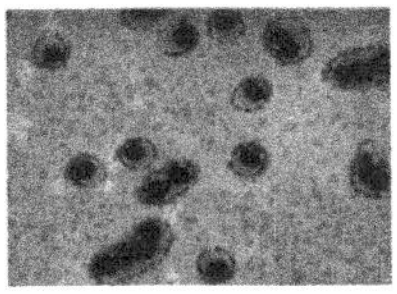	Reprinted under the terms of the STM agreement from [Ferguson, 2002a] Copyright (2002) American Chemical Society
Inverted Core–Shell	PMMA/PS(dark)	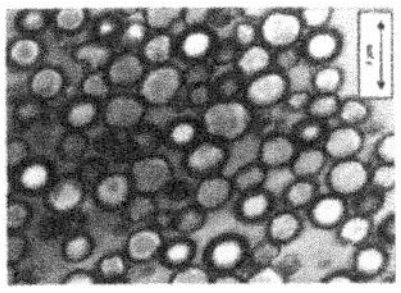	Reprinted with permission from [Jönsson, 2007] Copyright (2007) Elsevier Ltd
Hemispherical	PMMA/PS(dark)		Reprinted with permission from [Herrera, 2007] Copyright (2007) John Wiley and Sons, Inc.
Raspberry	PS(core)/PAN	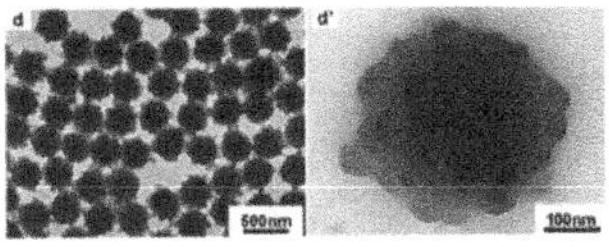	Reprinted with permission from [Huang and Liu, 2010] Copyright (2010) John Wiley and Sons Ltd
Hollow	Obtained by osmotic swelling	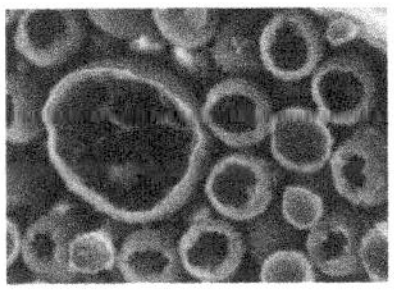	Reprinted with permission from [McDonald and Devon, 2002] Copyright (2002) Elsevier Ltd
Occluded (confetti) Particles	PS domains dispersed in natural rubber particle	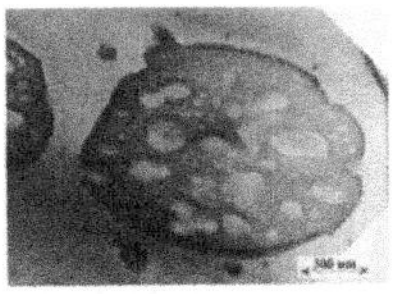	Reprinted with permission from [Schneider, 1996] Copyright (1996) John Wiley and Sons, Inc
Occluded (confetti) Particles	PMMA domains dispersed in PS matrix	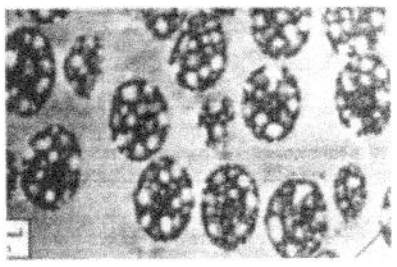	Reprinted with permission from [Jönsson, 1991] Copyright (1991) American Chemical Society

(a)

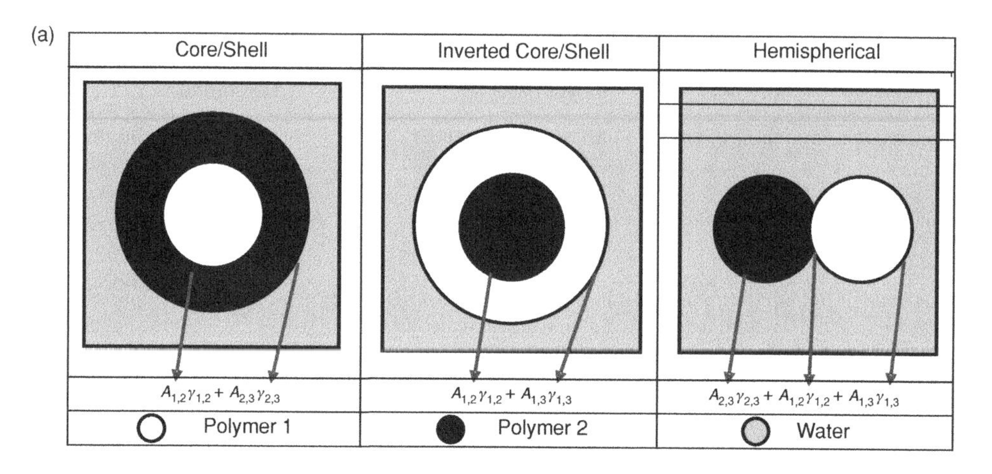

(b)

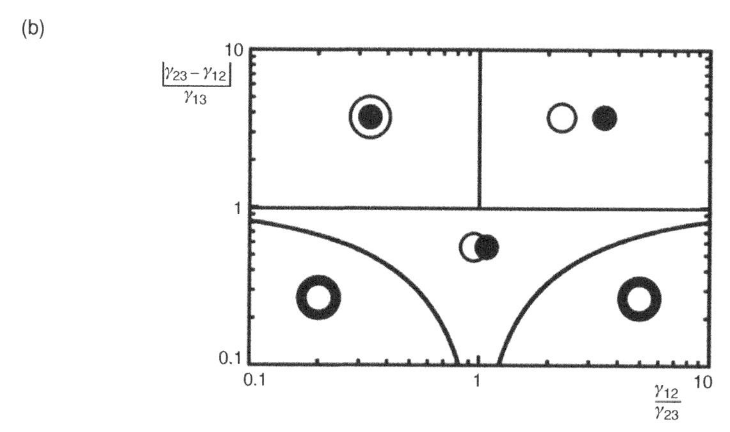

Figure 6.1 *In (a) the expressions for calculation of the interfacial energy for different equilibrium morphologies are given. In (b) a particle morphology map shows how the interfacial tensions can be used to determine the equilibrium particle morphology. Reprinted with permission from [González-Ortíz and Asua, 1995] Copyright (1995) American Chemical Society.*

Hard core–soft shell particles are used for high performance coatings, with the soft shell forming the film and the hard core imparting mechanical resistance to the film (Keddie, 1997). Soft core–hard shell particles are used as impact modifiers for brittle polymers, such as poly(vinyl chloride) (Wu *et al.*, 2004b) and polycarbonate (Parker *et al.*, 1989). Monodisperse core–shell particles can be used for photonic applications. The core contains optically active molecules chemically linked to the polymer. Assembly of core–shell particles by controlled drying leads to a periodic 3D structure of the cores that can be used for storage of optical data and security labeling (Gourevich *et al.*, 2004). The shell, that can be optically inert, provides mechanical properties to the film. Core–shell particles can be advantageously employed to create percolating structures of a non-film-forming polymer. Thus, electrical conductive films can be obtained from core–shell particles composed of a film-forming core and a conductive shell (Kahn and Armes, 2000). Core–shell particles are also used to protect the core from damage caused by the medium and for stimuli-responsive materials (Cayre, Chagneux and Biggs, 2011).

Hemispherical morphologies are formed when the two polymers have similar hydrophilicities and a relatively high polymer-polymer interfacial tension. Janus particles are hemispherical particles in which the two phases have a similar volume and different properties. The use of Janus particles as phase compatibilizers for blends has been proposed, but the advantages over the classical block copolymers are not evident. Janus particles have also been used as Pickering stabilizers (Tanaka *et al.*, 2010) and to study aggregation phenomena (Hong *et al.*, 2006).

Core–shell particles are used as an intermediate for the synthesis of hollow particles. The core is formed by a slightly crosslinked polymer containing ionisable groups (e.g., carboxylic acids) and the shell is a more hydrophobic polymer. The core is expanded by osmotic swelling of the functional groups with a suitable base at a temperature close to the glass transition temperature, T_g, of the polymer forming the shell. Upon drying, the water is evaporated, leaving a hollow particle. These materials are used as opacifiers in coatings (McDonald and Devon, 2002). Inorganic hollow particles can be synthesized by degrading the polymeric core of a core–shell particle having an inorganic shell (Caruso, Caruso and Möhwald, 1999a).

Table 6.1 presents a few cases of non-equilibrium morphologies. All of them result from the hindered movement of the clusters due to the high internal viscosity of the particles. Thus, occluded particles are formed when the second stage monomer polymerizes within the matrix formed by the first stage polymer and the viscosity of the matrix is high. These structures have found applications in toughening of polystyrene by forming a fine dispersion of rubber particles in the polystytene matrix, avoiding the aggregation of the rubber particles during polymer processing (Schneider, Pith and Lambla, 1996, 1997).

Raspberry and multilobed morphologies are non-equilibrium morphologies resulting from the high viscosity of the polymer forming the grains (lobes). This can be achieved by working under starved conditions and using polymers that are not soluble in their monomers (Huang and Lui, 2010). The raspberry-like particles are interesting due to the very large interfacial area that allows the collocation of specific chemical groups in a large quantity on the particle surface. Because multilobed particles have a larger hydrodynamic volume than spherical particles of the same volume, they are used as thickeners in the coating industry (Anderson and Daniels, 2003).

A mathematical model for development of the two-phase polymer–polymer particle morphology during the second stage of the emulsion polymerisation has been developed (González-Ortiz and Asua, 1995, 1996a, 1996b). The model is also applicable to miniemulsion polymerisation. The model accounts for phase separation leading to cluster nucleation that is assumed to be proportional to the excess of Polymer 2 in the Polymer 1-rich phase (matrix) with respect to saturation. The model also accounts for the polymerisation of Monomer 2 in clusters and the matrix as well as for the diffusion of polymers between the matrix and the clusters. A key component of the model is that it includes the motion of the clusters and the coagulation among them. The motion is described by means of the equation of the terminal velocity ($\mathrm{d}r/\mathrm{d}t$)

$$\frac{\mathrm{d}r}{\mathrm{d}t} = \frac{F}{\psi\,\eta} \tag{6.1}$$

Table 6.2 Polymerization variables that affect the internal viscosity of the particles.

Increase in the internal viscosity	Decrease in the internal viscosity
Low monomer concentration in the particle, i.e. feeding under starved-fed conditions	High swelling of the particles by the monomer(s) and, less often, by inert solvents
High molecular weight and/or crosslinking of the seed polymer	Low polymer molecular weight (substantial chain transfer to monomer and/or to a chain transfer agent)
High T_g of the swollen polymers	Reaction temperature higher than the T_g of the swollen polymers

where F is the net force acting on the cluster, ψ the friction factor acting on the cluster and η the viscosity of the Polymer 1-rich matrix. Although both van der Waals and Brownian forces act on the clusters, it was considered that the Brownian forces were negligible. The net force was then calculated as

$$F = -\nabla E \tag{6.2}$$

where E is the van der Waals energy of interaction. E can be estimated from the interfacial energies by using the relationships between the Hamaker constants and the interfacial tensions (Hunter, 1987).

The model was able to simulate well the evolution of the particle morphology during batch emulsion polymerisations of methyl methacrylate in polystyrene seeds, for which the experimentally observed particle morphology was reported in the literature (Jönsson *et al.*, 1991; Chen, Dimonie and El-Aasser, 1992a, 1992b). The model shows that the final particle morphology depends strongly on kinetic factors, and that the lower the particle viscosity and the slower the polymerisation rate, the closer the particle approaches to its equilibrium morphology. Some process variables that affect the viscosity of the particles are given in Table 6.2.

The non-equilibrium morphologies may evolve during storage if the T_gs of the phases are close to the storage temperature. It has been observed that during storage at room temperature for one year and a half of a latex of core–shell particles, in which the core is made of poly(vinyl acetate) and the shell of poly(butyl methacrylate), the particles evolved to the equilibrium morphology, with more hydrophilic poly(vinyl acetate) forming the shell (Zhao *et al.*, 2004). For coating applications, latexes are formulated before transportation and storage. A common practice to promote the film formation is the addition, during the latex formulation, of coalescing agents that plasticize the polymer particles and decrease the T_g. This procedure can reduce the time needed for the particles to reach the equilibrium morphology (Stubbs and Sundberg, 2011).

6.3.1 Effect of Grafting

The polymer–polymer and polymer–aqueous phase interfacial tensions play a key role in the development of the particle morphology as they strongly affect both the equilibrium morphology and cluster migration towards the equilibrium morphology. Modification of the polymer–polymer interfacial tension through grafting reactions is a way to tailor-make particle morphology (Rajatapiti, Dimonie and El-Aasser, 1995; Herrera *et al.*, 2006, 2007, 2010).

Thus, Rajatapiti *et al.* (Rajatapiti *et al.*, 1997) prepared a miniemulsion of butyl acrylate containing a methyl methacrylate macromonomer, which upon polymerisation led to the formation of the polybutyl acrylate seed with containing poly(butyl acrylate)-graft-poly(methyl methacrylate) copolymer. Polymerisation of methyl methacrylate on this seed led to composite PBA/PMMA particles. The authors showed that the morphology was strongly affected by the presence of the graft copolymer, which reduced the poly(butyl acrylate)–poly(methyl methacrylate) interfacial tension. For polystyrene–poly(methyl methacrylate) and polystyrene–poly(butyl acrylate), block copolymers produced *in situ* by controlled radical polymerisation (CRP) substantially modified the particle morphology (Herrera *et al.*, 2006, 2007, 2010). The strategy consisted in adding a small amount of a CRP agent during the formation of the seed so that some of the polymer chains were capped with the CRP agent at the end of the seed formation. Polymerisation of the second stage monomer in the presence of additional initiator led to the formation of some block copolymer chains that help to improve the compatibility between the two polymers. The practical importance of improving polymer–polymer compatibility by increasing the fraction of graft copolymer is evident in the case of waterborne polymer–polymer hybrids (e.g., alkyd–acrylic (Tsavalas, Gooch and Schork, 2000; Goikoetxea *et al.*, 2009), polyurethane–acrylic (Lopez *et al.*, 2011) and epoxy–acrylic (Iijima, Yochioka and Tomoi, 1992)) that have been developed in an attempt to combine the positive properties of both polymers, avoiding their drawbacks.

If the amount of graft copolymer is low, it is still possible to use the models developed for biphasic particles to predict the equilibrium particle morphology in the presence of graft and block copolymers (Herrera *et al.*, 2010). However, for high amounts of grafted copolymer, the system should be treated as a multiphase system (see below).

6.4 Two-Phase Polymer–Inorganic Particles

A broad range of waterborne dispersed particles can be synthesized by combining polymers and inorganic materials (Bourgeat-Lami and Lansalot, 2010). Inorganic materials are incorporated to improve mechanical and thermal properties (Negrete-Herrera *et al.*, 2006; Chen *et al.*, 2006; Diaconu *et al.*, 2007, 2009), to increase the barrier properties (Diaconu, Paulis and Leiza, 2008b) to provide hiding power (opacity) (Erdem *et al.*, 2000a, 2000b), to potentially block UV light (Erdem *et al.*, 2000c; Garnier *et al.*, 2012; Aguirre *et al.*, 2013) and to stabilize latexes (Binks, 2002; Pattamasattayasonthi *et al.*, 2011). Incorporation of quantum dots leads to a wide range of applications in areas like chemical sensors (Galian and Guardia, 2009), optoelectronics and photonics (LED) (Mattoussi *et al.*, 1998), solar cells (Gu *et al.*, 2005), photonic crystals (Woggon *et al.*, 2003)) and in a large number of biomedical applications (Bruchez *et al.*, 1998). Table 6.3 presents some examples of experimentally observed morphologies of polymer–inorganic particles.

Both emulsion and miniemulsion polymerisation can be used to place inorganic materials at the surface of the particles, whereas only miniemulsion polymerisation allows location of the inorganic material within the particles.

Table 6.3 *Illustrative examples of experimentally observed morphologies of two-phase polymer–inorganic particles.*

System	Method of synthesis	TEM image	Reference
PMMA particle armored with silica	Soap-free emulsion polymerization	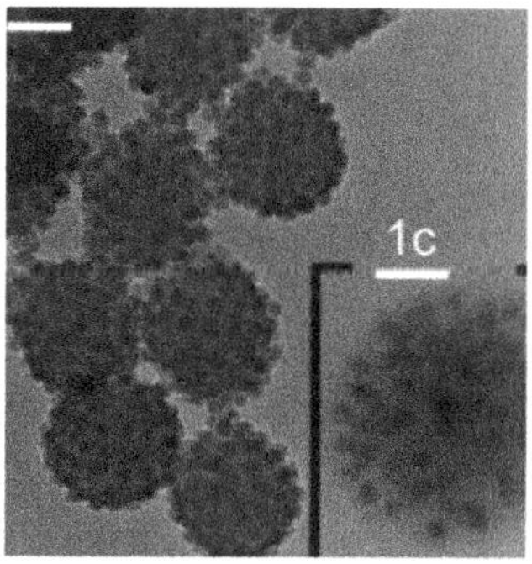 1c	Reprinted with permission from [Colver *et al.*, 2008] Copyright (2008) American Chemical Society
Magnetite-core/ PMMA-shell	Seed formed by organomodified magnetite, followed by monomer feeding	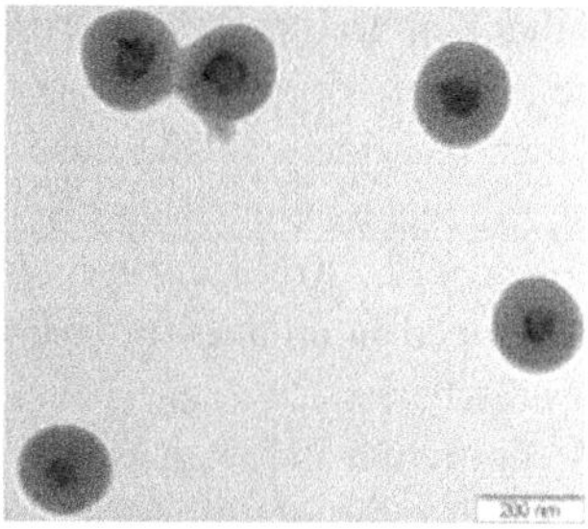 200 nm	Reprinted with permission from [Sacanna and Philipse, 2006] Copyright (2006) American Chemical Society
BA-co-MMA/organomodified Cloisite clay miniemulsion	Miniemulsion	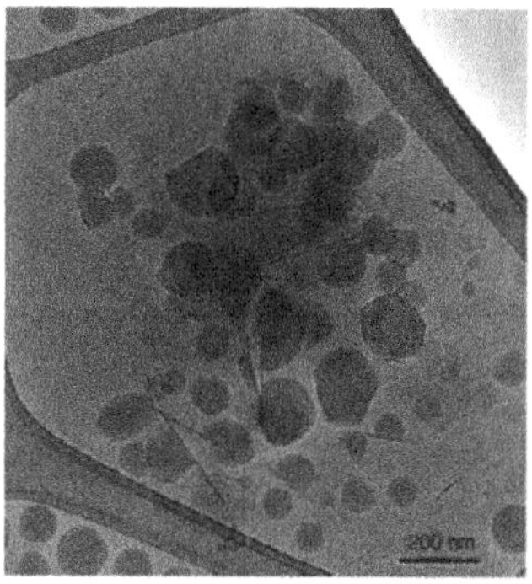 200 nm	Reprinted with permission from [Bonnefond *et al.*, 2012] Copyright (2012) Springer Science + Business Media
P(S-co-BA)-core/ laponite clay	Emulsion	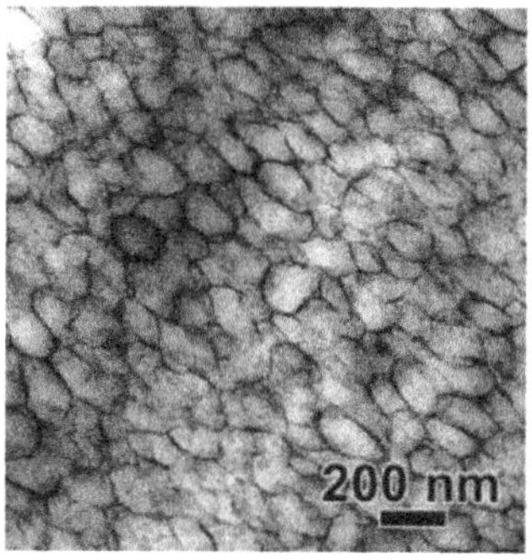 200 nm	Reprinted with permission [Negrete-Herrera *et al.*, 2007] Copyright (2007) Wiley-VCH

Table 6.3 *(Continued)*

System	Method of synthesis	TEM image	Reference
Anisotropic Fe-Silica/PS particles	Emulsion	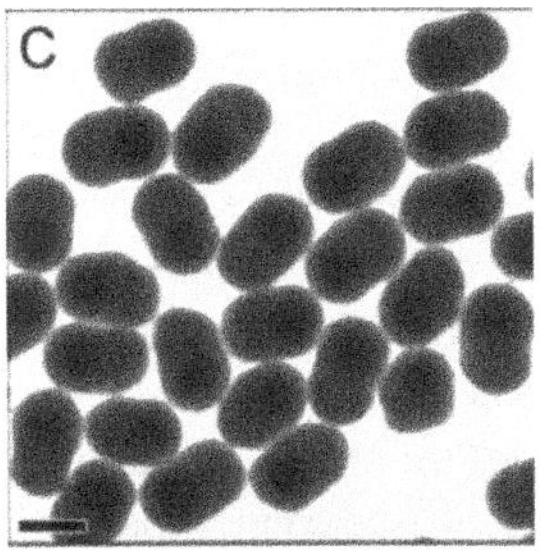	Reprinted with permission from [Ge *et al.*, 2007] Copyright (2007) American Chemical Society
Surfactant-free encapsulation of magnetite nanoparticles	Miniemulsion	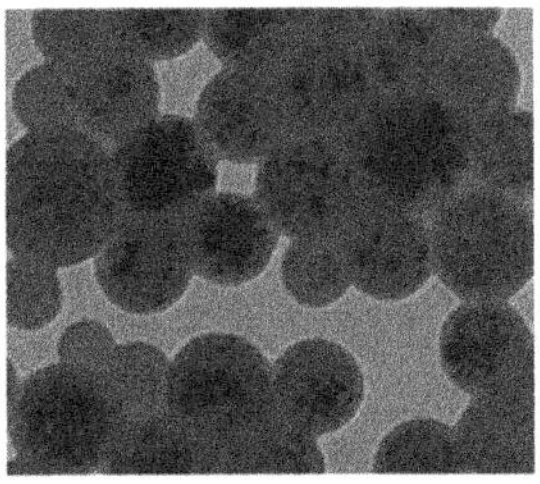	Reprinted with permission from [Ramos and Forcada, 2011] Copyright (2011) American Chemical Society
Asymetric super-paramagnetic nanoparticles	Miniemulsion	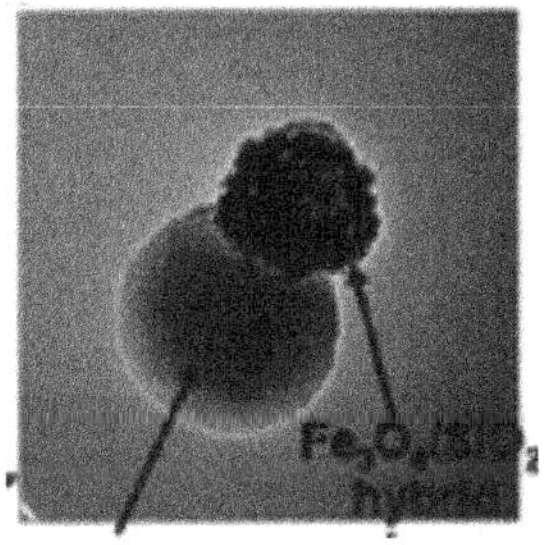	Reprinted with permission from [Wang *et al.*, 2011] Copyright (2011) American Chemical Society
Montmorillonite modified with cationic macromonomer	Miniemulsion	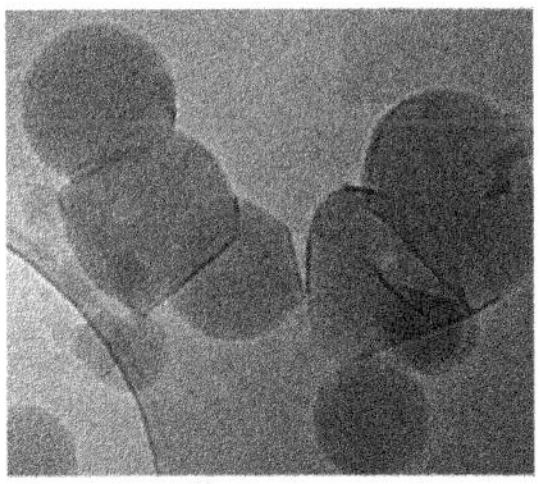	Reprinted with permission from [Bonnefond *et al.*, 2012] Copyright (2012) Springer Science + Business Media
Double modified Laponite	Miniemulsion	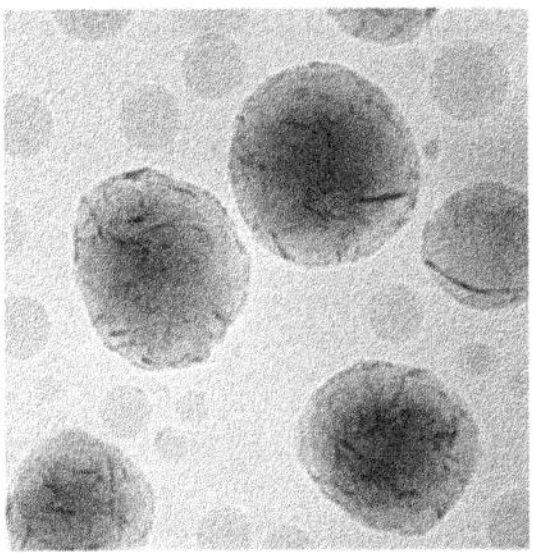	Reprinted from [Mellon, 2009] Copyright (2009) V. Mellon

Pickering (mini)emulsion polymerisation refers to processes in which no surfactant is used and the stability of the particles is provided by the inorganic material. The key aspect of these processes is to have a good adsorption of the inorganic material on the polymer particle. There are several ways to achieve this goal. Silica and clays, made more hydrophobic by lowering the pH, can stabilize particles in the emulsion polymerisation of hydrophilic monomers (MMA, VAc), but fail to stabilize hydrophobic monomers, such as styrene and vinyl pivalate (Colver, Colard and Bon, 2008; Colard, Teixera and Bon, 2010; Teixeira *et al.*, 2011). Salt has been used to enhance the adsorption of laponite on miniemulsion droplets/particles (Bon and Colver, 2007). Cationic monomers improve the adsorption of negatively charged silica (Tiarks, Landfester and Antonietti, 2001; Chen *et al.*, 2004). The combination of glycerol-functionalized silica and a cationic initiator allows high incorporation of the silica on the particles (Schmid, Tonnar and Armes, 2008; Schmid *et al.*, 2009).

Miniemulsion polymerisation is used to incorporate inorganic materials inside the polymer particles (Hu, Chen and Wu, 2011). The inorganic material should be surface modified in order to make it as hydrophobic as the monomer mixture. In addition, the modified in organic material has to be compatible with both the monomer mixture and its corresponding polymer. Compatibility with the polymer is often achieved by including reactive double bonds in the modified surface of the inorganic material (Diaconu *et al.*, 2009; Micusik *et al.*, 2010). For example, modification of montmorillonite clays with cationic surfactants and reactive cationic organomodifiers (like cationic macromonomer or macroinitiators) improves the hydrophobic character and the compatibilization of the modified clay with the monomer/polymer mixture. Hence, miniemulsion polymerisation carried out in the presence of this type of clay produces morphologies with the clay located at the particle surface (clay with low compatibility with the monomer and rather hydrophilic) (Diaconu, Paulis and Leiza, 2008a, 2008b), or with the clay embedded in a substantial amount between polymer particles, forming a dumb-bell like structure (Bonnefond *et al.*, 2012). Another important issue in the hybrid polymer/clay waterborne dispersions is the aspect ratio of the clay (Reyes, Paulis, Leiza, 2010). High aspect ratio clays, such as montmorillonte, can hardly be incorporated within the polymer particles, whereas smaller aspect ratio clays, such as laponite, have been successfully encapsulated (Voorn, Ming and van Herk, 2006). In the case when the clay (laponite) was double modified by using cationic surfactants to change the basal space and with reactive siloxanes at the edges, miniemulsion polymerisation of organic monomers resulted in morphologies in the majority of the clay platelets were exfoliated and located within the polymer particles (Mellon, 2009).

In the systems described above, particle morphology is also the result of the interplay between kinetics and thermodynamics. However, the predictions of the morphology are more complex than for two-phase polymer–polymer systems. Even the thermodynamic equilibrium morphologies are not well defined because the inorganic objects do not lose their identity to form a single object (see Table 6.3). Therefore, the interfacial surface area is poorly defined and the surface energy is hard to compute. For these systems, the methods developed for multiphase particle morphologies, that are discussed in the next section, are better suited.

6.5 Multiphase Systems

Multiphase systems may present advantages over two-phase systems. Thus, rubber–thermoplastic core–shell particles are useful to impart toughness to thermoplastic resins

(Sato and Tateyama, 1998). However, mouldability is poor if there is little thermoplastic in the shell. Mouldability is increased by growing the shell thickness, but in this case flexibility is sacrificed and elastic recovery decreases. Decreasing the glass transition temperature of the rubber to compensate flexibility leads to differences in refractive index and good transparency is not obtained. It has been reported that three-phase polymer particles containing at least two inner rubber phases and a thermoplastic phase in the outermost of the particles give a moulded product having excellent elastic properties and transparency, while maintaining a good mouldability (Naruse *et al.*, 2001). The properties of alkyd-acrylics (Tsavalas, Schork and Landfester, 2004; Goikoetxea *et al.*, 2009), PU-acrylic (Li, Chiu and Don, 2007) and epoxy-acrylics (Kawahara *et al.*, 2001) hybrid polymers are substantially improved by generating large amounts of graft copolymer, that is, a third phase. Three-phase systems formed by a rubber core including occluded polystyrene particles and a crosslinked poly(methyl methacrylate) shell are used to increase the toughening, the elongation at break and the Notched Izod impact resistance of a polystyrene matrix (Schneider, Pith and Lambla, 1997). Three-phase particles can also improve the damping properties, without losing the toughening increment of the polymer matrix (El-Aasser *et al.*, 1999). Reinforcement of three-phase polymer particles with inorganic materials results in a four-phase system.

The morphology of multiphase particles also results from the interplay between thermodynamics and kinetics. Thermodynamic equilibrium morphologies are obtained for systems with low viscosity particles (Goikoetxea *et al.*, 2009) and for long process times.

The prediction of the equilibrium morphologies can be conveniently performed using canonical Monte Carlo simulations (Reyes and Asua, 2010). For this purpose, the system is described in a coarse-grained way by a collection of spheres, each of them representing a fraction of a given phase. The number of spheres determines the detail in which the system is described. Because of the relatively small size of the waterborne dispersed polymers (often < 200 nm) it is computationally feasible to choose the number of spheres so that each of them represents one individual polymer chain. The water surrounding the particle is also represented by a collection of spheres. The equilibrium morphology is governed by the interaction between the spheres that is described by interaction potentials. Different interaction potentials (e.g., repulsive and Lennard-Jones) can be used and attention should be paid to relating the interaction potentials to measurable macroscopic properties, in particular, to surface and interfacial tensions. At the beginning of the simulation, the spheres representing the phases of the particle are randomly distributed and are surrounded by water spheres. Then, the system is allowed to equilibrate for a high enough number of Monte Carlo steps following the Metropolis algorithm (Metropolis *et al.*, 1953). In each step, a sphere is chosen at random and moved in a random direction and magnitude, and the energy of the new configuration is calculated to decide if this configuration is retained. There is no limit to the number of phases composing the particle and both polymer–polymer and polymer–inorganic systems can be handled. Figure 6.2 presents the morphologies calculated for three-phase particles.

The dynamic evolution of the particle morphology for multiphase waterborne systems can be simulated using stochastic dynamics. As in Monte Carlo simulations, the simulated system consists of spheres representing fractions of phases and water (Akhmatskaya and Asua, 2012, 2013). The movement of each of these spheres is described by the Langevin equation that states that the force acting on the sphere (left-hand side of Equation 6.3) results from the interaction with the other spheres, the viscous drag and the thermal energy

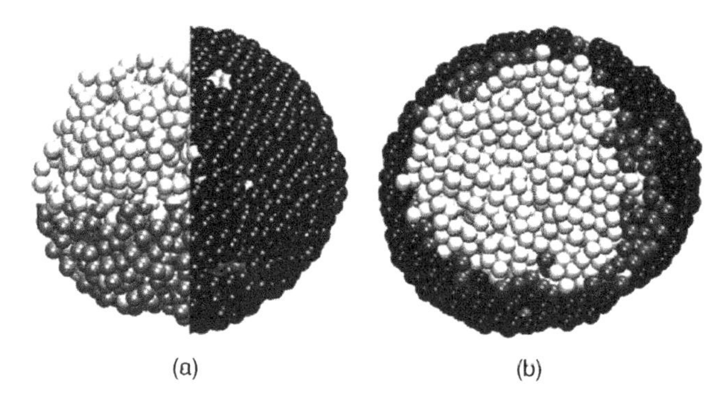

Figure 6.2 *Equilibrium morphologies of a three-phase system; Polymer 1 (in black) is less hydrophobic than Polymer 2 (in white), and graft Polymer 3 (in grey), in a volume ratio 0.4/0.4/0.2, respectively. (a) Visualization of the 3D structure cut at different layers and (b) cross-sectional view of the particle. Reprinted with permission from [Reyes and Asua, 2010] Copyright (2010) John Wiley and Sons, Inc.*

(respectively, the three terms on the right-hand side of Equation 6.3)

$$m\frac{\mathrm{d}^2 r}{\mathrm{d}t^2} = \nabla U - \psi m\frac{\mathrm{d}r}{\mathrm{d}t} + \sqrt{2\psi\, k_{\mathrm{B}} T\, m} R(t) \tag{6.3}$$

where m is the mass of the sphere, U is the interaction potential acting on the sphere, ψ is the friction factor, k_{B} is the Boltzmann constant, T is the temperature and $R(t)$ a random force. In comparison with the model developed by González-Ortiz (González-Ortíz and Asua, 1995, 1996a, 1996b) (Equation 6.1) this simulation approach removes the limitation of the terminal velocity and includes the effect of the Brownian motion.

The interaction between dissimilar phases (e.g., polymer–polymer, polymer–inorganic, polymer–water, monomer–inorganic) is modeled using a repulsive generalized soft sphere potential (Akhmatskaya and Asua, 2013)

$$U_r(r_{ij}) = \varepsilon_{ij}\left(\frac{\sigma}{r_{ij}}\right)^6 \tag{6.4}$$

where ε_{ij} is the interaction strength, σ is the size of the sphere and r_{ij} the distance between the spheres. A repulsive potential can also be used for inorganic–inorganic interactions because, in practice, the surface of the inorganic particles is modified to avoid aggregation (Akhmatskaya and Asua, 2013). An incomplete modification of the surface of the inorganic particles would lead to a lower repulsion potential or even to an attractive potential.

A 6–12 Lennard-Jones interaction potential is used to model the interaction between miscible phases. The parameters of the Lennard-Jones and repulsive potentials can be estimated from the surface and interfacial tensions using approaches described in the literature (Gloor *et al.*, 2005; López-Redón, Reyes and Orea, 2006).

The friction factor, ψ, relates to the internal viscosity of the particle that controls the rate at which the phases move within the particle, and it increases with the internal viscosity of the particle. High values of ψ correspond to high internal viscosity of the particle. This leads to a slow phase migration, which, for fast polymerisation rates, may result in a system in which the newly formed polymer remains in the place where it is formed. On the other hand, low values of ψ correspond to low particle viscosity that may lead more easily to the equilibrium morphologies. Because the viscosity increases as $\eta \sim \phi^5$ (ϕ being the volume fraction of polymer (van Krevelen, 1992)), the following equation has been used for the friction factor (Akhmatskaya and Asua, 2013)

$$\psi = \psi_0 \,(0.5 + 0.5X)^5 \tag{6.5}$$

where X is the fractional conversion of monomer and ψ_0 is the value of ψ at $X = 1$. The actual value of ψ_0 depends on the particular system considered, and can be related to the diffusion coefficient (Df) as

$$\psi_0 = \frac{k_B T}{m\, Df} \tag{6.6}$$

The diffusion coefficient can be estimated during film formation from latexes using fluorescence resonance energy transfer (FRET) methods (Oh *et al.*, 2003; Wu *et al.*, 2004a; Farinha *et al.*, 2005; Tomba *et al.*, 2009).

Polymerisation is simulated by converting the monomer spheres into spheres of a new polymeric phase, namely, a new type of spheres appeared in the system, but the total number of simulated spheres was constant during the simulation process. Polymerisation kinetics was accounted for by means of the rate of transformation of monomer spheres into spheres of polymer. The transformation of the monomer into the new polymer was achieved by gradually changing the values of the interaction potentials from those of the monomer to the characteristic values of the new polymer (Akhmatskaya and Asua, 2012, 2013).

The type of initiator affects the evolution of the particle morphology. When oil-soluble initiators are used, there is a uniform distribution of radicals within the polymer particles and the monomer spheres that are transformed into polymer are chosen randomly. If water-soluble initiators are used, the distribution of radicals may not be uniform because of the anchoring effect of the hydrophilic part of the entering radical to the surface of the particle. The distribution of radicals in the particle may be calculated using the available methods (Chern and Poehlein, 1987; de la Cal, Adams and Asua, 1990a; de la Cal *et al.*, 1990b) and the monomer spheres that are going to undergo polymerisation in a given time interval are chosen according to the radical concentration profile. More specifically, the monomer subparticles that are near to the surface of the particle have a higher probability of reaction.

The simulations can be carried out using the modern implementation of the Langevin dynamics on high performance computers, such as the free software GROMACS (Hess *et al.*, 2008). Ways to accurately integrate the Langevin equations over the long periods of time required for the polymerisation, have been proposed (Akhmatskaya and Asua, 2012, 2013).

Figure 6.3 presents the evolution of the particle morphology in a miniemulsion polymerisation aiming at producing a polymer–polymer–inorganic hybrid using an oil-soluble initiator. Figure 6.3a shows the morphology of the initial particle. Monomer 2 and Polymer 1 are uniformly distributed in the particle, which corresponds to a system in which Polymer 1 is soluble in Monomer 2. The inorganic particles are placed at the surface of the particle. Polymerisation of Monomer 2 (dull gray) leads to the formation of Polymer 2 (bright gray) that is not compatible with Polymer 1 (bright black). Therefore, phase separation occurs and the more hydrophobic polymer (Polymer 1) accumulates in the core of the particle. At the end of the polymerisation (Figure 6.3f), the particle presents a non-equilibrium

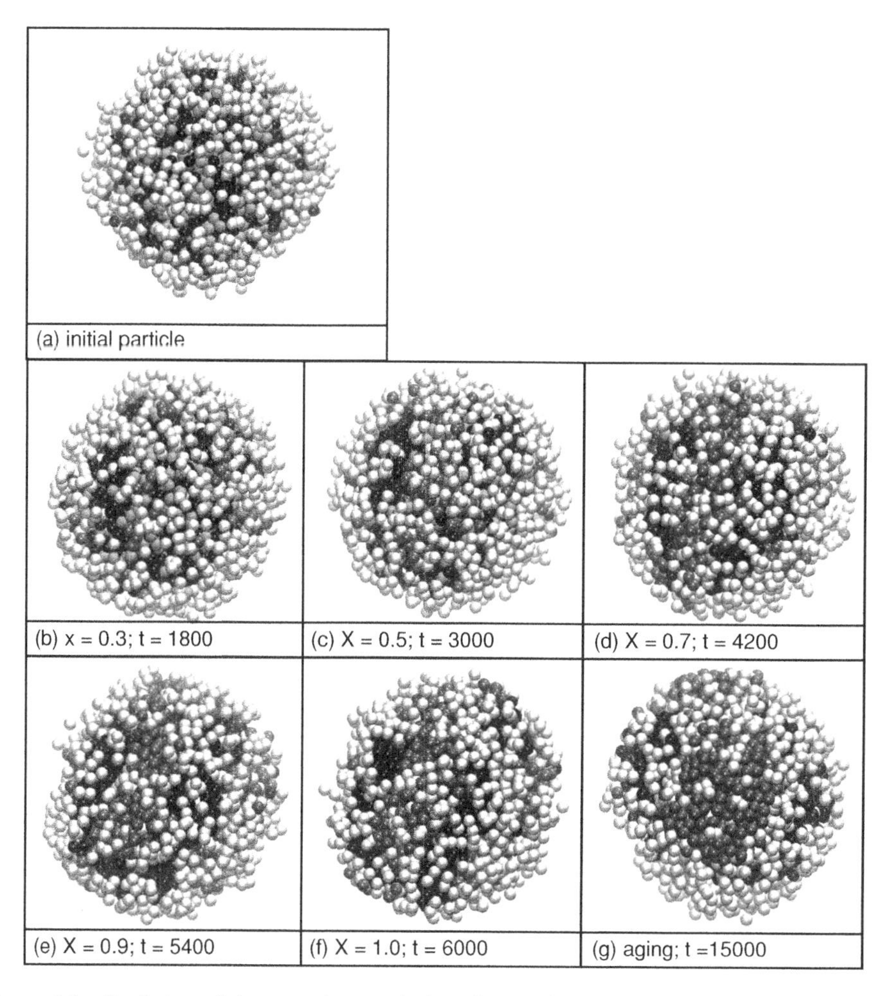

Figure 6.3 *Evolution of the particle morphology for a polymer–polymer–inorganic hybrid using an oil-soluble initiator. Legend: Polymer 1 (bright black); Monomer 2 (dull grey); Polymer 2 (bright grey); inorganic (white); X is the Monomer 2 conversion; t is the time in reduced units. The spheres of the water particles are not shown.*

morphology. Because of the high internal viscosity, particle morphology remains almost unchanged upon aging (Figure 6.3g).

Figure 6.4 shows the effect of the type of initiator (oil-soluble vs water-soluble) on the evolution of the particle morphology. It can be seen that when a water-soluble initiator is used, Polymer 2 accumulates in the region in which it is formed, namely, near the surface of the particle.

The method outlined above can also be used to simulate the effect of the formation of grafting on the evolution of the particle morphology for a polymer–polymer system. Figure 6.5 shows an example in which 20% of graft copolymer of composition 50/50 (Polymer 1/Polymer 2) is formed. It can be seen that, in the absence of grafting, the

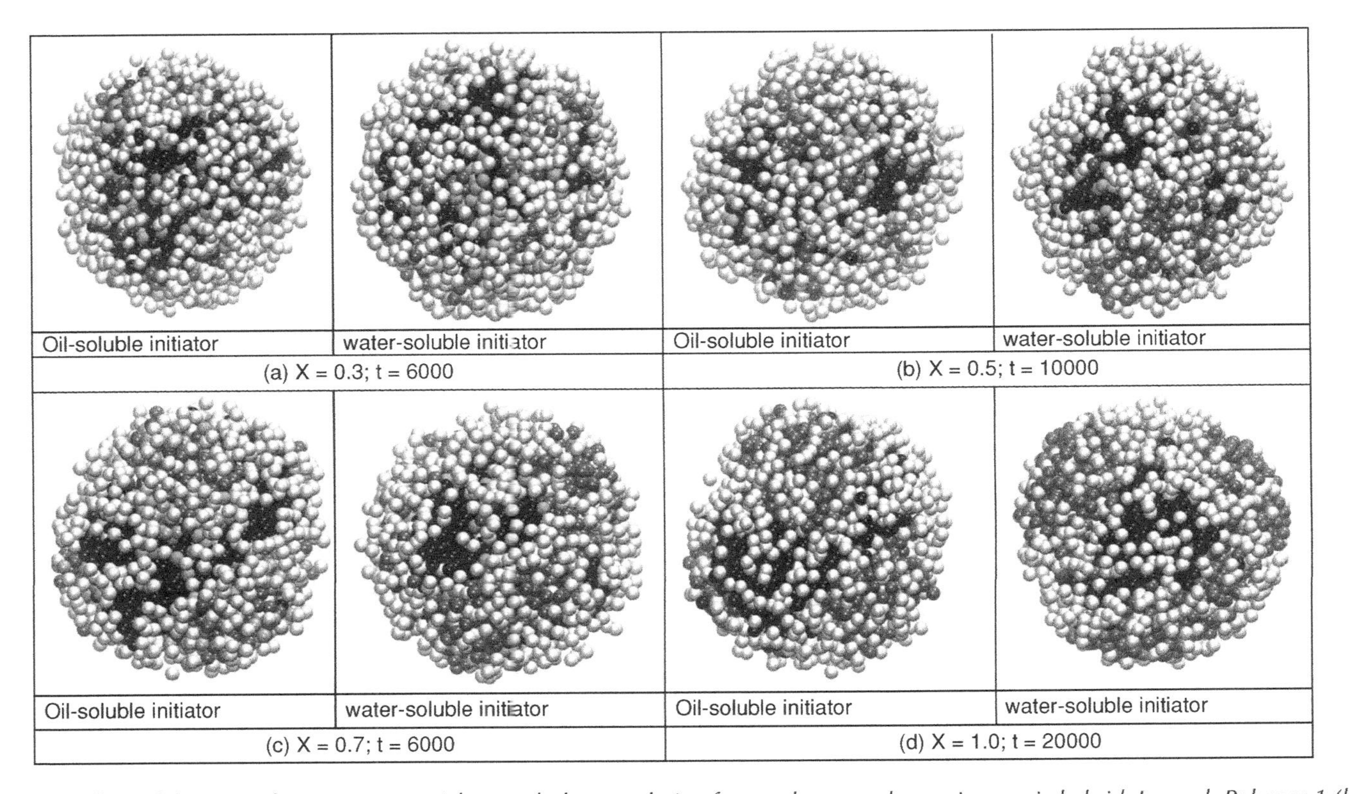

Figure 6.4 *Effect of the type of initiator on particle morphology evolution for a polymer–polymer–inorganic hybrid. Legend: Polymer 1 (bright black); Monomer 2 (dull gray); Polymer 2 (bright gray); Inorganic (white); X is the Monomer 2 conversion; t is the time in reduced units.*

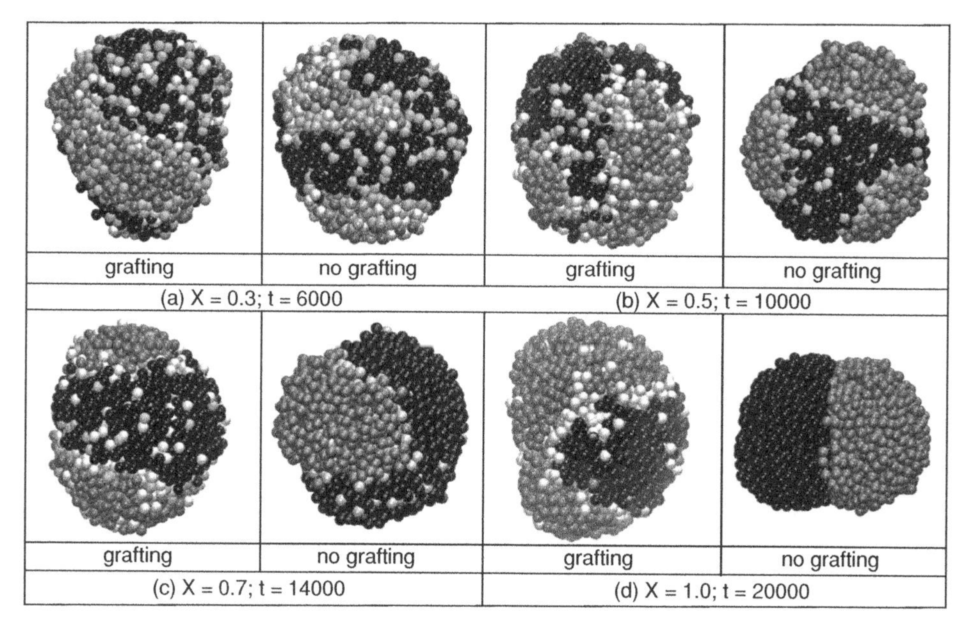

Figure 6.5 *Effect of the formation of grafting on the evolution of the particle morphology for a polymer–polymer system. Legend: Polymer 1 (bright black); Monomer 2 (dull gray); Polymer 2 (bright gray); graft polymer (white);* X *is the conversion rate;* t *is the time in reduced units.*

particle morphology evolves towards a hemispherical morphology, whereas in the case in which graft copolymer is formed, the morphology approaches core–shell morphology. This shows the potential of the modification of the polymer–polymer interfacial tension for tailor-made particle–particle morphology and the predictions closely reproduced the experimental results obtained for polystyrene–poly(methyl methacrylate) containing block copolymers produced *in situ* by means of controlled free radical polymerisation (Herrera *et al.*, 2007).

6.6 Effect of Particle Morphology on Film Morphology

Synthetic latexes are mainly used in applications (e.g., paints, adhesives, paper and coatings) that require the formation of a film (Asua, 1997). Film formation is described as consisting of three main processes (Keddie, 1997): (i) evaporation of water to achieve the close-packing of particles, (ii) deformation of particles to fill all the void space, and (iii) interdiffusion of the polymer across particle interfaces to fuse particle boundaries. The properties of the film depend strongly on the film morphology, which is determined by the particle morphology. This is illustrated in Figure 6.6 adapted from the work of Goikoetxea *et al.* (Goikoetxea *et al.*, 2012). These authors synthesized alkyd–acrylic waterborne hybrids with different particle morphologies by varying the formulation and polymerisation conditions. Because of the low internal viscosity of the polymer particles, equilibrium particle morphologies were obtained at the end of the process. The equilibrium morphologies were strongly affected by both the hydrophilicity of the alkyd used and the fraction of graft copolymer

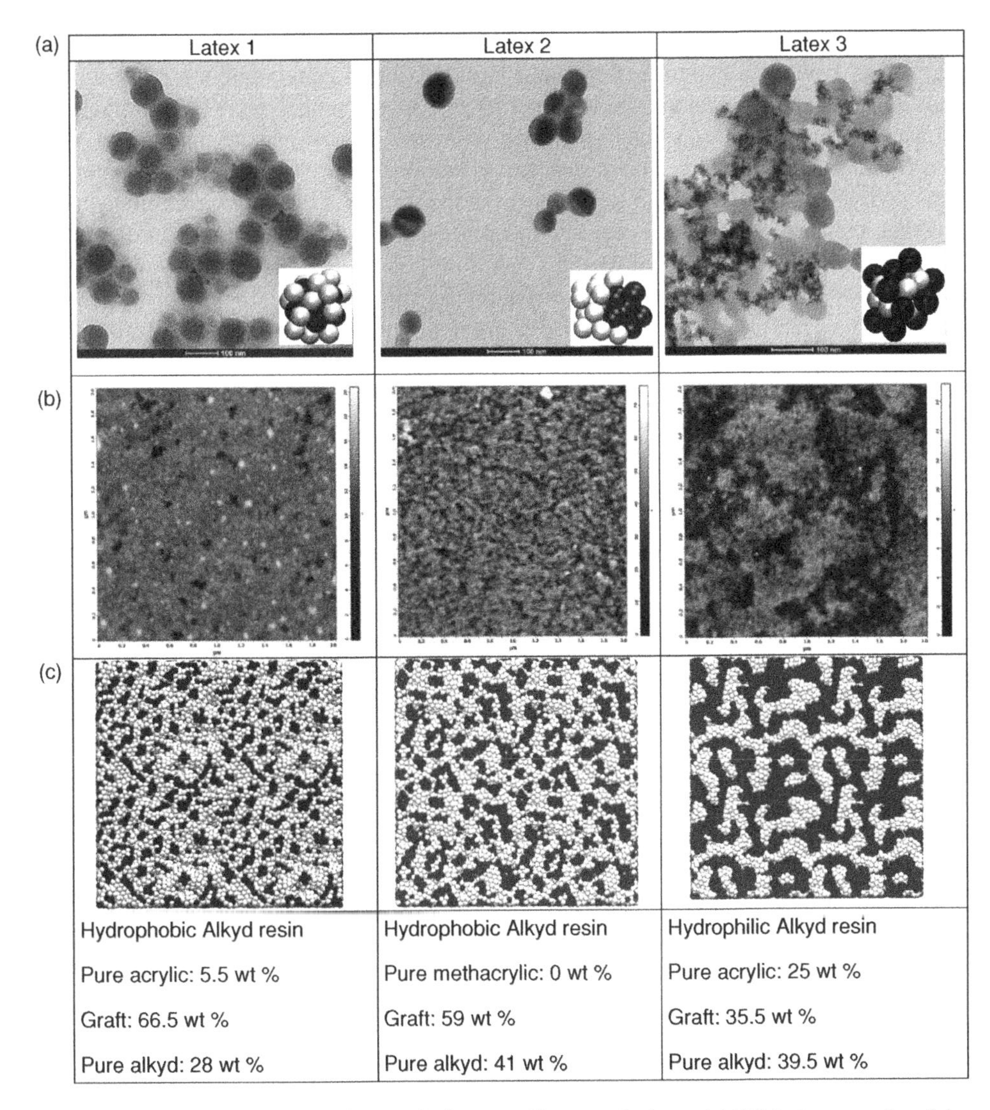

Figure 6.6 *Effect of the particle morphology on film morphology. (a) TEM photographs of the polymer particles; dark regions correspond to the alkyd-rich phase and bright regions to the acrylic-rich phase. The inserts within the TEM images are the particles used in the simulation of the film morphology (see Section 6.6.1). (b) AFM phase images (2 × 2 μm^2); the dark zones are related to the alkyd resin and the lighter zones are related to the acrylic-rich phase. (c) The morphology of the film/air interface obtained by simulation; the alkyd and acrylic rich phases are represented in black and white, respectively.*

formed during the polymerisation. Referring to Figure 6.6, Latex 1 has a core–shell particle morphology with the alkyd in the core and a high fraction of graft copolymer (66.5 wt%). Latex 2 has a hemispherical morphology and 59 wt% of graft copolymer. The morphology of Latex 3 is not apparent in Figure 6.6 because the low viscosity hydrophilic alkyd, that was outside the particle, spread upon sample preparation. Latex 3 has the lowest fraction of graft copolymer (35.5 wt%). The AFM images of the surface of the films cast from hybrids show that the particle boundaries are not apparent, which indicates that the particles have coalesced. In the phase image of the film cast from Latex 1 there are small domains that appear darker in the images and can reasonably be attributed to the alkyd phase. Latex 1 particles have an alkyd-rich core and it is possible that the AFM tip is penetrating the acrylic shell and detecting the alkyd core, but the possibility of some alkyd leaking from the particles cannot be ruled out. A higher fraction of the darker domains in the phase images is observed at the surface of Latex 2. Latex 3 shows an even higher fraction of the darker phase, which is likewise interpreted as being composed of alkyd. For Latexes 2 and 3, the mean size of these alkyd domains is much greater than that of the polymer particles (135 and 88 nm, respectively), which means that the free alkyd resin that is out and/or partially surrounding the particles has migrated to create larger domains. The alkyd aggregates formed in the Latex 3 film are larger than those formed with Latex 2.

Figure 6.6 shows that the morphology of the films depends on the polymer microstructure and morphology of the particles, and that there is a phase separation and migration to form domains that are larger than the size of the phases within the original composite particles.

The moment at which the phase migration occurs is not well defined, but one may speculate that, because the morphology of the waterborne hybrid particles is at thermodynamic equilibrium, they are not subjected to any significant driving force for phase migration while they are surrounded by water. Therefore, it seems reasonable to assume that phase migration occurs after water evaporation has been completed. In the absence of water, the driving force for phase migration is a minimization of the surface energy associated with the interfaces between the different polymer phases. Migration is favoured by the presence of low molecular weight alkyd resin.

The hybrids presented in Figure 6.6 contained up to three phases: alkyd resin, grafted copolymer and acrylic polymer. The higher the graft content, the better the compatibility between the acrylic polymer and the alkyd resin and, consequently, the lower the driving force for phase migration. Latex 1 had the higher fraction of graft copolymer and the alkyd resin occupied the core of the polymer particles that likely offered some resistance to alkyd migration. The combination of these effects resulted in a moderate phase separation. On the other hand, Latex 3 contained a lower fraction of grafted copolymer, a higher amount of the more mobile phase (alkyd resin) and the resin was located at the exterior of the particle. This led to a strong phase separation in the film cast from Latex 3. The properties of Latex 2 in terms of polymer microstructure and particle morphology were intermediate between Latexes 1 and 3 and, hence, the extent of phase separation was also intermediate.

The results discussed above show that during film formation, phase migration leading to the formation of aggregates increases when (i) the fraction of low molecular weight polymer (free alkyd resin) increases, (ii) the compatibility between the phases decreases (smaller fraction of grafted copolymer), and (iii) when the more mobile phase (alkyd resin) is located at the outer part of the particles. However, the relative importance of the contributions of the polymer characteristics (extent of grafting) and particle morphology on phase migration cannot be elucidated from these results.

6.6.1 Modelling Film Morphology

Mathematical modelling may help in understanding the relative importance of the polymer characteristics and the particle morphology on the film morphology, and also the application of this knowledge to other systems. A standard coarse-grained canonical Monte Carlo model aimed at predicting the effect of particle morphology and polymer properties on the morphology of the film has been developed (Goikoetxea *et al.*, 2012).

The model assumes that phase migration occurs after completion of water evaporation. At this moment, the film was considered to be formed by particles randomly distributed in a dense packing. Although the particles may contain three components (Polymer 1, grafted copolymer and Polymer 2), only two phases were considered in the model: one rich in Polymer 1 and the other rich in Polymer 2, the graft copolymer being distributed between these phases. Each composite particle was considered to be composed of a number of subparticles, each of them either Polymer 1 or Polymer 2. The number of subparticles used depends on the affordable computer time. Figure 6.6 shows an example in which the alkyd–acrylic particles of Figure 6.6 are represented by 25 subparticles; the black subparticles standing for the alkyd phase while the white ones correspond to the acrylic-containing polymer.

The initial configuration of the film is obtained by randomly distributing the hybrid particles (each of them composed of 25 subparticles) within a parallelepiped simulation cell without overlapping and maintaining the particle morphology. Once the initial configuration is obtained, the subparticles that formed the polymer particle are allowed to move individually according to the Metropolis algorithm (Metropolis *et al.*, 1953), which drives the system towards its equilibrium state. In the simulation presented in Figure 6.6, the interaction among subparticles was described by a Lennard-Jones potential. Repulsive potentials were used for the interaction of the subparticles with both the substrate and the air. Figure 6.6 presents a comparison of the simulated morphologies of the film/air interface (after 5×10^6 Monte Carlo steps) with the morphologies determined experimentally by AFM. It can be seen that the simulation predicts well the trends observed experimentally, that is, the size of the aggregates increased from Latex 1 to Latex 2 and Latex 3. Furthermore, Figure 6.6 shows that the final morphologies of the films (both experimental and simulated) correspond to metastable morphologies (in the simulation, increasing the number of Monte Carlo steps did not result in a significant change in the film morphology).

The model has also been used to gain some understanding about the relative importance of the polymer characteristics (extent of grafting) and the particle morphology on the film morphology. It was found that the effect of the phase compatibility on film morphology was stronger than that of the particle morphology (Goikoetxea *et al.*, 2012).

Acknowledgements

The financial support of Diputación Foral de Gipuzkoa, Basque Government (GV IT373-10), and Ministerio de Economía y Competitividad (Ref. CTQ2011-25572) as well as the support of the computing infrastructure of the i2BASQUE academic network are gratefully acknowledged for their financial support.

7

Colloidal Aspects of Emulsion Polymerisation

Brian Vincent
School of Chemistry, University of Bristol, UK

7.1 Introduction

Emulsion polymerisation involves the formation of polymer (latex) particles in the colloidal size range, namely a few 10s of nm to a few μm. As such they need to be stable to aggregation, both during formation and also during storage. There are, of course, some applications, where coagulation of the latex particles is a required step, for example, in the manufacture of rubber gloves, where film-forming latex particles are deposited on a steel former (e.g. by electrophoretic deposition) and then coagulated (and coalesced) in the presence of added electrolyte. In this chapter we will consider a number of questions pertinent to these issues. These include: (i) what determines the stability/aggregation behaviour of latex particles; (ii) what are the origins of the various interparticle forces; (iii) what is the nature of weak, reversible flocculation and colloidal phase separation; (iv) how may aggregate structure and strength be determined?

Before proceeding, a few general comments, mostly of a semantic nature, need to be made. The first point concerns the use of the word "stability", since this is a somewhat ubiquitous word in colloid science. It may be used in the context of several different types of breakdown processes. These different processes are related to different types of forces operating in a colloidal dispersion; these forces are listed in Table 7.1.

In this chapter we will only be discussing *aggregation* processes, and the various types of *interparticle* forces involved. Clearly, in some contexts, other types of forces and processes do need to be considered, for example, the sedimentation of large latex particles (in general, diameters >1 μm) may be a problem, on the timescale of hours or so. Latex

Chemistry and Technology of Emulsion Polymerisation, Second Edition. Edited by A.M. van Herk.

Table 7.1 *Breakdown processes in colloidal dispersions.*

Type of force	Type of breakdown process
external (e.g. gravity, centrifugal, electrostatic, magnetic)	particle migration (e.g. sedimentation or creaming), leading to a non-uniform concentration distribution.
inter-particle (repulsive or attractive)	aggregation
interfacial tension (arising from the imbalance of *intermolecular* forces at an interface)	coalescence or sintering; Ostwald ripening

particle coalescence has already been referred to and is an important feature of polymer film formation, but this topic is not directly pertinent to this chapter. Ostwald ripening, which is the process by which larger particles grow at the expense of smaller particles, through molecular transfer through the dispersion medium, is rarely encountered in latex dispersions, simply because of the very low solubility, in general, of polymer chains in the continuous phase.

Secondly, the word "aggregation" will be used as the generic word for the process by which particles come together in temporary or permanent contact. If the contact is temporary (i.e. the aggregation is *reversible*) then the word "flocculation" is used. If the contact is permanent (i.e. the aggregation is *irreversible*), then the word "coagulation" is used.

7.2 The Stabilisation of Colloidal Particles against Aggregation

The quantitative foundations of the theory of this subject were laid down in the so-called "Derjaguin–Landau–Verwey–Overbeek" or "DLVO" theory of colloid stability, following the pioneering work in Russia by Derjaguin and Landau (1941) and in Holland by Verwey and Overbeek (1948). The basic premise of this approach is that the stability of a (dilute) colloidal dispersion to aggregation can be described in terms of the pair-potential between the particles. The pair-potential is the potential energy (V) of any two mutually approaching particles, as a function of their separation (h). In the DLVO theory $V(h)$ is made up of two, supposedly additive contributions, an attractive part deriving from the van der Waals attraction between the particles, and a repulsive part, deriving from the overlap of the diffuse parts of the electrical double layers surrounding each of the (charged) particles. These two contributions will be discussed in more detail later. $V(h)$ may be directly related to the force, as a function of distance, $F(h)$, through Equation 7.1.

$$F(h) = -\frac{\mathrm{d}V(h)}{\mathrm{d}h} \tag{7.1}$$

Two common forms for $V(h)$ are illustrated schematically in Figure 7.1. They lead to two very different types of behaviour, as will be seen.

Figure 7.1a represents a typical DLVO total pair-potential (see Figure 7.4 later for an actual example). It has two principle features: (i) a maximum (V_{max}) at $h > 0$ and (ii) a minimum (V_{min}) at $h = 0$, that is, at particle contact. In this case the stability of the dispersion depends on the magnitude of V_{max}. If V_{max} is sufficiently large then the dispersion will be

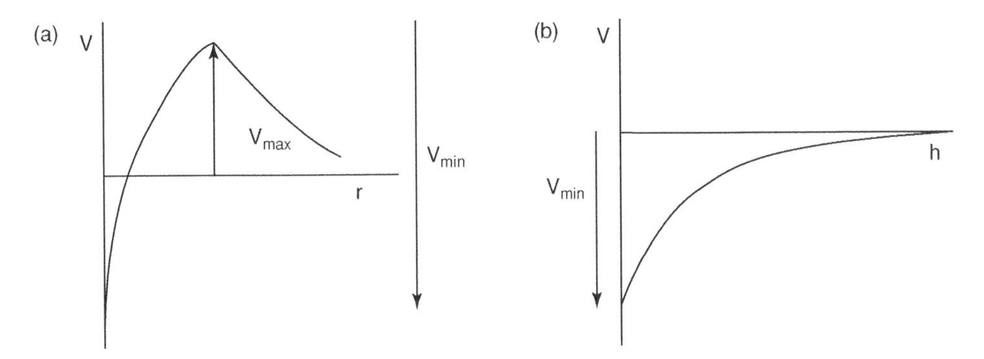

Figure 7.1 *Schematic of the pair-potential (V–h) for two approaching particles: (a) with an energy barrier (V_{max}) present; (b) with only a energy well (V_{min}) present.*

kinetically stable to aggregation (see Table 7.3 later). For lower values of V_{max}, where slow aggregation is observed, then such aggregation is said to be of the "reaction-limited" type, by analogy to chemical reaction kinetics. In general, V_{min} will be sufficiently large that the aggregation process will be irreversible, that is, *coagulation* is occurring. The corresponding aggregates (*coaguli*) which form will, in general, be large, strong and open-networked.

In Figure 7.1b there is no energy barrier; the only significant feature now is V_{min}. The stability now depends on just how large V_{min} is. This is a delicate question, which will be discussed in much greater detail later (see Section 7.4). Suffice to say here that, if V_{min} is sufficiently small, then the dispersion will be *thermodynamically* stable. Beyond some critical minimum value of V_{min} weak, reversible aggregation (i.e. *flocculation*) will occur. This may well lead eventually to colloidal phase separation, that is, to two co-existing colloidal phases, where one phase contains the weakly flocculated particles, and the second phase contains singlet particles. A close, molecular analogy would be the equilibrium between a liquid phase and its vapour. However, if V_{min} becomes sufficiently large, such that de-aggregation becomes unlikely, then the aggregation process again becomes essentially irreversible, and coagulation will now occur. The aggregation process is now said to be *diffusion-limited*, since the aggregation rate is governed solely by the mutual diffusion process of the two approaching particles. The boundary between reversibility and irreversibility cannot strictly be defined. Suffice to say that if, for example, $V_{min} < 5\ kT$ then the process is reversible; on the other hand, if $V_{min} > 10\ kT$ then the process may be considered irreversible (kT may be considered to be the thermal or kinetic energy of the two interacting particles).

In order to understand these two extreme forms of aggregation behaviour, which have their roots in the two types of pair-potential illustrated in Figure 7.1, one needs to understand more fully how these different forms for $V(h)$ originate. This requires a more detailed analysis of the various types of inter-particle forces which contribute to $F(h)$, and hence $V(h)$; this is given in Section 7.3 below. However, before proceeding, it should be pointed out that the DLVO theory is based, as mentioned earlier, on the premise that pairwise (two-body) interactions control the stability / aggregation behaviour of a dispersion. This is only the case for *dilute* dispersions. For *concentrated* dispersions, the situation is much more complex; multi-body interactions now become important. Again the boundary between *dilute* and *concentrated* in this context cannot be defined exactly. A working hypothesis might be to say that a dispersion becomes "concentrated" when the time between particle collisions becomes sufficiently small that it approaches the time-period when two particles can be said to be interacting, that is, $V(h) \neq 0$, that is, typically $\sim$ µs. In general, the "boundary" between dilute and concentrated will be somewhere in the range of particle volume fractions (ϕ), say $\phi = 0.01$ to 0.1.

7.3 Pair-Potentials in Colloidal Dispersions

Figure 7.2 represents two approaching, spherical colloidal particles. Each particle has a core (i.e. the actual particle, radius, a), surrounded by a sheath (thickness, δ). The sheath represents that region of the surrounding fluid which is structurally different from the bulk medium. All particles have such a region, due to the influence of the core particles on the surrounding medium. Exactly how these structural differences arise will be discussed later.

One may divide the interactions between two particles into two broad categories. The first category comprises those interactions which are due to the particle *cores*; they are generally attractive and exist even when $h > 2\delta$. The second category only occurs when $h < 2\delta$, and results from the overlap of the two sheaths around the particles, which necessarily leads to perturbations of these structurally different regions, especially in the overlap zone. The exact nature of these so-called "structural" interactions will also be discussed later. They may be repulsive or attractive.

7.3.1 Core–Core Interactions

The first category of interactions (i.e. between the cores) includes the van der Waals forces, in particular the dispersion forces. The simplest form for the van der Waals interaction, V_A (Hamaker, 1937) is given by Equation 7.2],

$$V_A = -\frac{\left(A_p^{1/2} - A_m^{1/2}\right)^2 a}{12h} \tag{7.2}$$

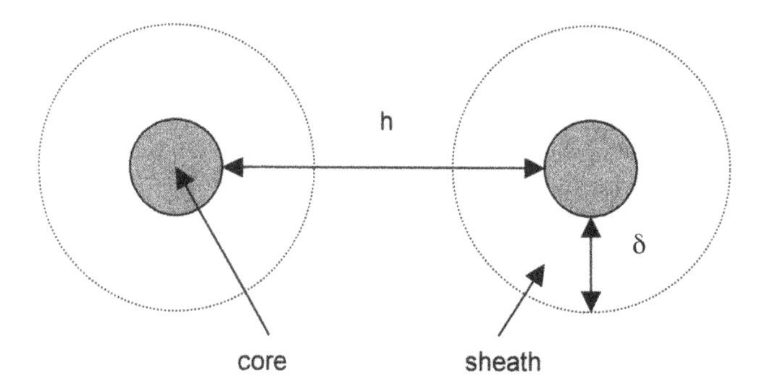

Figure 7.2 Schematic representation of the interaction between two colloidal particles.

Table 7.2 *Some typical values for A_p and A_m for various polymers and liquid. Reprinted under the terms of the STM agreement from [Israelachvili, 1991] Copyright (1991) Elsevier Ltd.*

Polymer	A_p (10^{-20} J)	Liquid	A_m (10^{-20} J)
poly(perfluroethylene)	3.8	water	3.7
polystyrene	6.6	benzene	5.0
poly(vinyl chloride)	7.9	cyclohexane	5.2

where A_p and A_m are the Hamaker constants of the particle and medium, respectively. Some typical values for A_p and A_m are listed in Table 7.2.

Thus, based on Equation 7.2, the value of V_A for two, 100 nm radius polystyrene latex particles in water, at $h = 0.3$ nm (about the diameter of one water molecule), is predicted to be $\sim$28 kT.

There may be other types of interactions in this category which arise when the particles are placed in an external magnetic of electric field; if the particles can respond to that field, they will develop a magnetic or electrical dipole moment. If the magnitude of the field is E, then $V(h) \sim E^2$. For example, Goodwin, Markham and Vincent (1997) demonstrated that the application of an (AC) electric field to dispersions of electrically-conducting polymer particles (namely, sterically-stabilised polypyrrole particles) in dodecane led to chain aggregation of these particles, as illustrated in Figure 7.3.

7.3.2 Structural Interactions: (i) Those Associated with the Solvent

As previously mentioned, the second general category of interparticle interactions, namely the "structural" interactions, arises from perturbations of the sheaths around the particles (see Figure 7.2) when overlap occurs, that is, for $h < 2\delta$. Even if the particles carry no charge or adsorbed layers (e.g. an adsorbed surfactant or polymer layer) the liquid in the region of the particle surface may well be different from the bulk liquid. For example, for *hydrophilic* particles in water (e.g. silica particles), the water is usually more "structured" (i.e. more H-bonded) than bulk water. This leads to a short-range repulsive ("solvation") force between two silica particles when the particles approach. On the other hand, for *hydrophobic* particles in water (e.g. polystyrene latex particles), the water molecules tend to be somewhat depleted in concentration close to the polymer surface. The thickness (δ) of this "depletion zone" is not well-defined, but is unlikely to be less than a few molecular thicknesses. The situation now is that, as two polystyrene particles approach, they will develop a hydrophobic *attraction*. The magnitude of these short-range hydrophobic interactions is difficult to calculate, but clearly the "contact" energy of two such particles will be significantly greater than the 28 kT contribution from the long-range dispersion forces, estimated above. Clearly also, the fact that PTFE particles in water may be readily coagulated (e.g. by the addition of electrolyte) implies that the contact energy between two PTFE particles must be largely due to these short-range hydrophobic attraction forces, since the dispersion force contribution, as estimated from Equation 7.2, will be negligible, based on the Hamaker constant values given in Table 7.2.

An important point to make here is that these "true", short-range hydrophobic interactions are not to be confused with the so-called "long-range hydrophobic interactions" that have

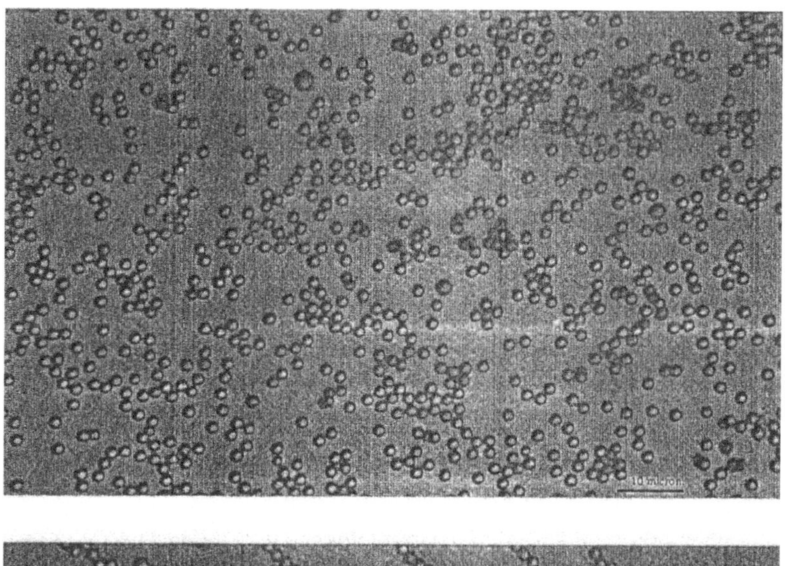

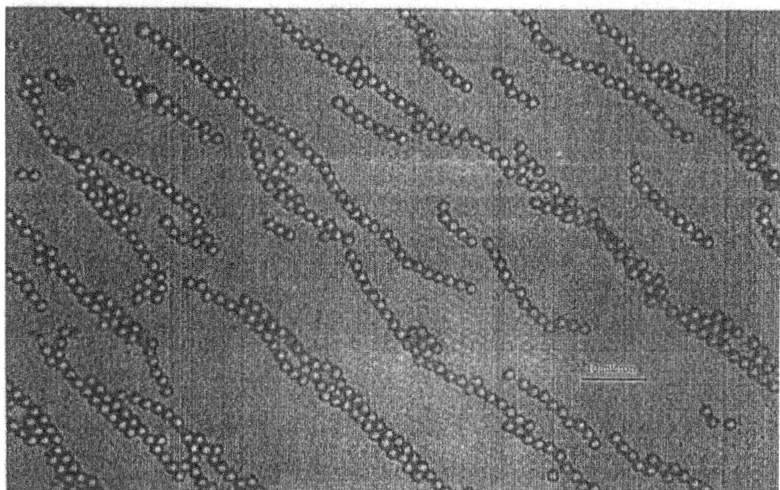

Figure 7.3 *Dispersions of electrically-conducting polypyrrole particle, in (a) the presence, (b) the absence of an applied electric field. Reprinted from [Goodwin* et al.*, 1997] Copyright (1997) American Chemical Society.*

been observed between hydrophobic surfaces in water, and are nowadays ascribed to the spontaneous formation of nano-bubbles between the two surfaces, particularly if the aqueous phase has not been de-gassed.

Even if the continuous phase is not water, there is likely to be a structurally different region of solvent near a polymer particle surface, associated with preferred molecular orientations (e.g. with the longer chain hydrocarbons) or local density differences. Again there may be short-range forces associated with the overlap of these structurally different solvent layers close to the surface. Whether they are repulsive or attractive will depend on the exact nature of the polymer particles and the solvent, in an analogous manner to aqueous dispersions of latex particles.

7.3.3 Structural Interactions: (ii) Electrical Double Layer Overlap

Polymer particles usually acquire a net surface charge. This could arise from: (i) deprotonation, in an aqueous environment, of hydrophilic end-groups, such as –COOH or $-SO_4H$, which migrate to the polymer/solvent interface during particle formation; (ii) co-absorption of ionic surfactants into the particle periphery during the polymerisation process, or their post adsorption onto the surface of pre-formed particles. Associated with this layer of (fixed) surface charge there will be an associated diffuse layer of charge, of net opposite (but equal) charge, made up of counter-ions near the surface. The thickness of this diffuse part of the electrical double layer will depend on the bulk electrolyte concentration (c_{el}), through the Debye parameter, κ, that is, $\delta \sim 1/\kappa$, where κ is given by Equation 7.3,

$$\kappa^2 = \frac{2e^2 N_A c_{el} z^2}{\varepsilon\varepsilon_o kT} \tag{7.3}$$

where N_A is the Avogadro constant, z is the valency of the counter-ions, ε is the dielectric ratio of the continuous phase (~80 for water and ~ 2 for an alkane) and ε_o is the permittivity of free space.

Some of the counter-ions may be "specifically adsorbed", that is, reversibly paired with charge groups at the surface. Such ions are considered to be adsorbed in a layer adjacent to the particle surface, called the "Stern layer". The electrostatic potential of particles in the Stern layer (ψ_d) is an important parameter, which may be approximated to the zeta-potential (ζ), as determined from electrophoresis measurements.

When, for two approaching charged particles, their diffuse layers of counter-ions overlap, there is an accompanying perturbation (i.e. an increase) in the ion concentration in the overlap region, resulting in a build-up of an excess osmotic pressure locally in that region. This is the simplest explanation of the origin of the repulsive electrical double layer interaction, V_E. However, the situation, in general, is more complex than this, in that other perturbations may occur. For example, there may be changes in the surface charge density (unlikely in the case of latex particles, but likely for say AgI particles) or changes in the distribution of specifically adsorbed ions between the Stern layer and the diffuse layer (certainly possible for latex particles). For this reason, the reader will find that there are many equations listed in the textbooks for V_E. The subtle differences depend on the exact nature of the double layer perturbations which occur in any given system during particle encounters, including the timescale of the interaction (for example, the situation during a rapid particle collision may not be the same as in a slower, direct force measurement). For simplicity, only the simplest equation, relating to the so-called "linear superposition" mode of interaction, will be given here (Verwey and Overbeek, 1948); this assumes that the counter-ion concentration profiles are additive, on overlap of two diffuse layers, with only solvent molecules being displaced into the bulk medium. This leads to Equation 7.4,

$$V_E = 2\pi\varepsilon\varepsilon_o a\psi_d^2 \exp(-\kappa h) \tag{7.4}$$

where all the symbols have been defined previously. The effect of the electrolyte concentration (c_{el}) on V_E enters directly in Equation 7.4 through κ (see Equation 7.3), and

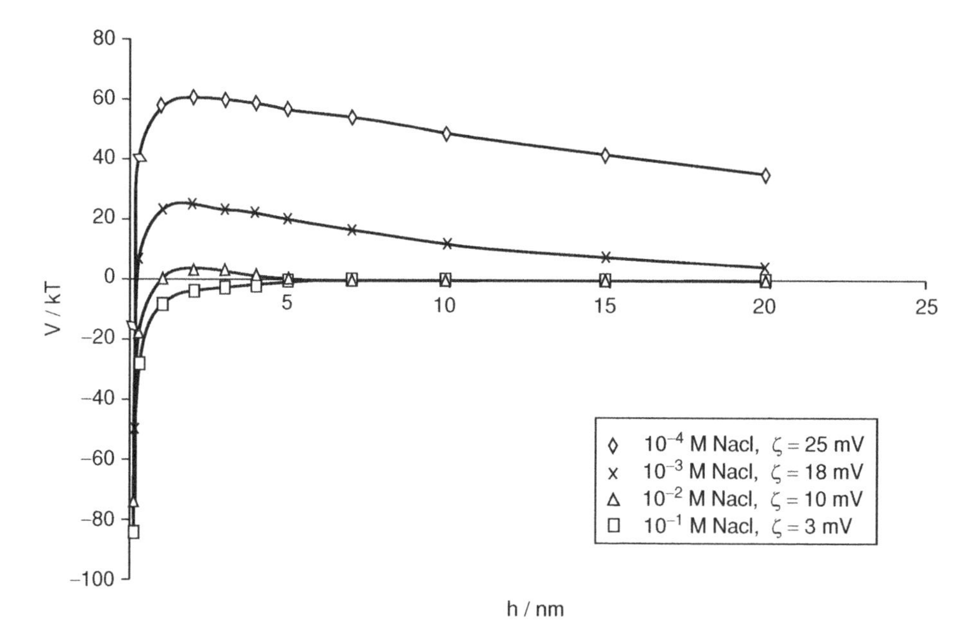

Figure 7.4 *DLVO-based calculations of V(h) for aqueous dispersions of polystyrene latex particles, based on Equations 6.2 and 6.4, at various NaCl concentrations, as indicated. The corresponding zeta potentials (ζ) were taken from Vincent (1992).*

also indirectly through ψ_d, since ψ_d decreases as c_{el} increases (as does ζ). We may now explore the DLVO theory more quantitatively, since its basic premise is that $V = V_A + V_E$. Figure 7.4 illustrates this for the parameters listed in the legend.

The presence of V_{max} and V_{min} are obvious from this figure. V_{min} is very deep, at all c_{el} values, implying irreversible coagulation, *provided* V_{max} can be surmounted within a reasonable timescale. As V_{max} decreases with increasing c_{el} so the rate of coagulation will increase. The value of k_D for which V_{max} just reaches zero is called the critical coagulation concentration ($c_{e,crit}$) for that system. For values of c_{el} beyond $c_{e,crit}$, the coagulation rate is second-order and diffusion controlled. Smoluchowski (1917) showed that the rate constant (k_D) for this process is given by Equation 7.5

$$k_D = \frac{4kT}{3\eta} \tag{7.5}$$

where η is the viscosity of the medium. Equation 7.5 implies that k_D should be fixed for a given solvent at a given temperature, and independent of the nature of the actual particles. Thus, for aqueous dispersions at 20 °C, Equation 7.5 predicts $k_D = 6 \times 10^{-18}$ m^3 s^{-1}. Experimental results for aqueous latex dispersions have been found to be close to this value, but, in general, about a factor of 2 to 3 times lower (attributed to hydrodynamic drag on the particles during a collision). For values of $c_{el} < c_{e,crit}$, that is, where $V_{max} > 0$, reaction-limited coagulation occurs, as discussed in Section 7.2. By analogy with the Arrhenius equation for chemical reactions, one could express (*very* approximately) the coagulation

Table 7.3 Values of $\tau_{1/2}$, for various values of a and V_{max}, for $\phi = 10^{-4}$.

a/nm	$V_{max} = 0$	$V_{max} = 3\ kT$	$V_{max} = 10\ kT$	$V_{max} = 30\ kT$
10	7 ms	140 ms	2.5 min	2400 a
25	109 ms	2.2 s	42 min	3.7×10^4 a
100	7 s	140 s	43 h	2.4×10^6 a
1000	1.9 h	38 h	4.8 a	2.4×10^9 a

rate constant, k, under these conditions by,

$$k = k_D \exp\left(-\frac{V_{max}}{kT}\right) \tag{7.6}$$

For a second-order rate process (such as coagulation), the half-life $\tau_{1/2}$ (i.e. the time taken for the initial number of particles, N_0, to be reduced by a factor of two) is given by, $\tau_{1/2} = 1/(kN_0)$. The volume fraction of particles,ϕ, is given by $\phi = (4\pi a^3 N_0/3)$. Hence, from Equations 7.5 and 7.6, and these relationships, one may readily show that $\tau_{1/2}$ is given by,

$$\tau_{1/2} = \frac{\pi a^3 \eta}{kT\varphi} \exp\left(\frac{V_{max}}{kT}\right) \tag{7.7}$$

In Table 7.3 some values for $\tau_{1/2}$ are given, as a function of a (10 to 1000 nm) and V_{max} (0 to 30 kT), for aqueous latex dispersions at 20 °C; the value of ϕ has been fixed (at 10^{-4}), but the effect of ϕ is readily calculated from Equation 7.7. It is noteworthy that the timescale for $\tau_{1/2}$, for this set of variables, varies from ~ ms to the age of the Universe (~4.5×10^9 a)! In terms of Figure 7.4, one can deduce that 100 nm polystyrene latex particles latex at 10^{-3} and 10^{-4} M NaCl should be indefinitely stable, and that the critical coagulation concentration for this latex is somewhere between 10 and 100 mM with NaCl, as was indeed observed experimentally (Vincent, 1992). Furthermore, in the context of growing latex particles, in an emulsion polymerisation process, Table 7.3 illustrates how the particles become much more stable to coagulation as both their size and their surface charge (and, hence, ψ_d and therefore V_{max}) increase.

7.3.4 Structural Interactions: (iii) Adsorbed Polymer Layer Overlap

If the particles carry an adsorbed polymer layer there will be a sheath (thickness, δ) around the particles, containing the polymer segments and solvent molecules. Some of the segments will be in contact with the interface, in trains, and others will be protruding into the sheath as loops and tails (Fleer *et al.*, 1993). In this regard adsorbed polymer layers somewhat resemble electrical double layers, which have both bound ions and free ions. If the particle surface is charged and/or the adsorbed polymer is a polyelectrolyte, then the sheath will contain both segments and ions. It is important to know how the adsorbed amount of polymer, Γ, and the corresponding sheath thickness, δ, vary with polymer concentration. Some illustrative forms of these dependences for an adsorbed, neutral homopolymer are shown in Figure 7.5.

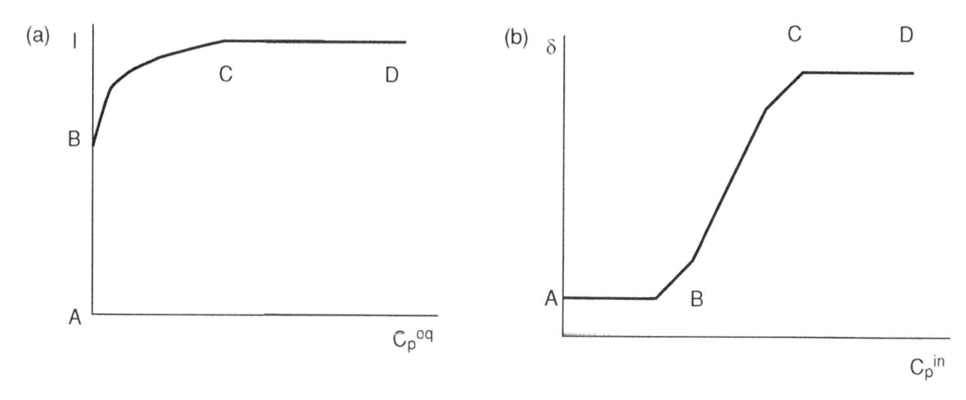

Figure 7.5 *(a) schematic adsorption isotherm for a homopolymer on latex particles: adsorbed amount (Γ) versus the equilibrium polymer concentration (c_p^{eq}); (b) adsorbed layer thickness (δ) as a function of the initial polymer concentration (c_p^i).*

In the region of point C, the two parameters Γ and δ both attain their maximum (plateau) values, that is, Γ_{max} and δ_{max}, respectively. If two particles, carrying such adsorbed chains, overlap, such that $h > 2\delta$, then there will be (as with electrical double layers) a local build-up of excess osmotic pressure in the overlap zone, leading to a repulsive force; this is termed the "steric interaction". Again, as with electrical double layer interactions, there are many equations listed for this interaction (Napper, 1983; Fleer *et al.*, 1993), since it depends on exactly how the segment density distribution in each polymer sheath is perturbed on overlap, and the timescale involved. The simplest assumption to make (as with overlapping electrical double layers) is the linear superposition one, which assumes simple additivity of the two segment density distributions upon overlap. In that case only solvent molecules are displaced into the bulk, and no changes in the segment density distribution occur. Moreover, if the segment density in each sheath is assumed to be *uniform* (rather than say exponential away from the surface, which is commonly found for adsorbed homopolymers, Fleer *et al.*, 1993), then one obtains an equation for V_S which is directly analogous to Equation 7.4 for electrical double layers, that is, Equation 7.8,

$$V_S = 2\pi a \varphi_S^2 \left({}^1\!/_2 - \chi\right) \frac{kT}{v} (2\delta - h)^2 \tag{7.8}$$

Here, ϕ_s is the average segment volume fraction in the sheath, v is the volume of a solvent molecule and χ is the Flory interaction parameter ($\chi = 0$, for an athermal solvent, and $\chi = {}^1\!/_2$ for a theta solvent); ϕ_s is related to Γ through Equation 7.9,

$$\varphi_S = \frac{\Gamma M}{\delta \rho_p} \tag{7.9}$$

where M is the molar mass of the adsorbed polymer and ρ_p is its mass density. Equation 7.8 indicates that V_S is positive (i.e. repulsive) if $\chi < {}^1\!/_2$, but negative (i.e. attractive) if $\chi > {}^1\!/_2$.

The assumption of additive overlap (and, hence, also Equation 7.8) can only be valid for small degrees of overlap, that is, when h is less than, but $\sim 2\delta$. At larger degrees of overlap, the loops and tails of the adsorbed chains must start to contract. This gives rise to

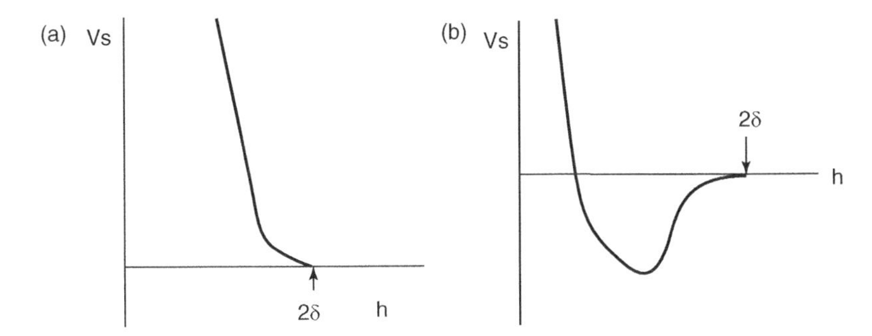

Figure 7.6 *Schematic representation of the steric interaction (V_S) as a function of particle separation (h): (a) in a good solvent for the stabilising moieties; (b) in a poor solvent for the stabilising moeities.*

an additional elastic contribution to the overall steric interaction, which is always repulsive, irrespective of the solvency conditions for the chains (Napper, 1983). For densely-packed chains (i.e. high ϕ_S), this elastic term may well dominate the mixing term, given by Equation 7.8. The general form of the total steric interaction is illustrated in Figure 7.6, for both good and poor solvent conditions.

In order to increase the "anchoring" strength of the polymer chains to the surface, and to avoid desorption during a particle collision, copolymers are often used. Some different types are shown in Figure 7.7. The "A" moeities are the anchoring ones and are essentially insoluble in the solvent ($\chi > {}^1/_2$), whilst the "B" moeities are in a good solvent environment ($\chi < {}^1/_2$).

Block copolymers (including non-ionic surfactants, which may be regarded as oligomeric block copolymers) are often used to stabilise polymer particles during emulsion polymerisation. This is particularly important in the early stages of the process when the nascent particles are small (and soft, because of imbibed monomer) and, as we have seen, charge-stabilisation is not always reliable. This is, of course, even more significant in the context

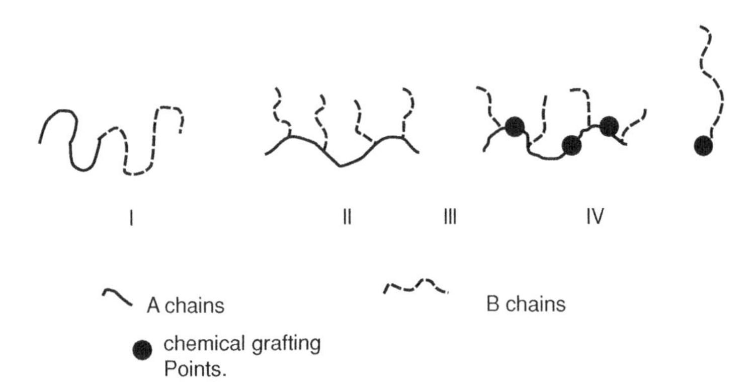

Figure 7.7 *Various types of polymeric stabilisers: (I) AB block copolymer, (II) AB graft copolymer, (III) "super-stabiliser" graft copolymer, (IV) terminally-grafted homopolymer.*

of non-aqueous media. Graft copolymers can also be used, but are perhaps better suited for stabilising the final hard, polymer particles. In that context, they may be added during the latter stages of the polymerisation, so that they migrate readily to the polymer/solvent interface, where the "A" chains become physically "buried". Another strategy is to effect what has been termed "super-stabilisation" (Barrett, 1975). This involves including a reactive group in the "A" chain, for example, glycidyl methacrylate, which can then react with specific functional groups on the particle surface; in this way "chemisorption" of the stabilising polymer is effected (see Figure 7.7, structure III). A more direct way of achieving this end is to attach a terminal reactive group to a homopolymer (e.g. by an anionic polymerisation route for the homopolymer) and then directly graft this homopolymer to a complimentary group on the polymer surface (Vincent, 1993), as shown in Figure 7.7, structure IV.

There are some other factors to consider in achieving effective steric stabilisation of latex particles. One is to ensure that the thickness (δ) of the adsorbed or grafted polymer layer is large enough, such that the range of V_S is sufficient to overcome the attractive forces (V_A). δ is clearly related to the molar mass of the "A" moeities. Secondly, the amount of the stabilising polymer added must be carefully controlled, such that Γ corresponds to region C of the adsorption isotherm (see Figure 7.5a). If the final value of Γ is between A and B, then the particle surfaces will have "bare patches" and polymer bridging may occur between particles, leading to bridging aggregation.

It is also important that not too much stabilising polymer is added, such that there is a significant excess concentration of free, unattached polymer present in the continuous phase, that is, around region D of the adsorption isotherm (se Figure 7.5a). If so, depletion flocculation may well arise. The mechanism of the depletion interaction may be understood from Figure 7.8. When two particles with their adsorbed sheaths, approach to within a distance $h < (2\delta + 2R_g)$, where R_g is the radius of gyration of the free polymer chains, then the free chains are restricted to some extent from fully entering the region between the particles; in effect an "exclusion zone", depicted by the shaded volume in Figure 7.8, is established. Within this volume there will be a lower segment concentration than in the bulk solution, such that an osmotic pressure gradient is established, which will pull solvent molecules for the exclusion zone into the bulk solution, pulling the particles together,

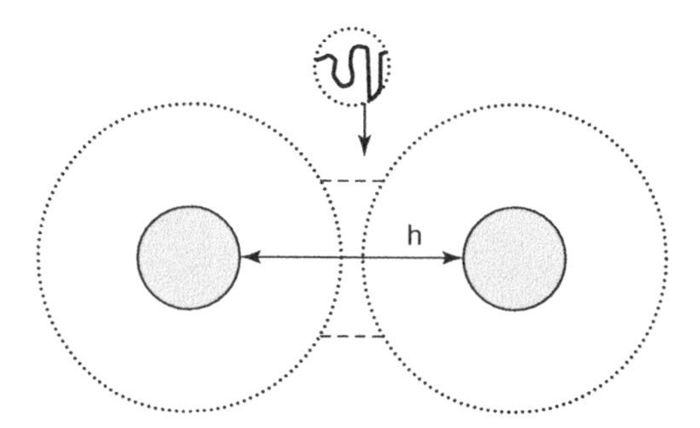

Figure 7.8 The mechanism of the depletion interaction between two particles in the presence of free polymer molecules, illustrating the "exclusion zone" in the region between the two particles.

and giving rise to an attractive interaction. A depletion interaction will also arise in the case of *non-adsorbing* polymer chains (i.e. for systems where the solvent molecules are preferentially adsorbed over the polymer chains). This case has been studied more often, and is sometimes referred to as "hard depletion". The *magnitude* of the depletion interaction depends on the concentration of free polymer (through the osmotic pressure), whilst its *range* depends on the molecular weight of the free polymer and its concentration; at high concentrations the range decreases (Fleer *et al.*, 1993). For those systems where an adsorbed or grafted layer is present on the particles, together with a large excess of free polymer in solution, "soft depletion" occurs. Jones and Vincent (1989) have modelled this soft depletion interaction. Depletion flocculation is a common occurrence in many latex systems. For example, it may occur in latex paints, which often contain free polymer as a viscosity modifier. In the drying film, as the concentrations of both the particles and the free polymer build up, so depletion flocculation may well occur, leading to a loss of gloss in the final coating.

7.4 Weak Flocculation and Phase Separation in Particulate Dispersions

As was demonstrated in the previous section (Section 7.3.4), breakdown of steric stabilisation can lead to weak, reversible flocculation. The main occurrences are summarised below:

1. The solvent for the stabilising moeities is changed from a good solvent to a poor solvent (see Figure 7.6). An example would be where poly(ethylene oxide) (PEO) chains are used as the stabilising ("B") moeities for latex particles, and the temperature is raised above the cloud point of the PEO chains (Cowell and Vincent, 1982).
2. The thickness (δ) of the adsorbed layer is not large enough, such that a significant attractive force exists beyond $h = 2\delta$ (Long, Osmond and Vincent, 1973)
3. Depletion interactions exist, due to the presence of excess free polymer.

In all these cases, the resulting weak, reversible flocculation is associated with a shallow energy minimum (V_{min}) in the pair-potential between the particles, as discussed earlier in Section 7.2. We now return to the question posed there, as to just how deep does V_{min} have to be, in order to see weak flocculation. Below, a simplistic, thermodynamic analysis of the situation is first presented (Cowell and Vincent, 1982), before referring to more detailed and advanced versions. As discussed in Section 7.2, provided V_{min} is not too great, say $< \sim$5–10 kT, flocculation leads to colloidal phase separation An example is shown in Figure 7.9, which is for silica particles dispersed in dimethylformamide, to which non-adsorbing polystyrene has been added (Zhou, van Duijneveldt and Vincent (2011)). Similar behaviour may be observed with polymer latex particles dispersed in water.

For flocculation equilibrium, one may write the classical thermodynamic condition,

$$\Delta G_f = \Delta H_f - T\Delta S_f \tag{7.10}$$

where ΔG_f, ΔH_f and ΔS_f refer to the (Gibbs) free energy, the enthalpy and entropy of flocculation, respectively. Both ΔH_f and ΔS_f are negative. At equilibrium, $\Delta G_f = 0$, and,

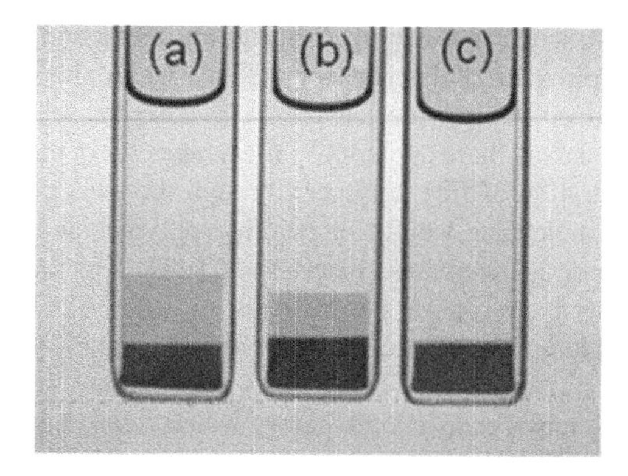

Figure 7.9 *Phase equilibrium in mixtures of two sizes of silica particles, polystyrene and DMF. In the three samples the polymer concentration and the* total *volume fraction of silica particles are both fixed, but the ratio (by volume) of the small: large particles decreases from (a) 3:1 to (b) 1:1 to (c) 1:3. In each case the top, transparent, fluid-like phase contains mostly polystyrene, the middle, more opaque, fluid-like phase (in a and b) contains mainly small particles, whilst the bottom (very opaque) solid-like phase contains a mixture of large and small particles.*

hence,

$$T\Delta S_{\mathrm{f}} = \Delta H_{\mathrm{f}} \tag{7.11}$$

One may also write down the classical Boltzmann relationship for the ratio of the concentrations of the particles in two "energy states", that is, the dispersed and flocculated states, c_{d} and c_{f}, respectively,

$$\frac{c_{\mathrm{d}}}{c_{\mathrm{f}}} = \frac{\varphi_{\mathrm{d}}}{\varphi_{\mathrm{f}}} = \exp\left(\frac{\Delta H_{\mathrm{f}}}{kT}\right) = \exp\left(\frac{zV_{\mathrm{min}}}{kT}\right) \tag{7.12}$$

where ϕ_{d} and ϕ_{f} are the volume fractions of particles in the dispersed and flocc phases, respectively, and z is the co-ordination number of particles in the flocc phase. From Equation 7.12,

$$\ln\frac{\varphi_{\mathrm{d}}}{\varphi_{\mathrm{f}}} = \frac{zV_{\mathrm{min}}}{kT} \tag{7.13}$$

Comparing Equations 7.11 and 7.13, one arrives at the two identities:

$$\Delta S_{\mathrm{f}} = k\ln\frac{\varphi_{\mathrm{d}}}{\varphi_{\mathrm{f}}} \quad \text{and} \quad \Delta H_{\mathrm{f}} = zV_{\mathrm{min}} \tag{7.14}$$

Figure 7.10 shows a plot of 7.12, in the form of V_{min} versus ϕ_{d}, for fixed z and ϕ_{f}. Both z and ϕ_{f} relate to the flocc structure, for example, for random close-packing $z \sim 8$ and ϕ_{f} is 0.64; for hexagonal close packing $z = 12$ and ϕ_{f} is 0.74. In Figure 7.10, the boundary between thermodynamic stability of the dispersion and colloidal phase separation is shown. This boundary line may be crossed either "vertically" or "horizontally", as indicated. The former case would correspond to the situation in Figure 7.9. Examples of the latter case

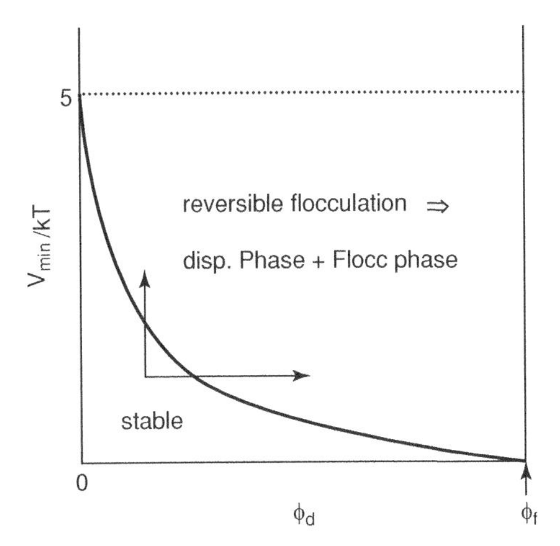

Figure 7.10 *A schematic plot of Equation 6.12, that is, V_{min} versus ϕ_d, showing the boundary between the (thermodynamically) stable region and the weakly flocculated region, where colloidal phase separation may be expected.*

(i.e. increasing ϕ_d at fixed V_{min}) could arise, for example, during evaporation of the solvent, or during sedimentation of the particles.

An aqueous, polymer particle system, which behaves similarly to the silica-nC_{18}/cyclohexane system, in the sense that the van der Waals forces may be tuned by varying temperature, has been described by Rasmusson, Routh and Vincent (2004). The particles were poly(*N*-isopropylacrylamide) (PNIPAM) gel microparticles, which carried surface $-SO_4^-$ groups from the initiator residues.

Figure 7.11 shows plots of the hydrodynamic radius (from dynamic light scattering) of the PNIPAM particles, as a function of NaCl concentration, at various temperatures in the range 25 to 50 °C. PNIPAM particles undergo a swelling/de-swelling transition around 32 °C. At low temperatures, with the gel microparticles in the heavily swollen state, the V_A term is negligible, so that high concentrations of NaCl can be tolerated (up to 1 M), without any aggregation being observed (since there is no driving force). Note that the decrease in size at higher NaCl concentrations is due to the fact that, at higher salt concentrations, water becomes an increasingly poorer solvent for PNIPAM. At higher temperatures, above 32 °C, with the gel microparticles now in the de-swollen state, the V_A term becomes more significant, and the particles flocculate at a given (critical) NaCl concentration. One difference, however, from classical latex coagulation behaviour on adding electrolytes, is that the observed aggregation behaviour in the PNIPAM case is weak and reversible, and, on standing, a flocc phase separates. This is because, even at higher temperatures (>32 °C), the PNIPAM particles retain a significant amount of water (maybe ~ 50% by volume), so that the van der Waals forces are still relatively weak, and hence the V_A term is small.

Although the thermodynamic analysis of weak flocculation and colloidal phase separation, given above, illustrates the basic principles, some of the details are incorrect, in particular for more concentrated dispersions. One missing feature is the prediction of an order/disorder transition in hard sphere dispersions (for which V_{min} is 0), where, at

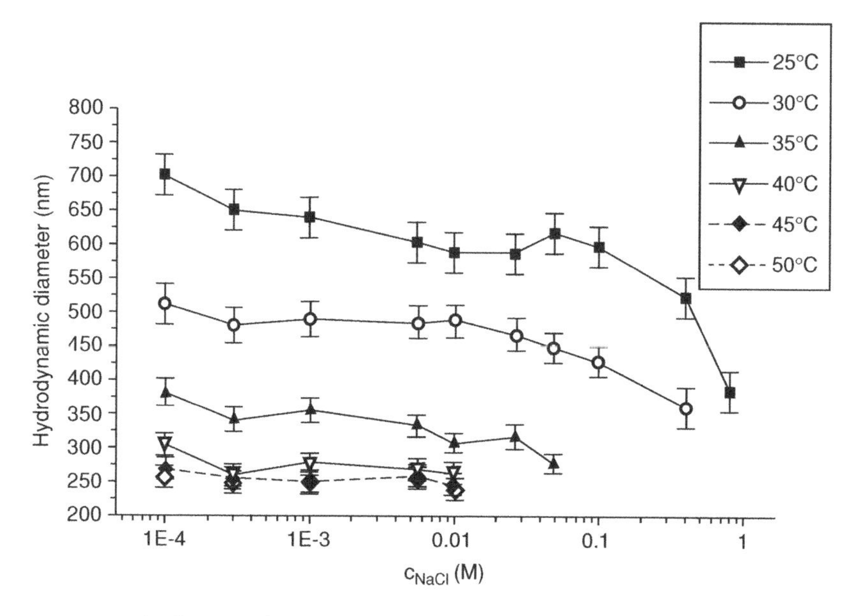

Figure 7.11 *Hydrodynamic diameter versus NaCl concentration, at various temperatures, as indicated, for PNIPAM gel microparticles. Reprinted with permission from [Rasmusson* et al.*, 2004] Copyright (2004) American Chemical Society.*

equilibrium, a colloidal crystal phase is predicted to co-exist with a disordered phase over a narrow range of particle volume fractions (ϕ) that is, $\sim$0.50 < ϕ < 0.55 (Dickinson, 1983). In molecular hard-sphere fluids this is known as the "Kirkwood–Alder transition", and is an entropy-driven effect.

The first experimental observation of this co-existence region was made by Pusey and van Megen (1986) with colloidal dispersions of poly(methylmethacrylate) (PMMA) particles, stabilised by poly(hydroxystearic acid) (PHS]) chains, dispersed in a closely refractive index matched solvent mixture of decalin and carbon disulfide. They demonstrated that, as ϕ gradually increased, one first saw a single disordered phase (ϕ, 0.5), then the predicted co-existence region ($0.5 < \phi < 0.55$) where colloidal crystallites co-existed with the disordered phase, and then a single-phase region ($0.55 < \phi < 0.58$) of just crystallites (which seemed to nucleate from the walls inwards). However, above $\phi \sim 0.58$ it became increasing difficult for these crystallites to form, because the required particle diffusion was increasingly suppressed. Instead, non-equilibrium, glass phases were obtained.

The question of what happens to the predicted equilibrium phase diagrams when a weak attractive force is introduced has been considered by a number of authors (e.g. Gast, Hall and Russel, 1983). The results are shown, schematically, in Figure 7.12.

Figure 7.12a is for systems where the range (L) of the attractive forces is much smaller than the particle radius (a); it shows a "gas + solid" co-existence region, which is an equivalent description to the "dispersed + flocc" phase region shown in Figure 7.10. The left-hand boundary in Figure 7.12a is similar in shape to that in Figure 7.10, at low ϕ values, but clearly deviates at higher ϕ values, particularly as $\phi \Rightarrow 0.5$.

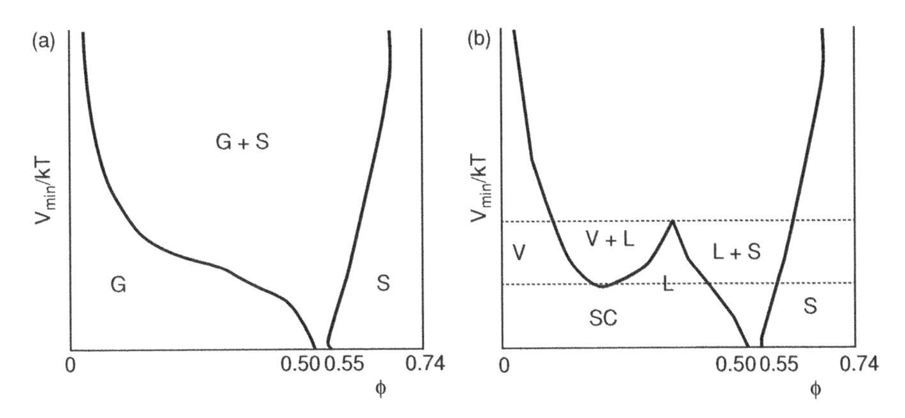

Figure 7.12 *Schematic equilibrium phase diagrams for colloidal particle having : (a) short-range interactions; (b) long-range interactions.*

For systems where $L \sim a$ (or greater), then, as Figure 7.12b indicates, it is possible to sustain a *liquid*-like colloidal phase. In fact, Figure 7.12b also describes the equilibrium phase diagram for spherical *molecules*, where the range of the van der Waals forces is comparable to the molecular size. The most straightforward way of generating colloidal systems, where $L \sim a$, is by utilising the depletion interaction, with small particles plus high molecular weight, non-adsorbing polymers. One such example is shown in Figure 7.13 for silica-nC_{18} dispersions in cyclohexane (at 20 °C), to which poly(dimethylsiloxane) [PDMS] has been added (Vincent *et al.*, 1988).

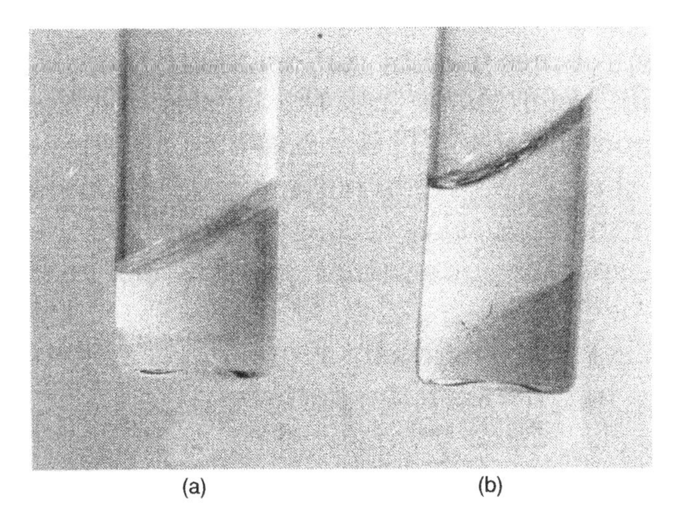

Figure 7.13 *An example of two co-existing colloidal phases, for silica-nC_{18} particles in cyclohexane, showing, (a)liquid–solid, (b) liquid–vapour equilibria. Note that the two tubes have been tilted to show the solid boundary in the former case and the mobile boundary in the latter case.*

So far, reference has only been made to *equilibrium* systems, but, as discussed earlier, at very high particle volume fractions, non-equilibrium states may appear, in this case glass states. Similarly, when systems are quenched rapidly into a two-phase, fluid/solid co-existence region, the equilibrium state for the solid phase ought to be a solid crystal, but very frequently colloidal gels are formed instead. Understanding such non-equilibrium behaviour is one of the challenges in modern colloid science.

7.5 Aggregate Structure and Strength

In previous sections, strong irreversible aggregation (coagulation) and weak, reversible aggregation (flocculation) have been discussed, and some aspects of aggregate structure relating to both have been referred to already. As has been described earlier (Section 7.4), the "openness" of a flocc can be described in terms of its (average) volume fraction (ϕ_f) or its (average) co-ordination number (z). Another parameter which may be used is the fractal dimension, d_f, which is defined in Equation 7.15

$$N \sim \left(\frac{R}{a}\right)^{d_f} \tag{7.15}$$

where N is the number or particles in an aggregate and R is some measure of its dimensions (e.g. its radius of gyration). There are a number of methods for determining d_f. Light scattering is often used, both static (small angle) and dynamic. For static light scattering, one may write,

$$I(q) \sim S(q) \cdot P(q) \tag{7.16}$$

where $I(q)$ is the intensity of the scattered light at a particular scattering vector, q. $S(q)$ and $P(q)$ are the structure factor and form factor for the particles, respectively.

$$q = \frac{4\pi}{\lambda} \sin \frac{\theta}{2} \tag{7.17}$$

where λ is the wavelength of the light used, and θ is the scattering angle. For length scales much larger than the particle radius (a), but smaller than the aggregate size, $P(q) \sim 1$, and then $I(q)$ essentially reflects $S(q)$, which is the Fourier transform of the pair correlation function, $g(r)$ in the aggregate. For fractal aggregate structures, it may be shown (Fernandez-Barbero and Vincent, 2000) that,

$$I(q) \sim S(q) \sim q^{-d_f} \tag{7.18}$$

Hence, a plot of log $I(q)$ versus log q, yields d_f.

Alternatively, if the hydrodynamic diameter (R_h) of the aggregates is measured as a function of time (t), using *dynamic* light scattering, then, under certain conditions, (Asnaghi *et al.*, 1992) the following relationship holds,

$$R_h \sim t^{1/d_f} \tag{7.19}$$

so a plot of log R versus log t, yields d_f.

The value of d_f for irreversibly coagulating dispersions depends on whether an energy barrier is present or not. For "reaction-limited" coagulation, d_f is, in general, >2, whereas

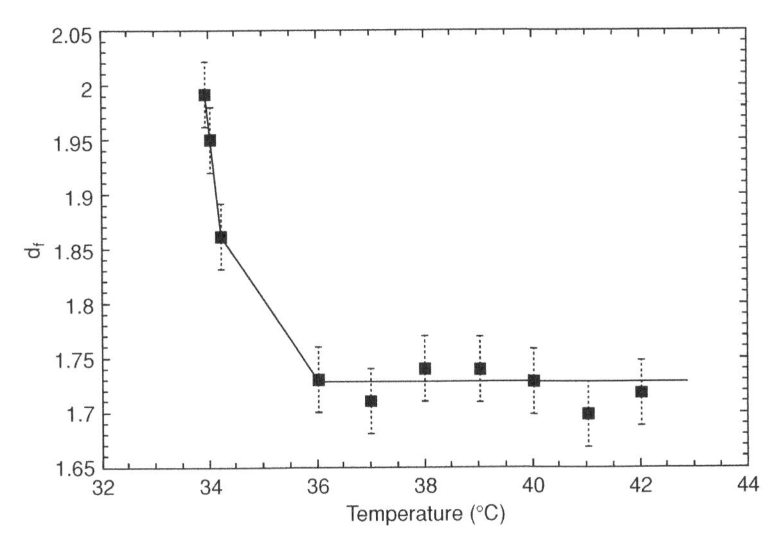

Figure 7.14 *The variation in the fractal dimensions (d_f), as a function of temperature (beyond the CFT), for dispersions of PNIPAM gel microparticles in 1 M NaCl solution. Reprinted with permission from [Routh & Vincent, 2002] Copyright (2002) American Chemical Society.*

for "diffusion-limited" aggregation d_f is ~1.7–1.8. The higher value in the former case is because particles may undergo several collisions before eventually surmounting the energy barrier (V_{max}) and "sticking", whereas, in the latter case, every collision results in the particles "sticking" where they touch.

For weak, reversible flocculation, d_f values >2 are often observed as well; in addition, some time-dependence may also be observed. This is because singlet particles can break away, and return to the flocc in a more favourable (i.e. a lower potential energy) position. Clearly, if a colloidal crystal is achieved, d_f approaches the maximum value of 3.

Routh and Vincent (2002) have made a study of the fractal dimensions of PNIPAM microgel aggregates in 1 M aqueous NaCl solution (see earlier, Figure 7.11), as a function of temperature, using small-angle static light scattering. The results are shown in Figure 7.14.

The critical flocculation temperature (CFT), below which the microgel particles are (thermodynamically) stable, is 34 °C. Just beyond 34 °C d_f is 2.0, indicative of weak, reversible flocculation. As the temperature is increased further, so V_{min} increases, and the aggregation becomes stronger. Eventually, (at ~36 °C) a value for d_f of 1.73 is reached, and becomes constant thereafter; this corresponds to irreversible diffusion-controlled aggregation. Rasmusson, Routh and Vincent (2004) confirmed, using dynamic light scattering with a very similar system, a value for of 2.0 ± 0.1 for d_f for floccs forming just beyond the CFT.

Some discussion has also been given earlier about aggregate strength. The parameter which best describes this is V_{min}. This is not a straightforward parameter to measure directly for polymer particles. Direct pairwise, force measurements between two polymer latex particles are not straightforward, although measurements between polymer particles and plates may be made using methods such as atomic force microscopy or total internal reflection fluorescence. Optical tweezers may be a possibility for directly measuring the force between two larger latex particles, say ~1 μm or so.

A number of more classical studies on latex particle assemblies have been reported in the past, using pressure-cell equipment. For example, Cairns and Ottewill used this method (Cairns et *al.*, 1976) to study the interaction of PMMA particles with grafted PHS chains, that is, a similar system to Pusey and van Megen's, discussed earlier, but with dodecane as the solvent. Napper (1983) has discussed how pair-potentials might be extracted from pressure / volume fraction data obtained using this method.

Rheology is another possibility for (indirectly) obtaining pair-potential information, but the main difficulty here in handling aggregated systems is their non-equilibrium nature in many cases. Consequently, any rheological measurements are often history dependent. It is easier, in principle, for weakly flocculated systems, where structures usually recover reasonably quickly, after any disturbance. Goodwin *et al.* (1986), who determined the shear modulus at a single frequency, and also Patel and Russel (1987), who measured the yield stress and also made small-amplitude oscillatory measurements, reported studies on weakly flocculated, polystyrene latexes. They both attempted to correlate their rheological data with the interparticle pair-potentials.

The determination of pair potentials for latex particles remains a serious challenge for the future.

8

Analysis of Polymer Molecules including Reaction Monitoring and Control

Peter Schoenmakers
Department of Chemical Engineering, University of Amsterdam, The Netherlands

To monitor, control, and optimize emulsion polymerisations, we need to be able to perform a variety of different measurements. The monomer conversion is a key parameter to monitor and control the reaction. A rapid response is required for real-time reaction monitoring.

The copolymer composition must be known to monitor and control copolymerisation reactions. Because various monomer-addition strategies allow various types of copolymers to be formed (random, gradient, block copolymers), we also must be able to characterize such products.

In this chapter we will consider on-line and off-line methods to measure the conversion. We will describe methods to determine the molar mass and the molar-mass distribution of polymers, to determine the chemical composition and the chemical-composition distribution and to determine two-dimensional distributions.

Methods for the detailed characterization of polymers will also be considered, as these are needed to study the principles of polymerisation (reaction mechanisms, kinetics), as well as to establish and interpret structure–property relationships.

Chemistry and Technology of Emulsion Polymerisation, Second Edition. Edited by A.M. van Herk.

8.1 Sampling and Sample Handling

8.1.1 Sampling

Withdrawing samples from the reactor is a major challenge, because the viscosity increases during the reaction. When using a sample loop, demixing and/or mechanical flocculation of the emulsion can occur, especially at high conversions (high solid contents). Guyot *et al.* (1984) designed a dilution cell to avoid flocculation in the loop used to transfer the reaction medium from the reactor to the injection port of the chromatograph.

In emulsion polymerisation the reaction mixture is necessarily very inhomogeneous. Initially, the emulsion consists of two liquid phases. During the reaction latex beads are formed whichare themselves heterogeneous.

8.1.2 Sample Preparation

In a latex system polymer may exist at three different locations, that is, in the aqueous phase or serum of the latex, at the surface of the latex particles and inside the latex particles. Besides the polymer, the latex contains also other ingredients, such as electrolytes and oligomers. If the composition of each of the three types of polymer present at these locations has to be known, latex isolation and cleaning is of crucial importance. El-Aasser (1983b) has summarized the methods of latex cleaning.

The most basic way of isolating a latex is to destabilize it, for example by adding an electrolyte to a charge-stabilized latex, followed by filtration and washing of the coagulate. The resulting latex is far from "clean". Oligomers and polymers are adsorbed on the surface, electrolytes can still be present. Therefore, extensive latex cleaning is needed when most of these ingredients have to be removed.

One of the more-often used techniques is that of serum replacement (Ahmed *et al.*, 1980), the latex particles are confined in a stirred cell by a filtration membrane. Washing with water cleans the latex, while the serum also becomes available for analysis. Monitoring, for example, the conductivity of the serum makes it possible to follow the cleaning process.

Dialysis (Everett, Gülteppe and Wilkinson, 1979) and hollow-fibre dialysis (McCarvill and Fitch, 1978a, 1978b) involve the use of regenerated cellulose dialysis tubing containing the latex. The driving force is the chemical potential of the ions. Replacing the water several times can introduce contamination with polyvalent cations. Dialysis is a slow process, while hollow-fibre dialysis and cross-flow filtration are relatively fast. Dialysis is not very efficient in removing oligomers and residual monomer.

Asymmetrical flow field-flow fractionation (Flow FFF) and, specifically, hollow-fibre flow FFF (HF_5) have the potential of combining latex clean-up and characterization. Capillary electrophoresis has seen significant advances in the last decade. It may potentially be used to separate and characterize charged latex beads and low-M_r charged components (stabilizers, surfactants and charged oligomers) without prior sample clean-up.

Other cleaning techniques include the use of ion-exchange resins for removing electrolytes (van den Hul and Vanderhoff, 1970), repeated centrifugation with replacement of the serum, contacting the latex with an activated carbon cloth, and gel filtration (see (El-Aasser, 1983b) and references therein). For the removal of residual monomer from latexes, steam stripping is sometimes used (Goodall, Hearn and Wilkinson, 1979).

Several surface-characterization techniques are used to monitor the latex-cleaning process, amongst these are conductometric and potentiometric titrations, infrared spectroscopy, electrophoresis and titration with a surface active substance (El-Aasser, 1983b and references therein). The removal of polyelectrolytes is especially difficult. van Herk *et al.* (1989) contacted a cationically charged latex with the active surface of silica to thoroughly remove polycations from the surface of latex particles that were subsequently used as catalytically active latex systems. Kong, Pichot and Guillot (1987) used combinations of surface end-group characterization techniques to reveal the influence of the monomer addition profiles on the distribution of SO_2^- and COO^--groups.

In applying any molecular characterisation technique to polymer chains obtained through emulsion polymerisation one has to realize that common additives in the form of surfactants, chain transfer agents and initiator molecules or fragments thereof can be present and might interfere with the characterization technique.

8.2 Monomer Conversion

In Table 8.1 a number of methods for measuring monomer conversion are summarized. Gravimetric analysis is arguably the most accurate method for measuring the amount of polymer formed and – in the case of homopolymers – for measuring the conversion. In the case of copolymers there are two degrees of freedom (amount of A converted and amount of B converted) and the measurement of a single number does not suffice. Measurement of the copolymer composition (see Section 8.4) in combination with gravimetry would be adequate.

Gas chromatography (GC) can be used to measure the amount of unconverted monomer(s). In certain GC systems, for example, when using a programmed-temperature vaporizer (PTV) with a packed liner, it is possible to inject an emulsion directly. Direct-injection GC can be performed either on line or off-line. The liner will need to be replaced regularly, but this constitutes a minor effort. The total amount of monomer can also be measured by dynamic-headspace ("purge-and-trap") analysis in an off-line measurement. Static-headspace analysis can be performed either on-line or off-line. However, this is an

Table 8.1 *Summary of methods for measuring monomer conversion.*

Method	Comments
Gravimetric analysis	• Accurate and precise, but laborious method • Most suitable for homopolymers
Gas chromatography	• Highly precise and (potentially) accurate method • Applicable for homopolymers and copolymers • Static-headspace analysis can be performed on-line
Size-exclusion chromatography	• Allows measurement of both polymer and monomer • No implicit assumptions
Spectroscopy (FTIR, NIR, Raman)	• Direct method for composition; Indirect method for conversion

equilibrium method and knowledge or assumptions is required on the relationship between the concentrations in the gas-phase and in the emulsion. In off-line headspace GC, the analysed amount of sample can be measured (weighed) accurately. In on-line applications of GC (static headspace or direct injection) this is not possible and an internal standard should be present in the reaction mixture. If samples are withdrawn from the reactor for subsequent off-line analysis by GC, such a standard can be added after weighing the sample.

Size-exclusion chromatography (SEC) can – in principle – be used to measure both the amount of polymer formed and that of unreacted monomer (Lousberg *et al.*, 2002). However, other low-molar-mass materials present in the mixture may interfere with the measurement of the monomer. There is a definite trend towards fast separations by SEC (Pasch and Kiltz, 2003; Popovici and Schoenmakers, 2005). Typical analysis times of 1 min or less can be achieved for relatively simple separations (e.g. separation of monomer from polymer). This greatly increases the attractiveness of SEC for on-line measurements.

Spectroscopic techniques can be applied for monitoring conversion. Nuclearmagnetic resonance (NMR) and Fourier-transform infrared (FTIR) can be applied as off-line techniques. However, spectroscopic techniques are arguably most attractive if they can be applied on-line. For example, attenuated-total-reflectance (ATR) FTIR has been applied to monitor the copolymerisation of butyl acrylate and vinyl acetate (Jovanovic and Dube, 2003). Raman spectroscopy, in combination with a self-modelling curve-resolution technique, has been used for the on-line monitoring of the anionic dispersion block copolymerisation of styrene and 1,3-butadiene (Bandermann *et al.*, 2001). A great advantage of using Raman and NIR methods is that optical-fibre technology can be used to physically separate the spectrometer from the reaction vessel. Despite the attractiveness of on-line spectroscopic methods, one should realize that the calibration is a significant bottleneck. In many cases, especially when using near-infrared (NIR) spectroscopy, the method is essentially a black box and extensive calibration using a primary method is required. The value of such black-box models depends on their robustness and on the time during which the model can be applied ("pay-back period").

8.3 Molar Mass

The molar mass of a polymer strongly affects a broad range of properties. For homopolymers it is the single most important characteristic. There are many ways to measure molar mass. It is important to realize that the various methods yield different average numbers, which – even if the measurements are perfect – can be substantially different for broadly distributed polymers. Indeed, the width of the MMD is most commonly characterized by a polydispersity index (PDI or D) that is the ratio of the weight-average and the number-average molar mass (PDI $= M_w/M_n$).

Table 8.2 provides a list of some of the many techniques that can be used for measuring molar-mass averages. To gain some idea on polymer molar mass, to monitor and control the (emulsion-) polymerisation process, and to compare different batches of sample, each individual method may suffice. However, to gain insight into the polydispersity of the sample polymer, two different averages must be obtained and, hence, two different measurements must be performed. Apart from the extra work involved, Table 8.2 also illustrates the difficulty of obtaining two accurate average values for any given polymer. Colligative-property

Table 8.2 *A selection of techniques for measuring molar masses.*

Technique	Result	Power[a]	Comments
Colligative properties (membrane osmometry, vapour-pressure osmometry)	M_n	1	Works best at low M
End-group analysis (titration, NMR)	M_n	1	Works best at low M
Viscometry	M_v	$(1/a+1)^b$	Works best at moderate to high M
Light scattering	M_w	2	Works best at high to very high M
Ultracentrifugation			Works best at high to very high M
• Sedimentation velocity	M_w	2	
• Sedimentation equilibrium	M_z	3	

[a] $M_k = \sum_M n_M M^x / \sum_M n_M M^{x-1}$

[b] a is the exponent in the Mark–Houwink equation, that relates the intrinsic viscosity to the molar mass: $[\eta] = K \times M^a$.

measurements and end-group analysis (the methods yielding M_n) are essentially only suitable for polymers of relatively low molar mass. In the case of high-M polymers, there are simply not enough molecules or end groups to allow accurate measurements. In contrast, light scattering yields a direct estimate of the weight-average molar mass for relatively large polymers, but smaller ones ($M_r < 10\ 000$) exhibit hardly any scattering. For large polymers, ultracentrifugation can provide some indication of the sample polydispersity ($\mathrm{PDI}' = M_z/M_w$) if both sedimentation velocity and sedimentation equilibrium can be measured.

It is also worth noting in Table 8.2 that, with the possible exception of end-group analysis, all methods require dissolution of the sample. In principle, end-group analysis can be performed in the solid phase by spectroscopic techniques (e.g. IR, NMR, see Section 8.3). However, obtaining accurate quantitative results is very difficult in practice. Once the polymer is dissolved, size-exclusion chromatography (Section 8.3.1.1) is usually the most attractive proposition, as it provides – in principle – the complete MMD.

(Simple) methods that provide an indication of the molar mass from the polymer melt (e.g. various methods related to the melt viscosity) are likely to retain a position of prominence in industrial practice. Very simple, low-cost methods, such as titration and osmometry, are also likely to prevail. More complex methods, such as light scattering, are increasingly replaced by (or combined with) size-exclusion chromatography.

8.3.1 Molar-Mass Distributions

In the previous section a number of methods have been identified for measuring molar-mass averages. However, we have discussed that it is difficult to use these methods for characterizing the width (and the shape) of the molar-mass distribution (MMD). Accurate MMDs are essential for the control, optimization and understanding of emulsion polymerisations.

8.3.1.1 Size-Exclusion Chromatography

Size-exclusion chromatography (SEC, aka gel-permeation chromatography, GPC, Table 8.3) is the bread-and-butter technique for measuring MMDs. SEC employs columns

Table 8.3 *Strengths and weaknesses of size-exclusion chromatography.*

Strengths	Comments
Flexibility	• In principle applicable to all soluble polymers • Great variety of columns available • Variety of detectors available
Information	• Complete MMD, plus information on low MM fraction (unreacted monomer, conversion).
Precise data	• Highly repeatable data allow comparison between different samples
Weaknesses	**Potential remedies**
Polymer solubility	• Dedicated solvents • High-temperature SEC
Polymer dissolution	• No good remedy for slow dissolution
Polymer adsorption	• Mobile-phase additives • High-temperature SEC
Limited loadability	• Use low concentrations
Calibration needed	• Universal calibration • On-line coupling to viscometry and/or light-scattering detectors • Calibration by on-line or off-line coupling with mass spectrometry
Copolymers	• Comprehensive two-dimensional liquid chromatography (LCxSEC) or LCxSEC//MALDI-MS

packed with porous particles. The pore size is a critical parameter. If the pore diameter is smaller than the (hydrodynamic) diameter of the polymer molecules in solution, the latter cannot enter the pores and they are totally excluded. If the pore sizes are much larger than the analyzed molecules, the latter can enter all the pores and are totally permeating. In the absence of any interactions between the analyte molecules and the surface of the packing material – a critical assertion in SEC – the extent of exclusion is directly related to the size of the polymer molecules. The extent of exclusion is reflected in the elution time or elution volume (the two properties are related by the flow rate; $V_e = F \cdot t_e$). For a homopolymer, the molar mass is directly related to size in solution and thus to the retention volume. This relationship takes the form shown in Figure 8.1 and is known as a calibration curve in SEC. The calibration curve is usually described by a mathematical function, often a third-order polynomial (Vanderheyden Popovici and Schoenmakers, 2002). Using the calibration curve, the SEC elution can be converted into an MMD.

SEC is by far the most popular technique for measuring MMDs. SEC is highly flexible in that it can be applied to all soluble polymers. It also yields much information. Apart from the complete MMD, information on the amount of unreacted monomer (and thus on the conversion) may be obtained (Lousberg *et al.*, 2002). There is a wide variety of SEC columns available, with narrow pore-size distributions for specific MM ranges or with broader distributions that yield approximately linear calibration curves across a wide range of masses. Different column materials are available for use with specific polymers and eluents.

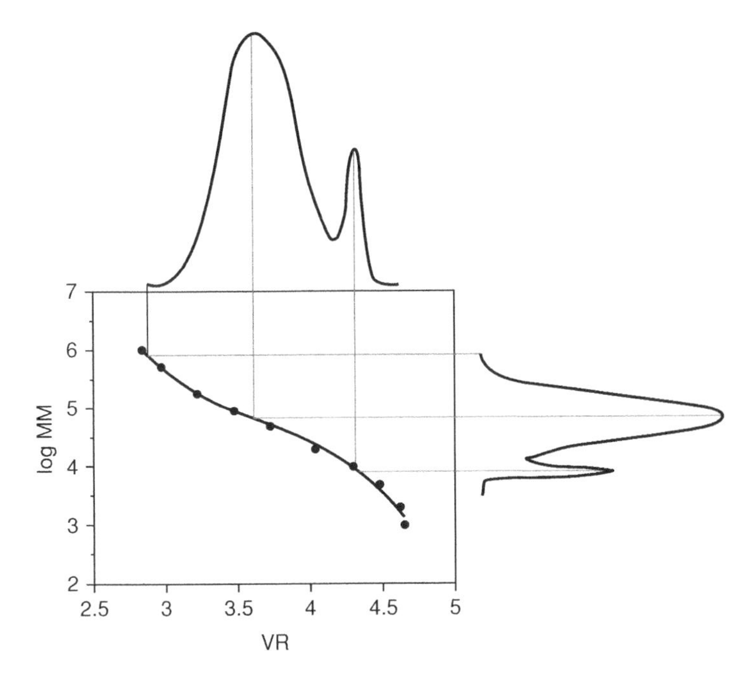

Figure 8.1 *Conventional calibration in size-exclusion chromatography. The calibration curve is constructed by measuring the elution volumes of a series of narrow standards, ideally with the same chemical structure as the analyte polymer. The elution profile of a polymer (top) can be transformed into a molar-mass distribution (right).*

Calibration is usually performed using a set of narrow standards. To obtain correct (accurate) MM data, these standards should be chemically identical to the analyte polymer. When appropriate standards are not available, relative data are obtained. These are usually inaccurate (biased), but precise (repeatable). Such relative data are often used for comparing different samples. To obtain absolute data, the principle of universal calibration may be used, which states that the hydrodynamic volumes – and therefore the SEC elution volumes – are equal for polymers for which the product of the molar mass and the intrinsic viscosity is identical. The intrinsic viscosity can be related to the molar mass if the Mark–Houwink coefficients of a specific polymer–solvent combination are known. By using the Mark–Houwink coefficients, relative data on the molar mass can be converted to absolute values.

Alternative ways to obtain absolute data from SEC include the use of viscometry detectors that allow the intrinsic viscosity to be measured on-line. Light scattering may also be used to determine absolute molar masses as a function of the elution volume. Finally, SEC may be calibrated using mass spectrometry (MS). For some polar polymers, on-line SEC–electrospray-ionization-MS has been demonstrated More commonly, SEC is combined off-line with matrix-assisted laser-desorption/ionization (MALDI) MS Commercial devices exist that make this latter combination highly convenient.

Despite the immense popularity of SEC, the technique still has some significant problems. Dissolving samples is not always easy – and almost invariably slow. Solubility can be enhanced by using specific solvents, which may be expensive, or by using elevated temperatures, which increases the risk of polymer degradation prior to or during the analysis. There is no good remedy for the slow dissolution of polymers. The higher the molar mass, the more time is typically needed to dissolve a polymer, with times in excess of 24 h indicated for polymers with $M_r > 10^6$. If the sample is not completely dissolved, incorrect data will be obtained. However, very long dissolution times are definitely unattractive. Enhancing the dissolution by vigorous shaking or ultrasonication is not recommended if polymer degradation (chain scission) is to be avoided.

Overloading occurs easily in SEC, but it can be avoided by injection of highly dilute polymer solutions. Concentrations of a few (1 to 5) mg ml^{-1} are common. For the largest polymers, for which overloading occurs most easily, the lower end of this range is recommended. Elution profiles in SEC are broadened by dispersion processes that occur within and outside the columns. When the peaks are significantly broadened, incorrect results (inflated polydispersities) will be obtained. In general, the effects of band broadening can be reduced by using long SEC columns or a number of columns in series, at the expense of long analysis times. There is a contemporary trend towards fast SEC using short columns (Pasch and Kiltz, 2003; Popovici and Schoenmakers, 2005). However, it must be understood that an increased speed almost inevitably results in reduced accuracy.

While extensive research has led to a number of elegant ways to determine accurate MMDs for homopolymers using SEC, the characterization of copolymers is still notoriously difficult. For homogeneous copolymers, that is, those with extremely narrow chemical-composition distributions (CCDs), combinations of SEC with viscometry and or light scattering may be used. In fact, such copolymers with a constant chemical composition may be treated very much like homopolymers. However, for copolymers that exhibit both an MMD and a CCD no easy solutions exist. Two-dimensional separation techniques are required and comprehensive two-dimensional liquid chromatography (LC $\times$ SEC; see Section 8.4.4) is developing into a routinely applicable technique. However, calibration of LC $\times$ SEC so as to obtain accurate results on (co-)polymer distributions is far from trivial.

8.3.1.2 Matrix-Assisted Laser-Desorption/Ionization

While size-exclusion chromatography is the most-popular technique for measuring MMDs, matrix-assisted laser-desorption/ionization mass spectrometry (MALDI-MS, Table 8.4) is the most promising and most-intensively-researched technique (Nielen, 1999; Montaudo and Lattimer, 2002; Murgasova and Hercules, 2003). The mass spectrometry of polymers has taken gigantic strides forward in the last 20 years. We now have ways to create and detect molecular (i.e. non-fragmented) ions of very large polymers. This gives us a direct and very accurate way to determine the molar mass of polymer molecules. Having said that, we must be aware that, although dramatically improved, the possibilities for analysing polymers by MS are still limited.

In mass spectrometry the molecules in a sample are first ionized, then separated according to the mass-to-charge ratio (m/z), and finally detected. The m/z ratio can be measured with high accuracy (typically within 0.1%). While qualitative information on molecular ions is often very accurate, accurate quantitative information cannot easily be obtained (Nielen and Buijtenhuijs, 2001; Murgasova and Hercules, 2002). MS suffers from discrimination (different sensitivities for the ionization, separation, and detection of different molecules)

Table 8.4 Strengths and weaknesses of matrix-assisted laser-desorption/ionization mass spectrometry.

Strengths	Comments
Mass range	• *M* from (about) 1000 up to 1 000 000, provided that the sample is very narrowly distributed
Accuracy	• Highly accurate molar masses can be obtained
Structural information	• High-resolution data provide information on end groups and (sometimes) on chemical composition
Sensitive	• MALDI-MS is an extremely sensitive technique, requiring very small amounts of sample
Weaknesses	**Comments**
Polarity range	• MALDI is not applicable to apolar polymers
Discrimination	• Small and polar molecules experience "positive discrimination"; Large and non-polar molecules are discriminated against
Ion suppression	• Molecules present in high concentrations tend to obscure molecules present at low concentrations
Poor quantitation	• Quantitative results (e.g. MMDs) are hard to obtain, because of discrimination, ion suppression, isotope patterns, and other reasons.
Poor repeatability	• Every spectrum ("shot") is different; Large numbers of spectra must be averaged to obtain reliable information. • MS sensitivity varies with time

and suppression. Small and polar molecules tend to be favoured over large and non-polar ones; more-abundant molecules tend to be favoured over less-abundant ones.

Ionization methods that yield (predominantly) non-fragmented ions are known as soft ionization methods. Such methods are essential for polymer analysis. Because any synthetic polymer consists of a variety of different molecules (different masses, and possibly also different compositions, end groups, etc.), significant fragmentation usually results in extremely complex spectra that cannot be judiciously interpreted. Two soft ionization techniques have emerged as the main tools in polymer analysis. These are electrospray ionization (ESI) and MALDI. ESI is especially useful in combination with LC or SEC separations of polar polymers of relatively low molar mass (see Section 8.4.2.3). MALDI has a significantly broader application range.

MALDI has been developed since 1970 (see Hillenkamp and Karas, 2000). However, its rise to popularity has occurred much more recently, especially in the domain of synthetic polymers. To perform MALDI experiments, a polymer sample is put on a sample disk ("target plate"), which can accommodate a large number of samples, typically in shallow "wells". The laser can be directed at a specific spot. The sample must be in contact with a matrix, which (i) absorbs laser light and (ii) donates a charge to the polymer molecules. It is not yet clear what the exact mechanism is that gives rise to the formation of molecular ions. "Matrix assistance" is arguably needed. However, it is still a matter of discussion (and research) whether desorption of the polymers occurs first, followed by ionization in the gas phase, or whether ionization precedes desorption. This dilemma is indicated by inserting a slash between desorption and ionization in the full name matrix-assisted laser-desorption/ionization. The matrix can be added before (using a pre-conditioned well plate), during (by post-column addition of a matrix solution), or after the sample deposition. Concomitant application of the sample and the matrix is likely to give the best results, because "co-crystallization" with the analyte is desirable (Scrivens and Jackson, 2000).

Remarkable results have been obtained with MALDI of synthetic polymers. It can be used for macromolecules with high polarities (such as proteins), but also with moderate polarities (such as polystyrene) or even low polarities, such as polyisobutylene. Useful spectra can be obtained from very high-M_r samples, provided that the polydispersity is sufficiently low (Schriemer and Li, 1996). However, MALDI-MS should not be considered a mature technique. A number of aspects still must be improved in the coming years:

- The repeatability of MALDI-MS experiments is generally poor. Some "shots" (laser pulses) may give no spectral sensitivity, while other shots on the same spot may yield beautiful spectra. In between these two extremes, the sensitivity of the MALDI-MS experiment is highly variable. Typical MALDI spectra are the accumulated (or average) result of a large number of shots. Local variations in spot composition and morphology are likely to play a role, so that more homogeneous spots may lead to greater precision.
- The selection of a suitable matrix for MALDI-MS experiments is still essentially based on trial and error. Generally, matrices that have been used successfully by others are suitable – but not necessarily optimal. The number of possible choices is increasing, enhancing the need for generally applicable matrix-selection rules.
- Larger ions are discriminated against (or small molecules are favoured). In addition, molecules present in large concentrations may suppress the ionization of trace constituents, an effect known as ion suppression. In a sample of a polydisperse polymer, all sample components are trace constituents.

Among various types of mass analysers (including magnetic-sector instruments, quadrupoles, ion traps), time-of-flight (ToF) systems have gained a prominent place in the field of polymer MS. ToF systems are eminently compatible with MALDI. MALDI produces ions in pulses, while ToF accept ions in pulses. As a result, ToF provides a very high sensitivity for the ions produced with MALDI. TOF systems also offer a high spectral resolution and accuracy and ToF systems are relatively simple and, lately, relatively affordable.

8.3.1.3 Other Methods

SEC and MALDI-MS are now by far the most important methods to investigate MMDs. However, neither technique is perfect. Lee and Chang (1996) have developed temperature-gradient interaction chromatography (TGIC) and they have demonstrated that it yields a much higher resolution than SEC. For narrowly distributed samples, the peak width in SEC is essentially determined by chromatographic band broadening (Popovici, 2004; Popovici, Kok and Schoenmakers, 2004), so that the apparent sample PDI is too high. Indeed, Chang *et al.* found much lower polydispersities for PS, PMMA, and polyisoprene standards (Lee and Chang, 1996; Lee *et al.*, 1998; Lee *et al.*, 2001a, 2001b). The usefulness of TGIC for the separation of complex polymers has been amply demonstrated (Chang *et al.*, 1999; Lee *et al.*, 1999; Park *et al.*, 2002). Apart from the high resolution, a significant advantage of TGIC is that a mobile phase of constant composition can be used. This creates the possibility of using various detectors, such as viscometry and light scattering, provided that thermal re-equilibration (cooling) takes place before the detector. Recently, there have been attempts to speed up TGIC analysis by using fast temperature gradients (Bruheim *et al.*, 2001). Fitzpatrick (Fitzpatrick *et al.*, 2004, Fitzpatrick, Staal and Schoenmakers, 2005) has demonstrated that interactive (gradient-elution) LC is equally applicable for separating polymers according to molar mass. Like TGIC, it can provide a much better resolution than SEC. An advantage is that this high resolution can be tuned for different polymers and for specific mass ranges, without the need to change (and buy!) different columns. This is also a disadvantage. LC separations require optimization for different polymers and the method-development effort is thus greater than for SEC. Because solvent gradients are used, LC is more restrictive as to the application of different detection devices than is TGIC.

Hydrodynamic chromatography (HDC) has the potential to yield rapid high-resolution separations of polymers according to molar mass. This potential has been demonstrated for packed-column HDC by Stegeman (1994), who has shown some fast and efficient separations on columns packed with very small particles (typically below 2 μm diameter). HDC is easiest and most attractive for the separation of very large molecules and small particles. The same is true for field-flow fractionation techniques. These are considered, therefore, for the characterization of latex particles.

Ultracentrifugation can yield MMDs. It can also be used in the preparative mode to obtain a series of specific fractions. The technique cannot be automated easily, but a number of samples can be analysed simultaneously. A significant disadvantage is the limited availability of the (rather complex and expensive) equipment.

8.4 Chemical Composition

8.4.1 Average Chemical Composition

We can consider both the overall chemical composition of (co-)polymers and the content of end groups or functional groups as aspects of the chemical composition (Phillipsen, 2004). However, the determination of the overall (average) composition and the end-group content often pose different requirements. The number of end groups is typically rather small, especially for high-M_r polymers. Many techniques are available for determining the average composition.

Low concentrations of highly polar (functional) end groups are still most commonly determined by titration. When applied for this purpose, titration is a simple and reliable technique which often outperforms complicated instrumental methods. In order to obtain the number of end-groups per molecule, the number-average molar mass must also be measured (see Section 8.3).

Spectroscopic techniques are highly appropriate tools for determining the average chemical composition, provided that (i) the spectral properties of the different co-monomeric units are sufficiently different, (ii) the end groups do not interfere with the measurement, and (iii) the unit absorptivity is independent of the molar mass and of the chain sequence.

UV–vis spectroscopy is a straightforward and reliable technique for determining the overall composition, if the different monomers absorb UV (or visible) light differently. Mathematical (multivariate) techniques can be used elegantly to obtain the various concentrations from the spectra, including the contribution from (UV-absorbing) end groups. However, such multivariate techniques are more commonly applied in combination with near-infrared (NIR) spectroscopy. Almost all polymers absorb in the near-infrared (wavelength range 0.8–2.5 μm). In comparison with the conventional mid-infrared (2.5–25 μm), NIR is much less sensitive and yields much less structural information. However, NIR and UV–vis signals are much more linear (signal vs concentration) as well as bilinear (additivity of the different contributions) than mid-IR signals. Also, the background of NIR is usually much lower and better defined than the background of mid-IR or UV–vis spectra.

Infrared spectroscopy is predominantly performed in the Fourier-transform mode, commonly abbreviated as FTIR. The great advantage of FTIR spectroscopy is the great number of measurement options (and accessories), that allow spectra to be taken conveniently from just about any kind of sample. Polymeric powders can be characterized by pressing them into the conventional KBr pellets, but also, without any sample preparation, by diffuse reflectance (DRIFT). Very thin films of polymers can be measured in the conventional transmission mode, but any kind of film (thick or thin), as well as large polymeric objects, can be measured by attenuated total reflectance (ATR). ATR probes can also be used to characterize solutions and emulsions. These and many other techniques combine to ensure that FTIR is still a very important technique for characterizing polymers. Quantitative chemical-composition data can be obtained from the relative intensities of different bands in the FTIR spectrum. However, this may be the weakest point of the technique. Because of nonlinearities, interferences in the spectrum and in the background and various other reasons, chemical-composition data obtained from FTIR are not as rigorously correct as one would hope – and perhaps expect.

In contrast, nuclear magnetic resonance (NMR) spectroscopy (Llauro *et al.*, 1995; Ando *et al.*, 2000) often provides rigorously correct quantitative data on the relative chemical composition. This is especially true for ^{1}H-NMR solution spectra, provided that interferences (e.g. from the solvent) are avoided. ^{13}C-NMR solution spectra can also be recorded in such a way that all signal intensities are proportional to the number of carbon atoms. The chemical-composition data are relative, in the sense that an internal standard is implicitly or explicitly used within the spectrum. Either the intensities of individual signals or the cumulative (integral) intensities of groups of peaks can be used in such relative computations. Solution NMR is much less convenient than FTIR spectroscopy, even though the demands on the quality of the solution are much less severe than in SEC. Agglomerated molecules (including emulsions with soft, swollen cores) are amenable with NMR, but not with SEC. As a result, the dissolution times can be much shorter in NMR. Although NMR spectra can also be recorded from solid samples, this is at the expense of spectral resolution and quantitative accuracy. Thus, for quantitative analysis solution NMR is the method of choice. Because NMR is a rather insensitive technique, it cannot be used to detect trace amounts of, for example, specific end groups or functional groups. For the same reason, NMR data are usually highly accurate, but not always highly precise. NMR spectra are affected significantly by the molecular microstructure. Thus NMR can be used to obtain detailed molecular information (see Section 8.4). However, the effects of, for example, the chain sequence on the spectrum may jeopardize quantitative measurements. For example, NMR yields much clearer, interpretable spectra for block copolymers than for random copolymers.

MALDI-ToF-MS (see Section 8.3.1.2) can be used to detect the total end-group mass. From this information, the nature of the end groups may possibly be determined. The procedure is fairly easy for homopolymers. A plot of the masses of a series of corresponding peaks (one monomeric unit apart) will yield the total mass of the end groups (plus possible adduct ions) as the intercept. Copolymers yield much more complicated mass spectra and a similar simple procedure does not yet exist, despite significant progress in this direction (Staal, 2005). Because of the high sensitivity of the technique, MALDI-ToF-MS is especially useful for obtaining qualitative information on end groups that are present in low concentrations in homopolymers. Again, obtaining quantitative information on the concentrations of different end groups is much more difficult.

8.4.2 Molar-Mass Dependent Chemical Composition

Composition drift, that is, variations in the polymer composition with variations in the molar mass, can be observed by combining SEC (separation according to molar mass) with methods that yield information on the chemical composition. On-line combinations are denoted with a hyphen (e.g. SEC-IR, SEC-NMR, SEC-MS) and are hence known as "hyphenated systems". Off-line systems can be denoted by one or two slashes (in this chapter we use the latter, e.g. SEC//FTIR, SEC//MS). Hyphenated on-line systems operated in the stop-flow mode can be denoted by a ⊖ sign (e.g. SEC⊖NMR).

The easiest hyphenated system consists of an LC instrument with a multiwavelength (e.g. diode-array) UV detector. Such a system is excellent for characterizing copolymers consisting of two or more types of monomeric units, all of which exhibit (different) UV activity. Unfortunately, this is hardly ever the case. A combination of a UV detector and a refractive index (RI) detector connected in series does, in principle, provide sufficient information for copolymers (two different monomeric units). However, the interdetector volumes and band broadening are a complicating factor, as are the different background and blank signals (solvent peaks) provided by the two instruments.

LC and SEC can be coupled with other spectroscopic techniques, such as (Fourier-FTIR or NMR spectroscopy, or with MS.

8.4.2.1 LC-FTIR

FTIR spectroscopy is an excellent tool for characterizing polymers. It yields clear information on the overall chemical composition and on the presence or absence of specific functional groups in the polymer molecules. The coupling of LC and FTIR has already been investigated and applied for about 25 years. FTIR has the inherent advantages of speed and sensitivity, so that it can be applied as an on-line, real-time measurement technique after separation by LC. However, a very large obstacle is the strong IR absorption by almost all common LC solvents. On-line LC-FTIR can be realized using a flow-cell interface, which is the most convenient coupling technique. LC-IR yields highly repeatable (Kok *et al.*, 2002) quantitative data. However, large parts of the spectrum may be completely obscured by solvent absorbance. The choice of solvents is very limited and gradient elution cannot be applied. LC-FTIR is compatible with SEC, which is an isocratic technique (constant eluent composition), but THF is not a suitable eluent.

To avoid the problem of solvent absorption, solvent-elimination techniques have been developed. Usually, these are off-line techniques that employ a convenient and largely automated interface between the two instruments. The sample is deposited on a substrate (e.g. an IR-transparent disk), which can then be transferred directly to the FTIR instrument. Solvent-elimination interfaces have matured over the years. Several different types are commercially available, as is the required software for recording and manipulating data (e.g. reconstructed chromatograms using the Gramm–Schmidt algorithm). Solvent-elimination interfaces are compatible with gradient-elution LC and not measuring in real time has the advantage that (small) peaks can be measured longer and irrelevant parts of the chromatogram (baseline) can be scanned quickly. However, it has proven difficult to perform accurate quantitative measurement using a solvent-elimination LC//FTIR interface. Making a deposition trace of constant quality (width, thickness and homogeneity) is a challenging task (Kok *et al.*, 2002) and even relative band intensities are not always correct. LC//FTIR is, therefore, more suitable as a problem-solving tool than as a method for routine quantitative analysis.

The off-line (Adrian *et al.*, 2000) and on-line (Kok, 2004) coupling of FTIR spectroscopy with comprehensive two-dimensional liquid chromatography (LCxSEC, see Section 8.4.4.) has already been demonstrated.

8.4.2.2 LC-NMR

The combination of LC and NMR is arguably the most attractive hyphenated system for polymer analysis, as well as in many other fields. NMR may yield a wealth of information on molecular composition *and* structure (e.g. chain regularity, branching, co-monomer sequence). Also, as mentioned in Section 8.4.1., NMR provides excellent opportunities for quantitative analysis. Thus, LC-NMR is a highly desirable proposition for polymer analysis.

Unfortunately, LC and NMR are also among the least compatible techniques. Some of the problems encountered in realizing an on-line coupling include interference from solvents and solvent gradients and the limited sensitivity of NMR. The latter is aggravated by the low concentrations in the effluent and by the desire to maintain short measuring times. One way to overcome the sensitivity problem is to use the stop-flow approach (SEC⊖NMR). However, when continuous distributions have to be characterized (as in the SEC of synthetic polymers) this is far less attractive than when a few well-separated LC peaks must be identified. Instead of using a stop-flow approach, a simple off-line combination may be far more attractive, as it allows a change of solvent and concentration of the sample by simple evaporation and redissolution.

Despite these obstacles, LC-NMR systems are increasingly available and increasingly applicable to real problems. SEC-NMR can be used to establish accurate MMDs for relatively low-M_r polymers by on-line measurement of the number-average molar mass (Ute *et al.*, 1998; Hatada *et al.*, 1988). However, SEC//MALDI seems a more attractive option for this application. Determination of the chain regularity is a strong aspect of NMR (see also Section 8.4.1.) and, in combination with LC or SEC, tacticity distributions can be determined (Kitayama *et al.*, 2000; Ute *et al.*, 2001a, 2001b). The chemical heterogeneity of high-conversion poly[styrene-co-ethyl acrylate] (Krämer *et al.*, 1999) and the functionality-type distribution of low-molar-mass polyethylene oxide (Pasch and Hiller, 1999) were studied by on-line SEC-NMR.

8.4.2.3 LC–Electrospray-Ionization–MS

In an electrospray (Montaudo and Lattimer, 2002) a flow of a liquid is dispersed into very fine droplets, while being subjected to a strong electric field. This ultimately leads to the association of intact molecules with one or more small cations or anions. Electrospray is a very soft ionization technique, that is, it yields virtually no fragment ions. This greatly simplifies the resulting spectra. On the other hand, multiple charges often occur on polar polymers, such as polyglycols. This complicates the spectra. Multiple ionization also allows larger molecules to be studied by ESI-MS. However, the technique is typically applied to study relatively low-molecular-weight polymers (or the low-molecular-weight fraction of a polydisperse sample).

When applying ESI-MS for the characterization of polymers, a high spectral resolution is beneficial. This allows the isotope pattern of multiply ionized peaks to be resolved. Even after a separation by, for example, SEC, copolymers may still give rise to very complex ESI-MS spectra, because many different molecules may elute at the same retention time. Again, high-resolution MS is desirable. MS-MS is also an interesting option.

While it is legitimate to perform ESI-MS separations off-line after separation and fraction collection, it is quite feasible to perform LC-ESI-MS on-line and this will doubtlessly be the way to go in the future. Optimum flow rates for an electrospray are of the order of 30 μl min^{-1} or less. This allows direct coupling (without the need for post-column splitting) to LC columns with inner diameters of 1 mm or less. This is quite feasible in combination with LC separations of polymers (Jiang, Lima and Schoenmakers, 2003a; Fitzpatrick *et al.*, 2004), but high-resolution SEC cannot easily be performed on narrow-bore columns (Popovici, 2004; Popovici, Kok and Schoenmakers, 2004).

In order to form a sufficiently large number of ions, ESI is most suitable for studying polar macromolecules. Fully organic (non-aqueous) eluents can be used, in which case a solution of a salt (often containing a polar solvent, such as isopropanol) is added between the column and the MS.

Electrospray is a form of atmospheric ionization, as is atmospheric-pressure chemical ionization (APCI). The latter technique is not quite as relevant for large (polymeric) molecules, because vaporization is required to a larger extent. Because of the greater analyte volatility required, APCI requires higher temperatures than ESI. This can lead to thermal degradation of polymers, but also of low-molecular-weight compounds, such as additives.

Several groups have studied the on-line coupling of MALDI and liquid chromatography. In most cases, peptides are the target analytes (Whittal, Russon and Li, 1998; Karger, Foret and Preisler, 2001). However, the application of on-line SEC-MALDI-TOF-MS for the separation of synthetic poly(ethylene glycols) has been discussed in (Zhan, Gusev and Hercules, 1999). The authors of this latter paper describe a seemingly simple interface that allows on-line interfacing of LC and MALDI. The effluent from the HPLC column is mixed with a solution of the matrix in a T-piece, the third leg of which is connected to the MALDI ionization chamber by a capillary tube, at the end of which a stainless-steel frit is glued. The LC effluent crystallizes together with the matrix on the MS side of the frit. A laser beam is used to effectuate the MALDI on the crystallized effluent. The idea is to continuously regenerate the interface through the combined actions of solvent flushing and laser ablation. However, so far the interface of (Whittal, Russon and Li, 1998) has only been used for short periods of time (5 to 10 s). This interface looks very promising, but the capacity of the vacuum system was said to constitute a limiting factor. The system will also need some changes before it can be applied in conjunction with non-aqueous eluents. For example, the PEEK tubing and epoxy resin may have to be replaced (Nielen and Buijtenhuijs, 2001).

Advances in the direction of on-line LC-MALDI-MS are obviously desirable and research is to be strongly encouraged. However, because the (off-line) application of MALDI for the analysis of synthetic polymers is still *im*mature (poor repeatability and robustness), the on-line coupling of SEC and MALDI is arguably *pre*mature.

The off-line SEC//MALDI coupling is at present a much-more realistic proposition. After the possibility of using MALDI for the accurate calibration of SEC systems was demonstrated in the 1990s (Nielen, 1998), we now see the emergence of suitable procedures and software for the determination of accurate MMDs by SEC//MALDI (Sato *et al.*, 2004).

8.4.3 Chemical-Composition Distributions

Just as the determination of MMDs requires separation of polymers according to molar mass (or molecular size in solution, e.g. by SEC) the determination of CCDs requires separation of polymers according to chemical composition (Stegeman, 1994). A distinction can be made between the overall chemical composition (monomer ratio) of a copolymer and the distribution of end groups or functional groups (functionality-type distribution) in either a homopolymer or a copolymer. In either case, separation must be based on the chemical composition of the polymer, not on its size. Although some other separation mechanisms exist (e.g. temperature-gradient interaction chromatography (Chang *et al.*, 1999; Chang, 2003) capillary zone electrophoresis and micellar electrokinetic chromatography (Oudhoff, 2004) "interactive" liquid chromatography (*i*-LC) is by far the most popular technique (Stegeman, 1994).

In *i*-LC the molecules of the analyte polymers interact with the mobile phase and the stationary phase in the column. A thermodynamic equilibrium arises, which is characterized by a distribution coefficient ($K = c_s/c_m$, where c is the concentration of the polymer in the indicated phase). The retention factor (k) is proportional to the distribution coefficient and to the phase ratio

$$k = \frac{V_R - V_0}{V_0} = \frac{t_R - t_0}{t_0} = K\frac{V_s}{V_m} \tag{8.1}$$

where V_R is the retention volume (and t_R the retention time), V_0 is the column hold-up volume (and t_0 the column hold-up time), V_s is the volume of stationary phase in the column and V_m is the volume of mobile phase. Retention is thus determined by the distribution coefficient, which in turn is determined by thermodynamic interactions, as is demonstrated by

$$RT \ln K = -\Delta G = -\Delta H + T\Delta S \tag{8.2}$$

where ΔG is the partial molar Gibbs energy associated with the transfer of one mole of analyte from the mobile phase to the stationary phase, ΔH is the partial molar enthalpy and ΔS is the corresponding entropy effect. Enthalpic (heat) effects arising from molecular interactions are reflected in ΔH. SEC is a strictly entropic process, that is, $\Delta H = 0$ and temperature has no significant effect on the elution volume.

Separation can be achieved if different parts of the molecule (different co-monomers, end groups, functional groups) exhibit different interactions with the mobile phase and the stationary phase in the column. We can rewrite Equations 8.2 for a homopolymer as follows

$$RT \ln K = -\Delta G = -p\Delta G_{\text{monomer}} - \Sigma\Delta G_{\text{endgroups}} \tag{8.3}$$

where we assume that the partial molar free energy is built up from p contributions of monomeric units (p being the degree of polymerisation) and the sum of all contributions from end groups (or functional groups). Because p is a large number for high-M_r polymers, reasonable distribution coefficients (and thus retention factors) can usually be obtained only if $\Delta G_{\text{monomer}} \approx 0$.

A special case is the situation in which $\Delta G_{\text{monomer}} = 0$. In this case the distribution coefficients and chromatographic retention factors are determined only by the functional groups and are independent of the chain length (p). This situation is known as critical chromatography (or liquid chromatography at the critical conditions) and it is eminently suitable for separating polymers based on functionality. Figure 8.2 shows an example of the separation of functional poly(methyl methacrylates) (Jiang, Lima and Schoenmakers, 2003a; Jiang *et al.*, 2003b). All PMMA molecules without OH groups are eluted around 4 min, irrespective of the molecular weight and (possible) other (weakly interacting) end

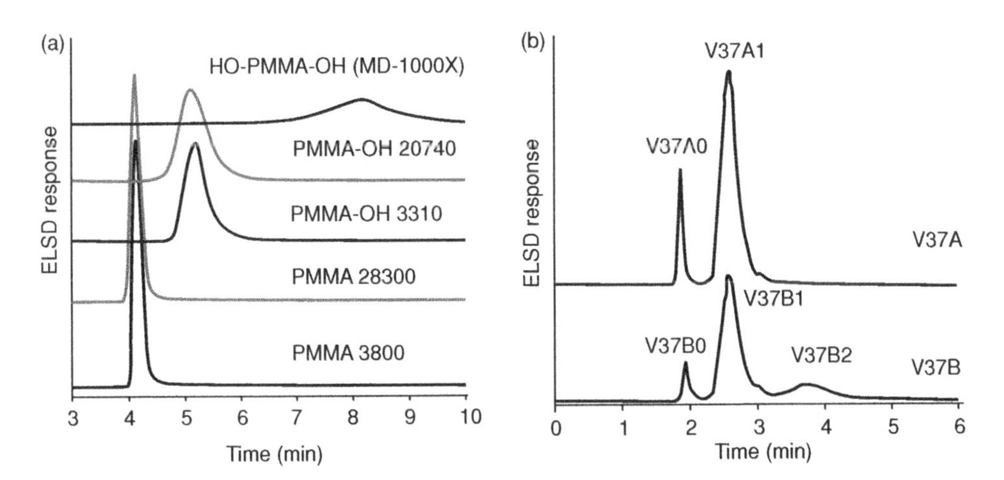

Figure 8.2 *(a) Separation of end-functional poly(methyl methacrylates) based on the number of end-groups. Column: 150 mm length × 4.6 mm i.d., home-packed with Hypersil silica (3-μm particles; 100 Å pore size); mobile phase: 43% acetonitrile in dichloromethane; temperature: 25 °C; flow rate: 0.5 ml min^{-1}; injection volume: 10 μl; sample concentration: 1 mg ml^{-1} in dichloromethane; detector: evaporative light scattering. (b) Separation of end-functional PMMAs prepared by RAFT polymerization. Mobile phase: 40% acetonitrile in dichloromethane. Other conditions as in Figure 8.2a. Reprinted with permission from [Jiang, 2003] Copyright (2003) Elsevier Ltd.*

groups present. Likewise, the polymers with one OH end group are all eluted around 5 min and the bifunctional ("telechelic") PMMAs are eluted around 8 min in Figure 8.2a. The elution profiles of real samples (Figure 8.2b) can be translated into a functionality-type distribution (FTD), provided the detector is suitably calibrated (Mengerink *et al.*, 2001; Peters *et al.*, 2002; Jiang *et al.*, 2004; van der Horst and Schoenmakers 2002). This particular example of OH functional polymers is very relevant to emulsion polymerisation because initiation of polymerisation by OH radicals can occur as well as hydrolysis of persulfate-derived end groups, leading to OH end groups. Note that the retention times in Figure 8.2b are different from those in Figure 8.2a, because a somewhat different mobile phase is used. However, separation according to the number of functional groups is achieved in both cases.

The separation shown in Figure 8.2 has proven to be quite robust. However, this is not usually the case for critical separations. Indeed, for carboxyl-functionalized PMMAs it proved much more difficult to achieve genuine critical chromatography (Jiang, Lima and Schoenmakers, 2003a; Jiang *et al.*, 2003b). Critical liquid-chromatographic separations are not always easy, but they can be highly rewarding, especially for determining FTDs.

For a copolymer Equations 8.2 becomes

$$RT \ln K = -\Delta G = -\sum_i p_i \Delta G_{\text{monomer},i} - \Sigma \Delta G_{\text{end groups}} \tag{8.4}$$

where the subscript i depicts the different monomeric units. Whereas it is difficult to achieve $\Delta G_{\text{monomer},i} = 0$ for one particular monomer (critical chromatography), it is impossible to find conditions that are critical for several different monomers simultaneously. Therefore, critical chromatography is much more useful for determining the FTDs of functional polymers than for determining the CCDs of copolymers. In the latter case, two options are open. One is to find conditions at which the separation is critical towards one type of monomer, while the second monomeric unit does not show any interaction, so that it is eluted under SEC conditions. Such conditions have been applied to block copolymers. The block for which critical conditions are maintained is made "invisible" and the separation reflects the block-length distribution of the second block. An example is shown in Figure 8.3.

Although separations such as in Figure 8.3 are not easy, and although there is some discussion on whether the A block can really be made "invisible" (Falkenhagen *et al.*, 2000; Lee *et al.*, 2001a, 2001b; Sato *et al.*, 2004), it is clear that the separation can be very useful for analysing block copolymers.

The alternative way to analyse copolymers is to resort to gradient elution. In this case the composition of the mobile phase is changed during time. At the initial composition, both monomers are highly retained (negative ΔG). When the eluent becomes stronger, the critical conditions for one of the monomeric units will be approached. At a later point in time, this will be the case for the second monomer. In this way, a blend of polymers can be separated into its constituents. Copolymers will be eluted according to their composition (Figure 8.4). In principle, gradient-elution liquid chromatography can be used to obtain CCDs of copolymers. Again, proper calibration of the detector is a significant issue.

i-LC separations can also be coupled to spectrometers and other highly informative detection devices. The situation is similar, but not identical to that described for SEC. In many cases, gradient elution is used, that is, the solvent changes as a function of time. In

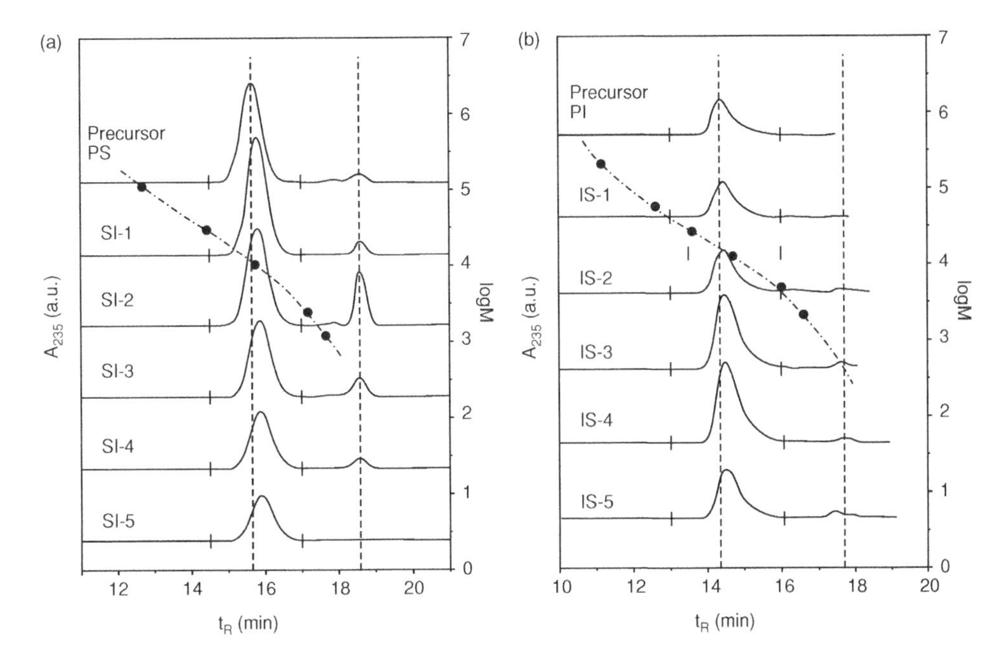

Figure 8.3 *Liquid chromatography of PS*-b-*PI copolymers at the critical conditions of the block of variable length. (a) Critical conditions for PI: three Nucleosil C18 columns in series (100 Å, 500 Å, 1000 Å pore sizes; 250 length × 4.6 i.d. mm each); mobile phase dichloromethane : chloroform 78:22 (v/v); temperature 47 °C; PS block* $M_w = 12.0$ *kg* mol^{-1}*; PI blocks (from top to bottom)* $M_w = 3.0$, *6.0, 11.1, 21.4 and 34.2 kg* mol^{-1}*, respectively. (b) Critical conditions for PS: three Nucleosil silica columns in series (100 Å, 500 Å, 1000 Å pore sizes; 250 length × 4.6 i.d. mm each); mobile phase THF: iso-octane 78:22 (v/v); temperature 7 °C; PI block* $M_w - 12.5$ *kg* mol^{-1}*; PS blocks (from top to bottom)* $M_w = 3.3$, *5.9, 13.5, 26.6 and 38.1 kg* mol^{-1}*, respectively. Reprinted with permission from [Jiang, 2003] Copyright (2003) Elsevier Ltd.*

that case hyphenation between *i*-LC and viscometry or light scattering is horribly difficult. Gradient elution is not used with such devices, so that it is much more difficult to determine the (average) molar mass as a function of the chemical composition ($M_r(\varphi_{pol})$) than it is to determine the average composition as a function of molar mass ($<\varphi_{pol}(M_r)$). In principle it is possible to combine viscometry or light scattering with critical chromatography. However, since the latter technique is more practical for relatively low molar masses and the detection devices are most suitable for high masses, this is not a good match.

Both LC-FTIR and LC-NMR can be applied in combination with solvent gradients. In both cases there are some complicating factors. In the case of FTIR, only solvent-elimination interfaces can realistically be used. Some authors have programmed the deposition conditions to obtain optimal results for gradient-elution LC-FTIR. However, as was mentioned earlier (Section 8.4.2.1), it is not easy to obtain accurate quantitative results on the copolymer composition using LC-FTIR. In LC-NMR a solvent gradient causes severe complications associated with the suppression of the solvent signal. While suppression techniques for

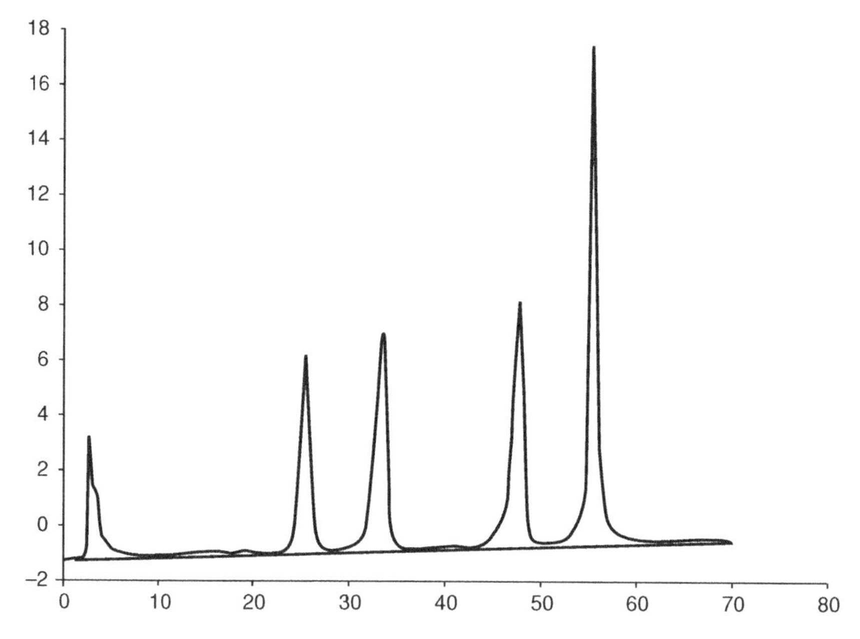

Figure 8.4 Gradient-elution liquid chromatography of a mixture ("blend") of a number of copolymers. Column: Supelcosil Discovery C18, 150 mm length × 2.1 mm i.d., particle size 5 μm, pore diameter 180 Å, temperature 25 °C, flow rate 0.2 ml min^{-1}, injection volume 5 ml, sample concentration 1.5 mg ml^{-1}, gradient from 5 to 95% THF in acetonitrile.

gradient-elution LC have been developed and successfully applied, the interferences in the spectrum become more serious than they are in isocratic separations. In either case, LC-FTIR or LC-NMR, the amount of additional information obtained is limited. The LC retention axis contains information on the polymer composition. The information present in the spectra is related to this. Although additional information on structural aspects may be obtained from both FTIR and NMR spectra, the two information dimensions are far from orthogonal.

This is fundamentally different for the combination of *i*-LC with MS, either on-line (most commonly using LC-ESI-MS) or off-line (most commonly using LC//MALDI-ToF-MS). In that case, the LC axis contains mainly structural information, while the MS axis provides information on the molar mass. A disadvantage of this combination is that fractions resulting from the *i*-LC separation are expected to be narrow in terms of their chemical-composition distribution, but may be quite broad in terms of their molar-mass distribution. In critical or pseudo-critical *i*-LC the very purpose of the separation is to obtain all different molar masses in a single narrow fraction. Such broad fractions are not really compatible with MS. As was mentioned in Section 8.3.1., biased results are anticipated from the MS analysis of broadly distributed samples.

Better results are anticipated if fractions that are both narrow in chemical composition and in molar mass are subjected to mass spectrometry. Such fractions can be obtained from two-dimensional separations (see next section).

8.4.4 Two-Dimensional Distributions

Just as the characterization of polymer distributions necessitates polymer separations, the characterization of two-dimensional polymer distributions necessitates two-dimensional polymer separations. Only if two distributions are fully independent do two separate one-dimensional separations suffice. This is the case if every chemical-composition fraction exhibits the same MMD and every molar-mass fraction exhibits the same CCD. Because this is not usually the case, one two-dimensional separation usually reveals (much) more information than two one-dimensional separations.

8.4.4.1 Comprehensive Two-Dimensional Liquid Chromatography

Two-dimensional liquid-chromatographic separations can be performed in the linear ("heart-cut") format or in the comprehensive mode. In the former case, one (or a few) fractions are isolated from the sample and these are subsequently subjected to a second separation. An advantage of this approach is that the specific fraction(s) can be subjected to two (lengthy) high-resolution separations. A great disadvantage is that only one or a few small fractions of the sample are extensively characterized. In comprehensive two-dimensional liquid chromatography the entire sample is subjected to two different separations. The word "comprehensive" is justified if the final (two-dimensional) chromatogram is representative of the entire sample (Schoenmakers, Marriott and Beens, 2003). The recommended notation for linear ("heart-cut") two-dimensional liquid chromatography is LC-LC, whereas comprehensive two-dimensional liquid chromatography is commonly denoted by LC × LC (Lee *et al.*, 2001a, 2001b).

In the case of polymer separations, the MMD is usually one of the distributions of interest. The second most important distribution is usually either the CCD or the functionality-type distribution (FTD). This implies that SEC and *i*-LC are attractive candidates for the two dimensions in comprehensive two-dimensional liquid chromatography. These two techniques can, in principle, be coupled in two different orders (either LC × SEC or SEC × LC, with the first dimension listed first). LC × SEC has a number of prevailing advantages (van der Horst and Schoenmakers, 2003). These include (i) the possibility to perform high-resolution (gradient) LC in the first dimension, (ii) the finite time of analysis in the second dimension, (iii) the greater choice of detectors (because the separation in the second dimension is isocratic), (iv) that the first dimension LC conditions can be changed without the need to re-optimize the seconddimension conditions, and (v) that the first dimension LC system is not easily overloaded. If the first dimension were SEC, the sample (fraction) transferred to the second dimension LC would be dissolved in a very strong solvent, creating a great danger of detrimental "breakthrough" peaks (Jiang, van der Horst and Schoenmakers, 2002). A disadvantage is that the resolution in the second (fast-SEC) dimension is limited, but the series of advantages prevails. Therefore LC × SEC is now the commonly employed technique.

If we are to maintain the separation (resolution) that has been achieved in the first dimension in the eventual LC×SEC chromatogram, we need to collect a large number of fractions. To maintain a reasonable overall analysis time, this implies that the second dimension separation should be fast and that the resolution that can be obtained in this second dimension is limited. There have been significant developments towards fast SEC in recent years (Pasch and Kiltz, 2003; Popovici, 2004; Popovici, Kok and Schoenmakers,

2004). Moderate-resolution SEC can be performed within 1 or 2 min. If we want to collect 100 fractions from the first dimension, this implies that typical LC × SEC analysis times are of the order of 2–3 h. Indeed, these are the analysis times commonly encountered in practice.

If we wish to transfer the entire first dimension fraction to the second dimension, then the first dimension column should have a much smaller internal diameter than the second dimension column. Either a "miniaturised" first dimension column can be used in combination with a conventional second dimension column, or a conventional first dimension column can be used in combination with a wide-bore ("maximised") second-dimension column. Both approaches have been successfully demonstrated. The first approach requires (very) much less solvent, produces correspondingly less waste, and is compatible with most existing LC detectors, including the molar-mass selective detectors (viscometry, light scattering) that are of great interest in polymer separations. The second approach puts fewer demands on the chromatographic (first dimension) system in terms of extra-column band broadening and it yields larger separated fractions for subsequent off-line analysis by other methods (e.g. NMR).

Figure 8.5a shows an outline of a typical LC × SEC system and Figure 8.5b shows an enlarged representation of the switching valve. A comprehensive two-dimensional liquid

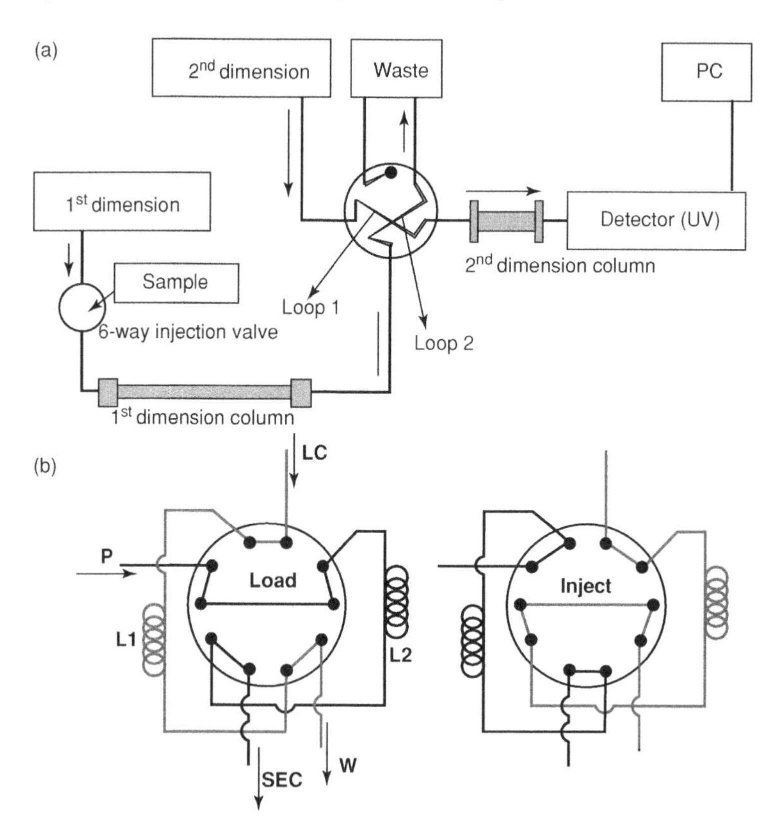

Figure 8.5 *Instrumentation for comprehensive two-dimensional liquid chromatography. (a) scheme of the complete instrument and (b) preferred configuration of a 10-port switching valve. Reprinted with permission from [Schoenmakers, 2003] Copyright (2003) Advanstar Communications.*

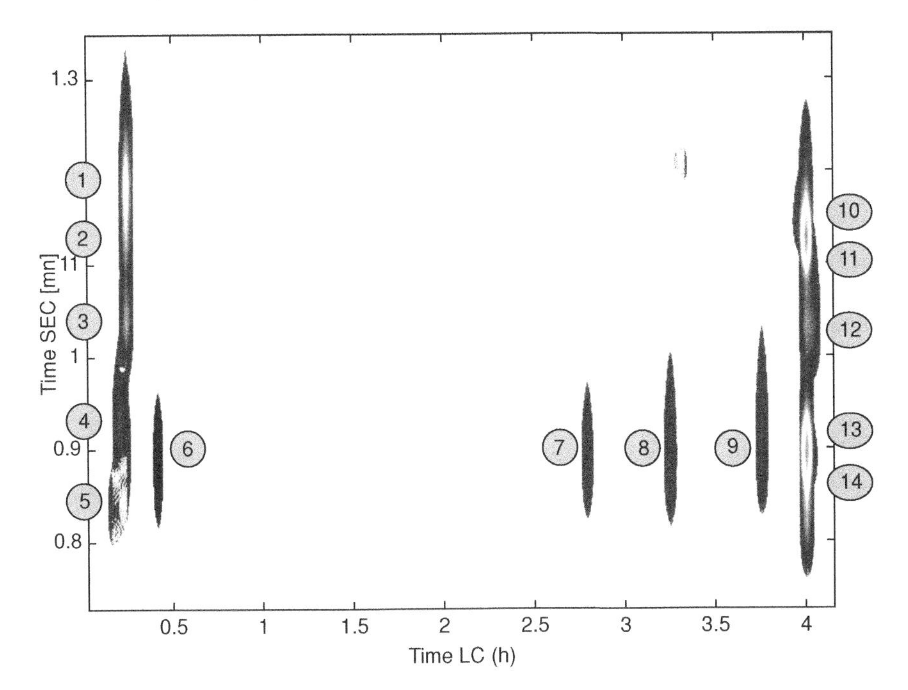

Figure 8.6 *LC × SEC-ELSD contour plot of a mixture of homo- and copolymeric reference materials: PMMA 2900 (1), 6950 (2), 28 300 (3), 127 000 (4), 840 000 (5); S-co-MMA 20% S (6), 40% S (7), 60% S (8), 80% S (9); PS 2450 (10), 7000 (11), 30 000 (12), 200 000 (13) and 900 000 (14); LC (first dimension): C18-column; flow 4 μl min^{-1}; gradient 5–70% THF in ecetonitrile 0–300 min (40 °C); SEC (second dimension): mixed-C-column; flow 0.6 ml min^{-1} THF. Reprinted with permission from [Lee, 2001b] Copyright (2001) American Chemical Society.*

chromatography system typically consists of two liquid chromatographs that are interfaced by means of a switching valve. In the case of LC×SEC the first dimension often features a gradient-elution system, that is, the composition of the eluent can be programmed during the run. The valve is configured such that while one fraction is being analysed, the next fraction is being collected (Figure 8.5b).

Figure 8.6 shows a contemporary example of an LCxSEC separation (Van der Horst and Schoenmakers, 2003). It shows the two-dimensional separation of a series of copolymer "standards" of known molar mass and chemical composition. The first (gradient-elution LC) dimension shows a high resolution, whereas the resolution in the second (SEC) dimension is adequate. The two separations are seen to be nearly orthogonal, that is, separation in the first dimension is (nearly) completely based on the chemical composition, whereas that in the second dimension is based on molecular size.

Comprehensive two-dimensional liquid chromatography has seen a strong increase in popularity and in the number of applications in recent years. LC × SEC has been applied to a large number of problems in polymer science. For example, the techniques has been used to provide a detailed analysis of polystyrene-poly(methyl methacrylate) diblock copolymers (Pasch, Mequanint and Adrian, 2002), to analyse well-defined star polylactides (Biela *et al.*, 2002), and to study the grafting reaction of methyl methacrylate onto EPDM (Siewing *et al.*, 2001) or onto polybutadiene (Siewing *et al.*, 2003).

The main bottlenecks for the proliferation of LC × SEC (and other two-dimensional polymer separations) are now the development of suitable calibration procedures and the associated software. Because suitable hardware for LC × SEC is already commercially available, significant progress is anticipated in this direction.

8.4.4.2 *Two-Dimensional Distributions from MALDI-MS*

The great advances made in the analysis of polymers by matrix-assisted laser desorption/ionization mass spectrometry are also reflected in new approaches for determining CCDs. Willemse *et al.* (2004) obtained "polymer fingerprints" from MALDI spectra of block copolymers of styrene and isoprene. The four fingerprints shown in Figure 8.7 represent four samples withdrawn from the reaction vessel during the synthesis of the second (polyisoprene) block. The distribution of the first block (polystyrene, horizontal axis) is seen to remain constant. A truly comprehensive two-dimensional distribution is obtained that can easily be converted in an MMD and a CCD (and the corresponding MMDxCCD) if desired. However, the representation as in Figure 8.7 is probably clearer for the purpose of distinguishing block copolymers from random copolymers.

The data shown in Figure 8.7 were verified by measuring the average chemical composition independently by NMR spectroscopy. The data agree very well with theoretical expectations based on random-coupling statistics. The authors concluded that, in this particular case, variations in the efficiency of ionization and mass-spectral analysis were insignificant. Despite this remarkable success, the authors warn that this is not necessarily the case for other polymers. Yet, it has been demonstrated that detailed information on complex polymers can be obtained from the MALDI-ToF-MS spectra of complex (co-)polymers and efforts in this direction are ongoing (Ando *et al.*, 2000).

8.5 Detailed Molecular Characterization

8.5.1 Chain Regularity

The stereochemical microstructure (tacticity) of polymers can greatly affect their thermal and mechanical properties. Therefore, the ability to investigate the tacticity is of great importance for understanding structure–property relationships. Also, studying the chain regularity may yield information on the monomer-addition process. NMR is the prime technique in this context (Pichot, Llauro and Pham, 1981; Marciniec and Malecka, 2003; Phillipsen, 2004). NMR has been applied to study the chain structure of many different homopolymers and even of some complex copolymers (Koinuma, Tanabe and Hirai, 1982; Van der Velden, 1983).

The tacticity distribution of polymers can be assessed by measuring their ease of crystallization (or, conversely, the ease of redissolution). This is routinely done for polyolefins by temperature-rising elution fractionation (TREF (Mingozzi, Cecchin and Morini, 1997; Boborodea *et al.*, 2004)), a technique which can also be combined with viscometry and light scattering detectors (Yau and Gillespie, 2001). In TREF a hot solution of a polymer is brought onto a column. The flow is then stopped (or reduced to a very low value) and the temperature is slowly reduced. Fractions with a greater degree of chain regularity

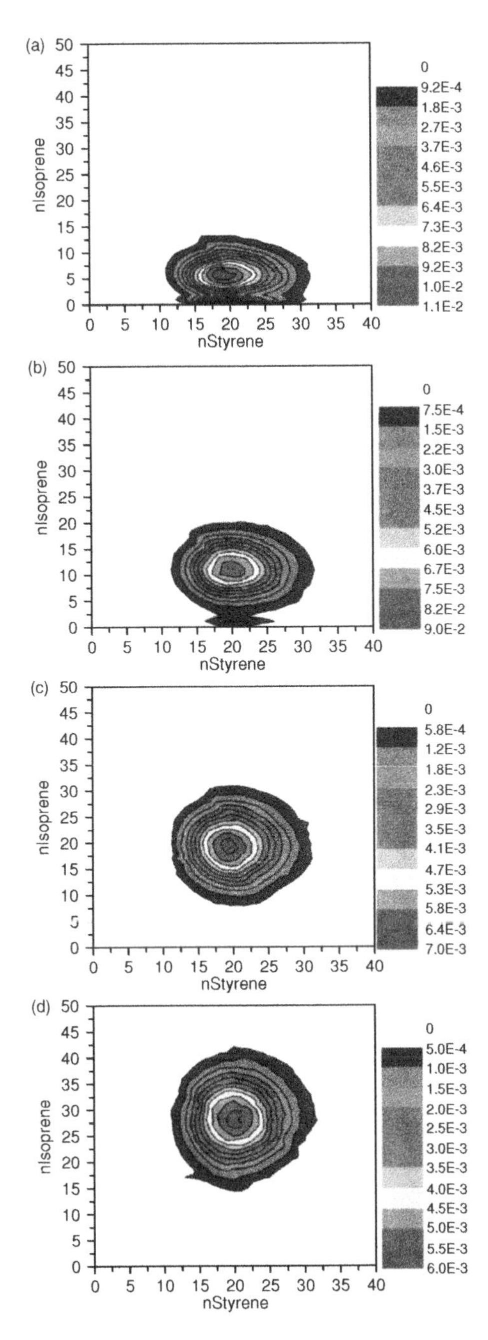

Figure 8.7 *(a–d) Copolymer fingerprints of the system polystyrene-*block*-polyisoprene corresponding to approximately 25%, 50%, 75%, and 100% conversion of the isoprene monomer. The number distributions of styrene and isoprene monomeric units are calculated from the MALDI-ToF-MS spectra of the copolymers. Reprinted with permission from [Siewing, 2001] Copyright (2001) Wiley-VCH.*

crystallize more easily. They are deposited first and eluted last when the temperature is increased again for the actual analysis. Many other polymers are not amenable to TREF, because they are readily soluble at room temperature. For such polymers, conventional *i*-LC procedures can be used, with the advantage that the lengthy temperature programming can be avoided (Berek *et al.*, 1994). *i*-LC methods have also been coupled on-line to NMR to study polymer tacticity (Hiller and Pasch, 2001; Ute *et al.*, 2001a, 2001b).

8.5.2 Branching

The average number of branches (per molecule) can be estimated if the (number-average) molar mass and the total number of end groups (per unit mass of polymer) can be measured. One of the techniques (e.g. VPO) from Table 8.1 can be used to determine M_n. The number of end groups can, in special cases, for example, be determined by titration. Knowledge of the polymerisation process is required to know which end-groups are present. For example, one end group often arises from the initiation reaction, whereas the other end group of a linear polymer chain is determined by the termination reactions.

NMR can be used to estimate the number of end groups, provided that the degree of branching is reasonably high. If a signal for the branch point can be assigned, ^{13}C-NMR may yield a direct estimate of the average number of branches per given number (usually 1000) of C-atoms. Britton *et al.* studied the branching and sequence distribution of copolymers of vinyl acetate and *n*-butyl acrylate prepared by semi-batch emulsion copolymerisation using ^{13}C-NMR spectroscopy (Britton, Heatley and Lovell, 2001).

For polymers of moderate to high molar mass the hyphenated systems with SEC and M_r-selective detectors (viscometry, light scattering) are extremely useful. If a branched polymer is compared with a linear reference material, the degree-of-branching distribution can be estimated either from the intrinsic-viscosity distribution (IVD) as obtained from on-line viscometry, or from the ratio of root-mean-square radii obtained using multi-angle light scattering (Grcev, Schoenmaker and Iedema, 2004). Viscometry is feasible for polymers with molar relative masses in excess of a few thousand. Measuring root-mean-square radii is based on the angular dependence of the scattered intensity, which can be measured reasonably well for polymers with relative molar masses exceeding about 50 000.

Frequently, viscometry and light scattering detection are combined in a single instrument (Mendichi and Schieroni, 2001; Wang *et al.*, 2004). This yields direct estimates of the IVD and the MMD, as well as of the Mark–Houwink constants. The root-mean-square radius cannot be obtained from such a “triple SEC” or “triple detector” system.

9

Particle Analysis

Ola Karlsson[1] *and Brigitte E.H. Schade*[2]
[1]*Division of Physical Chemistry, Lund University, Sweden*
[2]*Particle Sizing Systems, Waterman, Holland*

9.1 Introduction

The recent rapid progress within particle nanotechnology has expanded the number of stakeholders and broadened the fields of application for polymer or composite particles produced via emulsion polymerisation. Not only coating and adhesive applications but increasingly also applications in life science and various electronics contribute to the expansion. The need for analytical tools in the different fields is somewhat different but when the actual particles are to be analysed very much the same questions as previously remain to be answered and often the most obvious property to look for when particles have been produced is the particle size. Well-characterized latexes with respect to particle size, shape, surface properties and internal structure are a necessity to drive the development further. This chapter will deal with methods for particle size measurements, surface characterisation, particle shape and structure. However, the content in this chapter is also closely related to the discussions in several of the other chapters in the book and in Figure 9.1 is shown a schematic representation of a colloidal suspension containing latex particles. The chapters in the book covering the analysis and/or the origin of specified characteristics are indicated by notations in the figure. In addition to the polymer composition, its molecular mass and molar mass distribution (Chapter 8) and the colloidal properties such as stability and rheology (Chapter 7), the latex particle size and morphology (Chapter 6) have a large impact on the performance of the final products. Depending on what the important properties are in the final application there are many ways to characterize particles produced via emulsion polymerisation and we will provide assistance to choose appropriate analytical methods for solving common problems associated with particle analysis. In the first section, particle

Chemistry and Technology of Emulsion Polymerisation, Second Edition. Edited by A.M. van Herk.

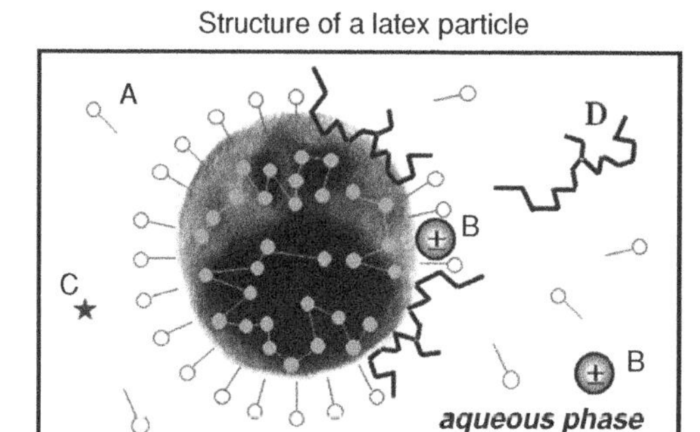

A	surfactant
B	charge
C	salt (initiator, buffer etc)
D	protective colloid
E	polymer

Structure parameters of latexes

colloid properties

- surfactant, type, amount (*ch. 7*)
- number of sphere charges (*ch. 7,9*)
- number of charges in serum (*ch. 8*)
- M_w of sphere chains (*ch. 7*)
- M_w of serum chains (*ch. 7*)

polymer properties

- chemical composition (*ch. 8*)
- T_g (*ch. 8*)
- M_w (*ch. 8*)
- M_w distribution (*ch. 8*)
- degree of crosslinking (*ch. 8*)
- grafting (*ch. 8*)

particle properties

- shape (*ch. 9*)
- size (*ch. 9*)
- size distribution (*ch. 9*)
- morphology (*ch. 9*)

Figure 9.1 *Schematic representation of a colloidal suspension containing latex particles.*

size analysis methods and solutions as well as challenges associated with that will be presented. The text will focus on the most frequently used techniques more in detail and briefly mention older or more specialized techniques for determination of particle sizes. Analytical aspects of latex particle shapes, structures and surfaces will be treated from Section 9.6 onwards.

9.2 Particle Size and Particle Size Distribution

9.2.1 Introduction

Knowledge of particle size or the particle size distribution (PSD) is important for reliably characterizing the quality and stability of a wide variety of particulate-based systems. Examples of chemical/physical properties of polymer emulsions affected by the PSD include: viscosity, suspension and emulsion stability, film uniformity and hardness, gloss, opacity, colour, thermal conductivity and others (Collins, 1991).

Particle size and the PSD rarely influence the properties of a particulate system in a straightforward way. For example, for paint pigments the projected area diameter is important, while for chemical reactants the total surface area is the most relevant parameter. The "type" of diameter measured depends on the method used. For a particle of irregular shape, no single number, or set of numbers, may adequately describe its physical dimensions. Instead, a fictitious "spherical-equivalent" diameter is often used to characterize its "size". Therefore, in referring to particle size or the PSD, one must realize which size characteristic best correlates with the final properties of the system of interest, and what uncertainties are introduced by the analysis and calculation method utilized.

The concepts of average particle size and PSD are discussed in Section 9.2, and the difficulty in obtaining a representative sample for particle size measurements is reviewed in Section 9.3. Approximately 400 methods for particle size analysis have been reported (Scarlet, 1982). The methods used most frequently for characterizing submicron particles made by emulsion polymerisation are reviewed briefly in Section 9.4. Several techniques used for analyzing latexes are compared in Section 9.5.

Most particle sizing methods are developed on the basis of uniform (monodisperse) spherical particles. Hence, non-spherical particles usually must be defined by their equivalent spherical diameter (ESD) – the diameter the particles should have, assuming they are spheres. However, the ESD is strongly dependent on the physical method underlying a given particle sizing technique. A small subset of the many different diameter values that can be used to define the "size" of a particle is shown in Table 9.1.

Table 9.1 *Definitions of particle size.*

Symbol	Name	Definition
d_s	surface diameter	The diameter of a sphere having the same surface area as the particle
d_v	volume diameter	The diameter of a sphere having the same volume as the particle
d_d	drag diameter	The diameter of a sphere having the same resistance to motion as the particle in a fluid of the same viscosity and at the same velocity
d_a	projected area diameter	The diameter of a sphere having the same projected area as the particle when viewed in a direction perpendicular to a plane of stability
d_f	free-falling diameter	The diameter of a sphere having the same density and the same free-falling speed as the particle in a fluid of the same density and viscosity
d_{St}	Stokes' diameter ($d_{St} = (d_v^3/d_d)^2$)	The free-falling diameter in the laminar flow region ($Re = 0.2$)
d_A	sieve diameter	The width of the minimum square aperture through which the particle will pass
d_{vs}	specific surface diameter	The diameter of a square having the same ratio of surface area to volume as $d_{vs} = d_v^3/d_s^2$ the particle
d_F	Feret's diameter	The distance between tangents on opposite sides of the particle along the direction of scanning across the particle
d_M	Martin's diameter	The distance between opposite sides of a particle measured on a line bisecting the projected area
d_c	perimeter diameter	The diameter of a circle that has a perimeter equal to the perimeter of the projected area of the particle
d_{max}	maximum diameter	The maximum dimension of the particle
d_{min}	minimum diameter	The minimum dimension of the particle

Table 9.2 Some mean diameter definitions.

Number, length mean diameter	$d_{NL} = \dfrac{\sum L}{\sum n} = \dfrac{\sum nD}{\sum n}$
Number, surface mean diameter	$d_{NS} = \sqrt{\dfrac{\sum S}{\sum n}} = \sqrt{\dfrac{\sum nD^2}{\sum n}}$
Number, volume mean diameter	$d_{NV} = \sqrt[3]{\dfrac{\sum V}{\sum n}} = \sqrt[3]{\dfrac{\sum nD^3}{\sum n}}$
Length, surface mean diameter	$d_{LS} = \dfrac{\sum S}{\sum L} = \dfrac{\sum nD^2}{\sum nD}$
Surface, volume mean diameter	$d_{SV} = \dfrac{\sum V}{\sum S}$
Volume, moment mean diameter	$d_{VM} = \dfrac{\sum M}{\sum V} = \dfrac{\sum nD^4}{\sum nD^3}$
Geometric, mean diameter	$d_g = \dfrac{\sum n \log D}{\sum n}$

9.2.2 Average Particle Diameter

In practice, dispersed systems are rarely monodisperse; rather, they contain a range of particle sizes, described by a PSD. The system can be characterized most simply through the use of an average, or mean, diameter. A variety of mean particle diameters can be defined, some of which are listed in Table 9.2.

9.2.3 Particle Size Distribution

Because the average particle size is often not adequate to characterize a sample, a particle size distribution (PSD), either volume- (i.e., mass-) or number-weighted, must usually be determined. In the former case the number of particles of a given diameter, d, is multiplied by the volume of a (spherical) particle of that size – that is, $\pi d^3/6$.

9.3 Sampling

An important, potentially difficult problem in particle size analysis is the need for a representative sample, requiring that dispersions be made homogeneous (e.g., by stirring) before taking a sample. However, depending on the mechanical stability of the system, excessive shear forces can induce agglomeration or flocculation of the particles. Attention also must be paid to possible adherence of particles to the walls of the container. After obtaining a representative sample, additional problems may arise from handling or treatment of the sample associated with the analysis. Most methods require that the sample be dispersed and diluted in a fluid. The selection of an appropriate fluid and, optionally, additives for wetting, dispersing and stabilizing the sample is important.

9.4 Particle Size Measurement Methods

Some popular methods of particle size analysis and their ranges of applicability are listed in Table 9.3. The methods used most often to analyze polymer emulsions, described below, are conveniently divided into three categories: (1) ensemble techniques (e.g., LD and DLS); (2) separation techniques (e.g., capillary hydrodynamic fractionation (CHDF) and disk centrifugation); (3) ultra-high separation techniques (e.g., SPOS and electrozone sensing).

Table 9.3 *Particle size methods and ranges.*

	Method	Range [μm]
Microscopy:	Electron microscopy	0.001–10
	Dark field microscopy	0.05–0.5
	Flow ultramicroscope	0.05–1
	Optical microscopy	>0.5
	Optical array	>0.3
Light scattering:	Classical:	
	Dissymetry	0.05–0.5
	Maximum–minimum	>0.2
	Turbidity	>0.3
	Polarization ratio	0.02–0.25
	Forward angle ratio	0.05–0.5
	Higher order Tyndall spectra	0.2–2.4
	Dynamic light scattering	0.003–5
Particle movement:	Sedimentation field flow	
	Fractionation	0.001–1.0
	Hydrodynamic chromatography	
	Normal HDC	0.03–1.5
	Capillary HDC	0.015–1.1
	Size exclusion chromatography	<1.5
	Disk centrifuge	0.08–6.0
	Ultracentrifuge	0.05–3.0
	Sedigraph	0.1–100
	Andreason pipette	1–100
	Electrozone sensing	0.2–1000
	Electro-acoustics	0.1–10
Others:	Membrane filters	0.01–100
	Soap titration	0.05–0.5
	Fractional creaming	0.05–1.0
	Permeametry	>1
	Gas adsorption	
	Sieving	20–5000
	SPOS	0.5–5000

9.4.1 Ensemble Techniques

Two ensemble techniques, both based on light scattering, are often used to characterize polymer emulsions. The first, usually referred to as "laser diffraction" (LD), is based on classical, or "static" (i.e., time-averaged), light scattering. The second, usually referred to as dynamic light scattering (DLS), is based on fluctuations in the scattered light intensity due to Brownian motion of the suspended particles. While both of these techniques yield information on the "polydispersity" of the sample (i.e., the range of particle sizes contained therein), they are usually used only to obtain a reliable measure of the mean diameter, in order to identify the "end point" in a manufacturing process (e.g., batch-process emulsion polymerisation). To ensure that the performance of a newly produced material will be acceptable, it is often sufficient to monitor only the mean particle size, comparing the values obtained from previous successful runs.

9.4.1.1 Laser Diffraction Technology

The technology referred to as LD in fact embraces two completely different physical methods – Fraunhofer diffraction and Mie scattering. Which technique is chosen depends on the particle size range of interest. For particles larger than 1.5–2 μm, a spatial pattern of diffracted light appears in the near-forward direction (small angles). The analysis is straightforward in the case of uniform-size particles, for which the diffraction pattern consists of concentric rings of light, where the angular spacing between rings (intensity maxima) varies approximately inversely with the particle diameter. For a typical polydisperse sample, the analysis is "ill-conditioned", and therefore challenging, because the diffraction pattern consists of a superposition of ring-like patterns, each having a different periodicity, requiring mathematical deconvolution to discover the constituent sizes and approximate volume-weighted PSD.

Analysis of smaller particles requires the use of Mie-scattering theory, which describes the angular variation of the scattered intensity due to the mutual interference of scattered light waves originating from different points within each given particle. For uniform particles the variation of the scattered intensity with angle depends not only on the particle size (and laser wavelength), but also on both the real and imaginary (absorptive) components of its refractive index, as well as the index of the solvent (typically water). As with Fraunhofer diffraction, the analysis of a polydisperse sample is also ill-conditioned and challenging, because the measured plot of intensity versus angle consists of a superposition of individual angle-dependent intensities associated with each particle size. The measured plot of intensity versus angle must therefore be "inverted", using an appropriate deconvolution algorithm, to obtain an approximate PSD. Because polymer emulsions are typically colloidal, dominated by submicron particles, Mie-scattering, rather than Fraunhofer diffraction, is usually employed.

Figure 9.2a and b show the PSD results obtained by LD/Mie scattering for a polymer emulsion, using two different assumptions for the imaginary component, b, of the refractive index, $n = a + bi$, where the real component, a, equals 1.48 in both cases. In the first case (Figure 9.2a) the absorptive component is very small, $b = 0.01$. In the second case (Figure 9.2b) it is considerably larger, $b = 0.2$, although still small on an absolute basis. Clearly, the shapes of the resulting PSDs differ significantly, demonstrating extreme sensitivity to

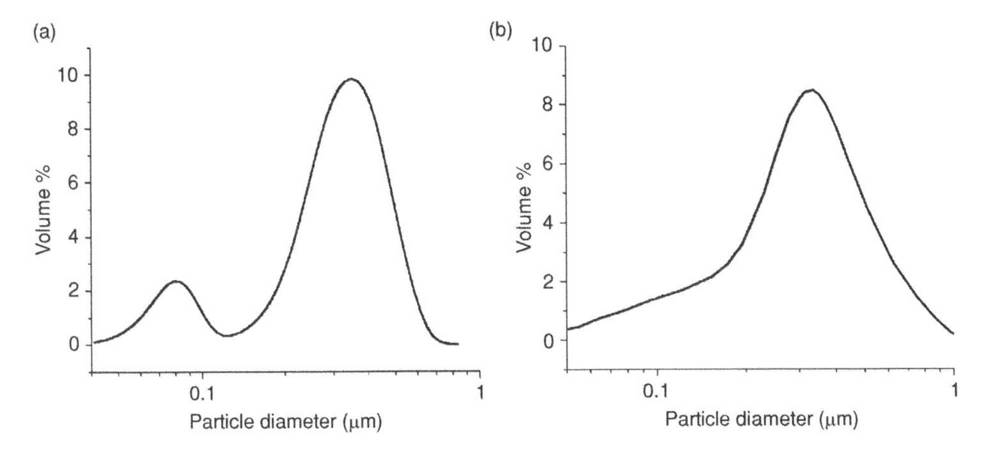

Figure 9.2 *(a) PSD obtained by LD/Mie scattering, assuming particle index* n $= 1.48 + 0.01i$. *(b) PSD obtained by LD/Mie scattering, assuming particle index* n $= 1.48 + 0.2i$.

the choice of the absorptive component, which in practice is unknown and often difficult to determine.

9.4.1.1.1 Mie Scattering Theory – A Brief Summary. According to classical (Mie) light scattering theory, both the magnitude and angular dependence of the intensity of light scattered by particles depend on their size. If the particles are very small compared to the laser wavelength (in the liquid), λ, there is no appreciable angular dependence. If the particles grow to approximately $1/10$ of λ, there is measurable enhancement of the intensity in both the forward and backward directions, shown schematically in Figure 9.3. If the particle size becomes comparable to λ, there is significantly more scattering in the forward than backward direction. Further increases in the particle size cause even stronger scattering in the forward direction, as well as intensity minima and maxima at larger scattering angles due to intra-particle interference, resulting in a unique particle size "fingerprint".

The angular pattern of scattering also depends on the shape of the particles. However, in the simplifying case of uniform (isotropic), non-adsorbing and non-interacting particles, the scattering behaviour is mainly determined by two parameters, α and m, defined by

$$\alpha = \pi d/\lambda \tag{9.1}$$

$$m = \mu/\mu_0 \tag{9.2}$$

where μ and μ_0 are the refractive indices of the particles and liquid, respectively. The full expression describing the angular dependence of the scattered light is given by Mie. However, for certain ranges of α and m the theory can be simplified:

1. $\alpha < 0.3$, $m \approx 1$; diameter $d \ll \lambda$ (Rayleigh region), (Allen, 1968)
2. $0.3 < \alpha < 3/(m-1)$, $m > 1$; diameter d comparable to λ (Lorentz–Mie region), (Pangonis, Heller and Jacobson, 1957)
3. $\alpha > 3/(m-1)$; diameter d significantly $> \lambda$ (Fraunhofer diffraction region).

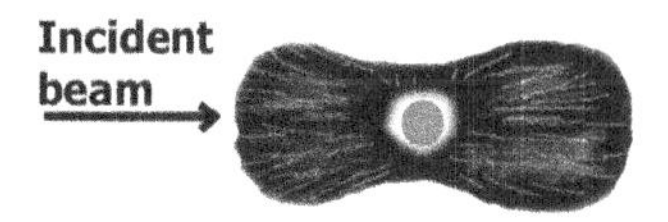

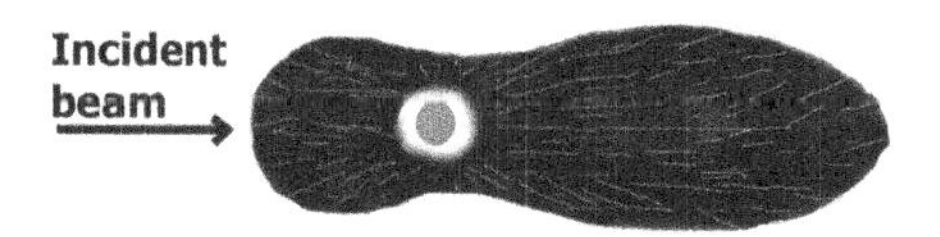

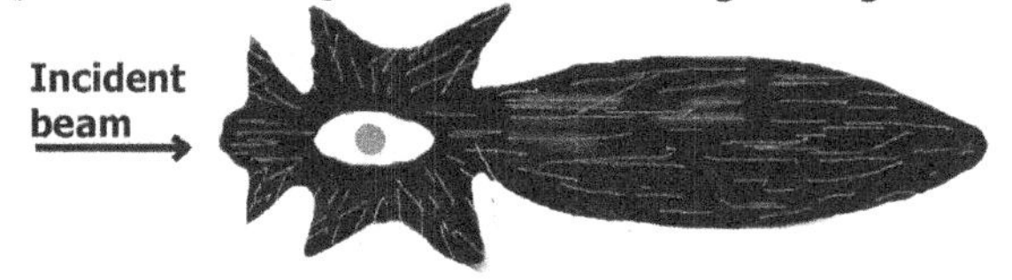

Figure 9.3 *Intensity versus angle light scattering patterns – a function of particle size.*

1. Rayleigh region
   The ratio of the intensity scattered at angle θ to that of the incident beam, I/I_0, can be described as a sum of vertically and horizontally polarized intensity ratios (Allen, 1968),

$$I/I_0 = |\alpha|^2 (2\pi/\lambda)^4 [(1 + \cos^2\theta)/2r^2] \tag{9.3}$$

   where r is the distance from the particle to the point of measurement. In the case of spherical particles of volume V, parameter α is given by,

$$\alpha = (V/4)[3(m^2 - 1)/(m^2 + 2)] \tag{9.4}$$

2. Lorentz–Mie region
   When $m > 1$ and d is comparable to λ – the case for most latex systems – the full Lorentz–Mie theory must be employed. The scattered intensity ratio I/I_0 is then given by,

$$I/I_0 = \lambda^2(i_1 + i_2)/8\pi^2r^2 \tag{9.5}$$

   where i_1 and i_2 are Bessel and Legendre functions, respectively. The results of the Lorentz–Mie theory are often multi-valued and cannot be expressed in closed form; however, useful values can be obtained from calculated tables and numerical evaluations (Pangonis, Heller and Jacobson, 1957; Stevenson and Heller, 1961).
3. Fraunhofer diffraction region.
   When $d \gg \lambda$, part of the light is diffracted. When $m > 1$, refraction of the incident light can be neglected and Fraunhofer diffraction theory applies. (A geometric shadow

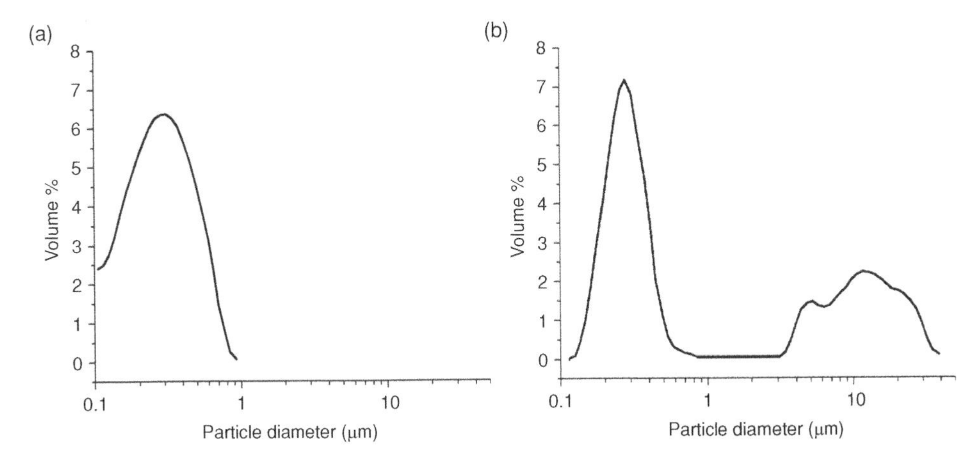

Figure 9.4 *(a) PSD obtained for a "good" (stable, relatively uniform) polymer resin using laser diffraction (LD). (b) PSD obtained for a "bad" (unstable, aggregated) polymer resin using LD.*

develops behind each particle.) The angle-dependent scattered intensity is given by,

$$I(\theta) = c\, I_0(\pi^2 d^2/16\lambda^2)[2J_1(\pi d\theta/\lambda)/(\pi d\theta/\lambda)]^2 \tag{9.6}$$

where c is a constant and J_1 is a Bessel function.

Typical PSD results obtained for a polymer resin using LD technology (combining Mie scattering and Fraunhofer diffraction) are shown in Figure 9.4a and b. In the case of a "good" (relatively monodisperse, unaggregated) resin (Figure 9.4a), the PSD consists of a single peak, having a median diameter of 283 nm. In the case of a similar, but colloidally unstable sample (Figure 9.4b), the PSD also contains a main peak, but with median diameter shifted upward to 368 nm. It also shows a second peak, centred at 15 μm. The combination of these features signals the presence of an aggregate population in the second sample, but only in a qualitative sense. Unfortunately, this result fails to describe accurately either the true shape of the PSD or the quantitative extent of the "tail" of aggregated, over-size particles actually present. As will be discussed later, a technique of much higher resolution is required to reveal the true PSD.

Advantages:

1. Fast measurement – usually just a few minutes
2. Easy to operate and interpret PSD results – useful for QC monitoring
3. Robust measurement – relatively insensitive to background contamination
4. No input sample parameters required for particles larger than 1.5–2 μm.

Disadvantages:

1. Ill-conditioned mathematical analysis – different "models" can be used to invert the data, often resulting in different PSD results and/or serious artifacts
2. Requires both real and imaginary (absorptive) refractive index parameters for sizing small particles – that is, below 1.5–2 μm

3. Measures many particles at once – insensitive to small amounts of coarse, over-size particles
4. Volume-weighted PSD – often insensitive to significant percentages of fine particles
5. Relatively poor resolution – unable to characterize highly polydisperse PSDs.

9.4.1.2 *Dynamic Light Scattering (DLS)*

A second ensemble technique is DLS – also called quasi-elastic light scattering (QELS) or photon correlation spectroscopy (PCS) – is based on analysis of the temporal fluctuations in the scattered intensity caused by Brownian motion, or diffusion, of the particles (Nicoli and Toumbas, 2004). In recent years it has become a popular technique for characterizing many submicron colloidal systems, including polymer latexes, because of its large size range (roughly 1 nm to 5 μm) and approximate independence from optical properties.

A laser beam is focused into a cell containing a stationary, dilute suspension of particles, which scatter light in all directions. A small fraction of the scattered light is detected at a fixed angle, θ (typically 90°), representing a superposition of waves originating from the various particles. The phase of each light wave depends on the instantaneous location of the particle in the suspension from which it originates (assuming, for simplicity, $d \ll \lambda$). Brownian motion causes the net detected intensity to fluctuate randomly due to the "random walk" of each of the particles. Nevertheless, there is a well-defined average "lifetime", τ, of the intensity fluctuations. In the simple case of just two particles, τ is the average time required for the two scattered waves to change from being in phase to out of phase with each other, due to a change in relative optical path length of $\lambda/2$. The mean lifetime τ is inversely related to the diffusion coefficient, D. In order to determine D from the fluctuating intensity, it is useful to construct the intensity autocorrelation function, $G^{(2)}(\delta t)$,

$$G^{(2)}(\delta t) = \langle I(t) \times I(t - \delta t) \rangle \quad (9.7)$$

for a large number of discrete times, δt, by means of a digital autocorrelator. The symbol $\langle \ \rangle$ indicates a sum of intensity products over many discrete time values, t. A large number (e.g., 10^5–10^7) of intensity values are sampled to obtain a statistically reliable value for $G^{(2)}(\delta t)$ for each value of δt. The maximum value of $G^{(2)}(\delta t)$ occurs at $\delta t = 0$, as the particles are perfectly "correlated", having had no time to move: $G^{(2)}(\delta t) = \langle I(t)\rangle^2$. Conversely, the smallest value occurs in the limit of very large δt, where the sampled intensities are essentially completely uncorrelated due to extensive Brownian motion: $G^{(2)}(\delta t) = \langle I(t)\rangle^2$.

In the simplest case of particles of uniform size, $G^{(2)}(\delta t)$ consists, after subtraction of the long-time "baseline" $\langle I(t)\rangle^2$, of a decaying exponential function,

$$G^{(2)}(\delta t) - \langle I(t) \rangle^2 = A \exp(-2Dq^2\delta t) \quad (9.8)$$

where A depends on the sample, optics and run time, and q is the "scattering wavevector",

$$q = (4\pi n/\lambda_0)\sin(\theta/2) \quad (9.9)$$

where n is the refractive index of the suspending liquid and λ_0 the laser wavelength in vacuum. The diffusivity, D, is easily obtained from the decay time constant of $G^{(2)}(\delta t)$, and finally the particle diameter can be obtained from the well-known Stokes–Einstein relation,

$$D = kT/3\pi\eta d \tag{9.10}$$

where k is Boltzmann's constant, T the temperature and η the shear viscosity of the solvent.

In the more usual case of a polydisperse colloidal system, containing a mixture of particle diameters, d_i, and corresponding diffusivities, D_i, Equations 9.8 is replaced by the square of a weighted sum of exponential functions, each decaying at a rate corresponding to a particular particle diameter contained in the PSD,

$$G^{(2)}(\delta t) - < I(t) >^2 = A\,[\Sigma\; f_i\; \exp(-D_i q^2 \delta t)]^2 \tag{9.11}$$

where each weighting coefficient, f_i, is proportional to the number concentration of the particles of diameter d_i, the square of the particle volume (i.e., $d_i^{\,6}$) and the square of the scattering intensity "form factor", which accounts for intra-particle interference.

A suitable mathematical algorithm must be used to "invert" Equations 9.11 to obtain the set of coefficients, f_i, constituting the intensity-weighted PSD, or PSD_I, from the "raw data", $G^{(2)}(\delta t)$. Typically two kinds of algorithms are employed in DLS-based instruments. The simplest is the method of cumulants, based on a least-squares fit of a polynomial (quadratic or cubic) in δt to the "reduced" autocorrelation function, $Y(\delta t)$,

$$Y(\delta t) = \ln[G^{(2)}(\delta t)/ < I(t) >^2 - 1]^{1/2} \tag{9.12}$$

which yields the z-average diffusion coefficient and normalized standard deviation (or variance) associated with the PSD_I. The volume-weighted quantity, PSD_V, is obtained from the PSD_I by dividing by the particle volume and the square of the intensity form factor. Relatively little scattering data are required to achieve stability in the computed PSD_I, given the "smoothing" of $G^{(2)}(\delta t)$ inherent in the polynomial fitting procedure, requiring only two, or at most three, parameters to be extracted from the data. This straightforward method is often effective when applied to simple systems that can be approximated by single-peak ("unimodal") distributions.

Alternatively, in the case of multi-modal PSDs, a second, more complex, algorithm is required to invert the $G^{(2)}(\delta t)$ data – typically a non-linear, least-squares regression technique. Several variations on this approach are commercially available. The inversion algorithm should have several attributes: (i) high accuracy and precision (e.g., for sizing monodisperse latex standards); (ii) highest size resolution (e.g., clean separation of close bimodals); (iii) robustness, or "stability", so that the computed PSD_I is relatively insensitive to the acquisition of additional scattering data. Figure 9.5 shows the PSD_I found by nonlinear least-squares regression analysis of the $G^{(2)}(\delta t)$ data for a bimodal mixture of polystyrene latex standard particles: 101/171-nm (1 : 2 vol). Considering the close spacing of the two sizes, the results are very good. When the same data are analyzed by the method of cumulants, an intensity-weighted mean diameter of 122 nm is obtained, with a normalized coefficient of variation equal to 0.31. Despite its favourable goodness of fit (low chi-squared value), this simplified single-peak PSD must be rejected in favour of the physically correct bimodal result shown in Figure 9.5.

Advantages:

1. Rapid analysis time (usually < 10 min)
2. Wide size range, 0.001–5 μm

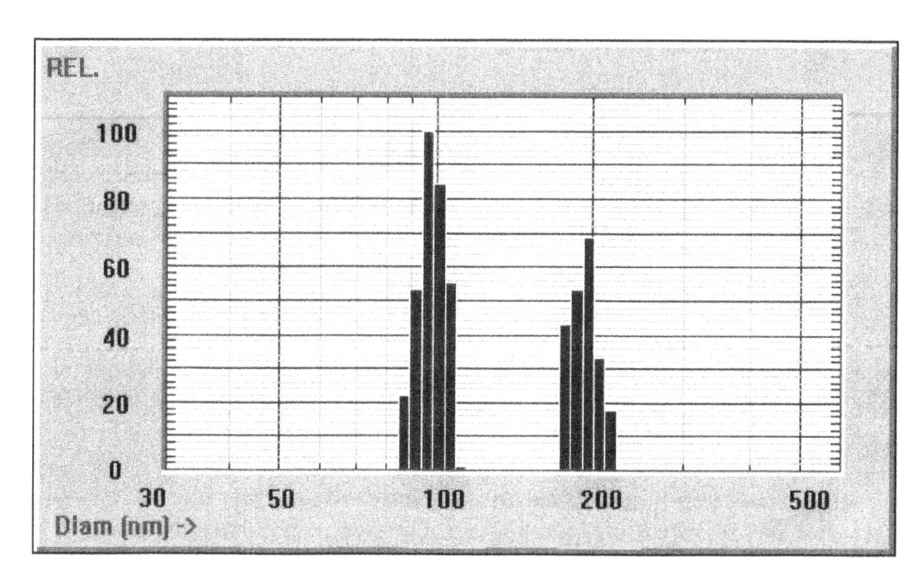

Figure 9.5 PSD obtained by DLS for a bimodal mixture of 101- and 171-nm polystyrene latex particles (2 : 1 vol).

3. Small sample volume (1 ml or less), minimal preparation/handling
4. Sample optical properties not required (first approx.)
5. Absolute method – no calibration required
6. Suitable for stability assessment
7. High precision and reproducibility (for nearly uniform particles).

Disadvantages:

1. Ensemble technique – inversion algorithm required (ill-conditioned)
2. Limited resolution – difficult to measure multimodal distributions
3. Analytical function assumed for PSD
4. Dust-free samples required – low tolerance for background contamination.

9.4.2 Particle Separation Methods

Particle separation methods can be divided into two groups: (1) those that use gravity or centrifugal force, in which the particles move relative to the suspending medium; (2) those that use chromatographic principles, where the particles move with the same velocity as the fluid.

9.4.2.1 Sedimentation Methods

Sedimentation is one of the oldest techniques, apart from sieving, for determining the PSD, relating the velocity of sedimentation due to gravitational or centrifugal force to particle size – the "hydrodynamic radius". While gravitational sedimentation is restricted to particles of larger masses, analytical centrifugation may be applied to particles as small as 50 nm.

Two versions of the technology are employed – disk and batch (cuvette) centrifugation. In both cases a sensor (single or multiple) detects a signal related to the concentration of the particles within the sensing volume. Light extinction is generally used, but X-ray attenuation is also useful for certain applications (e.g., low-Z).

Two standard operating procedures are used – line start (LIST), generally used for disc centrifuges, and homogeneous start (HOST), usually for cuvette centrifuges. In the LIST method a small sample is injected into the hollow centre of the spinning disk containing the suspending fluid. For best results the fluid should have an externally generated density gradient to avoid "negative" density gradients with increasing radius, yielding unstable sedimentation curves and distorted results, including peak broadening. In the HOST method the cuvette contains a well-mixed sample of the dispersion, having uniform concentration at the outset. Although the LIST method gives more precise PSD results, including better resolution, the HOST method is easier to operate in practice.

In the case of the LIST method the curve of particle concentration versus sedimentation time of the particles passing the detector (at a given position) represents a differential PSD. The relationship between the (spherical) particle diameter, d, and the time, t, required for it to move from the meniscus to the detector, in the absence of a density gradient, is given by Stokes' law,

$$d^2 = (18/\omega^2 t)\,[\eta/(\rho_s - \rho_f)]\ln(x/x_m) \tag{9.13}$$

where ω is the rotor speed, ρ_s the density of the colloidal particles, ρ_f the density of the spin fluid, x the distance from the rotor axis to the detector position, x_m the distance from the rotor axis to the meniscus of the spin fluid and η the viscosity of the spin fluid. In the case of the HOST method, the concentration dependent signal at a fixed position, x, within the sample cuvette is monitored as a function of time t, and the same relationship given by Equations 9.13 applies.

The intensity signal produced by the detector does not provide a direct measurement of the particle concentration, as it also depends on the particle mass and, in the case of light extinction, on the particle size itself. The signal can be related to the particle concentration using the theory of electromagnetic waves and particle interactions. The intensity, I, transmitted by a uniform dispersion illuminated by monochromatic light of intensity I_0 is given by Beer's law,

$$I/I_0 = \exp(-\tau L) \tag{9.14}$$

where L is the path length through the rotor fluid and τ characterizes the optical density of the sample. From this equation the extinction S can be obtained,

$$S = \tau L = \pi/4d^2\, N^* Q_{ext} L \tag{9.15}$$

where N^* is the number concentration and Q_{ext} the extinction efficiency of the spherical particles, a function of their size, refractive index and λ. For a differential PSD there is an equation representing a volume size distribution,

$$S/Q_{ext} = \text{const}\; d^3\, \delta N/\delta d \tag{9.16}$$

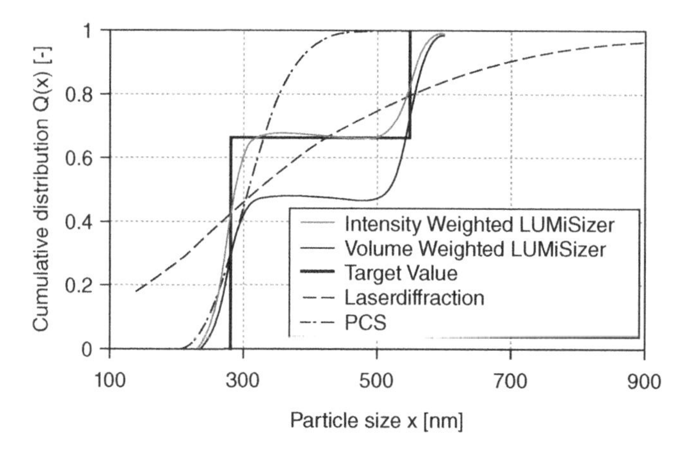

Figure 9.6 *Theoretical and experimental (intensity- and volume-weighted) PSDs obtained for a 280/540-nm (2 : 1 vol) mixture of silica particles by cuvette photocentrifuge (HOST).*

which is a function of the particle size, d, and the spectral characteristics of the detector. It can be transformed into an area size distribution using

$$d^3 \delta N / \delta d = d^2 \, \delta N / \delta \log d \tag{9.17}$$

Using this optical correction, obtained from S/Q_{ext}, an accurate PSD can be obtained.

Figure 9.6 shows the PSD obtained for a bimodal dispersion of 280 and 540 nm spherical latex particles (mixed 2:1 by volume), using a cuvette photocentrifuge based on the HOST approach. The theoretical PSD, the intensity-weighted PSD (not including the size dependence described by Equations 9.16) and the volume-weighted PSD (including Equations 9.16) are shown. The high size resolution, in contrast to other methods, is obvious.

Advantages:

1. Large, representative sample analysed
2. High resolution due to separation – effective for closely-spaced multimodals
3. Large size range – able to analyse particles typically from 0.05 to 300 μm in diameter
4. Any particle which can be centrifuged can be analysed
5. PSD calculations are straightforward – no calibration required
6. Only small sample volume required.

Disadvantages:

1. Particle size limit depends on the densities of particles and fluid and the viscosity of the fluid
2. Selecting a proper spin fluid and gradient can be tedious (disc centrifuges)
3. Hydrodynamic instabilities can arise if concentration gradients occur
4. Very broad distributions require long measurement times
5. Refractive indices (including absorbance) of particles and fluid are required to obtain volume-weighted PSDs.

9.4.2.2 *Chromatographic Methods*

9.4.2.2.1 Hydrodynamic Chromatography (HDC). The HDC technique for separating colloidal particles, introduced by Small *et al.*, (Small, 1974) originated from the observation that particles flowing through a packed column of uniform, impermeable beads are fractionated, emerging in order of decreasing diameter. Larger particles are excluded from the regions near the walls of the flow channels and, therefore, attain a greater average velocity than smaller particles, owing to the parabolic fluid velocity profile associated with laminar Poisseuille flow. Later, other workers developed a variant of this technique, called capillary hydrodynamic chromatography (CHDC), using a long thin capillary tube (Noel *et al.*, 1978; Mullens and Orr, 1979).

Originally, CHDC suffered from low efficiency in the submicron size range. This shortcoming was addressed by adding non-ionic surfactant to the eluent, which adsorbs on both the capillary wall and the particles, so as to reduce both the absolute and relative cross sectional area available to the particles (De Jaeger, Trappers and Lardon, 1986). However, to substantially augment the separation of submicron particles, much smaller diameter capillaries are required. Significant improvements were made by Silebi *et al.* (Dos Ramos and Silebi, 1989) who renamed the technique capillary hydrodynamic fractionation (CHDF), recognizing that, strictly speaking, HDC is not a chromatographic method. Figure 9.7 shows a representative PSD result obtained by CHDF for a trimodal mixture of latex particles – mean diameters of 43, 152 and 260 nm.

Advantages:

1. Rapid analysis time (15 min or less)
2. Small sample required
3. Universal – no limitation on chemical composition
4. Able to be automated (sample injection and data analysis)
5. Yields a size distribution
6. Attractive size range for latexes, 0.015–1.5 μm

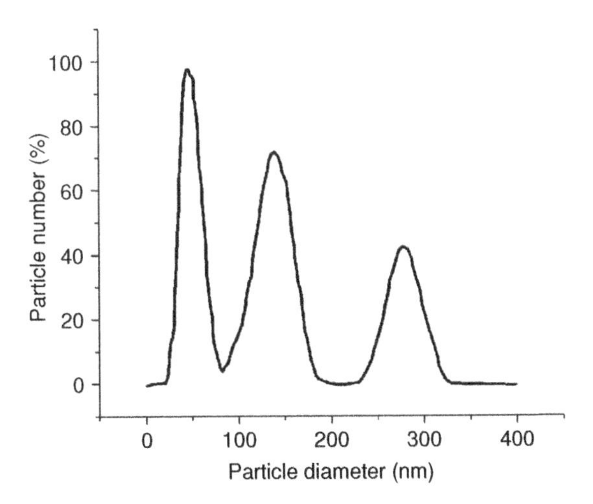

Figure 9.7 *PSD obtained by CHDF for a trimodal mixture of latex particles (43, 152 and 260 nm, 1 : 1 : 2 vol).*

Disadvantages:

1. Susceptible to column plugging – possible retention of particles (“soft” polymers)
2. Possible mass balance errors due to particle retention
3. Calibration of columns required
4. Relatively low resolution
5. Requires removal of over-size/outlier or agglomerated particles.

9.4.2.3 *Ultra-High Separation Methods*

Ultra-high separation techniques – specifically, single-particle optical sensing (SPOS) and electrozone sensing (ES) – are required for applications in which quantitative detection of subtle features of the PSD, often inaccessible by ensemble or normal separation techniques, is required. Examples include “tails” of either ultra-fine or over-size particles, undetectable using ensemble methods if their volume-fraction is sufficiently small. Knowledge of these features is often critical for assessing the stability and overall quality of process materials and final products. Unlike LD or DLS, the SPOS and ES techniques respond to individual particles passing through an active sensing zone, in which the signal consists of discrete pulses of varying heights, each corresponding to a single particle, requiring no mathematical inversion or other manipulation. A particle size is assigned to each measured pulse height, through use of a suitable calibration curve. The PSD is constructed with the highest possible resolution – one particle at a time.

9.4.2.3.1 Single Particle Optical Sensing (SPOS). A known quantity of concentrated sample suspension is diluted sufficiently that particles in a given size range pass individually through the active sensing zone of the sensor, thereby avoiding coincidences and distortion in the resulting number-weighted size distribution, PSD_N. The sensing zone is a thin, ribbon-like volume traversing the flow cell, with a thickness (20–40 μm) defined by a focused “sheet” of incident laser light and lateral dimensions (≈0.5–1 mm) fixed by entrance/exit windows and the remaining cell walls. Each particle entering the sensor traverses this zone, resulting in a signal pulse, the size of which depends on the physical sensing method employed – light extinction (LE) or light scattering (LS).

The LE method involves detection of the momentary decrease in light flux transmitted through the flow cell caused by the passage (transit time typically 10–20 μs) of a particle through the thin sensing zone. The physical mechanisms responsible for light extinction include refraction/scattering, reflection and diffraction. In general, the larger the particle, the greater the fraction of incident light that fails to reach the distant LE detector and hence the greater the pulse height, ΔV_{LE}, superimposed (negative-going) on the background signal level, V_0, in the absence of particles. For relatively large particles ($d \gg \lambda$), refraction dominates the extinction process. The particles behave like lenses, bending incident rays away from the beam axis, provided their refractive index differs sufficiently from that of the fluid. The size of each particle is determined from ΔV_{LE} by real-time interpolation of the calibration curve, constructed using standard particles of known size. The effective size range for a typical LE sensor is 1.3–400 μm.

By contrast, the LS method relies on the detection of light momentarily scattered by a particle when it passes through an optical sensing zone similar to, or the same as, that described above. The LS method is used to detect and size particles that are too small to be measured using the LE method. The signal consists of a brief pulse of height ΔV_{LS} superimposed on a background level that is ideally zero, given clean fluid and no other particles in the sensing zone. Pulse height ΔV_{LS} increases monotonically with particle size, provided the scattered light is collected over an appropriate range of sufficiently small angles. The sensitivity of the sensor is a function of laser power and the "contrast" of the particles – their refractive index compared to that of the suspending fluid. Unfortunately, the maximum measurable particle size is limited, due to the strong dependence of the scattering intensity on particle diameter, resulting in saturation of the detector/amplifier at large sizes. The effective size range for a typical LS sensor is 0.5–5 μm.

Fortunately, both high sensitivity and large dynamic size range can be achieved in a single sensor by using a novel, hybrid SPOS technique, herein called "LE+LS", which smoothly combines the separate LE and LS signals (Nicoli and Toumbas, 2004). The effective size range resulting from a typical LE+LS sensor is 0.5–400 μm. Figure 9.8a shows the PSD "tail", $d > 0.56$ μm, obtained for the "good" polymer resin depicted in Figure 9.4a. The volume fraction of polymer found in the size range 0.56–10 μm is 0.01%. By comparison, Figure 9.8b shows the PSD tail obtained for the "bad" (aggregated) resin seen in Figure 9.4b. The volume fraction of polymer in the same size range is fully 7.75%. Despite being unable to depict the "entire" PSD, the smooth decaying plot shown in Figure 9.8b clearly provides a much more accurate physical picture of the outlier particles in the unstable polymer compared to the PSD shown in Figure 9.4b, obtained using the LD technique. In addition, the SPOS technique yields accurate values for the volume fraction in the PSD tails.

Recent significant advances in the SPOS technique involve the use of an incident light beam that is much narrower than the lateral extent of the flow cell, resulting in a much smaller sensing zone than normally employed (Nicoli and Toumbas, 2004). There are two important consequences of this radical change in optical design. First, the maximum particle concentration (coincidence limit) increases from approximately 10^4/ml to 10^6/ml for the LE mode of detection and up to 10^7/ml for the LS mode. Hence, starting concentrated sample suspensions require much less dilution, and the diluent fluid can contain higher levels of contaminant particles without compromising analysis results – both important practical advantages. Second, given the much smaller cross-sectional area of the sensing zone, particles of a given size are capable of producing a much larger signal in either the LE or LS detection mode. The result in either case is a significant reduction in the threshold for particle detection: <0.6 μm in LE mode and <0.15 μm in LS mode.

The large increase in working concentration for the new SPOS technique results from the fact that only a small (but fixed) fraction of the particles flowing through the sensor are detected, yielding individual signal pulses. However, unlike the traditional technique, particles of uniform size give rise to a "spectrum" of pulses of widely varying heights, depending on the particle trajectories through/near the sensing zone. The raw data consist of a pulse height distribution (PHD) – number of pulses (particles) versus pulse height – which must be deconvoluted, using an appropriate algorithm, in order to obtain the desired PSD. Despite this process, the resulting PSD still possesses relatively high resolution, because the original signal still consists of individual pulses, each representing the response of the sensor to a single particle.

Figure 9.9 illustrates the high resolution that is achievable in the new LE design, showing the PSD obtained from a mixture of polystyrene latex standard particles of diameters 0.20,

(a)

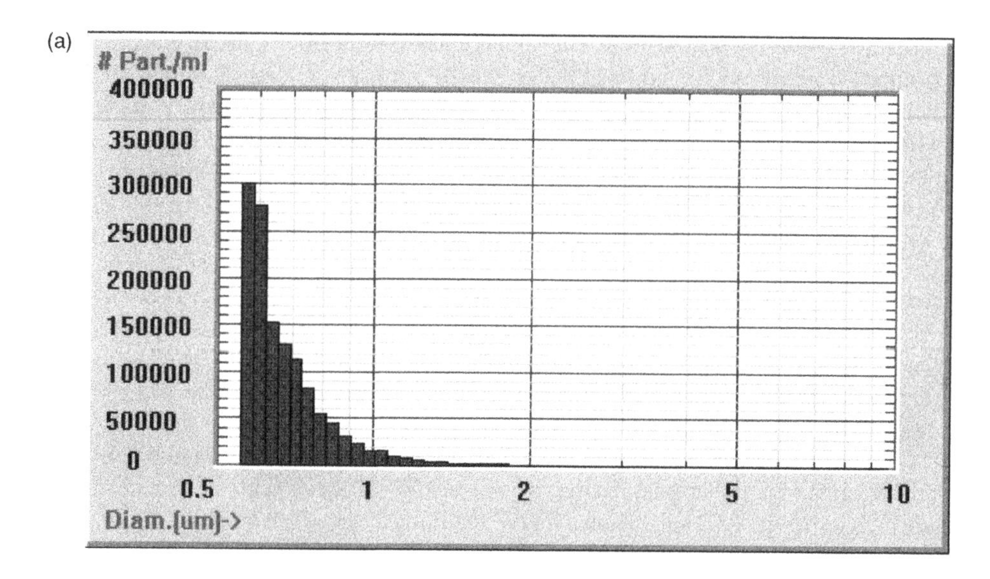

(b)

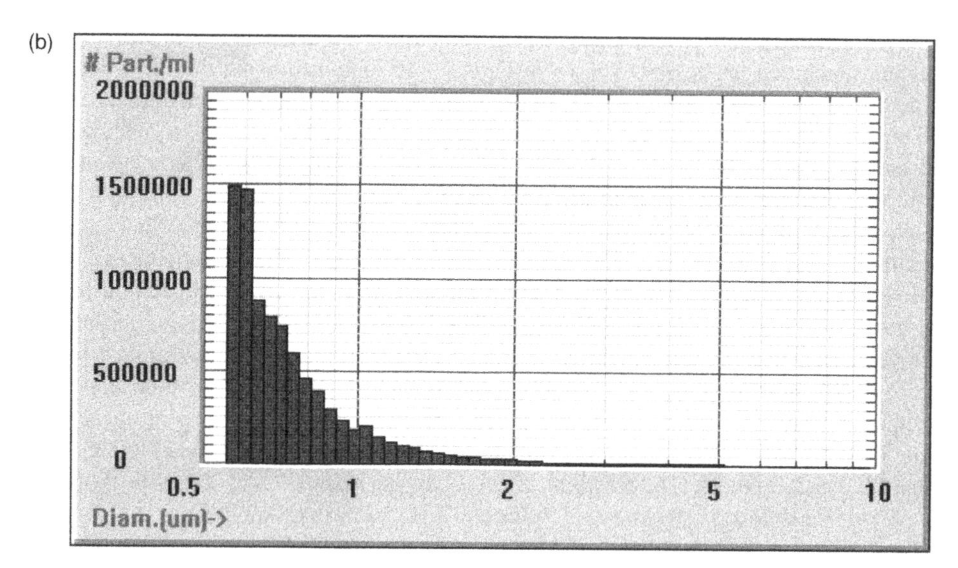

Figure 9.8 *(a) PSD obtained by SPOS (LE+LS) for a "good" polymer resin (see Figure 9.4a). (b) PSD obtained by SPOS (LE + LS) for a "bad" polymer resin (see Figure 9.4b).*

0.24, 0.30, 0.35, 0.40 and 0.50 μm. More than 400 000 particles were detected in 15 mL of sample suspension (2.6×10^6 particles/ml), resulting in cleanly separated, accurately sized peaks. Figure 9.10 shows a typical PSD obtained for a homogenized emulsion using the new SPOS technique in LS mode. Again, more than 400 000 particles were detected in 15 mL of dilute sample suspension (4.8×10^6 particles/ml), yielding a "real" PSD, with a peak/mode diameter of 0.21 μm.

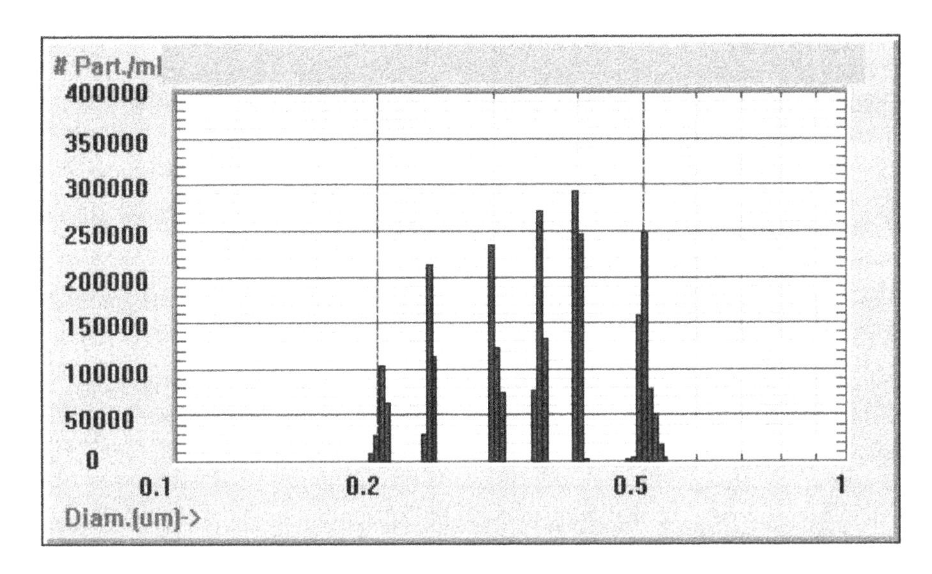

Figure 9.9 *PSD obtained by the new small-beam SPOS technique (LS method) for a mixture of six latex standard particles (0.20, 0.24, 0.30, 0.35, 0.40 and 0.50 μm).*

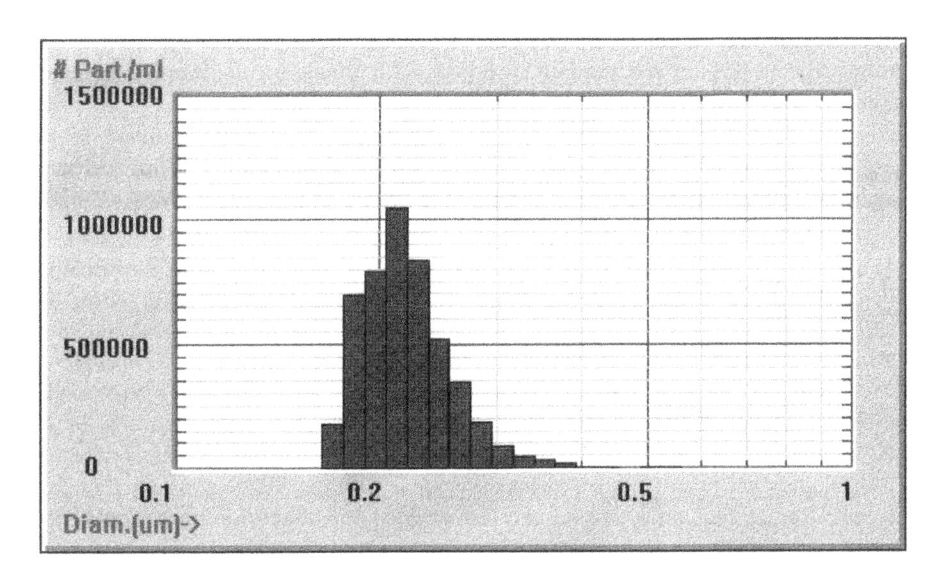

Figure 9.10 *PSD obtained by the new small-beam SPOS technique (LS method) for an aqueous emulsion made by homogenization.*

Advantages:

1. Highest resolution – single particles counted/sized at high speed
2. Relatively fast analysis time (typically < 5 min)
3. No segregation effects – all particles pass through the sensing zone (traditional SPOS)
4. Relatively large particle size range (e.g., 0.5–400 μm for LE + LS)
5. Relatively immune to clogging (relatively large flow channel)
6. Compatible with any fluid medium, aqueous or organic
7. True PSD – no inversion of raw data required (traditional SPOS)
8. Relatively independent of sample optical properties (LE method)
9. Optional high concentration – new SPOS technique (small beam)
10. Optional low particle size limit (<0.2 μm) – new SPOS technique (small beam).

Disadvantages:

1. Requires low particle concentration to avoid PSD distortion (traditional SPOS)
2. Requires very clean fluid for sample dilution (traditional SPOS)
3. Limited detection sensitivity – approx. 0.5 μm (traditional SPOS).

9.4.2.3.2 Electrozone Sensing (ES). A known quantity of concentrated sample suspension is diluted sufficiently that particles in a given size range pass individually through the sensing zone – in this case, a small orifice of known size, through which fluid and particles can flow. A vessel containing water and electrolyte, having this orifice located on its cylindrical surface, is immersed in a larger cylindrical vessel, also containing water and electrolyte. Electrodes are inserted into each body of fluid and a voltage applied between them, allowing a current to be established between the two fluids, by means of the orifice that connects them. Particles are suspended in the fluid located in the outer vessel and made to pass individually through the orifice into the inner vessel by applying suction to the latter. Whenever a particle passes through the orifice it displaces its own volume of water/electrolyte, causing the current between the two electrodes to decrease momentarily with respect to a background level (in the absence of particles). The height of the resulting negative-going pulse in current is measured and the particle volume determined from a standard calibration curve. The diameter of a sphere having the same volume is then determined and registered in a multi-channel analyzer. As in the case of the SPOS technique, the ES technique generates a signal consisting of individual pulses that correspond to single particles passing through the sensing zone. The main difference between the two techniques concerns the nature of the sensing zone and the physical methods of detection. Given the significant practical advantages associated with the SPOS approach compared to ES, the former technique has largely superseded the latter for most applications of interest.

Advantages:

1. Single-particle technique – therefore high resolution and sensitivity (like SPOS)
2. Yields absolute number and volume-weighted PSDs (like SPOS)
3. Yields true volume-weighted PSDs (unlike SPOS).

Disadvantages:

1. Requires samples to be dispersed in water and electrolyte (unlike SPOS)
2. Relatively small dynamic size range (unlike SPOS)
3. Orifice is small and therefore clogs easily (unlike SPOS)
4. PSD erroneous if the particles are porous or partially conducting (unlike SPOS)
5. Large/dense particles may sediment and fail to be counted/sized (unlike SPOS).

9.5 Comparison of Methods

There are two important reasons for comparing the performance of different particle sizing techniques/instruments for a particular application. First, it is necessary to establish their reliability or, in the case of certain instruments, to verify that they are capable of yielding substantially the same particle size or PSD. The degree of success depends not only on the techniques in question but also on the nature of the system to which they are applied. Second, it may become clear that the use of two (or more) different measurement techniques to characterize a given kind of material may yield additional information that might not be obtainable from a single analysis method and which, therefore, would otherwise be overlooked.

It is useful to appreciate that, broadly speaking, there are basically two reasons why particle size analysers are utilized. First, they are used to determine the optimal "end-point" of a manufacturing process, such as grinding/milling, homogenization and emulsion polymerisation. In this case the principal purpose of the analyser is to inform the user that it is time to stop the process, with the expectation that the key properties of the final product will closely resemble the benchmark specifications that were determined and optimized during previous production runs. Second, following termination of the process through end-point determination, particle size analysers are then often used to characterize a representative sample in greater detail, in order to ascertain that its "quality" – that is, performance specs and stability – meet or exceed requirements.

In the polymer field, by the nature of the production process, it is obviously important to be able to determine the end-point of the reaction. It is, therefore, necessary to choose a particle sizing technique that is appropriate and effective, one that can yield reliable and accurate measures of the mean diameter and standard deviation of colloidal, mostly-submicron emulsions. In production environments it is particularly important that the size analysis instrument be fast, easy to use and able to reveal reproducibly the trend of particle growth during the polymerisation reaction, so the operator can change production parameters based on these real-time results.

The ensemble techniques of laser diffraction (Section 9.4.1.1) and dynamic light scattering (Section 9.4.1.2) are capable of providing the required information in a relatively short time and, therefore, they are often used in both production environments and research facilities. Both the LD and DLS techniques typically yield enough information to determine whether the desired end-point was achieved, both during the reaction and after. However, DLS is often chosen over LD because the answers that it produces quickly and reliably – the intensity-weighted mean diameter and standard deviation (typically using the method of cumulants) – require no knowledge of particle parameters. By contrast, LD-based instruments require that both the real and imaginary refractive indices of the particles be known. As discussed earlier (Figures 9.2a and b), relatively small changes in one or both of these input variables can result in significant shifts in the mean diameter and possibly other features of the computed PSD.

There are situations in which a more detailed size "picture" of the emulsion particles is needed, where simple trending of the mean diameter is not adequate to describe the quality of the end product. For example, the polymer suspension may be multi-modal, or it may have a tendency to agglomerate over time. In such cases, methods other than LD or DLS are often needed to provide a more accurate representation of the true, underlying PSD. The LD and DLS methods can still be applied, but there are restrictions on their use and interpretation. For example, when peaks are too close to each other or agglomerates are present in insufficient quantity, the PSD results produced by LD or DLS may lack validity, owing to limitations in resolution. The fact that their data inversion algorithms, while very different, are inherently both "ill-conditioned" may give rise to serious artefacts in the PSD results that they produce – for example, missing peaks, extra peaks that should not be present, or peaks seriously shifted in size. Therefore, in a research or quality control environment it may be more effective to use a technique that produces PSD results with inherently better size resolution, such as the separation techniques described in Section 9.4.2.

The CHDF technique (Section 9.4.2.2.) is capable of providing very accurate size distribution information for dilute polymer emulsions containing more than one peak in the PSD. However, it cannot be used to analyse "soft" polymers, due to particle retention problems. This technique is used more frequently in research/QC labs than in production environments, because it requires significant operator expertise and experience.

By comparison, the HOST centrifugation technique (Section 9.4.2.1) is much more operator friendly and, therefore, is attractive for both R&D and production applications. It can measure relatively concentrated emulsions, which may be an important advantage in cases where an emulsion becomes colloidally unstable upon dilution. A potentially significant disadvantage of this technique is that one measurement may require several hours of centrifugation, making it less useful in production situations where fast decisions are required. On the other hand, in R&D environments it can prove to be very attractive, since multiple samples (e.g., 12) can be analysed at the same time.

End-point determination for polymer emulsions is not the only important use of particle size analysers in this field. Therefore, instruments based on other techniques, such as those described in Section 9.4.2.3, are also very useful for characterizing polymer-based products. For example, the techniques mentioned above – LD, DLS, CHDF and centrifugation (HOST or LIST) – are unable to give reliable information concerning the "tail" of oversize outlier particles in the distribution. Ultra-high separation methods, such as SPOS, can indeed describe this tail in quantitatively accurate detail. This feature, frequently involving less than 0.1% (by volume) of the dispersed phase, can be very important, because critical characteristics, such as stability against separation, filtration behaviour and surface appearance (e.g., gloss) are often influenced by very small percentages of particles residing in the tail of the PSD. The analogous conductivity-based method, ES, in theory can be used to provide the same kind of information for the PSD tail. However, in practice its disadvantages relative to SPOS, especially concerning susceptibility to clogging and much reduced effective counting rate, usually cause SPOS to be chosen instead of ES for elucidation of the PSD tail in polymer emulsions.

In theory, the SPOS technique can also be used to characterize the main, submicron portion of the PSD of polymer emulsions, rather than just their over-size tails. In practice, however, these ultra-high-resolution techniques are confined to measuring the latter feature of the PSD, because of their limited detection threshold (sensitivity) – typically 0.5 μm for commercially available SPOS sensors (LS method) of traditional design. However, the new high-concentration SPOS-LS technique is able to provide accurate quantitative information (# particles/ml) over most of the size range encompassed by the PSD – that is, down to 0.2 μm, or even smaller. This qualitative improvement in sensitivity and working particle concentration (of the order of 10^7 particles/ml) allows SPOS to become one of the most important tools for characterizing the quality and stability of submicron polymer emulsions.

9.5.1 Choice of a Method

As should now be clear, there is no single, universal particle sizing technique that can be applied most effectively to every system and fulfil every need. Therefore, it is necessary to have some guidelines for selecting the optimal method for a given application and need. This is not a simple task, as there are many factors to consider, such as measurement time, desired size range, composition and properties of the particles and cost of instrumentation. It is important to develop some understanding of the particle size problem before selecting a method. Therefore, it may be useful to start with an optical microscope, followed by one or more of the simpler techniques as described, to provide some simple initial screening tests. All techniques possess disadvantages or limitations and, therefore, it is important to be aware of these before selecting one or more methods for performing initial tests on the system of interest. The most important limitations of various particle sizing techniques are listed in Table 9.4. Furthermore, one should not overlook the problems associated with sampling and sample preparation, as discussed in Section 9.3.

Table 9.4 *Particle sizing techniques – summary of properties.*

	Size Range (μm)	Particle parameters	Dilution	Measurement time (min)	On-line
LD	0.05–5000	Refractive indices	All	5–10	Yes
DLS	0.001–1	No	All	5–15	Yes
HOST	0.05–300	Refractive indices	All, or concentrated	30, - >	No
LIST	0.01–50	Refractive indices	All	30, - >	No
CHDF	0.01–3	No	Special eluent	10–20	No
SPOS Traditional	0.5–5000	No	All	5–10	Yes
SPOS (LS)	0.15–3	Refractive indices	All	5–10	Yes
ES	0.6–200	NO	electrolyte	5–15	No

9.6 Particle Shape, Structure and Surface Characterisation

9.6.1 Introduction to Particle Shape, Structure and Surface Characterisation

Structure performance relationships as well as an understanding of how the kinetic and thermodynamic driving forces for different latex particle morphologies in multi-component particles (Chapter 6 and Chapter 11) will be affected by variations in the process parameters are necessary to obtain in order to develop new high end latex and nanoparticle products for use in industrial applications. In addition to the morphological aspects, the nature of the particle surfaces is essential, for example, in film formation, interactions with pigments in coatings and with tissue in life science applications. The work scheme in Figure 9.11 provides guidelines for how to approach the analysis when a latex or composite particle is being characterised with respect to shape, structure and/or surface. A good starting point to get up to speed on fundamental terminology in microscopic techniques is the introductory chapters in "Polymer Microscopy" (Sawyer, Grubb and Meyers, 2008). Terms introduced there are, for example, magnification, resolution, contrast and depth of field, which all have to be understood in order to fully grasp the scope of this chapter.

Latex particles are usually spherical but if this is not the case, microscopic techniques are the only methods that can give information about the particle shape and the actual size. The possibilities for analysis of latex particle shape are closely related to the microscopic methods used for studies of latex particle morphologies and the shape and morphological analysis can, with advantage, be performed simultaneously. However, it is usually much more difficult to perform automated image analysis on multicomponent latex particles since the combination of varying contrast attributed to the different polymer phases within the particles and the often occurring multi-occluded or multi-lobed particle structures (Chapter 6) makes it difficult to define the particles.

Morphologies of multicomponent latex particles are complex to analyse properly and structural misinterpretation often occurs due not only to artefacts arising from the methods used but also to the initial sample treatment and preparation. Common for all the techniques used to distinguish between different particle structures is that the technique must be sensitive to variations in the polymer composition and its distribution within the particles, for example, contrast differences in a microscope or variations in the adsorption area for surfactants on a particle surface. When a particle structure analysis is to be performed it is always good to estimate what the expected thermodynamic equilibrium latex particle morphology would be based on the minimum free energy (Torza and Mason, 1970; Berg, Sundberg and Kronberg, 1986; Sundberg *et al.*, 1990; Chen, Dimonie and El-Aasser, 1991a, 1991b; Durant, Carrier and Sundberg, 2003). Even though kinetic restrictions (Chapter 6) are often present, the thermodynamic equilibrium morphology calculation provides the morphology that would be expected when there are no diffusional restrictions for the polymers in the particles, that is, when incompatible polymer phases in multicomponent particles are fully phase separated. When the observed particle morphology deviates from the expected morphology at thermodynamic equilibrium the reaction kinetics should, if they are available, be included as a complement to the particle morphology analysis in order to achieve a better appreciation of the origin of the experimentally obtained morphology (Cal *et al.*, 1990; Chen *et al.*, 1993, González-Ortiz and Asua, 1996a, 1996b; Dimonie *et al.*, 1997; Stubbs *et al.*, 1999a; Karlsson, Karlsson and Sundberg, 2003a; Karlsson,

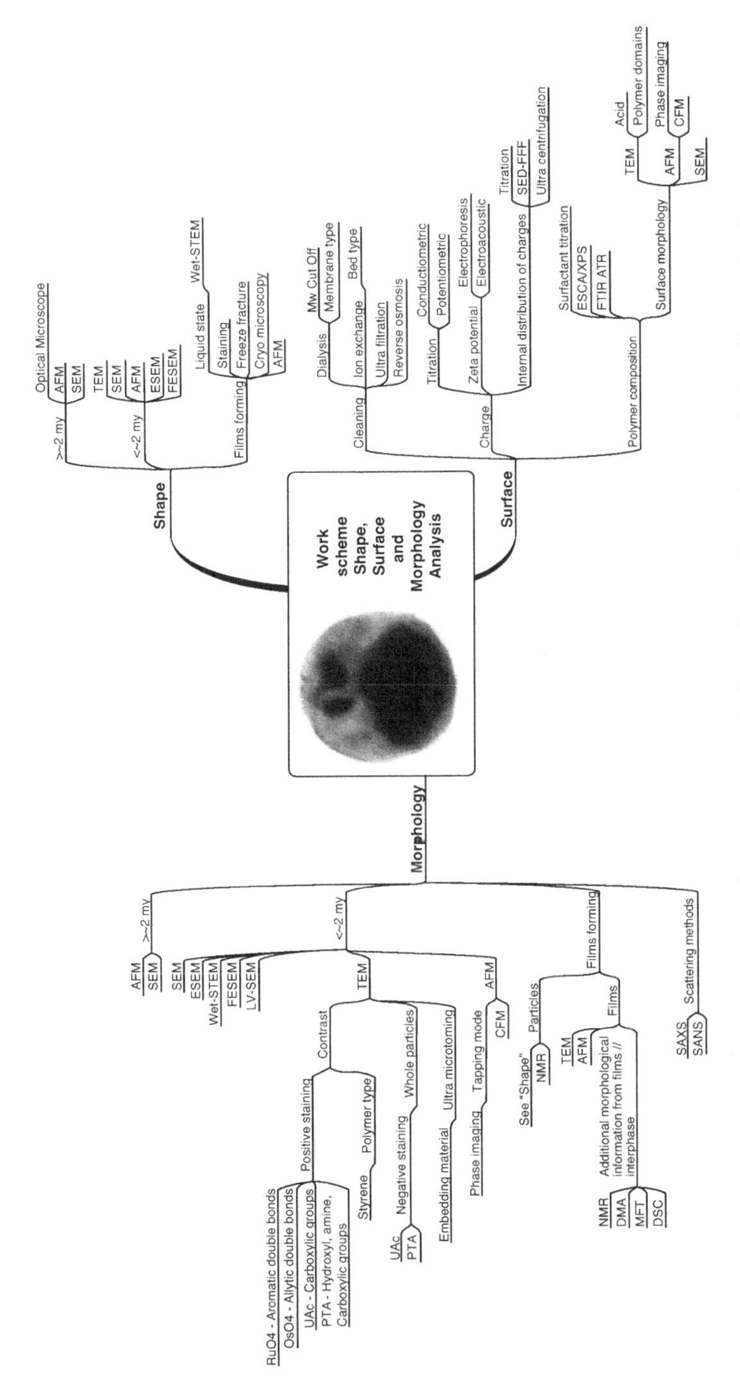

Figure 9.11 *Guidelines on how to approach the analysis when a latex is being characterised.*

Hassander and Colombini, 2003b; Karlsson *et al.*, 2003c; Stubbs *et al.*, 2003b; Stubbs *et al.*, 2010). Good examples of the complexity of structural analysis of latex particles are shown by (Stubbs and Sundberg, 2005b, 2008a) who coordinated an interlaboratory study where six independent laboratories participated in analysing the particle structure of latex particles having multicomponent particles using nine different analytical methods. Each analytical test was performed by at least two independent laboratories and the results from all the groups were compiled. It was concluded that the use of complementary analytical techniques largely improved the ability to make an accurate determination of the particle structure since each method by itself was insufficient and microscopy data were required to make definite conclusions about the particle morphology. In addition to the analytical methods used for whole particles, morphological studies of films formed from multicomponent particles are often a good and sometimes easy accessible source of additional morphological information (Hassander, Karlsson and Wesslen, 1994; Juhué and Lang, 1995; Hagen *et al.*, 1996; Karlsson, Hassander and Colombini, 2003b; Kirsch *et al.*, 2004; Colombini *et al.*, 2005; Goikoetxea *et al.*, 2012).

9.6.2 Classification of the Samples

It is well known to anyone in the field that it is easy to produce a large number of samples of synthesized latexes in a short time period. However, it is a challenge to analyse these samples in a correct and timely fashion. Very often the analyses of the synthesised samples are the bottleneck that holds back the speed of development in a laboratory. It is, therefore, of paramount importance that a clear strategy on how to analyse the samples is employed. Regardless of which type of analysis is needed there are a few questions that are useful to answer before the particle characterisation is initiated. (i) Is the dispersion film forming at room temperature? (ii) Is the particle size in the analytical range of an optical microscope, that is, larger than approximately 2 μm? (iii) Does the sample polymer contain any stainable chemical groups? The outcome of these questions will determine the extent of sample preparation needed and, by knowing which techniques are available for a sample, pitfalls can be avoided and a lot of effort and time saved.

9.6.3 General Considerations – Sample Preparation If the Latex is Film Forming

Unless a microscopic technique can be used, which allows the sample to remain in the liquid state, for example, *optical microscopy (OM)*, colloidal probes in *atomic force microscopy (AFM)*, *environmental* or *low vacuum scanning electron microscopy (ESEM)* (Uwins, 1994) and *wet scanning transmission electron microscopy (wet-STEM)* (Bogner *et al.*, 2005; Faucheu *et al.*, 2009) the preparation and analysis of film-forming dispersions present a number of problems, where the size, microstructure and composition should remain unchanged throughout the treatment. The sample preparation is often tedious and difficult to perform but there are useful techniques that in most cases give reliable results (Shaffer, El-Aasser and Vanderhoff, 1987). *Freeze drying* is an effective and relatively easy procedure for drying aqueous dispersions without the risk of formation of agglomerates or films in the drying process. The specimen is quickly frozen and cooled to –196 °C (liquid nitrogen) and the sample is placed on a cold surface about –80 °C in a chamber which is then evacuated to a high vacuum when the ice is then sublimed. The samples can then be analysed in a *cryo transmission electron microscope* (Talmon, 1987; Van Hamersveld *et al.*, 1999; Wittemann *et al.*, 2005; Crassous *et al.*, 2006) or sputtered with carbon or gold and studied in SEM (Katoh, 1979; Watanabe, Seibel and Inoue, 1984). *Cryo SEM* is not commonly used but was, for example, used in a study of particles having concentration-dependent morphologies (Weber *et al.*, 2011). In *cryo-ultramicrotomy* (Cobbold and Mendelson, 1971) the specimen and the knife are cooled to sub-zero temperatures and sections of suitably shaped specimens may be cut without preliminary embedding. *Freeze fracture* can be employed on dispersions to reveal particle shape (Stubbs and Sundberg, 2005b), where the samples are imbedded in a polymer resin and quickly frozen and then split into two pieces. If the samples are well below their glass transition it is possible to induce brittle fracture, which can reveal internal structure in special cases. The frozen fracture faces contain topographical information, and are examined either after sputter coating in a *scanning electron microscope* (*SEM*) or indirectly by replicas in the *transmission electron microscope* (TEM) (Winnik *et al.*, 1993). However, *staining* followed by TEM is by far the most common technique to analyse the shape and internal structure when the particles are film forming (Hassander, Karlsson and Wesslen, 1994). Another frequent way to solve the problem with film-forming structured particles is as mentioned above to choose an analytical method that, by analysing the formed film, will give some answers regarding the particle morphology.

9.7 Discussion of the Available Techniques

9.7.1 Optical Microscopy (OM)

The optical microscope is a basic tool for studying details of structure and shape down to particles of about 2 μm, and in special cases down to 0.5 μm. Particles below 2 μm will often be detected but cannot be measured due to diffraction. Nearly all samples can be examined, including liquid samples, and it is the only method by which individual particles can be observed directly. Additional problems arise from the short depth of field in the microscope. Colloidal dispersions can simply be studied by placing a drop on an object glass and covering with a glass slip. If the particle concentration is too high artefacts may arise from particle agglomeration and there will be an over-representation of large particles. To minimise this error it is common to keep the coverage of particles on the viewing surface to less than 5%. However, Weber *et al.*, 2011 used OM in combination with other microscopic techniques to report on concentration-dependent particle morphologies. Optical microscopes can be equipped with numerous accessories for the study of physical characteristics in colloidal suspensions and the limitations of the technique caused by low specimen contrast can be improved by the use of methods such as *phase contrast* and *dark-field illumination*.

Dark-field illumination can also be used to observe particles with sizes down to 50 nm, through observing the scattered light cone. This is a powerful way of directly looking at a latex in the original serum and one can see, for example, whether the latex is monomodal or bimodal. Also the onset of coagulation can be observed with dark-field microscopy.

9.7.2 Atomic Force Microscopy (AFM)

Scanning tunnelling microscopy (*STM*) is a collective term for a family of techniques and *scanning probe microscopes* (*SPM*) are derived from STM. New SPM techniques are continuously developed but they all share the same principal operation, in which a sharp local probe is scanned across a sample surface and the interactions between the probe and the sample surface are registered. The different SPM types are then characterised by the nature of the interactions. Two excellent reviews (Tsukruk, 1997; Mcconney, Singamaneni and Tsukruk, 2010) cover the basic principles and the applicability of various SPM techniques for probing polymer surfaces, for example, contact, dynamic, force modulation, friction force, chemical force, electrostatic, adhesion, and thermal modes are illustrated. *Atomic force microscopy* (*AFM*) belongs to a subgroup of SPM called *scanning force microscopy* (*SFM*). AFM is very useful in determining the surface structure of almost any material and has found widespread use in the analysis of structured polymer materials, similar to the SEM, but at a much higher resolution (Ozgur and Natalia, 2008). AFM provides the possibility to study samples in the liquid state or the dry state at ambient conditions. No vacuum is needed and normally no difficult sample preparation is involved. To study surface topography by AFM, a small, sharp tip on the end of a flexible cantilever is scanned across the sample. The van der Waals forces and quantum mechanical forces due to overlapping between the wavefunctions of the sample and the tip at short distances are always present and as the tip is moved with a piezoelectric scanner in the *xyz*-directions a laser measures the deflection of the cantilever. Depending on the sample the resolution varies between 10^{-6} to 10^{-10} m (Ohnesorge and Binnig, 1993). Height image data obtained by the AFM is three-dimensional and the usual method for displaying the data is to use colour mapping.

9.7.2.1 AFM Techniques

One common problem arising when AFM experiments are run in air is that there might be a 2–50 nm thick contamination layer on the sample surface, which usually includes water that also can condense in the gap between the tip and the surface, even if it is absent elsewhere (Thundat *et al.*, 1993). If a contamination layer exists, capillary forces will pull the tip towards the surface with a strength that can be greater than the van der Waal forces, and will depend on the sample, the humidity and the tip shape. Contamination can be observed as hysteresis in the force–distance curve, since the tip will remain wetted up to larger separations as it is withdrawn from the surface, but can be avoided by operating in vacuum, in dry gas, or in a liquid.

In *constant force mode* or *height mode* the positioning piezo responds to changes in the interaction force and alters the tip–sample separation distance to restore a predetermined force value. This mode of operation is commonly used to obtain topographical images. For studies of flat samples at high resolution *constant height* or *deflection mode* is often used. However, uneven samples may easily damage the probe when this mode of operation is used. The basic principles of operation, that is, *constant height mode* and the *constant force mode* can then be applied to all operation methods of the AFM. The most common AFM method is *contact mode*. Two other important methods are *non-contact mode* and *tapping mode*. In *contact mode* the tip and sample remain in close contact as the scanning proceeds so that the deflecting force is repulsive. The specimen topography is measured by sliding the probe tip across the sample surface and the measurement is influenced by lateral forces. Soft samples can be damaged and may also result in distorted images. In *non-contact mode* the cantilever oscillates out of contact above the specimen surface at such a distance that the repulsive intermolecular forces are exchanged by attractive forces acting on the cantilever. The topography is then measured by sensing the Van der Waals attractive forces. However, *non-contact mode* generally gives low resolution and the capillary forces arising from the presence of the contaminant layer may cause the probe to jump towards the surface, which then leads to misinterpreted images. *Tapping mode* measures topography by tapping the sample surface with the cantilever near or at its resonance frequency with typical amplitudes between 20 and 100 nm (Zhong *et al.*, 1993). Because the cantilever taps the surface for a very small fraction of its oscillation period it is prevented from being trapped by adhesive meniscus forces from the contaminant layer (Kühle, Sorensen and Bohr, 1997; Knoll, Magerle and Krausch, 2001; Ledesma-Alonso, Legendre and Tordjeman, 2012). Since the contact time is dramatically reduced frictional forces are also eliminated, which enables analysis of soft samples. *Phase imaging* is a variant of tapping mode where the phase shift between the driving voltage for the oscillations of the cantilever driving piezo and the optically detected oscillations is measured (Tamayo and García, 1998). The image contrast is derived from image properties such as stiffness and viscoelasticity, which can be used to map distributions of different materials (Leclere *et al.*, 1996; Vanlandingham *et al.*, 1997; Bar *et al.*, 1997; Magonov, Elings and Whangbo, 1997). It is possible to chemically map the sample by using *chemical force microscopy* (*CFM*), which uses a modified tip to customise its interaction with the sample. CFM is capable of providing rich information with lateral resolution down to a single molecular group (Janshoff *et al.*, 2000; Friddle *et al.*, 2007). By scanning the sample with the cantilever long axis perpendicular to the scan direction the torsion of the cantilever is used to derive variations in the friction between the tip and sample (Vancso, Hillborg and Schönherr, 2005).

Interpreting AFM images is sometimes difficult and complex since it is hard to know whether the AFM was run in contact or non-contact mode (Bar *et al.*, 1997). However, most artefacts generated in AFM analysis are caused by the tip (Rynders, Hegedus and Gilicinski, 1995; Van Cleef *et al.*, 1996), where the tip shape and sharpness mainly affect the resolution and, for example, may cause broadening of steps or adsorbed spheres.

9.7.2.2 AFM Applications with Latex Samples – Shape – Surface Structure – Morphology

AFM is a well-recognised tool to study films, film structure and various aspects of film formation of dispersion polymers. The direct use of AFM for studying single particle morphologies has not been widely employed, partly because not all dispersions are suitable for morphology studies by means of AFM. In order to be analysed, the latex particle morphology should either have at least one accessible low-T_g phase in order to reveal internal morphologies in the case of hemispherical or partly engulfed morphologies or to have a high-T_g core with a second phase collected in some arrangement around the first phase, for example, core–shell or raspberry-like. Early reports of morphology studies of single particles of polybutyl acrylate (PBA) and PMMA in various ratios studied by AFM were performed by (Butt and Gerharz, 1995; Sommer *et al.*, 1995; Gerharz, Butt and Momper*et al.*, 1996). Where Butt and Gerharz (1995) used contact mode and focused their study on the film formation properties in relation to single particle morphologies and

Sommer *et al.* (1995) used both contact mode and tapping mode to correlate the particle morphology with respect to the second stage polymerisation kinetics. Schellenberg *et al.* (1999) used phase imaging in tapping mode to study the structure and film forming ability of core–shell particles consisting of a soft liquid like poly(2-ethylhexyl methacrylate) cores and crosslinked PBA shells. Stubbs and Sundberg (2005b, 2008a) compared particle morphologies of structured latex particles obtained by AFM with seven other techniques. Pfau, Sander and Kirsch (2002), Kirsch *et al.* (2002) used tapping and phase mode to obtain both the particle morphology and the oriented adsorption of single deposited PBA/PMMA structured latex particles. An interesting combination of AFM and confocal laser scanning microscopy was used to study the surface nanostructure of natural rubber latex particles (Nawamawat *et al.*, 2011). A closely related application of AFM is wetting of a substrate by single latex particles. From the ratio between the initial particle diameter and the diameter after spreading, these types of investigations can give the interaction parameters between a latex polymer and a specific surface (Granier, Sartre and Joanicot, 1993). Lau *et al.* (2002) further developed the technique by also taking the elastic properties into account and Dreyer *et al.* (2011) used AFM to study the spreading kinetics of individual functionalized vinyl acetate-co-ethylene polymer nanoparticles on inorganic substrates and it was found that the kinetics underwent a transition from a fast initial regime to a slower regime, which was dependent on the wettability of the substrate. Figure 9.12 is an AFM image obtained in tapping mode and shows a single latex particle after spreading on oxidised silicon wafer. The polymer was a styrene-*co*-butadiene-*co*-acrylic acid, p(S/B/AA), 33/61/6 wt.% with a T_g of 6 °C and the original particle size was 102 nm. The estimated contact angle from the AFM image for this latex particle was 22° (Engqvist *et al.*, 2007).

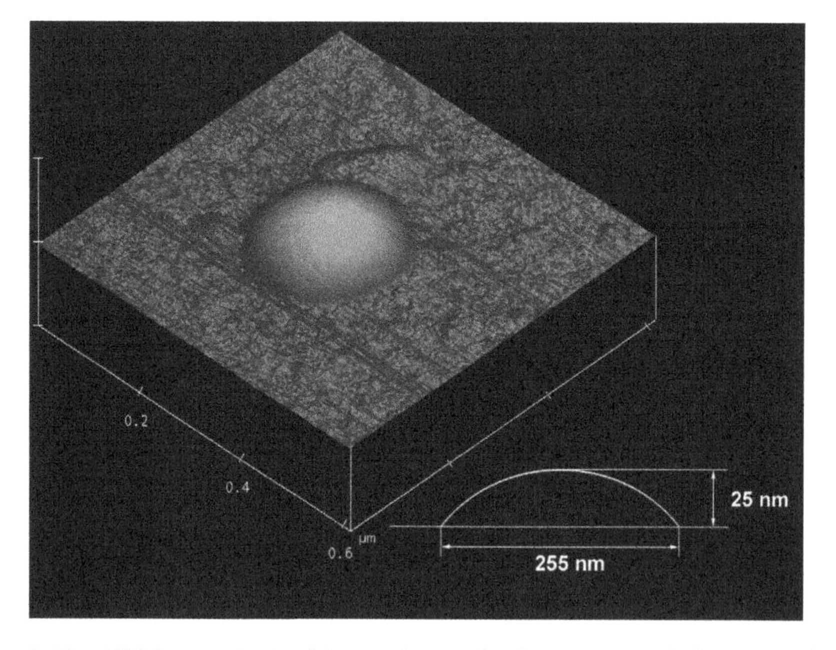

Figure 9.12 *AFM image obtained in tapping mode showing a single latex particle after spreading on oxidised silicon wafer. The polymer was a styrene-co-butadiene-co-acrylic acid, p(S/B/AA), 33/61/6 wt.% with a* T_g *of 6 °C and the original particle size was 102 nm.*

However, many of the published papers where AFM was used for studies of latexes also involve the combination of film formation and film morphologies and, in some cases, also particle morphologies. Even if these types of studies not exactly qualify for particle studies they often also give good indications of the particle morphology (Butt and Gerharz, 1995; Gerharz, Butt and Momper, 1996; Keddie, 1997). In more recent studies AFM has been combined with an ensemble of other techniques such as NMR (Schantz *et al.*, 2007), *X-ray photoelectron spectroscopy* (XPS), *dynamic contact angles* and *quartz crystal micro-balance* with dissipation monitoring (QCM-D) (Liang *et al.*, 2012). One main end application of a dispersion having structured particles is as a coating and Rynders, Hegedus and Gilicinski (1995) studied particle coalescence in waterborne coatings. Butt and Kuropka (1995) reported that the gloss of a paint film made from core–shell particles was strictly correlated to the surface roughness of the corresponding pure latex film. Schuler *et al.* (2000) studied acrylic composite particle morphologies with TEM and correlated the particle morphologies with film morphologies obtained by AFM. The particle and film morphological information was then used to interpret properties of architectural coatings. A dhesion of pressure sensitive films has also been successfully analysed using AFM and, for example, the properties of "tackyfying" core–shell resins made via mini-emulsion were correlated with film structures obtained using AFM combined with force spectroscopy (Canetta *et al.*, 2009). AFM has also been used to study the exclusion of surfactants from latex films (Butt, Kuropka and Christensen, 1994; Juhué and Lang, 1994a, 1994b). Comparisons on films made from dispersions having polymerisable surfactants and conventional surfactants showed a decreased amount of surfactant on the film surfaces when polymerisable surfactants were used (Lam *et al.*, 1997; Hellgren, Weissenborn and Holmberg, 1999) and differences in the rate of film formation (Schantz *et al.*, 2007).

9.7.3 Electron Microscopy

The theory behind the electron microscopic techniques and the general considerations are well covered in the book by Watt (Watt, 1997) and more specific electron microscopic polymer problems are thoroughly discussed by (Sawyer, Grubb and Meyers, 2008). The major advantage shared with the *OM* and *AFM* is the possibility of direct observation of the samples and subsequent imaging. Internal structural analysis of latexes by means of electron microscopic methods, that is, *TEM* and to a certain degree also *SEM*, have been performed for at least 35 years. The artefacts caused by using microtome cross-sections introduce sampling errors, and other disadvantages commonly encountered in electron microscopy, such as the effects of high vacuum and high temperature due to the electron beam radiation on the polymer sample, can alter the original sample properties. The sample preparation is, therefore, very important and, unfortunately, also difficult because much care has to be taken not to manipulate the sample. Since the samples are made from different polymers the phases normally do not differ much in contrast. Techniques for attaining particle phase structure will be emphasised in this section and most of the discussion will be about the commonly used electron microscopic techniques TEM and SEM and versions thereof.

9.7.3.1 Scanning Electron Microscopy (SEM)

The size of objects studied by means of SEM is normally larger than when using TEM but detailed images of a sample surface with considerable depth of focus can be obtained on solid specimens, which result in information regarding size, shape and structure. Basically there are two major categories of electron emitters used in SEM, that is, *thermionic emitters*, which emit electrons as they are heated and *field emission guns*, which emit electrons by a potential field, often referred to as being a "*cold source*". The spot size is smaller and the accelerating voltage is lower when a field emission source is used and, since the resolution is dependent upon the size of the area from which the signal is emitted, this results in much greater resolution than with a conventional SEM using a thermionic emitter. For example, high-resolution *field emission SEM* (FESEM) was used by (Teixeira-Neto and Galembeck, 2002) to examine submonolayers of poly(styrene–acrylamide) latex particles.

The second major factor that affects resolution in the SEM is the signal to noise ratio that exists due to such factors as primary beam brightness, condenser lens strength and detector gain. The various types of signals produced from the interaction of the electron primary beam with the specimen include *secondary electron emission* (*SE*), *backscatter electrons* (*BSE*), *Auger electrons*, *characteristic X-rays*, and *cathodluminescence*. SE gives topographic information and, for examining surface structure, the most widely utilised signal is the secondary electron emission signal, resulting in a maximum resolution of 10 nm (Watt, 1997). In order to increase the number of SE that are emitted from the sample, non-conductive specimens like polymers are normally sputter-coated with an approximately 10–20 nm thick layer of gold or palladium. Secondary electrons are used in the *environmental SEM* (*ESEM*), which enables one to image specimens that are not under vacuum, and wet or even uncoated living specimens can be imaged. With no prior potentially damaging sample preparation it is possible to image dynamic processes, such as wetting, drying, absorption and curing at high magnification (Uwins, 1994; Keddie, 1995 #367). When the particles are submerged in the liquid phase the top surface of the liquid is imaged with rather poor contrast in classical ESEM. Wet-STEM is a rather new technique that allows transmission observations of wet samples in an ESEM (Bogner *et al.*, 2005, 2007). Wet-STEM is performed as STEM in an environmental SEM with a resolution of a few nanometers. It has been employed for morphological studies of grafted natural rubber latex studies (Bogner *et al.*, 2008) and the location of surfactants in films made from alkyd acrylic hybrid particles (Faucheu *et al.*, 2009). In *low voltage SEM* (LV-SEM) the electron beam voltage is decreased so that the beam energy is at the charge equilibrium point and the technique has been applied successfully to determine latex morphology (Gaillard *et al.*, 2007).

Contrast due to BSE increases with increasing atomic number of the specimen. In samples having uniform topology, contrast and thus imaging with an approximate resolution of 1 μm can be obtained by using BSE if the sample is composed of two or more different elements which differ significantly in their atomic numbers. This can also be applied to selectively stained polymers (Ohlsson and Törnell, 1990). In polymer blends, one common method to achieve topographical differences is to selectively chemically etch one of the polymer phases but this is normally not employed on particles prepared via emulsion polymerisation. However, dispersion polymerised particles have been analysed in this fashion (Young, Spontak and DeSimone, 1999). They studied the morphology of multicomponent polystyrene/poly(methyl methacrylate) (PS/PMMA) particles prepared in supercritical carbon dioxide by SEM and TEM after the PS was extracted by cyclohexane from the particles. Another attractive way of achieving simultaneous sample cross-sectioning and analysis is *focused ion beam SEM* (*FIB-SEM*), which was employed for a study on hollow latex particles that were sectioned using FIB and then the internal particle structure and the particle collapse mechanism were analysed in the SEM (Beach *et al.*, 2005).

9.7.3.2 *Transmission Electron Microscopy (TEM)*

In a TEM the specimen is illuminated by an electron beam in an analogous way to light in optical microscopes. The main use of the TEM for colloidal particles is to examine in nanometre detail the particle size, shape and the internal structure of multi-phase polymeric and composite materials. By selectively staining one of the phases it is also possible to reveal the composition of specimens in ways that cannot be examined using other equipment or techniques. TEM builds an image by way of differential contrast due to electron scattering and is dependent on the atomic number of the material in the specimen, the higher the atomic number, the greater the degree of scattering and thus the contrast. Materials for TEM must be specially prepared to thicknesses which allow electrons to transmit through the sample ($\sim$100 nm), much like light is transmitted through materials in conventional optical microscopy. The resolution increases when the sample thickness decreases and those electrons that pass through the sample go on to form the image while those that are stopped or deflected by dense atoms in the specimen are subtracted from the image. Because the wavelength of electrons is much smaller than that of light, the optimal resolution attainable for TEM images is many orders of magnitude better than that from an optical microscope. Additionally, a hundred-fold increase in depth of field also gives an advantage over optical microscopy.

9.7.3.2.1 Sample Preparation for TEM. In order to examine a sample by means of TEM it must be: (i) completely free from water or other volatile components, (ii) able to remain unchanged under high vacuum conditions, (iii) stable in the electron beam and have regions of both electron opacity and electron transparency. Single whole latex particles are studied by applying a drop of diluted dispersion on a TEM grid coated with a supporting film, for example, polyvinyl formaldehyde (formvar), carbon or nitrocellulose. Staining, embedding, and sectioning are sample preparation methods often used and most latex samples are subjected to one or a combination of these. Organic–inorganic hybrids or composite particles very often have a slightly easier sample preparation due to the higher electron density of the inorganic parts (Reculusa *et al.*, 2004; Bourgeat-Lami and Lansalot, 2011).

Embedding: The sole purpose of embedding media for electron microscopy is to enable the object of interest to be cut sufficiently thin for the microscope to develop its full resolution. The best embedding medium permits thin sectioning with the least damage during the preparation and gives the least interference during microscopy. Usually the sample is placed in a mould chosen so that the finished block will easily fit into the microtome and desirable embedding medium properties are the possibility to polymerise near room temperature with a minimum of shrinkage. In addition, the final polymer should be mechanically stable to radiation while transparent to the passage of electrons, the resin should not chemically alter the sample or extract components. Finally, the resin should cut well and be hydrophilic enough to allow lubrication by water against the knife when microtomed. Used embedding media include acrylic resins, that is, methacrylates, polyester resins and epoxy resins, of which the latter is the most widely used class since it is relatively stable in the electron beam. The hardness of the embedding medium and the sample should match in order to achieve the best possible result in the microtome and variations can be made in the recipe for each medium in order to change its hardness. In multicomponent samples the various polymer phases can have different solubility in the embedding resin and extraction of sample components can be avoided by proper choice of embedding medium or by staining prior to the embedding (Stubbs and Sundberg, 2005b, 2008b).

Sectioning: Ultra-thin sections of the order of 50–100 nm are cut from a polymerised block containing the sample and the embedding medium by using an ultramicrotome and a freshly cleaved glass edge or a diamond knife. Sectioning is one of the most widely used methods in the preparation of polymers for electron microscopy but is perhaps also the most difficult skill in electron microscopy to master, experience is required to obtain useful sections of polymer specimens. Microtomy permits the observation of the actual structure in a bulk material such as films but the morphology of single particles can also be revealed. The steps involved in sample preparation for ultramicrotomy could, depending on the type of sample, include: pre-staining, drying (if needed), embedding and curing, trimming and sectioning and post-staining. Objects larger than the thickness of the sections are not possible to completely observe in one piece and in the case of multicomponent particles different projections of the particles may appear as different particle morphologies, depending on the sample thickness and how the particles are sectioned (Jönsson *et al.*, 1991).

Sectioning Imperfections. There are three major artefacts that commonly occur when sectioning. These are *knife marks*, *compression*, and *chatter*. *Knife marks* appear as scratches and are caused by either a dull or dirty knife edge and characteristically run along perpendicular to the edge of the knife. Knife marks can range from tiny thread-like lines that are barely noticeable to gaping fissures and/or holes. *Compression* is usually the result of cutting sections that are too thick or having a dull knife edge and occurs due to the stress that is placed on the embedding resin during the sectioning process (Studer and Gnaegi, 2000). Less compression can be achieved by using an angle of the knife edge that is less than 30° and the cutting speed and section thickness can be varied. In multicomponent particles, where the hardness of the sample polymer phases varies, it is difficult to properly match the hardness of the embedding medium, which may result in compression of the sample. Figure 9.13 shows a TEM micrograph of an unstained film made from structured particles having PS-*co*-PBA occluded cores with a T_g of 80 °C and film-forming PMMA-*co*-PBA shells with a T_g of 20 °C. The dark styrene-containing cores are elliptical and clearly compressed in the direction perpendicular to the cut. *Chatter* is caused by vibration and produces a repetitive array of dark areas that run parallel to the edge of the knife. These marks tend to blend into one another and resemble waves on the sea. The most important factor to avoid chatter is to ensure that vibrations are kept to a minimum, but if chatter continues to be a problem, changes in the cutting speed and thickness settings may help to minimise it.

Staining: Polymer specimens are mainly composed of carbon and hydrogen and additional elements that have relatively low atomic weights and in this way do not differ significantly from each other in a multicomponent sample. However, increased resolution and contrast can be achieved by staining the sample, which generally implies the incorporation of electron-dense, heavy-metal atoms into the polymer to increase the material density and thus increase electron scatter. In *positive staining* a specific chemical group in the polymer reacts with the staining agent and the higher contrast can be located to specific domains in the sample. *Negative staining* is preferably used with latex particles not having reactive groups and that not are stable in the microscope or that are film forming. By the use of a single staining method or by combinations it is possible, by positive

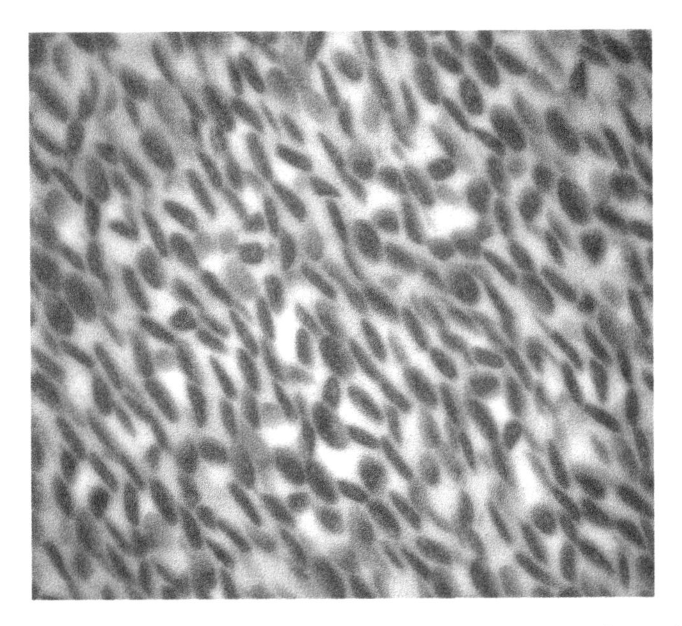

Figure 9.13 *TEM micrograph of an unstained film made from structured particles having PS-co-PBA occluded cores with a T_g of 80 °C and film forming PMMA-co-PBA shells with a T_g of 20 °C. The dark styrene containing cores are elliptical and clearly compressed in the direction perpendicular to the cut.*

staining, to achieve increased phase contrast to reveal internal particle structures, and sometimes to fixate film forming latexes by increasing the T_g of the soft phase or, by negative staining, to study particle size and shape (Shaffer, El-Aasser and Vanderhoff, 1987; Karlsson, Hassander and Wesslen, 1995; Sawyer, Grubb and Meyers, 2008). From the 1950s until today there has been a continuous development of staining methods for latexes. Brown (1947) used *bromination* of double bonds to preserve particle shape in TEM and Stromberg, Swerdlow and Mandel (1953) published one of the first attempts to compare various sample preparation methods for studies of particle size and shape of PS-*co*-PB latexes by looking at *particle shadowing*. In the 1980s, when the studies of multicomponent latex particles became more frequent, the need for new analytical methods advanced the area further (Shaffer, El-Aasser and Vanderhoff, 1987).

Among the many staining methods available there are some methods that are frequently used for studies of colloidal particles and these will be discussed in the following sections.

In *negative staining*, the staining agent is added to a diluted solution containing the particles and a drop of the sample is then allowed to dry on a coated TEM grid. When the liquid phase disappears the staining agent will assemble around the particles and will appear as dark rings of the same diameter as the particles in the microscope due to the increased electron opacity. *Uranyl acetate* (*UAc*) can be employed as a negative stain for latex shape and particle size measurements (Mahl, 1964; Karlsson, Hassander and Wesslen, 1995). Another more often used negative stain is *phosphotungstic acid* (*PTA*), which Shaffer, El-Aasser and Vanderhoff (1983) used in combination with a cold stage in the microscope for a study of multicomponent latexes in order to limit flattening and aggregation. In this PTA was used both for negative staining and positive staining, which was achieved with hydroxy, carboxy and amine groups.

Polymer phases containing residual allylic double bonds, for example, poly isoprene or poly butadiene can be *positively stained* with *osmium tetraoxide* (*OsO_4*). The reaction is generally done in the vapour phase and leaves the osmium atom as a bridge between the reacted sites (Kato, 1966). To avoid distortion of film-forming latex particles during drying, staining can be done in the liquid phase by adding OsO_4 solution to a diluted latex dispersion (Karlsson, Hassander and Wesslen, 1995). (Grancio and Williams, 1970) used OsO_4 staining and ultra microtomed sections of PS latexes copolymerised with small amounts of PB to discuss the locus of polymerisation in a PS latex particle and proposed the first core–shell model of a latex particle. Later Lee (1981) presented one of the first reports on a PS-PB core–shell and occluded multicomponent particles where the PB phase was stained with OsO_4. Another commonly used positive staining agent is *ruthenium tetroxide* (*RuO_4*), which is more reactive than OsO_4 and besides allylic double bonds RuO_4 will also react with double bonds in phenyl rings and with amine groups (Lee and Vandeelkes, 1973; Trent, 1984). The reaction is performed in the gas phase and RuO_4 splits the double bond but the resulting RuO_2 is not bound to the polymer. There are several reports in the literature of the use of RuO_4 in morphological studies of multicomponent particles containing styrene in one of the phases and if the experiments are carefully designed this is normally a good way of attaining structural information (Cho and Lee, 1985; Dimonie, El-Aasser and Vanderhoff, 1988; Chen, Dimonie and El-Aasser, 1991a; Stubbs *et al.*, 1999a).

In addition to studies of whole or sectioned particles, films made from structured latexes may also be stained and examined to reveal particle structures (Hassander, Karlsson and Wesslen, 1994; Hagen *et al.*, 1996). The sectioned films can be stained either before embedding or after the microtoming but in principle the same methods are used for staining of films. UAc is normally used for negative staining but it has been used as a positive stain to reveal distribution of carboxylic acid groups in films prepared from latexes containing carboxylic groups (Zosel *et al.*, 1987, 1989; Richard and Maquet, 1992). Furthermore, the stability of different bonds in the electron beam varies to a great extent but cross-linking and chain scission normally occur. PS for example, has high stability due the possibility to delocalise the electrons in the aromatic rings (Sawyer, Grubb and Meyers, 2008) and if a multicomponent sample contains a styrene-rich phase it might be possible to study the particle structure without staining due to the stability of PS in the electron beam (Talmon, 1987; Jönsson, Hassander and Törnell, 1994; Karlsson, Hassander and Colombini, 2003b).

9.7.4 Indirect Analysis of Particle Morphology

Apart from the direct microscopic techniques the interfacial structure and volume and thereby the phase mixing or phase separation can be studied in structured particles and films made from multicomponent latex particles. By comparing the size of the ΔC_p transitions in *differential scanning calorimetry* (*DSC*) for multicomponent particles to the transitions for the same pure bulk polymer, it is possible to estimate the amounts of the seed and second stage polymer that are present in the interfacial volumes between the pure polymer phases within the phase separated particles (Hourston *et al.*, 1997; Stubbs and Sundberg, 2005a). Similar approaches have been made using *dynamic mechanical analysis* (*DMA*) to study interphases in such films (Richard and Maquet, 1992; Dos Santos, Bris and Graillat, 2009; Karlsson, Hassander and Colombini, 2003b). Typical values obtained for the interphase thickness are of the order of 3–8 nm, which also has been observed in multicomponent latex films using *nuclear magnetic resonance* (*NMR*) (Hidalgo *et al.*, 1992; Nelliappan *et al.*, 1995; Spiegel *et al.*, 1995; Landfester and Spiess, 1998; Ishida *et al.*, 1999). In *minimum film formation temperature* (*MFFT*) measurements the shell polymer determines the MFFT to a large extent and the technique has therefore have been used to estimate latex particle structures (Matsumoto, Okubo and Imai, 1974; Morgan, 1982). *Small angle neutron scattering* (*SANS*) (Goodwin *et al.*, 1980; Bottle, Lye and Ottewill, 1990; Wignall *et al.*, 1990; Zackrisson *et al.*, 2005) and *small angle X-ray scattering* (*SAXS*) (Hergeth, Schmutzler and Wartewig, 1990; Ballauff, 2001; Bolze *et al.*, 2003) have also been used to study colloidal dispersions and the structure of multicomponent latex particles.

9.7.5 Surface Characterisation

The surface properties of the latex particles have a large impact on the final application properties and surface characterisation of particles is increasingly important. Historically, latex dispersions have been applied as model colloids (Hearn, Wilkinson and Goodall, 1981) and, therefore, required well-characterised surfaces, but as the sophistication of new coatings increases the latex particle surfaces also become more important from an industrial perspective. In addition to these applications the utilisation of latex particles in electronics, pharmaceutical and biomedical applications has also contributed to the development of new surface characterisation methods. The surface engineering, that is, variations in size, surface charge, chemical functional groups and surface hydrophobicity, of latex particles as colloidal carriers has, for example, been demonstrated to provide opportunities for site-specific delivery of drugs (Illum and Davis, 1982), and the number of new applications where surface modifications and surface synthesis methods of colloidal particles are applied is increasing at a high rate. Thus it will be outside the scope of this book to go into some of the very specialized techniques used to address certain properties and only the more common surface analytical methods will be reported on. The most obvious reason to study the surfaces of latexes is the origin of latex stability (Chapter 7). Surface groups that often are of interest are either sulfate groups, mainly originating from the initiator, or carboxy groups, either coming from carboxylic acids used as co-monomers or derived from secondary reactions during polymerisation. These reactions can, for example, originate from dissolution of carbon dioxide in the water, hydrolysis of ester groups of methacrylate and acrylate esters or oxidation of surface sulfate groups (Hul and Vanderhoff, 1972; Vanderhoff, 1981; Rasmusson and Wall, 1999). Functionalised latexes having special groups, such as epoxy, amine and oxyethylene, in the particle shells and on the surfaces have also been subjects for surface characterisation (Odeberg *et al.*, 1996, , 1998). The relative level of surface-bound PEO on latex particles has been analysed using *X-ray photoelectron spectroscopy* (*XPS*) combined with either *hydrophobic interaction chromatography* (*HIC*) (Dunn *et al.*, 1994) or *static secondary ion mass spectrometry* (*SSIMS*) (Brindley *et al.*, 1992). For the determination of the actual amount of PEO chains per unit surface area *sedimentation field-flow fractionation* (*sed-FFF*) has proven to be a useful tool (Li and Caldwell, 1991).

9.7.6 Cleaning of Latexes

Before the surface characterisation can be performed, in many cases it is necessary to clean the latex from, for example, surfactants, buffer, residual initiator and oligomers or polymers that are adsorbed to the particles or are soluble in the aqueous phase. The matter that is further analysed is either the cleaned latex dispersion or the aqueous phase collected after cleaning (or both). The three most commonly used cleaning techniques are mixed bed ion exchange, dialysis and serum replacement via ultra filtration and these methods are discussed in detail by (Vanderhoff *et al.*, 1970; Stenius and Kronberg, 1983; El-Aasser, 1983b). Before the latex cleaning is started, thorough cleaning of the cleaning material itself must be done in order to avoid contamination of the dispersion. Ion exchange is fast but can sometimes lead to problems with coagulation and separation of ion exchange beads and the dispersion. Dialysis is not fast but large volumes of dispersion can be placed in cylindrical membranes in tanks with running deionised water and be left for weeks in order to be cleaned. If high solids content after cleaning is a requisite, dialysis is the preferred method of cleaning. It is also recommended that high M_w cut off of the dialysis membrane is used in order to guarantee a good result. Serum replacement is slow but can be useful for small volumes of dispersion that need to be well cleaned.

9.7.7 Analyses of Particle Charge

The most frequently used methods to analyse charges on particle surfaces are *conductometric titration*, *electrophoresis* and *potentiometric titration* (Vanderhoff, 1981; Hearn, Wilkinson and Goodall, 1981; Rasmusson and Wall, 1999). *Conductometric titrations* give information about the nature of the charged surface groups and, in some cases, the total number of surface groups can be determined (Hul and Vanderhoff, 1972; Kawaguchi, Yekta and Winnik, 1995). For a negatively charged surface the proton concentration is always higher at the surface than in the bulk, which leads to ionisable surface sites often not being fully dissociated at all pH values and, therefore, a potentiometric titration will also give information about the pH dependence of the surface charge density (Healy and White, 1978). The distribution of buried charges, that is, carboxylic acid groups, between the particle surfaces and inside the latex particles has been the subject of many studies (Muroi, 1966; Hoy, 1979; Nishida *et al.*, 1981; Zosel *et al.*, 1987) and the lower the polymerisation pH, the more carboxylic acid will be copolymerised in the hydrophobic interior of the particles (Dobler *et al.*, 1992). A closely related subject is the swelling behaviour of carboxylated latex as a function of pH, which has been studied by combinations of *ultracentrifugation* and *conductometric titration* (Bassett and Hoy, 1980), *photon correlation spectroscopy (PCS)*, *viscometry* and *ultracentrifugation* (Bassett *et al.*, 1981; Bassett and Hoy, 1981), *field-flow fractionation (FFF)* and *sed-FFF* (Ratanathanawongs and Giddings, 1993) or by a combination of *PCS* and *sed-FFF* (Karlsson, Caldwell and Sundberg., 2000).

9.7.8 Additional Techniques Used for Latex Particle Surface Characterisation

Conductometric anionic surfactant titration has proven to be reliable in determining the surface fraction of a polymer phase in multicomponent particles from the individual adsorption areas for the surfactant on the polymer surfaces at saturation. The adsorption of surfactant on multiple polymer surfaces is determined by the activity of the surfactant in the water phase and, therefore, it is possible to calculate the partitioning of surfactant between two different polymer surfaces in a latex since the partitioning is dependent on the nature of the polymers and their relative surface areas (Stubbs, Durant and Sundberg, 1999b; Stubbs *et al.*, 2006). *XPS* has been used to analyse residual sulfate groups on polystyrene latexes (Stone and Stone-Masui, 1983) and also to determine the surface fractions of polymer phases in multicomponent particles (Jönsson, Hassander and Törnell, 1994; Arora *et al.*, 1995). As a consequence of the growing field of synthesis of composite particles consisting of inorganic matter and polymer the use of *XPS* and *Fourier transform infrared spectroscopy* (*FTIR*) for surface analysis of particles has increased significantly (Lin *et al.*, 2005; Zhang *et al.*, 2006; Makarov *et al.*, 2007; Liang *et al.*, 2012). *Ultra centrifugation* can be considered as an indirect surface analysis tool since it is regularly used for separation of latex particles from the aqueous phase and then the phases are further analysed, for example, determination of proteins immobilised on the surface of PS latex (Basinska and Slomkowski, 1991).

9.7.9 Zeta Potential

The layer of counterions surrounding a charged particle is called the diffuse double layer and the concentration of counterions in the diffuse double layer is a function of the distance from the particle surface. When a charged particle moves with respect to the surrounding liquid, that is, electrophoresis, there is a plane of shear between the two phases and the electric potential at the plane of shear is called the zeta potential, ζ. This is the experimentally measured quantity computed from the electrokinetic motion of particles. However, even if the zeta potential is not exactly the surface potential, Φ_0, it is the value used for surface potential in calculations of electrostatic stabilisation in the DLVO theory. Because the zeta potential determines the net interparticle forces in electrostatically stabilised systems it is a measure of relative stability (Stenius and Kronberg, 1983). Values far from zero signify stability and values approaching zero signify instability, which typically can lead to flocculation or coagulation. Experimental methods often used for zeta potential measurements of suspensions, emulsions, or macromolecules in solution are *electrophoresis* and *electroacoustic* techniques (Hidalgo-Alvarez *et al.*, 1996). There are two *electroacoustic techniques* available, which both provide non-intrusive measurements of the zeta potential of particles at concentrations over 40% by volume (Hunter, 1998; Dukhin *et al.*, 2000; Dukhin and Goetz, 2001). In c*olloid vibration potential* (*CVP*) a high-frequency acoustic signal is applied to the dispersion and the electric double layer is polarised back and forth, thus generating an electric field in phase that can be measured as a voltage, called the colloid vibration potential. In *electrokinetic sonic amplitude* (*ESA*) a high-frequency electric field is applied to a dispersion and the particles vibrate electrophoretically in the applied field, generating an acoustic signal from which the zeta potential can be calculated.

10

Large Volume Applications of Latex Polymers

Dieter Urban,[1] Bernhard Schuler,[2] and Jürgen Schmidt-Thümmes[1]
[1]BASF AG, GMD, Germany
[2]BASF AG, ED/DC, Germany

10.1 Market and Manufacturing Process

10.1.1 History and Market Today

The first large volume application of synthetic emulsion polymers was using styrene-butadiene copolymers (SBR) instead of natural rubber (NR). In 1941 the United States of America were cut off from the natural rubber supply due to war activities in South East Asia. At that time the NR consumption was 600 000 tons per year, in particular to make tyres. In a joint effort from government, universities and industry a manufacturing process was developed (similar to the Buna S process developed 1929 in Germany (ACS, 2001; Azom.com 2002; PSLC 2000) and 15 plants were built to produce monomers and polymers, the first started in 1942. In 1945 about 900 000 tons synthetic rubber were produced to serve the civil and military demand. After World War II new civil applications were found to utilize the installed capacity of SB emulsions. One very successful product was vulcanized foam made of SB-emulsions, which was used for foam backing of tufted carpets, for automotive seating and furniture upholstery. Due to cost reasons SB-foam has been increasingly replaced by polyurethane foam.

The first acrylic emulsion polymer was already introduced in 1931 by the German I.G. Farben Industrie for leather finishing, but the large volume applications did not occur until the 1960s when new waterborne adhesives and paints were introduced.

Polyvinyl acetate dispersions were produced on an industrial scale in the 1930s and used as wood glues and later for high-speed packaging. In the 1950s latex house paints were introduced.

Chemistry and Technology of Emulsion Polymerisation, Second Edition. Edited by A.M. van Herk.

The demand for SBR, acrylics, polyvinyl acetate copolymer dispersions and other emulsion polymers increased strongly in the 1950s and 1960s. In paints and adhesives, emulsion polymers are used to reduce the emissions of organic solvents by substituting traditional solvent-borne polymers. This is one of the most important driving forces to use emulsion polymers. In other application areas, like paper coating and carpet backing, emulsion polymers substituted natural materials like starch or they have been used directly from the beginning. Today, the worldwide demand for emulsion polymers is about 10 million metric tons (dry), 16% of them are used for paper and paperboard, 17% for paints and coatings, 17% for adhesives and sealants and 5% for carpet backing. These four applications take almost 60% of the worldwide production of emulsion polymers and will be brieflyly described within this chapter.

10.1.2 Manufacturing Process

Large volume applications, like paints, adhesives, paper coating and carpet backing, do not mean that one single emulsion polymer fits all respective application needs. The opposite is true, a huge variety of different emulsion polymers is used to fulfil the sometimes contradictory requirements of each application. This means that the equipment should enable the manufacturer to produce a huge variety of emulsion polymers. On the other hand, strong competition is a continuous force to reduce manufacturing cost, meaning that the specific investment cost needs to be as low as possible.

To balance the requirements of high versatility and low manufacturing cost for the production of emulsion polymers the discontinuous semi-batch process is widely used. Semi-batch means that initially only a portion of the water, monomers and emulsifiers is charged into the reactor, polymerisation is started and the remainder of the ingredients is added over a period of time until the desired filling volume is reached. The most common temperature range for emulsion polymerisation is 60–100 °C. The reactors used are normally agitated stainless steel vessels, ranging in size from 20 to 100 m^3. After the end of monomer addition non-reacted monomers are further polymerised, often using a redox initiator system. Other volatile organic compounds like monomer impurities or by-products from polymerisation are removed, most commonly by steam distillation. Afterwards any coagulum is removed by filtration and post additions of other ingredients may be made along with final adjustment of the latex properties, such as pH and solids content (Taylor, 2002).

10.2 Paper and Paperboard

This section will cover the applications of emulsion polymers in the paper industry. Today this industry consumes a volume of circa 1.6 million metric tons of latex (dry) worldwide, mostly in surface sizing and paper coating. In Western Europe about 3% of the paper industry's consumption of emulsion polymers goes into surface sizing and 97% into paper coating.

10.2.1 The Paper Manufacturing Process

In 2010, the world production of paper and paperboard totalled approximately 380 million metric tons and is expected to grow to about 500 million metric tons in 2020 with an annual growth rate of 2.6% (BASF SE).

The main raw material used to make paper is wood. Depending on the quality demands for the final paper, wood typically will be processed into two modifications:

- Cellulose from which lignin, resins and incrustations have been removed by the refining process to leave a high-grade cellulose fibre that is particularly well suited for paper manufacturing.
- Mechanical pulp, which is produced from wood that has been ground or refined by mechanical means. This type of pulp will lead to lower quality papers due to the presence of incrustations and wood fragments.

Especially in Europe, recycled paper has become increasingly important as a raw material for paper manufacturing. Modern technology combined with appropriate process chemicals enables this secondary raw material to be used not only for paperboard, but also for high-quality paper. In 2010 about 70% of the fibres used for paper making in German paper mills came from recycled papers. For board making the portion of recycled fibres was even >80%. Several paper mills are nowadays producing newsprint paper based on 100% recycled paper as fibre base.

The proportion of chemical additives used as process chemicals in paper manufacturing is about 3%, a surprisingly small amount compared to the other constituents, such as recycled paper, cellulose, and pigments. Of this 3%, synthetic additives comprise only about one third so that, overall, synthetic additives make up only about 1% of the total content of a paper. The two most important groups of synthetic additives are synthetic binders (50%) and sizing agents (25%). Whereas synthetic binders are based on emulsion polymers, sizing agents can be monomer or polymeric. In the latter case, they will also be produced by emulsion polymerisation.

The largest part of the paper and board produced today is for printing purposes. The requirements that these materials must meet include:

- High degree of surface uniformity and smoothness
- High opacity and high strength
- Good optical properties of which brightness and gloss are the most important
- Good printing properties, such as print gloss and print evenness.

In order to fulfil these requirements the following two processes of surface treatment can be applied after the manufacture of the paper web

- Surface sizing
  on-line in paper or board machines
- Coating.
  on-line or offline in paper machines, always online in board machines

The use of emulsion polymers in the paper industry is essentially restricted to these two processes, which are described in more detail in the following sections.

10.2.2 Surface Sizing

Surface sizing is the application of small spots of hydrophobic emulsion polymers onto the paper surface in order to prevent the water-based printing ink being spread over the paper surface. This improves the brilliant appearance of the print. The surface sizing agents are applied in combination with starch and without pigments. The application will be online to the paper machine by either a size press or a film press. In relation to the paper mass usually 3 to 5 wt% of starch and 0.1 to 0.25 wt% of sizing agents, each calculated as solid, will be applied. Starch enhances the strength of the paper, the surface sizing agent renders hydrophobic properties to the paper sheet, thereby reducing the water absorbency of the paper. Thus the penetration and spreading of water-based print colours are controlled and the loss of strength in the wet state is reduced.

For surface sizing mostly polymeric sizing agents are used. The most important product classes are acrylic copolymer dispersions stabilized by protective colloids. The particles of the sizing agent consist of a hydrophobic polymer core and a hydrophilic shell formed out of the protective colloid.

The composition of the polymeric core influences the hydrophobic properties, glass transition temperature and binding strength of the polymer. The hydrophilic shell is highly swollen in water and normally carries either an anionic or cationic charge. It renders stability to the dispersions during storage and against the high mechanical stress during application. It also plays an important role in the interaction between starch and sizing agent. The protective colloid acts as a compatibilizer between starch and hydrophobic polymer core, allowing the polymer particles to spread onto the surface of the starch film. After drying a halftone-like screen (raster), consisting of areas of hydrophilic character (starch) and hydrophobic barriers (polymer) will be formed on the surface of the paper. Whereas the hydrophilic areas allow a fast dewatering of the printing ink, mostly to the interior of the paper sheet, the hydrophobic points prevent a spreading parallel to the paper surface (Schmidt-Thümmes, Schwarzenbach and Lee, 2002).

10.2.3 Paper Coating

Paper coating is the most important surface finishing process for paper in terms of both the amount of paper that is coated and the quantity of emulsion polymers consumed in the coating process. The method involves coating the surface of the paper with a water-based pigmented coating colour. Typically between 5 and 30 g (dry) m^{-2} of coating are applied, the upper limit only being used in board coating. The average coating layer thickness will be between 2 and 12 μm. The emulsion polymer used in the coating colour formulation fixes the individual pigment particles together and helps the entire pigment layer to adhere to the surface of the paper. Emulsion polymers are also added as so-called co-binders in order to improve the processability and/or runnability of the coating colour.

A coating is typically applied onto paper and board for printing or packaging applications. Coating paper or board increases the homogeneity of the surface and considerably improves its optical characteristics, such as gloss, smoothness, brightness, and opacity. Compared to the open, porous structure of a natural paper, coated papers show a much more homogeneous and closed surface, leading to more uniform ink receptivity and better ink holdout than uncoated papers. The much smoother surface of a coated paper is a particularly significant factor when printing individual dots, especially when using the rotogravure process.

10.2.3.1 Coating Process

A number of different coating equipments exist for applying the coating colour onto the base paper. Figure 10.1 illustrates the basic principles of the most common coating methods.

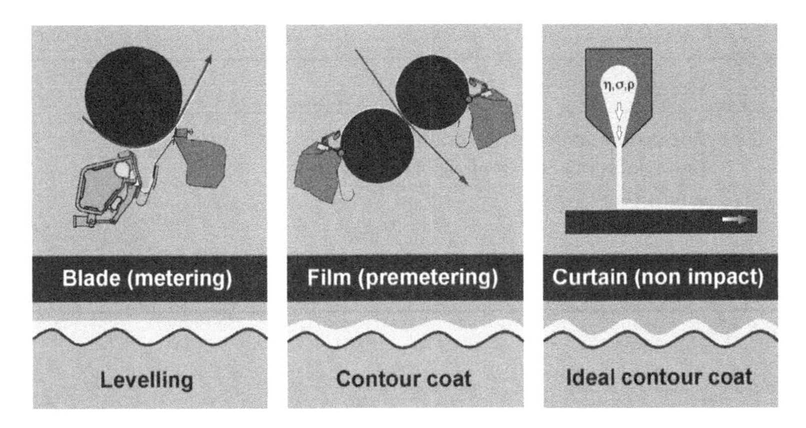

Figure 10.1 *Principles of different coating methods.*

In blade coating, as a first step a layer of coating colour is applied to the paper sheet. The sheet will then pass under a blade, which removes an excess of applied material and thereby creates a very smooth surface. Due to inhomogeneities of the base paper, surface blade coating leads to fluctuations in the coating layer thickness over the width and length of the paper sheet.

This can be avoided by contour coating processes, like the film coating process. In this process the coating colour is pre-metered to a cylinder surface. When the paper sheet passes over the surface of the cylinder a fraction of the coating colour will be transferred to the paper surface. In this process a much more homogeneous coating colour thickness will be obtained compared to blade coating. Surface smoothness, however, is lower than in blade-coated papers. In addition, the film press process allows the simultaneous application of coating on both sides of the paper or board.

New impact-free coating processes, like spray and curtain coating, which are currently being introduced to the mills, will result in even better contour coating profiles.

In practice, often more than one coating layer is applied, especially if higher coating weights are required. In this case different coating methods are very often combined, for example, a film press for a good fibre covering of the base coat and a blade for a high smoothness of the top coat surface.

It is apparent that the various coating methods place different demands on the rheological properties of the coating colour. These requirements must be taken into account when formulating a coating colour for a particular application.

Typical coating colour compositions and the major components of a coating colour will be described in the following section.

After each coating step the paper is run through a drying section. Here the water brought into the web by the coating colour will be evaporated by either hot air or IR radiation, or a combination of both. Drying intensity has to be properly adjusted since a too intensive drying can lead to migration phenomena in the coating, resulting in bad printability. At the end of the drying section the paper web will have a remaining water content between 4 and 8%.

As a final step, the paper will be smoothed in the calendering section. Calendering involves subjecting the paper surfaces to high temperatures and pressures in order to create

Table 10.1 *Typical coating colour composition for sheet-fed offset papers in Europe.*

Part	Components
80	fine ground calcium carbonate
20	fine kaolin clay (high gloss clay)
12	emulsion polymer
0.5	co-binder
0.5	curing agent, like epoxy resins or ionic compounds (e.g. zirconium acetate)
0.5	optical brightener, like sulfonated stilbene derivates

Solid content 65–70%.

a smooth, glossy surface. A distinction is made between supercalendered and soft-nip calendered papers. In the supercalender the paper is run over a high number of nips, typically formed out of twelve cylinders in a supercalender stack, with temperatures on the cylinders up to 100 °C and pressures in the nip up to 300 kN m^{-1}. Supercalendering leads to the smoothest and glossiest paper surfaces. In the soft-nip calender process, the number of nips is kept low, and higher temperatures and lower pressures are used compared to the supercalender process. The advantages of the soft-nip calendering are that it can be performed "on-line", that is, immediately after the coating process and that the bulk of the paper does not decrease in height as much as in supercalendering. By varying temperature and pressure in a controlled manner, a very broad range of gloss levels can be achieved. In future there will be a clear trend towards the use of soft-nip calenders.

10.2.3.2 Major Components of a Coating Colour

The major components of a coating colour are:

- Inorganic pigments to cover the surface of the base paper
- Co-binder and thickener for controlling the processing properties
- Binder (water-soluble or disperse systems or a combination of the two).

Table 10.1 shows a composition of a coating colour for sheet-fed offset typical for European coating mills.

The main constituents of a coating colour formulation are the inorganic pigments, which serve to cover the surface of the base paper and thus to improve its optical and printing properties.

Coating pigments, therefore, have to fulfil the following requirements:

- High purity
- High brightness and opacity
- High refractive index
- Good dispersibility and desirable rheological properties
- Amount of binder required should be low.

In nearly all cases, not only one but a combination of several pigments is used in coating colour formulations. Kaolin clay and calcium carbonate are the most commonly used pigments. There are a great number of different types in each of the two pigment groups:

the calcium carbonate grades being distinguished mainly by particle size, while the plate-like kaolin clays are classified according to their so-called aspect ratio (ratio of surface diameter to thickness) and particle size.

The pigments used in the preparation of coating colours are prepared as slurries. These are aqueous suspensions, which by using dispersing agents such as tetrasodium pyrophosphate or sodium polyacrylate can have a solid pigment content of higher than 70%.

Co-binders and thickeners are added to adjust the rheology of a coating colour.

Pumping, transfer and, most particularly, the actual coating process require certain rheological properties of the coating colours. Low-shear and high-shear viscosities (shear rates of 10 to $>10^6\ s^{-1}$) and water retention values are highly important parameters. For example, in film coating applications, the thixotropic behaviour of the coating colour is particularly important, whereas shear-thinning flow at high shear rates is important for all blade-coating techniques.

Typical amounts are 0.1–3 parts of co-binder or thickener to 100 parts pigment and approximately 12 parts binder. Besides emulsion polymers, described in greater detail below, other substances are used as co-binders and thickeners. These include natural products, such as starch, and synthetic water-soluble polymers, such as polyvinyl alcohol and carboxymethyl cellulose.

In contrast to the emulsion polymers used as binders, those employed as co-binders and thickeners contain large fractions of hydrophilic (typically carboxyl-rich) monomers. This high degree of hydrophilic units means that the particulate nature of the dispersion is lost when the acidic dispersion ($pH < 7$) is added to the alkaline environment of the coating colour formulation ($pH > 7$).

The resulting structures, which range from massively swollen polymer networks to polymer chains dissolved in the aqueous phase, influence the rheology of the coating colour in a complex manner (Figure 10.2).

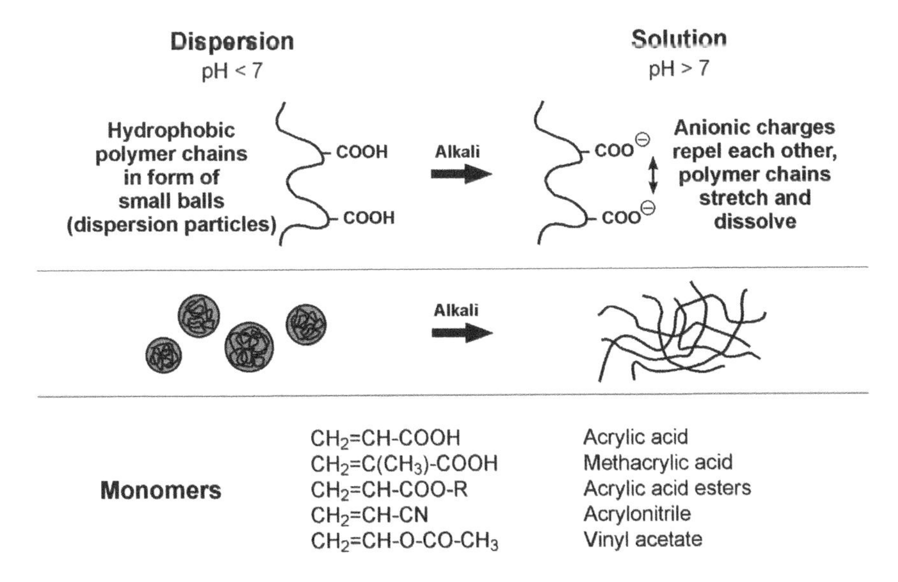

Figure 10.2 Chemical structure of synthetic co-binders.

All effects induced by the co-binder and thickener in the coating colour are very strongly dependent on the shape factor, charge distribution, and size of the pigments used, as well as on the solid content of the formulation. Choosing the right thickener or co-binder for a coating colour which is to be formulated for use in a particular type of coating machine is a complex task that requires good product knowledge and a considerable degree of practical experience.

Historically, natural products like starch and casein were used as binders for coating colours. Partly because of their high price and inconstancy in quality, partly because natural binders cannot be added directly to the pigments but must first be pre-processed, these products have been substituted by synthetic binders to a high degree. Whereas starch has kept some significance until today, mostly for pre-coats, the use of casein is restricted to (low volume) processes for extremely glossy papers (cast coating).

The overwhelming portion of binders used nowadays, however, are emulsion polymers based on

- Styrene and butadiene
- Styrene and butyl acrylate
- Polyvinyl acetate
- Acrylates
- Vinyl ester and acrylic ester
- Ethylene and vinyl ester.

These synthetic binders are mostly modified with functional monomers, such as vinylic acids, vinylic amides, acrylonitrile, and so on to improve the colloidal and rheological properties of coating colour formulations and the printing and/or converting characteristics of coated papers and paperboards.

Some of the most important properties are:

- Binding strength (dry pick strength)
- Water resistance or wet pick strength
- Print gloss
- Brightness (reflection of visible light)
- Opacity (hiding, opposite to transparency)
- Smoothness
- Stiffness (more important for light-weight papers)
- Water absorption capacity (the capacity of the paper to absorb water, thus permitting the transfer of inks to moist surfaces)
- Ink absorption capacity (the capacity of the paper to absorb ink and to prevent ink being transferred from the freshly printed areas to the rubber blanket of the following printing station)
- Print uniformity
- Blistering in web offset process (Due to high temperatures in the drying section of the printing machine the humidity of the paper web will evaporate instantaneously. Unless the coating has a high porosity, this will lead to "blisters" on the printed paper surface.)
- Gluability (packaging board)
- Varnish gloss (overprint varnishes in packaging board).

The extent to which a coated paper and board needs to fulfil the various requirements listed above depends on the printing process to be used.

When choosing or developing a suitable binder for one of the various printing processes, one generally focuses on those four parameters whose effect on binder properties is sufficiently well known. These are:

- Nature of the constituent monomers
- Glass transition temperature
- Particle size and particle size distribution
- Molecular structure of polymers.

As mentioned before, the binders used in coating colour formulations are based on combinations of different monomers. The most common combinations are styrene with butadiene or acrylic esters and vinyl acetate combined with ethylene or acrylic esters. An important difference between styrene–butadiene binders and styrene–acrylic ester binders is the tendency of the binder to yellow under the influence of UV radiation or heat. Products containing a butadiene-based binder are considerably more susceptible to yellowing due to the much greater fraction of double bonds in the polymer. Acrylic ester copolymers are significantly less prone to thermal or UV-induced yellowing and these are the copolymers of choice for the production of high-quality, long-life prints.

The glass transition temperature of a polymer is determined by the amounts of its different monomer constituents. Paper used in offset printing contains binders whose glass transition temperature lies between 0 °C and 30 °C. The high smoothness and compressibility required for paper grades used in the rotogravure printing process are achieved by using binders with a much lower glass transition temperature (<0 °C). Figure 10.3 shows the typical dependence of dry and wet pick strength, stiffness, gloss, porosity, and evenness of offset printing on the glass transition temperature.

Particle size and particle size distribution are influenced by the kind and amount of emulsifiers and protective colloids used. These components are added to stabilize the

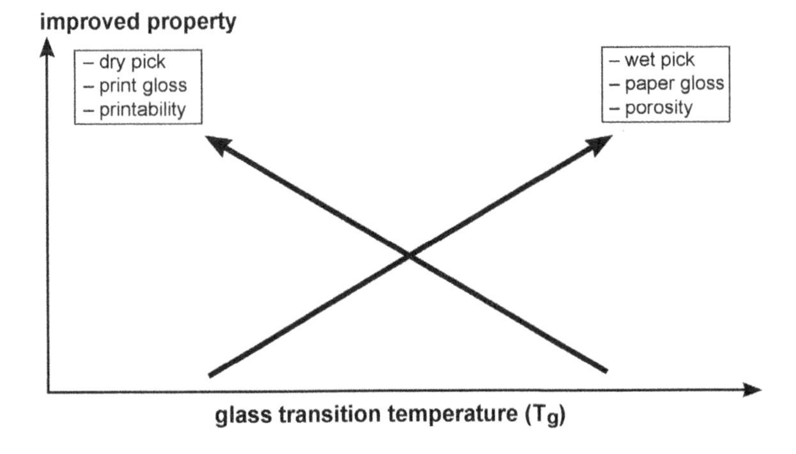

Figure 10.3 Dependence of paper properties on the glass transition temperature of the binder.

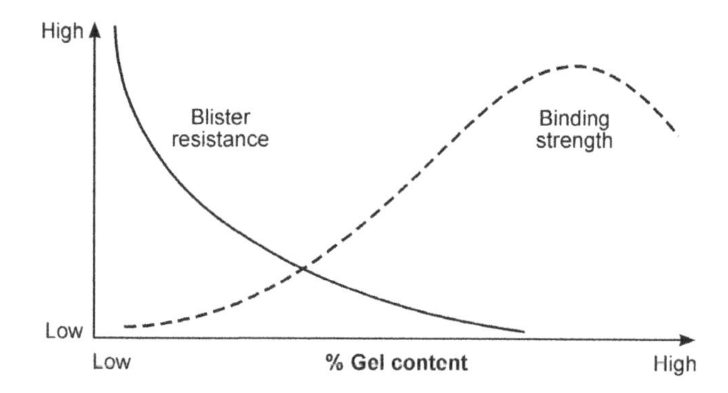

Figure 10.4 *Relationship between blister resistance and binding strength for styrene–butadiene binders.*

dispersion during processing, thus making it both able to be conveyed, metered, filtered, and so on, and stable during storage. Variations in the emulsion polymerisation process also have a major effect on the size and size distribution of the polymer particles. Typically, binders used in the paper coating process have particle sizes of between 100 and 300 nm. Both the viscosity of the coating colour and the wet pick strength of the coated paper are strongly dependent on particle size.

In contrast to the other possible monomer components, butadiene possesses two double bonds, both of which can act as polymerisation sites. Binders based on a styrene–butadiene combination, therefore, have a more cross-linked and branched polymer structure. The extent of cross-linking affects the binding strength, the print gloss and the degree of blistering, which is a highly significant parameter in web offset printing. Unfortunately, binding strength and blister resistance tend to oppose one another and cannot, therefore, be optimised by the choice of binders alone (Figure 10.4).

A very similar dependence is observed with the styrene–acrylate binders. In this case, binding strength and blister resistance show a mutually opposed dependence on the relative molecular weight of the polymers.

The polymer structure, and thus the desired balance between binding strength and blister resistance, can be controlled in the two classes of binders by careful adjustment of the polymerisation conditions and by the addition of so-called chain transfer agents.

More detailed information on paper coating can be found elsewhere (Booth, 1970; Dean, 1997; Schmidt-Thümmes, Schwarzenbach and Lee, 2002).

10.3 Paints and Coatings

Paints and coatings are applied to a variety of different substrates, for example, metal, masonry, wood or engineering plastics. The corresponding objects range from buildings and large industrial complexes, furniture and automobiles to specialized areas like concrete roof tiles or vessels. Paints and coatings thereby fulfil mainly two different functions. First, they protect the substrate from damage through external factors like UV-radiation, moisture, temperature or chemicals. Thus, they contribute largely to the preservation of

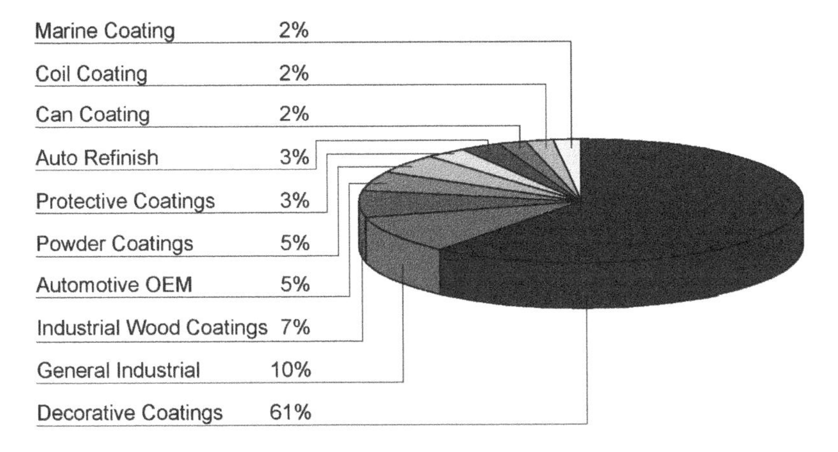

Figure 10.5 *European market for coatings – paint sales 2009: 6500 kt.*

the value of objects. Secondly, they make up for the visible appearance of the surface and improve optics and aesthetics. In the past most of the coating applications have utilized solvent-borne polymers as binder. However, over the last decades emulsion polymers have increasingly replaceded them. The change was driven by concerns over environmental pollution and the toxicity of solvents as well as by practical reasons like ease of use and clean up. As a prerequisite, emulsion polymer technology made a tremendous leap and provided solutions that not only caught up with the performance of solvent-borne systems, but frequently set the performance standards and lead the market in many areas.

Paints and coatings represent a large field of application for emulsion polymers. In Western Europe the total amount of coatings sold was 6.500 kt in 2009 (Figure 10.5). With 61%, architectural coatings represent by far the biggest segment, corresponding to 510 kt of dry polymer.

10.3.1 Technology Trends

The most important trend in the paint and coatings industry is a move to more environmentally friendly coating materials. Increasing environmental awareness of the end users, commitment of the industry, as well as laws and regulations drive this. The primary target of the efforts is the reduction of so-called volatile organic compounds (VOCs) to the lowest level possible for a given application. The first and biggest step can be done by switching from solvent-based to water-borne coating formulations. In several application areas the use of a low level of coalescing agents in water-based formulations that also contribute to VOCs could initially not be avoided completely, for performance reasons. However, in the last few years innovative emulsion polymers that even make those small amounts unnecessary have started to penetrate the market . The abandonment of film-forming agents first became an issue for paints and coatings used indoors . People wanted to paint their home or official buildings, such as schools or hospitals, without the necessity to move out until the paint had completely dried and the VOCs had evaporated and been sufficiently diluted. Nowadays, VOC levels of paints for exterior applications have become an issue too, since the organics contribute to the greenhouse effect and smog in the cities. The industrial coatings industry, that recycle solvents from coating processes, for example, for automobiles or furniture, are also seeking increasingly for low VOC solutions. The substitution process is still ongoing and represents the most important focus of research and development activities of paint manufacturers and raw material suppliers.

10.3.2 Raw Materials for Water-Borne Coating Formulations

Generally, coating compositions consist of quite a few different raw materials. Emulsion polymers as a binder as well as pigments and extenders are the most crucial and basic constituents. They usually make up the major part of a coating. However, they cannot be formulated into a paint that fulfils the required application properties. Additives have to be employed to adjust flow behavior, ensure a proper dispersing of the pigments, improve film formation, control foam formation, prevent microbiological attack and so on (Bielemann, 2000).

The ratio of the different raw materials has a big impact on the characteristics of coatings. It is frequently characterized by the pigment volume concentration (PVC). This is the ratio of the volume of pigments and extenders to the total dry volume of all raw materials, including pigments and extenders. Coatings with low PVC have a higher binder content than those with a high PVC where pigments and extenders are more prevailing. PVC is an important quantity because it relates to many of the performance properties of a dry paint film. In low-PVC coatings the binder dominates the properties. It forms a continuous phase in which the pigment is distributed (Figure 10.6a). The surface is closed and smooth. Therefore coatings with low PVC (up to 25%) are chosen to formulate high gloss or semi-gloss paints. As the amount of binder decreases the increasing surface area of pigments and extenders can no longer be wetted completely at a certain point. This point is called the critical pigment volume concentration (CPVC) (Bierwagen and Rich, 1983). The exact value of the CPVC depends on the specifics of the coating formulation and is usually

Figure 10.6 Electron micrographs of paint films with low (a) and high (b) pigment volume concentration (PVC).

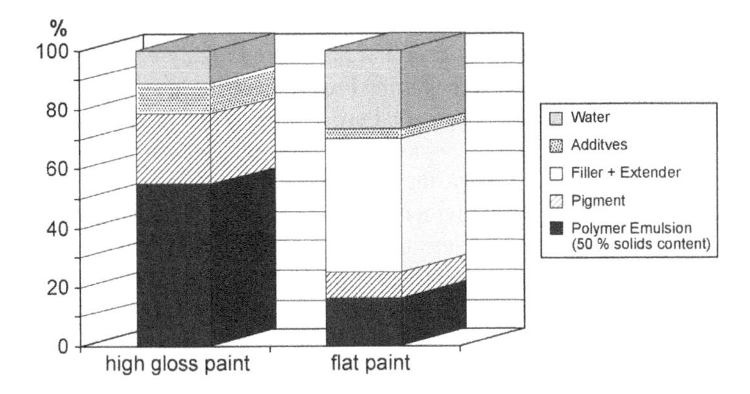

Figure 10.7 *Volume ratio of raw materials of a typical flat and high gloss paint.*

observed at PVCs between 45 and 60%. Above the CPVC the dry coating begins to develop small air voids between the solid components of the film and the properties, like porosity, elasticity or water absorption, change abruptly. The surface looks rather rough with pigments sticking out (Figure 10.6b).

Paints above the CPVC are flat and are used predominately for interior coatings. Figure 10.7 shows the volume ratio of the raw materials for different types of coatings, a flat interior paint and high gloss paint.

10.3.2.1 Emulsion Polymer Binders

The binder holds the coating together and makes it adhere to the substrate. It provides many of the performance features needed for specific coating applications for example, toughness and elasticity to resist mechanical impact, like scratching, abrasion or yield stress; stability against chemicals, water resistance and so on. In the market different types of emulsion polymers are established (Padget, 1994):

- copolymers of styrene and acrylic esters (styrene acrylics)
- copolymers of methacrylic esters and acrylic ester (pure acrylics)
- homopolymers and copolymers of vinyl acetate.

During the manufacturing of paints shear stress is applied through dispersing and pumping process steps. In addition, pigments and extenders can release multivalent cations that destabilize the colloidal paint system. To improve the stability of emulsion polymers usually minor amounts of monomers, like unsaturated acids (e.g. acrylic acid, methacrylic acid, itaconic acid), are copolymerised in addition. In the alkaline paint they are deprotonated. The negative charge sitting on the particle surface increases the resistance against agglomeration. The choice of emulsifier plays an important role in this respect, too. Quite often mixtures of different types are used to optimise the overall stability of emulsion polymers during production and processing.

In principal, any type of binder can be employed for any application, but in reality the particular properties of the different classes mean that each dominates specific areas. For exterior applications on mineral substrates, like architectural paints or textured finishes, usually styrene acrylics are preferred. They have the highest resistance to saponification and thus do not undergo hydrolysis if coated on not fully cured highly alkaline substrates, like concrete or lime cement. In addition, they bring low water absorption, good adhesion to the substrate and high pigment binding capacity. Pure acrylics are used especially in low PVC applications for example, clear-coats, varnishes or high gloss paints. Since these coatings contain only little or even no pigment they have to demonstrate their low susceptibility to UV-degradation. Homo- and copolymers of vinyl acetate are in general the most cost efficient type and dominate the price-sensitive segment of interior paints.

A very important characteristic of a binder is the temperature at which it forms a clear and homogeneous film. This temperature is called the minimum film-forming temperature (MFFT). It can be determined experimentally with a special apparatus. This consists of a support that provides a temperature gradient along its length. The emulsion polymer is drawn down on it. After equilibration the transition from a cracked to a clear film determines the MFFT. The MFFT of an emulsion polymer is usually a few °C lower than its glass transition temperature (T_g). One reason is that a small amount of water dissolved in the latex particle acts as a plasticizer. Most coatings are applied under ambient conditions (either on a job site or in a factory) at typical temperatures between 5 and 40 °C. For obvious reasons the MFFT of paints and coatings should be lower than the temperature at the application side. The low end of the temperature range is defined by architectural coatings used outdoors that should still form a neat paint film under unfavourable conditions.

The MFFT of paints can be temporarily lowered by use of a coalescing agent. This works by partitioning into the emulsion polymer particles, disrupting the packing of the polymer chains and thus lowering the MFFT. After film formation the coalescent evaporates. This principle is employed for certain applications where hard polymers are required. For example, wood coatings for door or window frames should not stick to each other when brought in contact by closing. In technical terms the coating should show sufficient block resistance. Unfortunately, coalescents contribute largely to VOCs that are emitted after application. As already stated the clear trend in the coatings industry is directed towards environmentally friendly low-VOC coatings. Thus, in the last years new kinds of emulsion polymers have been developed that allow realization of the contradictory requirements of low MFFT on the one hand and good block resistance on the other without use of any coalescent: the so-called multiphase particles (Schuler *et al.*, 2000; Kirsch *et al.*, 2002).

They consist of at least two different polymers one with low MFFT and one with high MFFT. Figure 10.8a shows an electron micrograph of a two-phase latex particle. The soft phase was stained and is partly engulfed by the hard phase. This picture only represents one possibility of how domains of the different phases can be arranged. The particle structure can be varied largely by the choice of monomers and process conditions (Rudin, 1995).

Figure 10.8b displays the surface of the latex film determined by atomic force microscopy. The dark areas represent the soft phase. Since its MMFT was adjusted to 0 °C it easily forms a film, even at low temperature without coalescent. The hard domains, represented by the bright areas, are perfectly distributed in the soft phase matrix. Since they stick out of the film they impart block resistance. With multiphase particles a technology was developed to achieve high performance low-VOC coatings.

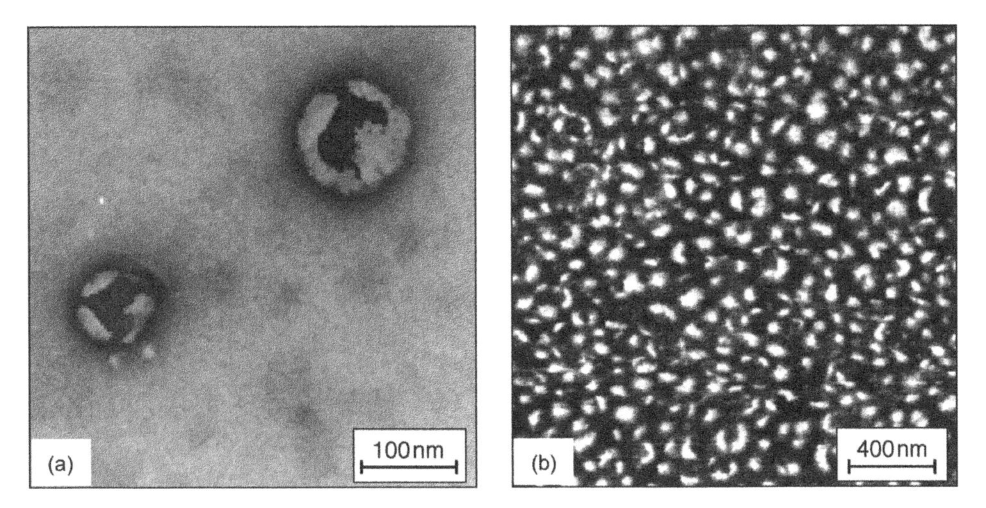

Figure 10.8 *Two-phase particles: electron micrograph (a) and atomic force micrograph (b).*

10.3.2.2 Pigments and Extenders

Pigments and extenders are part of almost all coating formulations (Stoye and Freitag, 1998). Only in specific segments are pigment-free clear coats employed. Pigments and extenders govern many coating properties, depending on their nature, structure and surface characteristics. Pigments provide hiding and colour properties of the coating. They also help to protect the binder from degradation by UV-light. Pigments can be either inorganic materials, such as titanium dioxide (TiO_2) and iron oxide, or organic materials, such as phthalocyanine blue and carbon black. By far the most important type is TiO_2, a white pigment. Extenders are generally white, too, but have a much lower refractive index than TiO_2. The ratio of the refractive index of the binder and the pigment or extender determines the hiding of the coating. Thus, extenders do not contribute much to it but they provide a low cost way to increase the solids level of a coating formulation and thereby impart higher film build of the coating to, for example, fill small cracks. A variety of extender materials is commonly used, such as calcium carbonate, clay, feldspar or talc.

10.3.2.3 Pigment Dispersants

Dispersants support the dispersing process during the manufacturing of coating formulations based on emulsion polymers. They facilitate the wetting and break down of agglomerates of extenders and pigments to single particles during milling or grinding processes. Predominantly, polymeric species with low molecular weight and high levels of acid functionality are employed. Important types are based on acrylic or methacrylic acid, optionally copolymerised with other monomers. They are neutralized with bases, such as ammonia, sodium or potassium hydroxide or organic amines. The dispersants act by adsorbing to the surface of the pigments and extenders. They provide a strong anionic stabilization under the basic pH conditions of a coating formulation by increasing the Coulomb repulsion of the particles. Besides the polymeric dispersants, polyphosphates are used as dispersing

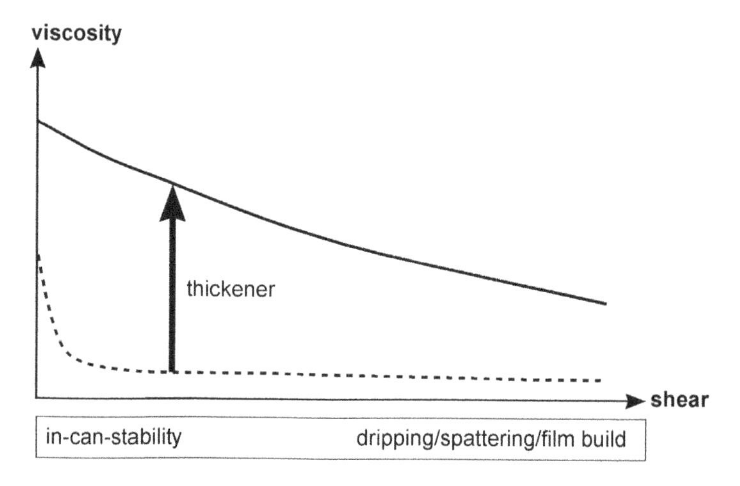

Figure 10.9 *Effect of thickeners on the rheology of paints.*

aids. They work by chelating metal ions in the coating formulation and thus reducing water hardness, which prevents agglomeration of pigment/extender particles.

10.3.2.4 Thickeners

Thickeners are used to provide coating formulations with the desired application rheology. A typical emulsion polymer shows a very unfavourable rheological profile (Figure 10.9). The viscosity at low shear rates is too low to prevent sedimentation of pigments and extenders during storage. Due to distinct shear thinning the viscosity drops even further at higher shear rates, leading to dripping and spattering during application. The low viscosity also results in low film build when applied by brush or roller, or unfavourable spraying behaviour. A formulation with an optimised thickener package corrects the rheology profile and provides storage stability as well as superior application properties.

There are three main classes of thickeners, which are commonly used: Cellulosics (modified natural products, usually hydroxy ethyl cellulose (HEC)), HASE (hydrophobically modified alkali-swellable emulsions) and HEUR (hydrophobically modified ethylene oxide urethanes).

Cellulosics are relatively high molecular weight water-soluble polymers, which thicken by raising the viscosity of the water phase of the coating.

HASE thickeners are polyacrylate dispersions that contain a higher amount of carboxylic acids, predominantly acrylic acid or methacrylic acid. Upon neutralization during the paint manufacturing they swell and thicken the paint. Due to additional hydrophobic functional groups attached to the polymer backbone they can interact with the hydrophobic part of surfactants and with hydrophobic entities of other thickener molecules, forming a kind of micellar structure in the water phase. In addition they can interact with hydrophobic domains on the surface of emulsion polymer binders. These extensive hydrophobic interactions improve the efficiency of the HASE thickener and help the coating formulation to resist volume exclusion flocculation, an undesirable aggregation process of the colloidal system associated with high molecular weight, non-associating polymers. HASE thickeners are cost-effective raw materials. However, the high acid content can lead to an increased water sensitivity of the coating.

HEUR thickeners are synthesized from diisocyanates, polyethylene glycols and long-chain alcohols. Thus, they are also capable of undergoing hydrophobic interactions. However, in contrast to HASE thickeners they do not carry a charge. Because of this, they proved to be particularly useful for applications where water resistance or barrier properties are important. With HEUR thickeners it is possible to develop water-based paints that come close to the Newtonian flow behaviour of solvent-based alkyd-systems, only.

In coating formulations often a combination of different types of thickeners is chosen to optimise rheology and application properties, respectively.

10.3.2.5 Others

As already stated, coalescents are employed in coating formulations to improve film formation of high-MFFT binders. More hydrophobic types like white spirit or Texanol®[c] (2,2,4-trimethyl-1,3-pentanediol-diisobutyrate) are highly compatible with the polymer and plasticize to a much higher extent than more hydrophilic solvents like ethylene glycol or propylene glycol. The latter slow down the evaporation of water additionally and retard the film formation, increasing the open-time of the coating.

Furthermore, defoamers are used to prevent or to dissipate foams. Foam can be formed during manufacturing and application of coatings. Water-based formulations are especially susceptible due to the presence of surface-active molecules like emulsifiers. If the bubbles are stabilized and long-lived they can interfere with the efficiency of manufacturing or leave unwanted voids in the dried film. Defoamers destabilize and break up foam in the liquid coating. Most commonly, silicon and hydrocarbon-based dispersions or emulsions are used.

Biocides and preservatives are used to prevent microorganisms from growing. If contaminated, aqueous formulations can change viscosity, form a separate liquid phase or even coagulate. Additionally, gas an with unpleasant odour might be formed. Typical products used are thiazolinones. In addition to in-can preservation of the liquid paint, so-called film preservatives are used to hinder microbiological attack of exterior coatings. They have to show low water solubility to restrict leaching and guarantee a long-lasting effect. Examples are carbamates, triazines and also thiazolinones.

10.3.3 Decorative Coatings

Since decorative coatings that are often also called architectural coatings represent the biggest application field for coatings the most important segments are briefly discussed.

10.3.3.1 Interior Paints

Interior paints can be described as coating materials that are not exposed to weather factors or UV-radiation. They are predominantly matt paints used to coat surfaces like plaster, wood chip wallpaper and the like. Because an interior paint mainly does not need to protect the substrate against moisture it can be formulated with low binder content. Therefore interior paints are usually formulated above CPVC yielding an open porous structure.

The quality of paint is rated by:

- hiding power of the wet and dry film
- easy and trouble free application
- one-coat application
- wet-scrub resistance as a measure for cleanability
- solvent content.

Moreover an attractive price/performance ratio is expected. Even though the amount of an emulsion polymer binder in an interior paint is rather low it has to contribute to a considerable extent to its properties. Its ability to wet and bind pigments and fillers should be distinct. In addition the binder should impart hydrophobicity to the paint film to allow cleaning of the surface with aqueous detergents without pronounced swelling and, consecutively, disintegration of the coating. Especially in this segment low-VOC paints have a very high market share. The end-user wants to avoid the burden of solvents in his home. The whole set of requirements in this segment is fulfilled best by styrene acrylics with MFFT around 0 °C.

10.3.3.2 Exterior Paints and Textured Finishes

Exterior decorative coatings are used to provide aesthetic and protective features to exterior walls of buildings or other objects like fences, decks or roofs which are mostly made from mineral substrates or wood. The coating has to protect the surface against weathering or UV-radiation and should keep a pleasant appearance over a long time. The most important requirements for an exterior decorative paint are:

- high UV-resistance
- water vapour permeability
- low water absorption
- good adhesion to the substrate.

Because an exterior coating should protect the surface against moisture it has to prevent liquid water or water vapour from penetrating through. Paints that are formulated below the critical PVC forming a non-porous coating achieve this feature. UV-resistance of an exterior paint can be imparted by styrene acrylics or pure acrylics (Baumstark, Costa and Schwartz, 1999). Only in low PVC coatings can pure acrylics demonstrate clear advantages due to their inherent higher UV-stability. To guarantee adhesion also to difficult substrates like, for example, wood or metal, emulsion polymers for exterior coatings can carry special functional groups such as amino, acetoacetoxy, phosphate, siloxane or urea groups that form specific interactions with the surface.

If coated on wood, architectural paints and thus the binder have to provide sufficient elasticity to cope with enormous changes in dimensions due to variation in temperature and moisture without cracking.

10.3.3.3 Elastomeric Coatings

Elastomeric coatings as well as emulsion gloss paints discussed below are smaller but nevertheless important segments of decorative paints. They are regarded to be high quality systems and impose special challenges to the binder.

Elastomeric coatings are applied in thick layers on masonry substrates. Because of their mechanical properties they can bridge cracks in the substrate and stretch and shrink with thermally driven building movements. They also prevent the penetration of rain into the substrate by sealing cracks, thus improving the durability of the masonry material. Polymer emulsions used for elastomeric wall coatings usually have T_g below −10 °C to provide a sufficiently elastomeric character even at lower outside temperature. The challenge for the design of an appropriate binder is to not impart tackiness and therefore high dirt pick up. This can be achieved, for example, by sophisticated cross-linking systems that work superficially.

10.3.3.4 Emulsion Gloss Paints

Emulsion gloss paints give the surface a shiny, brilliant and appealing look. Gloss is determined by the intensity of the reflection of light at the coating surface. The most important factors for this are the smoothness of the surface and the refractive index of the coating. Thus, high gloss paints have to be formulated with a high binder content and low PVC, respectively. Special application fields like window and doorframes require a high block resistance of the coating at the same time. It should be possible to close and reopen windows and doors without stickinging after short drying times. The already discussed multiphase particles can fulfil the requirement. The soft phase forms a film and imparts the elasticity; the hard phase contributes to the hardness and non-tackiness of the coating.

10.3.4 Protective and Industrial Coatings

Industrial coatings or protective coatings are defined as coatings that are applied to products produced in an industrial production process or to large industrial structures such as bridges and factories. The number of applications in this area is huge, ranging from coatings for furniture, coil coating, marine coating to OEM coating and refinish coating for automobiles. The requirements in these areas are generally more demanding than those for decorative coatings, regarding, for example, hardness, scratch resistance or water resistance. Because of this, the substitution of solvent-borne binders by water-borne ones has not been so pronounced as in the decorative area. However, innovations in polymer synthesis and improved formulation know-how have enabled waterborne finishes to increasingly gain volume in many areas of protective and industrial coatings. Nevertheless, the challenge for people in research and development for emulsion polymers still persists to come up with new technologies that allow for further replacement.

10.4 Adhesives

Reducing the emissions of VOCs is the main driving force for using waterborne adhesives. Until the 1960s most synthetic polymers used for adhesives were dissolved in organic solvents. So, huge amounts of organic solvents were released during application. In contrast, emulsion polymers are free of organic solvents; they are environmentally friendly, because no solvents are emitted. Therefore, they have replaced polymers dissolved in organic solvents and this substitution process is still ongoing. A special advantage of emulsion polymers is that high molecular weight polymers can be produced along with high solids content and a low viscosity. This combination cannot be achieved with polymer solutions based on organic solvents. Emulsion polymers are also easy to handle because of their low viscosity and because they can be diluted with water. For these reasons, acrylic dispersions were introduced in Western Europe to substitute solvent-based flooring adhesives in the early 1960s and this process was continued in the 1970s for other applications, such as pressure sensitive adhesives (Urban *et al.*, 1995).

Today about 1700 000 tons of emulsion polymers (dry weight) are used per year worldwide for adhesives. Polyvinyl acetate and ethylene/vinyl acetate copolymers are the biggest volume (circa 800 000 t), mainly used in packaging and as wood glue. Polyacrylate emulsion polymers (acrylics) are second in volume (circa 500 000 t) and mainly used for pressure sensitive adhesives. The demand for styrene butadiene (SB) dispersions used for adhesives is about 300 000 t per year (Freedonia, 2011).

10.4.1 Design of Emulsion Polymer Adhesives

Adhesives must have two key properties: they have to "wet" solid surfaces and adhere to them and they have to have internal strength, be cohesive. The right balance of adhesion and cohesion always needs to be considered when designing an emulsion polymer. The requirements are sometimes very different, depending on the application. The adhesive for a protective film for example must have high cohesive strength and very low adhesion, so that it can be removed without leaving a residue, even after long bonding times. By contrast, a packaging tape must stick immediately and durably, needing both very high adhesion, even after brief contact, and high cohesion. Monomer composition, molar mass and cross-linking are the main parameters to design the desired balance of cohesion and adhesion. Some examples of pressure sensitive adhesives (PSA) should illustrate this.

10.4.1.1 Monomer Composition

The adhesion force after brief contact is called tack (Zosel, 1985, 1986). Table 10.2 shows the tack of some acrylic homopolymers. The higher the amount of carbon atoms (C_n) in the side chain the higher the maximum tack, and the lower the temperature T(max tack) at which the maximum tack occurs. Low T_g monomers give soft and tacky polymers, in particular polyethylhexyl and polybutyl acrylate have the best tack at room temperature (Druschke, 1987).

10.4.1.2 Molar Mass and Cross-Linking

The molar mass and cross-linking of the polymer also affects the adhesion properties. All monomer units having activated H-atoms in the α-position to a C=O or C=C double bond

Table 10.2 *Maximum tack of polyacrylates.*

	C_n	max tack (J m^{-2})	$T_{(max\ tack)}$ (°C)	T_g [°C]	$T_{(max\ tack)}-T_g$ (°C)
Poly (methyl acrylate)	1	8	50	16	34
Poly (ethyl acrylate)	2	18	30	−14	43
Poly (n-butyl acrylate)	4	50	22	−46	68
Poly (ethylhexyl acrylate)	8	100	18	−62	80

(like acrylic acid and esters, butadiene, vinyl acetate) are accessible to radical transfer. This leads to branching and cross-linking. Above a certain degree of cross-linking the polymer is no longer soluble in organic solvents like toluene or tetrahydrofuran and can be separated from the rest that is soluble. Insoluble polymer structures are called gels and the gel content increases the cohesion of an adhesive. The soluble part is called a sol and is characterized by measuring the molar mass. With increasing molar mass the cohesion also increases due to entanglements of polymer chains.

Both molar mass and cross-linking can be reduced using chain transfer agents (CTA). Let us consider an acrylic with high amounts (0.6 pphm) of CTA. This polymer has no cohesion (shear strength) and almost no adhesion (peel strength). The fracture pattern of a peel test will show cohesion failure, which means that polymer is left on both sides of what was peeled apart. By decreasing the CTA (0.3 pphm) the molar mass will go up and so will the peel strength due to increasing entanglements of polymer chains. The shear strength is still zero the fracture pattern is still cohesive. Reducing the CTA further there will be a maximum of peel strength still showing cohesion failure. With a certain gel content the shear strength also increases. When then a little bit less CTA is used the cohesion failure from the peel test turns into an adhesion failure where no major amount of the adhesive is left on the substrate from which it was peeled off. The internal strength of the polymer is now higher than the adhesion to the substrate. At the change of the fracture pattern the peel strength shows unsteadiness and drops to a low value. From this point the further decrease in CTA increases the shear strength strongly and decreases the peel strength. The amount of CTA is an important tool to design the required balance of adhesion and cohesion (Zosel, 1991, 2000; Urban and Egan, 2002b).

Molar mass and cross-linking can be influenced in addition by the amount of initiator, by the process conditions (e.g. monomer-starved conditions versus monomer-flooded conditions) and by using cross-linking monomers like divinyl benzene, butanediol diacrylate or allyl methacrylate.

10.4.1.3 Functional Monomers

Functional monomers are used to improve the adhesive properties. One important functional monomer is acrylic acid. It increases strongly the cohesion of PSA and improves compatibility with fillers like $CaCO_3$ for construction adhesives.

In some adhesive applications it is necessary to have a high adhesion at the beginning and improve the cohesion by a curing step at room temperature. Some curing reactions are listed in Figure 10.10.

Carboxylic acid groups in combination with divalent metal ions like Ca^{2+} or Zn^{2+} give cross-linking by chelate formation. Hydroxyalkyl acrylates are used when water-dispersible, multifunctional isocyanates are taken for cross-linking. Monomers with carbonyl groups in the side chain are utilized for a curing mechanism with water-soluble acid dihydrazides to form a dihydrazone. An emulsion polymer containing dihydrazide in the water phase is stable and has a long shelf life. Only when the film has formed does cross-linking occur between the particles to build up cohesion. Additionally, a special advantage of dihydrazide-containing dispersions is the good adhesion to corona-treated polyolefins, since corona treatment generates carbonyl groups at the surface (Urban *et al.*, 1995).

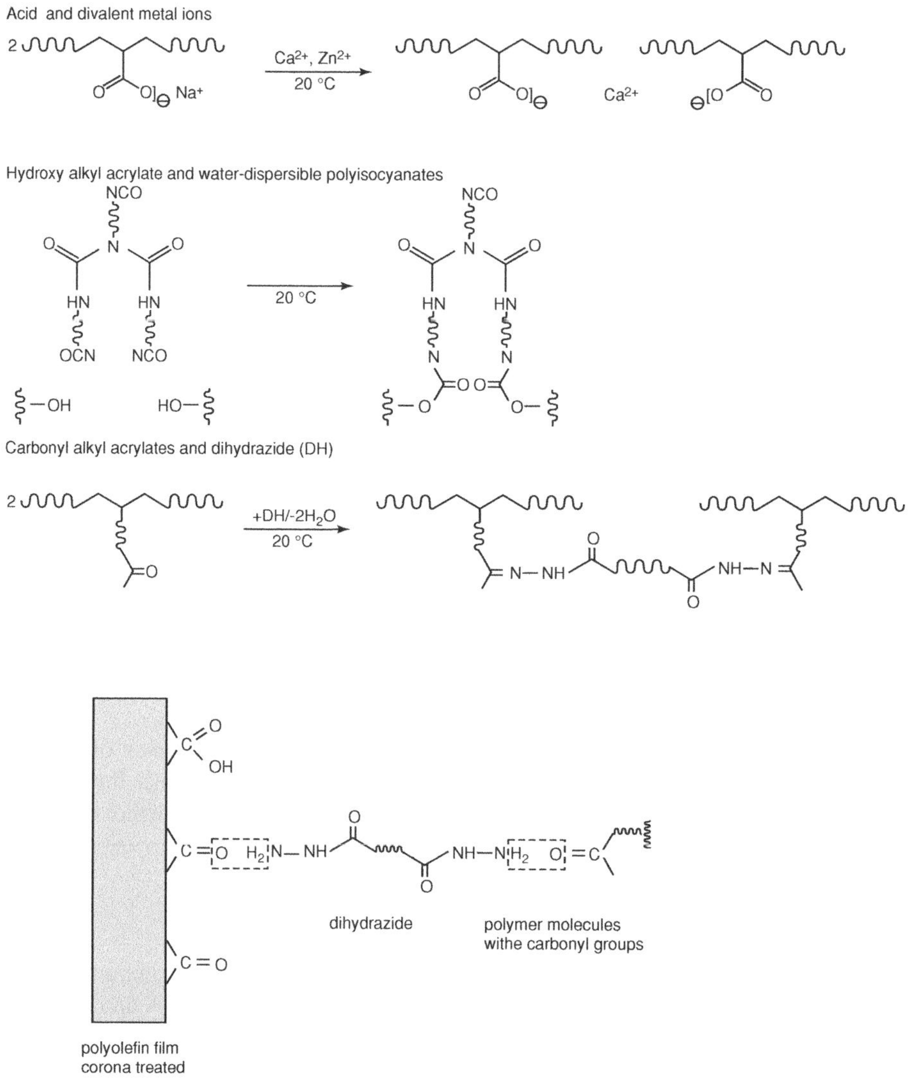

Acid and divalent metal ions

Ca^{2+}, Zn^{2+} 20 °C

Hydroxy alkyl acrylate and water-dispersible polyisocyanates

20 °C

Carbonyl alkyl acrylates and dihydrazide (DH)

$+DH/-2H_2O$ 20 °C

dihydrazide

polymer molecules withe carbonyl groups

polyolefin film corona treated

Figure 10.10 *Functional monomers and their chemical reactions for cross-linking at room temperature.*

10.4.1.4 Particle Size Distribution

The viscosity of an emulsion polymer depends on the solids content and is strongly influenced by the particle size distribution. Figure 10.11 illustrates that monomodal particle size distributions show a steeper increase in viscosity at a relatively lower solids content than bimodal or wide particle size distributions where solids contents of 70% are obtained in conjunction with low viscosity of about 200 mPa s. In a model of hard spheres with two different diameters, a maximum packing fraction is obtained when the volume fraction of

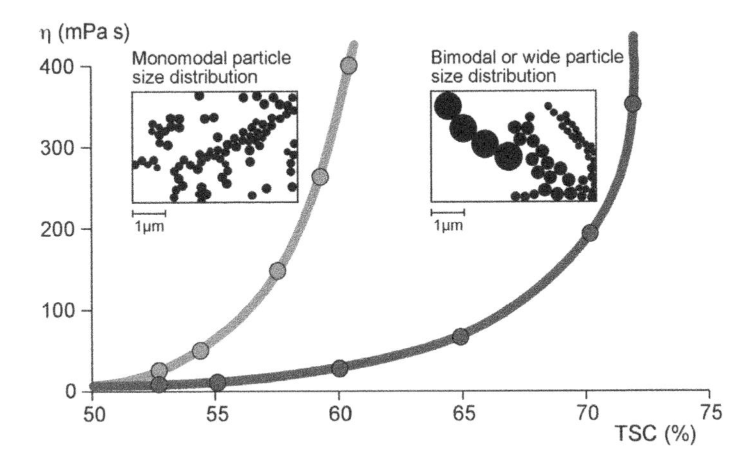

Figure 10.11 *Viscosity as a function of total solids content. Reprinted with permission from [Urban* et al., *1995] Copyright (1995) Exxon Chemical Europe Inc.*

the small particles is between 20 and 30% of the total volume of the spheres. The small particles are located within the voids formed by the large particles in a cubic packing pattern (Urban *et al.*, 1995).

The particle size distribution of an acrylic emulsion polymer with 70% solids content is shown in Figure 10.12: 30% (by weight) of particles with 200 nm diameter and 70% (by weight) of 850 nm particles result in a viscosity of 250 mPa s at a shear rate of 250 per second. This is not a mixture of two emulsion polymers but is produced in a single run by a special polymerisation process. High solids dispersions have the advantage that less energy is needed for transportation and evaporation of water, reducing the cost for a certain dry polymer amount to be applied.

Generally, emulsion polymers used for adhesives are produced and supplied with solids content between 50 and 70%, they have particle size distributions between 100 and 3000 nm,

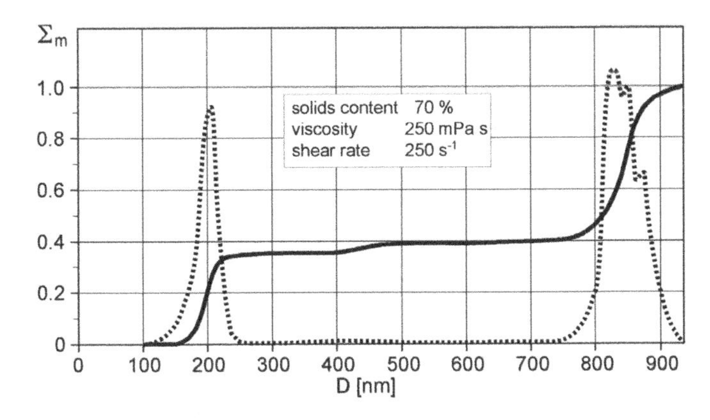

Figure 10.12 *Bimodal particle size distribution measured by analytical ultracentrifuge, D is the diameter. Reprinted with permission from [Urban* et al., *1995] Copyright (1995) Exxon Chemical Europe Inc.*

viscosities between 10 and 5000 mPa s and glass transition temperatures between –60 and +40 °C. The emulsion polymer itself is considered to be the most important raw material for making an adhesive. However, formulation additives are important as well to optimize the processing and application properties of an adhesive. Most emulsion polymers are formulated with tackifiers, plasticizers, thickeners, surfactants, antifoaming agents, fillers, pigments, biocides, antioxidant and curing agents to give a suitable adhesive.

10.4.2 Formulation Additives

Most emulsion polymers are formulated with additives to improve their performance from a technical and a commercial point of view. Some formulation additives are mentioned below.

10.4.2.1 Tackifying Resins

Resins are used, for example, to improve the tack of PSA or the green strength of a flooring adhesive. They must be compatible with the polymer and improve the polymer flow characteristics but generally a better flow is accompanied by less cohesion. Derivatives of gum rosins and of abietic acid are commonly used for acrylics, whereas hydrocarbon resins based on petroleum oil derivatives are preferred for tackifying SB emulsion polymers. To retain the "solvent-free" feature offered by emulsion polymers, resin dispersions or resin melts are used instead of resin solutions.

10.4.2.2 Plasticizers

In contrast to resins, which normally increase the glass transition temperature of the polymer, plasticizers decrease the T_g and make the polymer softer. This results in faster wetting of a surface and increases the initial peel strength. This occurs at the expense of cohesion and heat resistance. Common plasticizers are phthalates, like dioctylphthalate. Polypropylene glycol alkyl phenyl ether is a special polymeric plasticizer, which is extremely well compatible with acrylics and does not migrate.

10.4.2.3 Thickening Agents

Thickeners increase the viscosity, the water retention, the mechanical stability and the compatibility of the emulsion polymer with electrolytes and fillers. Viscosity adjustments are needed for optimum coatability of PSA, water retention improves, for example, the open time of a flooring adhesive and improved stability prevents coagulum formation during formulation. Since thickening agents in general are hydrophilic they reduce the water resistance of the adhesive. Thickeners of natural origin, for example, gelatin, casein and alginates are used as well as synthetic thickeners, for example, polyacrylic acids, polyvinyl alcohols, polyurethanes and polyvinylpyrrolidones.

10.4.2.4 Surfactants

Additional surfactants are used to improve, for example, the colloidal stability of the adhesive formulation in construction applications or the coatability of PSA. Surfactants reduce the surface tension and are used as wetting agents for the non-polar surface of a siliconized release liner. They naturally increase the tendency of foaming. The sodium salt of a sulfosuccinic acid ester is a very effective surfactant, which spreads very quickly onto newly generated surfaces (during coating) to stabilize them.

10.4.2.5 Antifoaming Agents

Surfactants used during emulsion polymerisation or in formulation often cause foaming, resulting, for example, in coating defects of PSA. This is prevented by addition of antifoaming agents. Antifoams are usually derivatives of aliphatic hydrocarbons or silicones. Some of them tend to migrate into the polymer during storage of the adhesive and thus are no longer available at the liquid/air interface where they are needed. Therefore, it is recommended to add the antifoam just before a process step in which foaming is expected.

10.4.2.6 Filler and Pigments

Fillers like $CaCO_3$ or SiO_2 are rarely added to PSA, since they strongly reduce the tack. But flooring adhesives, ceramic tile adhesive and sealants may have filler contents from 20 to 70%. Small inorganic filler particles evenly distributed reinforce the adhesive composition and increase cohesion to a certain level but too high filler levels will reduce the strength of the adhesive bond. Thixotropic fillers like SiO_2 reduce, for example, the sag of ceramic tile adhesives.

$CaCO_3$ is most widely used since it also reduces the formulation cost considerably. TiO2 is used when a white appearance and hiding power of the adhesive or sealant is required.

10.4.2.7 Biozides

To minimize the risk of bacterial, yeast or fungal contamination small amounts of biozides are added, like benzisothiazolinones or isothiazolinones.

10.4.3 Adhesive Applications

Emulsion polymer adhesives are used worldwide in many different applications, packaging being the biggest in volume (Ita, 2002). Here we focus on three areas in order to demonstrate the diversity: pressure sensitive adhesives, laminating adhesives and construction adhesives.

10.4.3.1 Pressure Sensitive Adhesives

Pressure sensitive adhesives (PSA) are highly viscous, visco-elastic materials, which adhere to virtually all surfaces when a small pressure is applied. They are permanently tacky and have an adequate cohesion.

Products based on pressure sensitive adhesives are protective films, self-adhesive labels and tapes. Protective films are made from flexible plastic material (PP, PE, PVC) coated with pressure sensitive adhesives (5–15 g(dry) m^{-2}) and they are used to protect, for example, painted or polished metal surfaces, lacquered furniture surfaces, acrylic sheets from scratching and soiling during manufacture, shipping and installation.

Labels are pieces of paper, plastic films or metal foil coated with pressure sensitive adhesives (15–30 g(dry) m^{-2}), which adhere to any solid surface after removal from a release liner. Labels are typically used to convey various kinds of information, examples are address labels, barcodes, price stickers and so on.

Tapes are flexible substrates (paper, plastic films) coated with pressure sensitive adhesives (20–100 g(dry) m^{-2}) and wound up to a roll. Examples are packaging tapes, double-sided adhesive tapes, masking tapes, medical tapes, electrical insulation tapes, office and household tapes.

Self-adhesive articles are usually produced by coating the emulsion polymer adhesive onto a substrate (siliconized release liner, film webs) and then drying it. (Türk, 1985, 1993) To apply the adhesive to the substrate roll or die, coating devices are used. By using a roll, first the roll takes up the emulsion polymer from a casting box and then transfers it to the substrate. With an optimized coating equipment web speeds up to 1500 m min^{-1} can be achieved. Pressure sensitive adhesives can also be directly applied to the substrate web using a slot die coater or a curtain coater. The advantage of die coating is that no foam is formed and the applied shear forces are lower than with roll coating (Willenbacher *et al.*, 2003).

10.4.3.2 Laminating Adhesives

Laminating adhesives are used to bond polymer films permanently to other films or to rigid materials in industrial manufacturing processes. According to the application field in industry a distinction is made between film-to-film lamination for flexible packaging, glossy film lamination for finishing print products, and technical lamination applications in automotive and furniture assembly processes (Urban and Egan, 2002b).

Flexible packaging materials are multi-layered structures increasingly used for packaging of food like cheese, bacon or potato chips. The multi-layer film for vacuum-packed coffee, for example, consists of polyethylene so that the pack is heat-sealable, an aluminum foil layer for aroma retention and light barrier, and a polyester film for mechanical strength and good printability. Film laminates are produced by coating the aqueous adhesive (circa 3 g(dry) m^{-2}) onto one side of the primary film, drying, and then laminating a second film onto the dried adhesive layer under heat and pressure. In low-performance laminates, dispersions are employed as the only adhesive component. If additional boiling resistance or sterilization capability is required, 3 to 5% of a suitable curing agent is added, for example, a water-dispersible polyisocyanate. This cross-linking agent (see Figure 9.10) does not just improve cohesion and heat resistance, but also significantly increases adhesion to corona-treated films (Fricke and Maempel 1994).

Glossy film lamination is the covering of print products with a transparent plastic film in order to protect them from scratching and soiling and to improve the brightness of the printing inks. Examples include book covers, advertising and packaging materials. A preferred emulsion polymer used for glossy film lamination wets the glossy plastic film very well and improves bond strength by a cross-linking reaction after evaporation of the water. This takes place at room temperature using a ketone-dihydrazide chemistry (see Figure 10.10) designed into the emulsion polymer (Fricke and Maempel, 1990).

Technical lamination in the automotive and furniture industries is used to bond decorative films onto preformed hard substrates, for example, printed PVC films on dashboards or medium density fibre boards. Adhesives based on polyurethane dispersions are mainly used, which provide excellent adhesion and extremely high bond strength.

10.4.3.3 Construction Adhesives

Modern construction adhesives are almost emission free and very easy to apply, which has widened the "Do It Yourself" applications considerably. Two examples are mentioned here, floor-covering adhesives and ceramic tile adhesives.

Floor-covering adhesives are used for gluing down flexible floor-coverings, like carpet or vinyl sheet. In Europe acrylics are predominantly used, formulated with tackifying resins and calcium carbonate as filler. In North America usually high solids content styrene butadiene emulsion polymers are applied, formulated with naphthenic oil and clay as filler (Urban and Egan, 2002b). In either formulation the inorganic filler content is between 25 and 50% and the polymer/resin ratio is about 1.

Ceramic tile adhesives are used for thin-bed applications on wall and floor, gluing tiles onto flat surfaces. Non-cementitious adhesives (mastics) are made of acrylic emulsion polymers (ca. 20%) mixed with silica sand and calcium carbonate as fillers (ca. 80%). One-component cementitious ceramic tile adhesives are made of cement (ca. 35%), redispersible polymer powders (1–25%) and fillers like silica sand (Lutz and Hahner, 2002). The redispersible polymer powders are made from emulsion polymers, which are spray dried. Predominantly, vinyl acetate copolymers are used.

10.4.4 Adhesive Test Methods

There are many test methods available for adhesives, but most important are the peel test, the shear test and the emission test. The peel and shear tests are used to determine the strength of an adhesive bond, usually by destroying it in a well defined way and measuring the forces needed. The emission test is used – mainly for flooring adhesives – to determine the VOCs released from the adhesive. These three test methods are described here brieflyly.

10.4.4.1 Peel Strength

The most common adhesion test is measuring the peel strength (Druschke, 1987), in which the force which occurs on peeling the adhesive layer off a substrate is determined. This test can only be applied using flexible adhesive layers, like paper labels, tapes, flexible floor coverings and so on. For pressure sensitive adhesives test strips of a certain width are rolled onto test panels using a roller with a defined weight. The test panels may be from stainless steel, glass, or plastic materials (e.g. HDPE) with a surface of defined roughness. After a dwell time the peel measurement is carried out using a tensile testing machine at constant peel rate and at a peel angle of 90° or 180°. Flooring adhesives are applied to plywood or cement board (according to regional requirements) using a trowel spreader, and after a certain time a 5 cm x 30 cm floor-covering strip is laid on the adhesive and pressed down again with a roller. After a defined time the floor-covering strip is peeled off using the tensile testing machine. The average peel force during this operation divided by the width must exceed a certain minimum level.

10.4.4.2 Shear Strength

To measure the shear strength a force is applied parallel to the adhesive bond. For floor covering adhesives the force per area, which results in fracture of bonded samples, is registered. For pressure sensitive adhesives the time required for a certain area of self-adhesive material to slide off a standard surface in a parallel direction to the surface with a constant load of 0.5 or 1 kg is measured. Shear strength is determined by destroying a bonded area and is taken as a measure of the cohesive properties, whereas peel strength is the force applied to a bonded length and represents the adhesives properties, in particular when adhesion fracture occurs. In either measurement it is important to know the fracture pattern in order to assess the properties correctly and to be able to improve the adhesive properties.

10.4.4.3 Emission Measurements

As already stated above, the main driving force for using waterborne adhesives is, to reduce the emissions of VOCs. However, depending on the manufacturing process and the raw materials used, there are volatile low molecular weight by-products in emulsion polymerisation, for example, Diels–Alder products of butadiene, alcohols from hydrolysis of acrylic esters, and residual monomers. To determine such VOCs, which are in the range of 10 to 5000 ppm, a chamber method for emission measurement has been established and is used in particular for flooring adhesives. In a stainless steel test chamber, samples of adhesives (besides adhesives all kind of materials can be tested as well) are sealed under well-defined temperature and humidity conditions. Purified air is flowing through the chamber and all VOCs are collected onto adsorption tubes containing suitable adsorbents. After certain times (e.g. 24 h, 240 h) the amount of adsorbed VOCs is determined by desorbing at higher temperature and analyzing them by gas chromatography (GC-MS coupling) or liquid chromatography. Important parameters are loading of the chamber (e.g. 0.4 m^2 m^{-3}) and the air exchange rate (e.g. 0.5 or 1.0 chamber volumes per hour). This test is used to identify and measure VOC emissions during processing (processor protection) as well as long-term VOC emissions (consumer protection). For flooring adhesives in Europe very low emissions are standard today, meaning that less than 500 μg m^{-3} of total VOCs are released after 240 h.

Emission measurements using this chamber method are also applied to tufted carpet which is described in the following section.

10.5 Carpet Backing

Carpet backing is – after paints, paper and adhesives applications – the fourth largest bulk application of emulsion polymers. About 480 000 tons per year (dry) are used worldwide, mainly as binder for the backing of tufted carpets. The function of a carpet backing is to anchor the pile fibres in place and to improve the dimensional stability of the tufted material.

The tufted carpet industry had its beginning in the late nineteenth century, when a Dalton (Georgia, USA) woman, Catherine Evans Whitener, produced bedspreads by sewing thick cotton yarns with a running stitch into an unbleached muslin base cloth, and cutting the surface loops of the yarn so they would fluff out. After tufting, the material was washed in hot water to cause the muslin to shrink around the tufts to hold them in place mechanically. She sold the first bedspread in 1900, and generated so much interest that a thriving cottage industry started. In the early 1930s multi-needle tufting machines and looms of greater width were developed in order to meet the demands for more bedspreads. Tufting machines for producing carpet appeared in the late 1940s, and by the late 1950s, carpet affordable to virtually every home owner in the USA, was being produced in 12 feet width (3.66 m), using nylon fibres, and jute as a secondary backing cloth. In the 1960s this technology was introduced into Europe. This resulted in woven carpet production in Europe declining sharply by the 1970s, and the establishment of Belgium, The Netherlands, Germany and Great Britain as Europe's major tufted carpet manufacturing countries (Blanpain, 2002).

10.5.1 Carpet Backing Binders

Initially starches and natural rubber latex were used as binders for improved tuft bind. They were replaced in the late 1950s by carboxylated styrene-butadiene dispersions (XSB). XSB emulsion polymers are very cost effective and easy to formulate, and they have become the work horse of the carpet backing industry and are almost exclusively used today.

Carboxylated styrene-butadiene dispersions have a solids content of about 52%, a pH between 7.5 and 9.0, and a monomodal particle size distribution with an average diameter of about 150 nm. With this small particle size, the instantaneous conversion during emulsion polymerisation is high, which leads to low concentrations of by-products, like 4-phenyl-cyclohexene (4-PCH) and 4-vinyl-cyclohexene (4-VCH) generated by Diels–Alder reaction of styrene and butadiene. 4-PCH and 4-VCH concentrations are required to be below certain limits, since they have a very low odour threshold.

The polymer composition is typically about 65% styrene and less than 3% carboxylic acid, for example, acrylic or itaconic acid, the rest being butadiene.

10.5.2 Carpet Backing Compounds

Manufacturing of residential carpets (used in private homes) uses typically two types of backing; primary and secondary backing. The purpose of primary backing is to securely fix the tufted fibres onto the fabric. The pre-coat compounds used here consist of XSB latex binder (10–15% dry on dry), calcium carbonate as filler (90–85% on dry), and small amounts of polyacrylate thickener and additional surfactants. The coating compound has viscosities between 5 and 15 Pa s and is directly applied onto the back-side of the tufted material (primary backing). The second compound is applied by means of a pan and lick roll directly to the woven or non-woven fabric, which is laminated to the already applied primary backing. Drying of both backing compounds is carried out after lamination. The second compound is more like an adhesive and has reduced filler load, ranging from 80% to even 0% (on dry). The viscosities are between 5 and 10 Pa s. This secondary backing improves the dimensional stability, which is particularly needed when the carpet is not glued down.

For commercial carpets (used in hotels, business offices and industrial buildings) a secondary backing for dimensional stability is not needed since they are directly glued onto the floor. They are coated with a high strength compound referred to as unitary coating. Unitary coating compounds consist of XSB emulsion polymers (33–40% dry on dry), calcium carbonate (67–60% on dry), and very small amounts of polyacrylate thickener and additional surfactants. The coating compound has viscosities between 5 and 10 Pa s. The coating speeds go up to 60 m min^{-1} (Blanpain, 2002).

10.5.3 Application Requirements

Carpet backing compounds are essential for the long-term performance and the aesthetic value of the installed carpet. Properties like high tuft-lock, minimum pilling and fuzzing, dimensional stability, water resistance and low odour increase the value of tufted fiber arrangements, whatever their pile density and color pattern. Within 60 years the tufted carpet business has grown from nothing to a more than $20 billion business per year. The continuous improvement of XSB emulsion polymers used as binders for carpet backing contributed to this growth. The combination of high binding strength, soft hand, good run ability and low emissions of VOCs have been important to support this growth. In particular the VOC emission of carpets was discussed in the late 1980s and some years later studies were published, connecting 4-phenyl-cyclohexene (4-PCH) emitted from carpets to adverse health effects. As a result of these allegations the Styrene Butadiene Latex Council (SBLC), the trade association of US latex producers and the EPA (Environmental Protection Agency) undertook extensive animal toxicological testing to investigate whether there was a link between 4-PCH and adverse health effects. As a result the EPA declared 4-PCH to be an "unremarkable chemical" (Fed Reg 1990; Haneke, 2002). However, 4-PCH is responsible for "new carpet odour" and this issue had to be addressed. As a result, the SBLC member companies have reduced VOC emissions by 95% since the late 1980s. With the present low VOC XSB emulsion polymers and a proper drying technique, carpet manufacturers today can produce odour free carpet (Blanpain, 2002).

Acknowledgements

The authors would like to express their sincere thanks to the following colleagues for friendly assistance in writing this chapter and for critical checking of the manuscript: G. Auchter, O. Aydin, R. Baumstark, HJ Fricke, S. Kirsch, K.-H. Schumacher, E. Schwarzenbach, M. Taylor, J. Tuerk, A. Zettl.

11

Specialty Applications of Latex Polymers

Christian Pichot,[1] *Thierry Delair,*[2] *and Haruma Kawaguchi*[3]

[1]*Saint-Priest, France*

[2]*Laboratoire des Matériaux Polymères et des Biomatériaux, Université Claude Bernard Lyon 1, France*

[3]*Graduate School of Engineering, Kanagawa University, Japan*

11.1 Introduction

For many years, emulsion polymers have hugely and increasingly been used in a broad range of applications where bulk, colloidal and surface properties of the dispersed material play a very important role. Nowadays, polymer particles, mainly in the subcolloidal size range, have found new applications to more advanced fields, such as information technology, electrical, electronic and optical devices, as well as in the life sciences (biomedical, biotechnological and pharmaceutical applications). The development of such latexes with innovating properties relies on the preparation of these emulsion polymers with a wide variety of polymers, particle shapes and sizes, morphologies and functionalities thanks to the use of numerous free radical and controlled radical heterogeneous polymerisation processes. Moreover, many physical and physico-chemical techniques allow the precise and complete characterization of the polymer colloids which are produced (Chapters 8 and 9).

A major advantage of the emulsion polymerisation technique is that, through the choice of the recipe and the polymerization strategy, it is possible to carefully adjust both macromolecular and colloidal properties of the obtained latexes, which is quite versatile in view of the variety of applications. For specialty applications, the control of particle size, particle size distribution, surface morphology, surface chemistry and functionality and so on, is

Chemistry and Technology of Emulsion Polymerisation, Second Edition. Edited by A.M. van Herk.

indeed of paramount importance. To that end, emulsion polymerisation has proven its use in the synthesis of functional latex particles (Arshady, 1999; Kawaguchi, 2000).

The scope of this chapter is to review specialty applications of latexes. After a short discussion on the common criteria that should be fulfilled for the design of polymer latex particles, we describe the main preparation methods suitable for producing them. Then, we will describe various specialty applications, dealing first with those concerning catalysis, information technology and so on, then with those related to the life sciences, that is, the biological, pharmaceutical and biomedical domains.

11.2 Specific Requirements for the Design of Specialty Latex Particles

Latex particles for specialty applications are generally high value dispersed material because either they are produced with a sophisticated formulation (sometimes requiring expensive monomers) or they will be used for supporting high cost species, like magnetic materials, biomolecules (proteins, nucleic acids, etc.) or drugs. Contrary to polymer latexes whose applications rely on bulk properties (coatings for instance) and are, therefore, produced on a large scale, latex particles for specialty applications are usually produced on a small scale.

Consequently, when dealing with latex particles for specialty applications, numerous variables should be considered with regard to macromolecular, surface and colloidal properties. Table 11.1 provides a non-exhaustive list of several important properties which should be carefully controlled, depending on the type of application. Appropriate techniques which can be used for the characterization of the particles are also included.

11.2.1 Nature of the Polymer

Polystyrene latexes are most extensively studied in general and, therefore, are also used in many specialty applications due to their appropriate polymer properties: (i) density slightly larger than one thus ensuring efficient dispersibility of the particles in aqueous dispersions; (ii) high T_g imparting robustness to particles when submitted to stress and avoiding coalescence; (iii) hydrophobicity suitable for biomolecules adsorption (especially antibodies) although some denaturation might occur. However, using copolymerisation of two or more monomers allows one to tune the properties much better for example, polarity, T_g values, functionalities, and so on.

Different from hard and hydrophobic polystyrene particles, soft and hydrophilic particles have attracted attention since the late twentieth century, mainly due to their possible stimuli-sensitive properties. Particles composed of highly hydrophilic polymer should be crosslinked to maintain the spherical shape in an aqueous medium. Not only hydrodynamic size but also interfacial properties of particles, such as mobility, sorption properties and so on, can be controlled by choice of monomer and designed morphology of the particles. In the case of polymer particles for life science applications, it is obvious that only a limited number of polymers is suitable, for example degradability might be desired and biocompatibility. This explains the interest in polyesters such as poly(lactic acid), poly(glycolic acid) (and their copolymers) and polycaprolactone. Synthetic non-toxic polymers of low MW that are bioresorbable, like polycyanoalkylacrylates and poly(acrylic acid)s, are also used.

Table 11.1 Variables of interest in the design of functionalized latex particles.

Property	Characterization techniques	Comments
Shape of particle	Electronic microscopy techniques	Mostly spherical but others can be conceived
Particle size and size distribution	Electronic microscopy, QELS, CHDF, FFF, etc.)	20 to 1000 nm
Particle morphology	NMR, TEM with staining techniques, SANS	Copolymers, core–shell, composite, porous, etc.
Nature of interface	Electronic microscopy, Neutron scattering, XPS, etc.	bare, hairy, etc.
Surface charge density	Conductometry, potentiometry	Presence of weak and strong acidic groups
Surface potential	Electrophoresis, acoustophoresis	Negative or positive values amphoretic systems
Colloidal Stability against electrolyte, temperature (T), shearing	Coagulation kinetics (UV, QELS, etc.)	Critical values for coagulation (ionic concentration, T, etc)
Functionality		
reactive surface groups	Chemical reaction, colorimetry, fluorescence, NMR, etc.	carboxyl, aldehyde, amine, thiol, epoxy, etc.
sensitivity to stimulus (T, pH, ionic strength, etc.)	QELS, electrophoresis	polyalkyl(met)acrylamide (thermal sensitivity)
color, fluorescent	Visible, UV spectrometry	High sensitivity limit
magnetic	Magnetization measurements	Iron oxide nanoparticles
biodegradability	NMR or spectroscopic techniques	Polyesters, polycyanoacrylates, etc
filmability	Measurement of T_g	Polyacrylates

11.2.2 Particle Size and Size Distribution

Both the particle size and distribution are very important parameters to be controlled, depending on the type of envisioned application. As will be shown later, particle size is usually controlled by the latex recipe, mostly by the presence and nature of surface-active species (see also the discussion on transparent latexess in Chapter 3). Specific surface area is directly proportional to the inverse power of the particle radius and could range from several to hundreds of $m^2\ g^{-1}$. Control of size distribution is often required for the sake of reproducibility. Generally, getting a fixed number of particles within the early stage of a polymerisation and preventing aggregation of particles in the later stage are the necessary conditions in order to obtain monodisperse particles.

11.2.3 Particle Morphology

As shown in Figure 11.1, various particle morphologies can be designed depending on the polymerisation process of choice. Depending on the number of polymer phases constituting the particle and/or the presence of water-soluble soluble chains, lipids, or

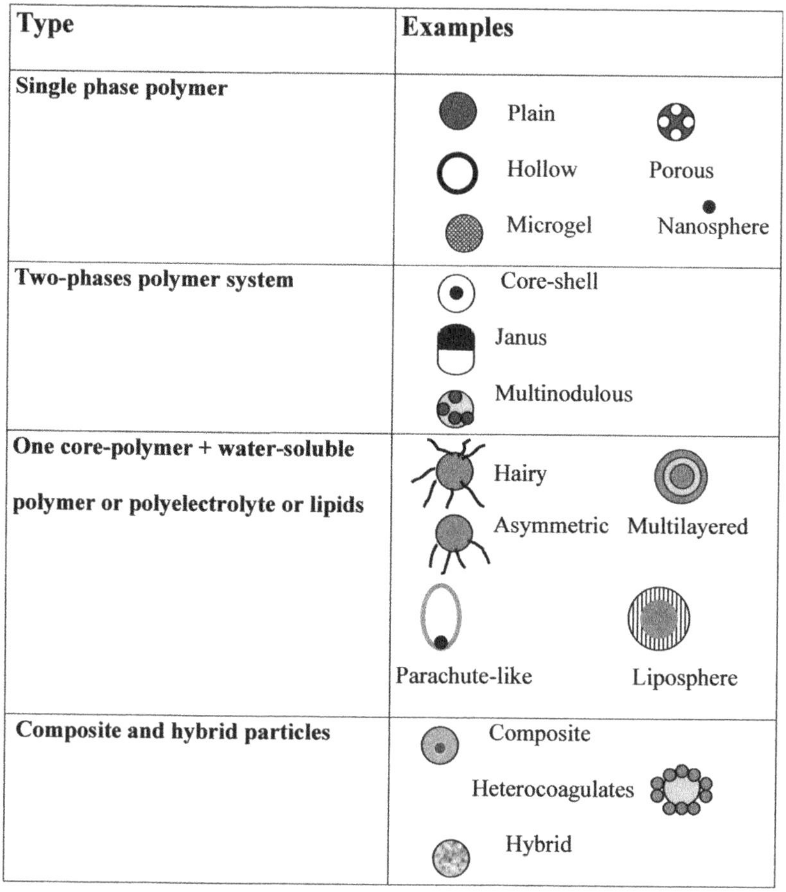

Type	Examples
Single phase polymer	Plain, Porous, Hollow, Microgel, Nanosphere
Two-phases polymer system	Core-shell, Janus, Multinodulous
One core-polymer + water-soluble polymer or polyelectrolyte or lipids	Hairy, Multilayered, Asymmetric, Parachute-like, Liposphere
Composite and hybrid particles	Composite, Heterocoagulates, Hybrid

Figure 11.1 Various particle structures and morphologies.

of an inorganic phase four categories of structured particles can be distinguished. Most of these morphologies can now be theoretically predicted based on thermodynamic and kinetic aspects (Stubbs, 2003a, 2003b and Chapter 6). A dynamic modelling approach was developed for the prediction of equilibrium morphologies of multiphase waterborne systems (González-Ortiz and Asua, 1995; 1996a; 1996b). Such particles are not only nano-objects of interest for academic research, but many of them exhibit attractive potentialities in various applications, especially for the immobilization/encapsulation of drugs or biomolecules.

11.2.4 Nature of the Interface

In many specialty applications, the interfacial properties are most important and need to be controlled. Surface chemistry and morphology (flatness, hairiness, polarity, presence of ionic charges or of hydrophilic polymer chains) should be tailored to the application.

11.2.5 Surface Potential

It is usually difficult to get direct information on the surface potential of polymer particles but alternatively a value of the zeta potential can be deduced from the electrophoretic mobility behaviour of the particles. Such knowledge is of paramount importance to predict the colloidal stability of the prepared polymer colloids when the pH and ionic strength of the aqueous medium are modified or when they are put in contact with other colloids, especially if those are of opposite sign. The electrophoretic behaviour of hairy particles does not necessarily give information on zeta potential but may offer a rough estimation of the layer thickness (Ohshima *et al.*, 1993).

11.2.6 Colloidal Stability

As colloids, latex particles are thermodynamically unstable and special care should be paid to impart long-term metastable stability, which means to provide an efficient energy barrier, which can be electrical or steric in nature. In many applications, especially those related to pharmaceutical and biomedical ones, polymer particles should often be able to withstand severe conditions (e.g. electrolytes, presence of biomolecules, temperature, shearing, etc.) which could induce reversible or irreversible flocculation. In those cases, steric stabilization is often brought through the use of polyethylene oxide containing molecular or macromolecular amphiphiles.

11.2.7 Functionality

Depending on the envisioned applications, the end-product polymer colloids should offer one or several specific functionalities which can be incorporated during the synthesis (by adjusting the recipe) or after preparation via a postpolymerization reaction. The first relies on the surface incorporation of functional groups able to interact with various (bio)organic or inorganic substrates. Classical reactive groups (carboxy, aldehyde, amine, thiol, epoxy, activated ester, etc.) are quite useful and control of their surface incorporation is challenging. It will be briefly reported in the next section how this can be done. Other functionalities can be imparted to the particles: magnetism (in order to rapidly separate the particles from a fluid medium as will be detailed later); coloration or fluorescence (detection by optical methods). It could be convenient to take advantage of stimuli-sensitive microgel particles which are able to change their structure or size in response to a given external stimuli (temperature, pH, electrical field, etc.), a property which mostly depends on the nature of the polymer (Pelton and Chibante, 1986; Pelton, 1999; Kawaguchi, Yekta and Winnik, 1995). Finally, for *in vivo* applications (drug delivery, gene therapy), one needs to select polymers exhibiting biocompatibility, bioresorbability, non toxicity, and so on.

Functionalized particles can be associated with plain surfaces in order to create two-dimensional assemblies of particles, also named *self-assembled monolayers* (SAMs). The development of such organized structures using well-defined colloidal particles and molecularly smooth surfaces (natural mica, silica wafer) was studied fundamentally (Zhang *et al.*, 2010; Li, Josephson Stein, 2011). Various techniques can be used to modify the surfaces, especially in order to control the deposition of latex particles. Figure 11.2 illustrates that chemisorption of functionalized latex particles onto silica wafer can lead to a non-compact hexagonal arrangement. Covalent bonding is provided by surface phenyl boronic groups.

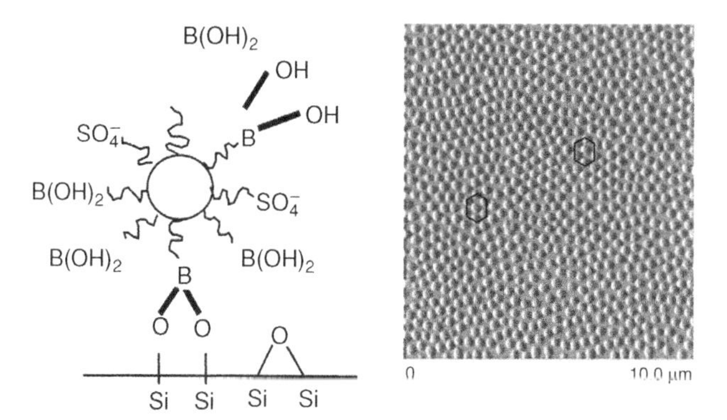

Figure 11.2 AFM micrograph of functionalized polyN-ethylmethacrylamide particles chemically adsorbed onto silica wafe. Reprinted from [Hazot 2001] Copyright (2001) P. Hazot.

11.3 Preparation Methods of Latex Particles for Specialty Applications

Many manufacturing methods are available for the synthesis of latex particles exhibiting appropriate functionalities and three main approaches can be followed: (i) polymerisation in heterogeneous media; (ii) modification of preformed particles; (iii) formulation of colloidal dispersions from pre-formed polymers.

11.3.1 Radical-Initiated Polymerisation in Heterogeneous Media

Due to the versatility of many radical-initiated (co)polymerisation processes in heterogeneous media, a large amount of polymer colloids for fine applications can be prepared using those techniques. When one wants to prepare monodisperse particles, it is obvious that emulsion polymerisation based techniques can be advantageously performed. Various strategies can be followed according to the type of particle which is envisioned: surface functionalized, hairy or composite particles.

11.3.1.1 Surface Functionalized Particles

Due to the broad interest in such particles, not only for specialty applications, a huge amount of work has been carried out in order to elaborate surface functionalized latex particles. For that purpose, it is possible to use molecular or macromolecular species (bearing the functionality) along with the polymerisation protocol (batch, semi-continuous, core–shell, shot-growth, etc.) (Pichot, 1995; Pichot and Delair, 1999; Pichot, Delair and Elaissari, 1999).

In Table 11.2 the main approaches which can be followed are summarized.

11.3.1.2 Hairy Particles

The production of such particles usually results from the emulsion copolymerisation of a hydrophobic monomer, like styrene, with a water-soluble monomer, like acrylic acid. Differences in the water-solubilityies of the two monomers together with disparate reactivity

Table 11.2 *Various strategies for producing functionalized latex particles.*

Origin of surface functional groups	Chemical structure	Surface groups	References
Initiator-derived	– Persulfate salts	$ROSO_{3-}$,M^{+}, ROH, RCOOH	Goodwin *et al.*, 1978 McCarvill and Fitch, 1978a,b
	– Redox systems	Persulfate/bisulfite/iron	Goodwin *et al.*, 1978
	– Azo derivatives	Amidine, Carboxylic Hydroxy	Guillaume, Pichot and Guillot, 1990 Filet *et al.*, 1995
Functional monomer or macromonomer	Carboxylic acids(acrylic and methacrylic acids)	COOH	Blackley, 1983
	Sodium styrene sulfonate,	$SO_3^-M^+$	Juang and Krieger, 1976
	Sodium 2-Sulfoethyl(or propyl) methacrylate		Liu and Krieger, 1978
	Vinylbenzylisothiouronium	$SC^+(NH_2)NH_2$,Cl^-	Delair *et al.*, 1994
	Chloride, 4-Vinylbenzyl chloride	$-CH_2Cl$	Verrier-Charleux *et al.*, 1991
	p-Formylstyrene	–CHO	Charleux, Fanget and Pichot, 1992
	Acrolein	NH_3^+, Cl^-	Basinska, Slomkowski and Delamar, 1993 Ganachaud *et al.*, 1995
	Dimethylaminoethyl methacry ate	Py+, X^-	Zhang, Zha and Fu, 2007
	Acrylamidoethyl(or propyl) methacrylate, hydrochloride 4-Vinylpyridinium		Ohtsuka, Kawaguchi and Hayashi, 1981
	Hydroxymethacrylate	–OH	Tamai, Fujii and Suzawa, 1987
	Glycidyl methacrylate	$-(CH-O-CH_2)-$	Magnet *et al.*, 1992
	Methacrylate-ended PEO	PEO	Ito, Cao and Kawaguchi, 2002
	Fluorinated alkyl methacrylate tab and then –F		Storsberg and Ritter, 2002
Reactive surfactant	Inisurf	Many chemical groups	Guyot and Tauer, 1994
	Surfmer	– idem	Schoonbrood, Unzue and Beck, 1997 Cauvin *et al.*, 2003
	Transurf	– idem	Dos Santos,Bris and Graillat, 2009
Chemical modification of surface groups	Benzyl chloride groups	$PhCH_2SH$, OH, CO_2H, NH_2, CHO	Delair, Pichot and Mandrand, 1993; Delair *et al.*, 1994
	Ester groups (acrylates)	COOH	Fitch and McCarvill, 1979a
	Sulfates	OH	Fitch and Watson, 1979b

ratios lead to the preparation of particles having a core–shell structure with the hydrophobic polymer in the core and the water-soluble polymer in the shell layer. It was also found that precipitation polymerisation of alkyl (meth)acrylamide, like *N*-isopropyl acrylamide in the presence of a crosslinker, allowed the formation of hairy particles (Pelton and Chibante, 1986). However, one disadvantage of those particles is that the hairy layer is poorly defined. It was also found that the use of hydrophilic macromonomers (like methacrylate-terminated PEO macromonomers (Ito, Cao and Kawaguchi, 2002)) and block copolymers (Riess, 1999) in emulsion polymerisation allowed production of highly stable hairy particles.

11.3.2 Modification of Particles and Related Methods

Modification of particles would be considered when the particles obtained by heterogeneous polymerisation do not possess the desired properties or functions. Modification can be done by chemical reaction or physical treatment. For example, a hydrophobic surface of particles can be converted into a hydrophilic surface by poly(ethylene glycol) chain grafting or by albumin adsorption. A comprehensive review was presented on the modification of particle surfaces (see Arshady, 1999). Here, methods for surface grafting to form hairy particles and the synthesis of hybrid particles (organic and inorganic) is focused on.

11.3.2.1 Hair Formation on Preformed Particles

The above-mentioned hairy particles prepared by general radical polymerisation are apt to have poorly defined hairs. It is worth mentioning that polymer nanoparticles having a well-defined surface of grafted polymers can be prepared by controlled radical polymerisation techniques such as NMP, ATRP and RAFT (see Chapter 5 for the meaning of these abbreviations). Various strategies can indeed be considered to produce such particles: (i) Grafting-from techniques applied to both polymeric or inorganic particles; (ii) Grafting-onto methods using preformed polymer chains produced by controlled radical polymerisation; (iii) heterogeneous polymerisation in dispersed media using amphiphilic polymers or macromonomers prepared either by anionic polymerisation (polyethylene oxide containing species) or by controlled radical polymerisation (Ferguson *et al.*, 2002a). Table 11.3 illustrates some examples of such polymer particles following these strategies.

The density of hairs on the particles can be controlled by the number of growing sites in the grafting-from method to form hairy particles from sparsely covered so-called mushroom particles to densely covered brush-like particles. The latter exhibits unique properties like that of rigid particles (Ohno, 2010).

11.3.2.2 Composite Particles

The association of inorganic and organic materials in dispersed form, especially colloidal particles, has received increasing interest in recent years. It was indeed shown that inorganic materials with nanoscale dimensions could exhibit improved behaviour with regard to electrical, magnetic and optical properties when associated with polymers to give novel materials (Bourgeat-Lami, 2002; Karg and Hellweg, 2009). Other merits of the association of inorganic nanoparticles with polymer microparticles are the control of nanoparticle function by the surrounding polymer and the synchronization of functions of inorganic and polymeric materials.

Table 11.3 *Various strategies to produce well-defined hairy particles by CRP techniques.*

Strategy	Example	Reference
Grafting from	Polymerization of *N*-isopropylacrylamide onto PS-polyvinyl benzylchloride particles	Tsuji and Kawaguchi, 2004
	Polymerization of hhydroxyethyl acrylate and 2-(methacryloyloxy) ethyl trietylammonium chloride onto functionalized particles (ATRP)	Manuszrak-Guerrini, Charleux and Vairon, 2000
Grafting onto	*N*-acryloylmorpholine-*N*-acryloxysuccinimide copolymers onto PS particles	D'Agosto *et al.*, 2003
Aqueous emulsion polymerization in the presence of preformed amphiphilic polymers	PS-b-PEO	Mura and Riess, 1995
	PMMA-polyacrylic acid	Rager, Mayer and Wegner, 1999
	Poly 2-(dimethylamino)ethyl methacrylate-b-PMMA	Li *et al.*, 2003
	PS-b-poly(vinylbenzyltrimethyl ammonium chloride)	Jaeger *et al.*, 1999
	PS-b-polyacrylic acid	Burguière, Chassenieux and Charleux, 2003
	PS-b-poly(vinylbenzylchloride)	Save *et al.*, 2005

The synthesis of such composite (nano)particles can follow three main methods: (i) assemblies of preformed organic and inorganic particles; (ii) chemical reaction of organic precursors in the presence of inorganic particles and vice versa; (iii) simultaneous reaction of both inorganic and organic precursors which then leads to hybrid particles. Among the many inorganic colloidal particles which were combined with polymers, it is worth to mention iron oxide, silica, titanium dioxide, and so on (see also Section 3.7.1. and Chapter 6).

Due to their interest in numerous biomedical applications, the preparation and characterization of magnetic particles have been investigated over the three last decades, however, with results which were not always fully satisfactory, especially with regards to the encapsulation efficiency of the preformed ferrofluids. The elaboration of magnetic colloids has been challenged by many research groups and many detailed reviews have been published (Elaissari *et al.*, 2003). A special issue of Advances in Polymer Science is devoted to the preparation of hybrid latex particles with the aid of (mini)emulsion polymerization (van Herk and Landfester, 2010).

Briefly, two major strategies have been envisioned, one using pre-formed non-magnetic particles, the other incorporating the magnetic material during the polymerisation process leading to the particle formation. For the first strategy, a pioneering report (Ugelstad *et al.*, 1993) relied on the precipitation of iron oxides within pre-formed monodisperse porous polymer, leading to magnetic composite particles of over 2 μm in size. More exactly, ferric and ferrous ions were introduced into the pores and thenoxidized *in situ* by treating them with alkali. In the same strategy, another concept consisted in the hetero-coagulation of iron oxide nanoparticles onto latexes in the 500 nm range, followed by encapsulation to avoid leakage of the magnetic material (Furusawa, Nagashima and Anzai, 1994; Sauzedde, Elaissari and Pichot, 2000). The latex particles obtained by this approach had a magnetic material content of less than 30 wt%, which may prove insufficient when high magnetization rates are required. The second strategy was pioneered by Daniel, Schuppiser and Tricot (1982) who synthesized polystyrene magnetic particles by dispersion of inorganic magnetic materials in an organic mixture of monomer and initiator, emulsified in water, followed by polymerisation. The resulting particles were fairly polydisperse, accompanied by free inorganic nanoparticles and polymer particles, but the magnetic material content was 40–50 wt%. An improvement in the size distribution was obtained by adding a miniemulsion of magnetic material to a second miniemulsion of styrene. After polymerisation, particles in the 40 to 200 nm range were obtained with a maximum magnetic content of 35 wt% (Ramirez and Landfester, 2003). Finally, a method based on the polymerisation of styrene within the submicronic droplets of a highly stable (using a polymer amphiphile) magnetic emulsion was developed, the particle distribution was controlled by the distribution of the initial magnetic emulsion. The magnetic latexes obtained were highly magnetic in nature with up to 60 wt% iron oxide (Montagne, 2002). As shown in Figure 11.3, a nice core–shell particle morphology was obtained, indicating that iron oxide was fully encapsulated. Several metal nanoparticles containing polymeric microspheres were prepared by introducing metal ions into the gel particles followed by *in situ* reduction of the metal ions to form metal nanoparticles distributed in the microspheres (Kumacheva, 2004).

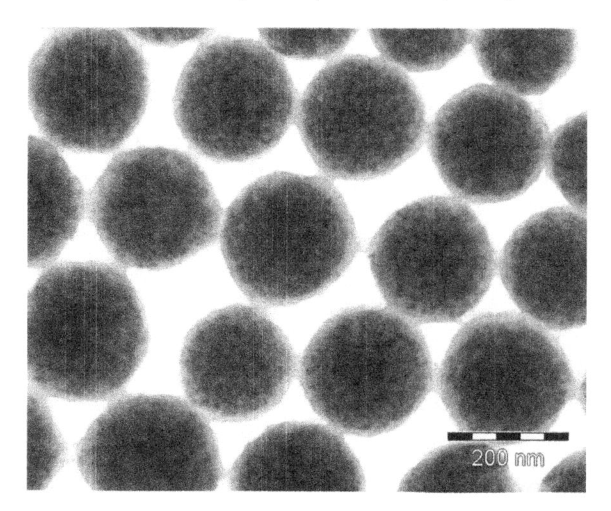

Figure 11.3 *Submicronic magnetic latex obtained by emulsion polymerization of styrene and divinylbenzene onto a ferrofluid emulsion stabilized by polyacrylic acid based amphiphilic surfactant and initiated by potassium persulfate. Reprinted from [Montagne, 2002] Copyright (2002) F. Montagne.*

In that domain, it is also worth mentioning the development of polypyrrole-based composite particles with various organic (polyacrylics) or inorganic (silica) (Armes, 1998) where the conductive polymers can be incorporated either in the core or in the shell.

11.3.3 Formulation of Colloidal Dispersions from Pre-Formed Polymers

The technologies involving the formation of particles from pre-formed polymers are widely used in the field of *in vivo* delivery of drugs, proteins or DNA. Obviously, none of these methods is based on emulsion polymerisation but *in vivo* applications are important for colloids, in terms of scientific and technical challenges, added value, and improvement of life for the human being.

The chemical nature of the preformed polymer will depend on the strategy of administration. For the peroral route, polymers do not need to be degradable as they will be eliminated via the digestive track. Hence, copolymers like poly(alkyl (meth)acrylates) or poly(meth)acrylic acids) can be used (Schmidt and Boodmeier, 1999). For injections, administration via the parenteral route, the polymer to be injected should be biodegradable and, therefore, the typical preformed polymers are poly(hydroxy acids), such as poly(hydroxybutyric acid), poly(α-caprolactone), poly(glycolic acid) or poly(D, L- lactic acid). The most widely used polymers are poly(glycolic acid) (PGA) and poly(D, L-lactic acid) (PLA) or the copolymers poly(glycolic-co-lactic acid) (PLGA). Most of these polymers are obtained by ring opening polymerisation of a cyclic precursor, ε-caprolactone, glycolide or lactide (Panyam and Labhasetwar, 2003).

Polyelectrolytes have been used in the formation of particles and the most promising structures are from the natural polymers such as chitosans (Sinha *et al.*, 2004) and alginates (Gombotz and Wee, 1998) which are water-soluble and respectively positively and negatively charged polysaccharides.

11.3.3.1 Formulation of Organo-Soluble Polymers into Particles

The simplest method to obtain particles is the solvent displacement method developed by Fessi *et al.* (Fessi, Puisieux and Devisaguet, 1987). The polymer, dissolved in a water-miscible organic solvent, is slowly added to an aqueous solution containing a surfactant or a non-ionic stabilizer, such as poly(vinyl alcohol) or PEO-PPO-PEO triblock copolymers. On contacting with water the solvent in the drops diffuses into the aqueous phase and the water-insoluble polymer precipitates in the form of colloidal particles, under the condition that the polymer concentration be low enough to prevent the formation of aggregates.

Even though the preceding method is quite attractive by its ease of use, it may appear somewhat limited in its range of particle size. Hence, methods based on the formation of emulsions were developed. Oil-in-water (o/w) emulsions can be formed, the oil droplets containing the polymer and the aqueous phase a stabilizer. The elimination of the solvent from the mixture induces a controlled precipitation of the polymer into particles (Singh and O'Hagan, 1998). This is a general methodology in which solvents quite different in chemical nature and physico-chemical properties (regarding water miscibility for instance) can be used, provided they are volatile enough to be easily eliminated. Moreover, this approach is used for the encapsulation of hydrophobic active molecules within the core of the particles. The formation of double emulsions (w/o/w) leads to the production of capsules with the cavities filled with an aqueous solution of a hydrophilic active compound like a peptide or a protein (Delair, 2003).

Supercritical fluid technology is an attractive method for producing particles in high yield free from traces of solvent. The supercritical fluid of choice is CO_2 because of its low critical temperature (31 °C), natural abundance and relative environmental friendliness. This technology has been reviewed by Richard (Richard and Deschamps, 2003). According to the comprehensive works by Desimone *et al.*, supercritical CO_2 was especially efficient in the syntheses of fluoropolymer particles (Du *et al.*, 2008)

Apart from synthetic polymers, particles can be obtained from fat, or solid lipids, using similar approaches to those reported above (Müller, Mäder and Gohla, 2000).

11.3.3.2 Formulation of Water-Soluble Polymers into Particles

Polymers from natural origin, like polysaccharides for instance, are of great interest in pharmaceutical sciences since they are considered generally as safe materials. The formulation of these biopolymers into particles often occurs in an aqueous medium and as such the following processes can be regarded as environmentally friendly, and safe for human health.

Ionic gelation relies on the formation of particles by cross-linking polyelectrolytes with multivalent, often inorganic, ions. For instance, alginates have been complexed with calcium salts (Gombotz and Wee, 1998) and chitosans with polyphosphates (Janes, Calvo and Alonso, 2001).

Colloids can be also obtained by formation of polyelectrolyte complexes of a polyanion and a polycation, as reported for synthetic (Buchhammer, Mende and Oelmann, 2003) and natural polymers (Cui and Mumper, 2001). In this strategy of particle formation, counter ion components are not limited to polymers but can also comprise ionic surfactants (Trabelsi, Raspaud and Langevin, 2007). Ionized cellulose derivatives dissolved in water were assembled with oppositely charged surfactants to form microspheres. Not only electrostatic forces but also hydrophobic forces are found to contribute to the formation of micro-aggregates (Guillot *et al.*, 2003).

11.4 Applications

In this section we give an overview of the various specialty applications of polymer latex particles. For illustration, Table 11.4 provides a non-exhaustive list of the numerous applications of these polymer colloids showing that a broad range of domains is presently covered.

11.4.1 Non-Biomedical Applications

The outstanding colloidal properties of latex particles which can be tuned in a wide range of size, shape, surface charges and functionalities make them very attractive in quite a lot of applications. Before reviewing some of them, it should be first remembered that for a long time latex particles have played a very important role in two domains: (i) as unique models in academic research in the field of colloid science or for testing polymerisation mechanisms in many heterogeneous polymerisations in dispersed media; (ii) as standards of calibration of sizing methods, such as TEM, SEM and AFM; dynamic light scattering, hydrodynamic chromatography, capillary hydrodynamic fractionation, field flow fractionation and so on.

Table 11.4 *Major applications of latex particles in specialty applications.*

Main application	Current application	Type of polymer particle
Calibration	Calibration of electronic microscopes and other sizing methods	Monodisperse PS particles
	Identification and enumeration of lymphocytes, bacteria, viruses	
Electronic or photonic devices	Measurement of colloidal forces	Charged latex particles, organic/inorganic hybrids or composites
Magnetic recording media	Information storage	Charged latex particles
Optoelectronics, lithography	Self-assemblies of nanoparticles	Charged-polystyrene particles
Catalysis	Oxidation and hydrolysis of small molecules	Anionic or cationic styrene-vinylbenzylchloride based copolymer particles
Diagnostic	Immunoassays	Magnetic particles
	NMR imaging Lab on a chip	Conducting particles (polypyrrole)
	Clinical diagnostics	Polystyrene particles
Purification and separation	DNA, proteins, virus separation	Thermally-sensitive particles
		Magnetic particles
Biomedical research	Immobilization of enzymes	Hydrophilic particles
Cosmetics	Shampoos, capillary products	Film-forming polymer particles
	nail varnishes, make-up products	(acrylics)
Therapy	Embolization, hemoperfusion	Polystyrene particles
	Drug delivery	Hollow polyester particles
	Immunotherapy	Polyester particles
	Vaccination	Polyester particles
	Chemotherapy, gene therapy	Polycyanocrylates

11.4.1.1 Polymer Colloids as Catalysts Supports and for Metal-Complexing

The study of Fitch *et al.* (1981) should be first mentioned. They investigated the hydrolysis of a series of alkyl acetates using simple sulfonated-charge polystyrene latex particles and found a significant increase in the reaction rate (almost eight times faster than under homogeneous conditions).

Polymer colloid particles can constitute appropriate alternatives for conventional heterogeneous catalyst supports. They indeed provide high surface area and can be produced in a variety of size, compositions and surface functionalities. The use of functional monomers can offer specific binding sites for catalysts. It might be expected that such polymer colloids would reduce mass transport and diffusion limitations together with favouring the reaction to occur at the particle surface. Such an area of research has been pioneered and developed by Ford and co-authors (Ford *et al.*, 1992) who investigated various classical reactions in order to understand where and how chemical reactions can proceed in a colloidal environment. They prepared slightly crosslinked cationic or anionic latexes by emulsion polymerisation of styrene, divinylbenzene with either vinylbenzylchloride or acrylic acid . The cationic ionic exchange latex was used as a support for cobalt phthalocyaninetetrasulfonate for the oxidation of various mercaptans, styrene or alkenes and high activities were observed. The anionic carboxylate latex as a support of *o*-iodobenzoate was used for the hydrolyses of various small molecules.

The synthesis of metal-complexing nanoparticles with an average size of 15 to 25 nm consists of a two-step procedure starting first with the microemulsion polymerisation of a mixture of styrene and vinylbenzylchloride in the presence of a cationic surfactant, then followed bypost-functionalization with a selective macrocyclic ligand, tetrazacyclotetradecane (cyclam) (Larpent *et al.*, 2003). The authors showed the ability of these cyclam-functionalized nanoparticles to complex copper and, in addition, they observed that the particles exhibited a "solution-like" complexation behaviour, demonstrating the accessibility of the immobilized ligand.

Metal nanoparticles such as Au, Ag and Pt nanoparticles are excellent reducing catalysts. The details of catalytic activity in polymeric particles were carefully studied in terms of the reduction of nitrophenol using sodium boron hydride as a reducing reagent (Wunder *et al.*, 2011). The results were clearly explained with the Langmuir–Hinshelwood model. This means that the adsorption is the critical process in the total reaction.

11.4.1.2 Magnetic Recording Media

Magnetic storage plays a very important and key role in audio, video and computer development. The stability of recorded data against thermal decay has become an important objective for judging the performances of magnetic systems. Continued growth of storage densities in the presence of thermally activated behaviour requires innovations in the recording system. Recent innovations are self-assemblies of nanocomposites or nanoparticles considered for recording media. In nanocomposite materials, an annealing process transforms a multilayered structure into a non-magnetic matrix containing magnetic particles. Monodisperse and non-interacting particles were found promising to reduce noise and decay rates (Moser *et al.*, 2002).

11.4.1.3 Latex Particles for Electronic and Optoelectronic Devices

Although there are some valuable individual optoelectronic particles, such as multilayer polymer particles with periodic modulation in refractive index for use as a resonator (Alteheld *et al.*, 2005), many of latex particles for the electronic and optoelectronic applications are assembled into one-, two-, or three-dimensionally ordered structures spontaneously or with the aid of some external forces. The ordered structure is classified into two categories, loosely-packed and closely-packed colloidal crystalline arrays (CCAs). A one dimensional array of particles is realized by spontaneous assembling of Janus particles or by electro-rheological fluids under an electric field. A two-dimensional array of particles can be observed when a dispersion of suitable concentration is poured on a substrate and dried under the condition that the particles have sufficient interparticle repulsive force (Pelton, 1999)

Loosely packed three-dimensional CCAs are also formed due to the repulsive forces of the charged mono-dispersed particles (Fitch and McCarvill, 1979; Fitch and Watson,

1979; Fitch, Gajria and Tarcha, 1979). Equilibrium in self-organization is characterized by a spacing distance (responsible for the Bragg diffraction) which depends on the ionic strength of the continuous medium (Okubo, 1993). Optical interaction of the particle array with incident light was observed in the case of many polymer colloids in the submicron size, which is comparable to the wavelength of visible light.

For example, as the particle size is changing, it causes a variation of the surface topographic structure which in turn exhibits optical coating effects such as anti-reflective coating effect (Koo *et al.*, 2004). Loosely packed colloidal crystals whose matrix was composed of stimuli-sensitive gel as a sensor of stimuli (e.g. temperature, glucose, metal, etc.) can be used as sensors. Because the lattice spacing is changing depending on the extent of the stimulus, consequently the optical spectra change (Weissman *et al.*, 1996; Muscatello, Stunja and Asher, 2009). Zhou and Hu proposed a novel method to make a film of microgel colloidal array using self-crosslinking during the drying process (Zhou *et al.*, 2009).

Close-packed CCAs are generally formed by centrifuging particle dispersions. CCA formation is sometimes assisted by some additives, such as surfactants liquid crystals and fibres, or guided by light, electric fields, magnetic forces and so on (Li and Matyjaszewski, 2011).

It was reviewed that CCAs could find innovative applications as a photonic material (Texter, 2003). Exhibiting periodic structures very analogous to those of simple salts such as NaCl, they were proposed as photonic band-gap materials, that is, for instance as narrow bandwidth filters.

There is a need for the near future to increase information storage (video and multimedia editing, imaging and storage archives), however, these applications require high storage capacity. Novel nanocomposite materials have been proposed for use in high-density optical memory storage. They consist of submicron latex particles comprising a hard core containing a covalently bonded photosensitive compound and a soft shell containing an inert resin and playing the role of a continuous phase of the matrix. These particles are arranged into a three-dimensional array and heated at a temperature at which the shell resin melts. The core particles are closely ordered with a concentration depending on the particle size. This ordered structure was proposed to be used in various applications, especially in optical memory storage devices. Each core particle, due to the presence of a photosensitive compound, is able to store a single bit of information. Writing information can be achieved by irradiating specific particles with a laser beam to induce photobleaching of the photosensitive compound (Siwick *et al.*, 2001).

11.4.1.4 Polymer Colloids as Templating Materials

Some two-dimensional colloidal crystals have attracted much attention because they were a valuable device for colloidal lithography (Zhang *et al.*, 2010). Colloidal lithography is a convenient, inexpensive and repeatable nano-fabrication technique which can be applied as masks and templates for evaporation, deposition, etching, and imprinting. Colloidal particles can be produced in a wide variety of materials, sizes and shapes. Dynamically tunable microlens arrays were developed using self-assembled temperature-sensitive micro hydrogels (Kim, Serpe and Lyon, 2005). Adsorption of charged particles onto oppositely charged surfaces by electrostatic interactions leads to nanostructures in which size, spacing (close-packed and non-close-packed) and shape can be varied. A systematic study has been reported by Hanarpp and coworkers who investigated the adsorption of three different negatively charged polystyrene particles onto flat titanium oxide substrates (Hanarpp *et al.*, 2003).

Templating of close-packed photonic arrays can be used to produce structured porous material by filling the interstices of the colloidal crystal with an organic or inorganic phase (Velev, Lenhoff and Kaler, 2000).

11.4.1.5 Latex Particles as Colloidal Materials for Micromanipulation

Latex particles can be used for micromanipulation by taking into account that some of their colloidal properties (particle size and dielectric behaviour) can be affected by the action of a laser beam or an electric field.

Considering the first case, new techniques have indeed been developed in the last ten years which use light instead of mechanical means to manipulate an object. Optical tweezers, first developed by Ashkin (Ashkin, 1997), also called *optical traps*, are three-dimensional traps that use a focused laser to trap and manipulate neutral objects, such as small dielectric particles. The basic principle of optical tweezers is as follows: thermal fluctuations of small particles which are determined by their environment (mostly temperature and medium viscosity) can be altered by external forces acting on the fluctuating particles. It can be interesting to investigate the position fluctuations of the particle interacting with its environment. That could be done by trapping the probe with applying forces arising from the momentum transferred from the laser beam onto the specimen. It is then possible to record the trajectory of the particle through interferometry. From the histograms of particle positions, the interaction potential and the corresponding forces acting on the particle can be deduced. In conventional optical tweezers, a measure of the force is recorded together with a specific direction after the trap whereas the photonic force microscope provides a complete three-dimensional potential (Rohrbach, Florin and Stelzer, 2001). Figure 11.4 shows the basic set-up of such a technique along with two examples of applications.

Optical tweezers have been involved in numerous applications using latex particles, such as those concerning colloidal and interface science or in the life sciences, where latex beads serve as supports for immobilizing biomolecules. This technique is becoming an elegant common technique to realize manipulation of polymer colloids and, more generally, micron-size objects (Masuhara, 2003).

Regarding the dielectric phenomenon of particles, its principle refers to the polarization and associated motion induced in abiotic and biotic particles by a non-uniform electric field. The phenomenon stems from the difference in the magnitude of the force experienced by the electrical charges within an unbalanced dipole. In the case of charged and neutral particles, the convergence of the field lines results in an uneven charge alignment and formation of an induced dipole moment. The consequence is an imbalance in force upon the particle, which makes it migrate toward the region of greatest field intensity, such as an electrode. The dielectric properties of dispersed materials are frequency dependent, in turn depending on the components and polarizability of the particle. Therefore the polarity and magnitude of the dielectric force will vary as a function of the field frequency. Such dielectric frequency spectra data can be used to favour the separation of particles from media and from complex mixture of particles.

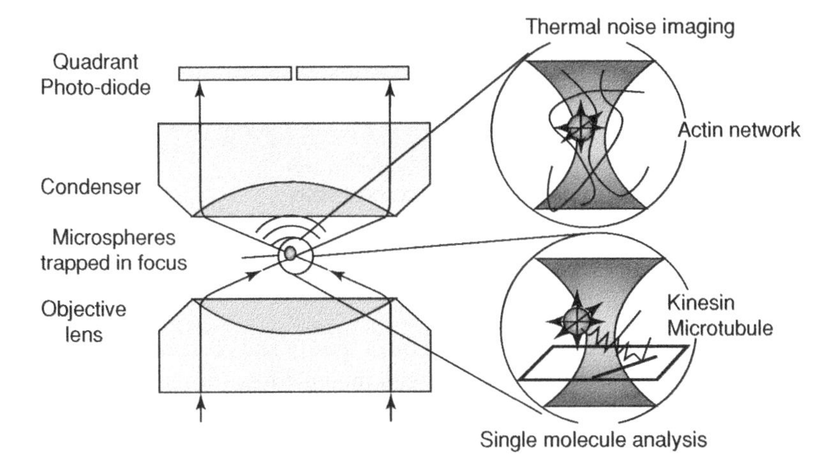

Figure 11.4 *Principle set-up of the phototonic force microscope and enlargement of two applications regarding the mechanical properties of biological molecules.A highly focused beam enables optical trapping of microspheres. The unscattered and the forward scattered lightis collected by a condenser lens. The resulting interference pattern is detected by a quadrant-dipole in the back focal plane of the condenser and provides the 3D-position of the particle. The probe fluctuates within the focal region due to Brownian motion and thus scans its local environment.* Thermal noise imaging*: the particles scan around the object and generate a negative image (of e.g. an actin network).* Single molecule analysis*: a kinesin molecule is attached to a microtubule and a microsphere; the stifness of the spring-like molecule is probed by the fluctuating microsphere.*

This process was applied to latex particles as well to cells and viruses (Green and Morgan, 1999; Hugues and Morgan, 1998).

11.4.2 Biological, Biomedical and Pharmaceutical Applications

This section is devoted to the use of colloids in conjunction with molecules of biological relevance. Therefore, the colloids will be a tool to purify or concentrate biomolecules or organisms, like cells or viruses, to perform a bio-assay or to carry and deliver a bioactive molecule in, respectively, biological, biomedical or pharmaceutical applications. The applications feature a high added value, with usually a low production scale as compared to other industries –paint or paper coating for instance. Moreover, the cost of the carrier is rarely a limiting factor as compared to the price of a recombinant protein, for example, or to the benefit brought to a patient by an improved delivery of a drug or a safer and faster result in diagnostics.

11.4.2.1 Biological Applications

Very often, a biological molecule of interest is contained in a complex mixture, such as a culture medium for recombinant proteins, cell components for the extraction of DNA from cells, whole blood for viruses, or food samples when looking for bacteria in the food

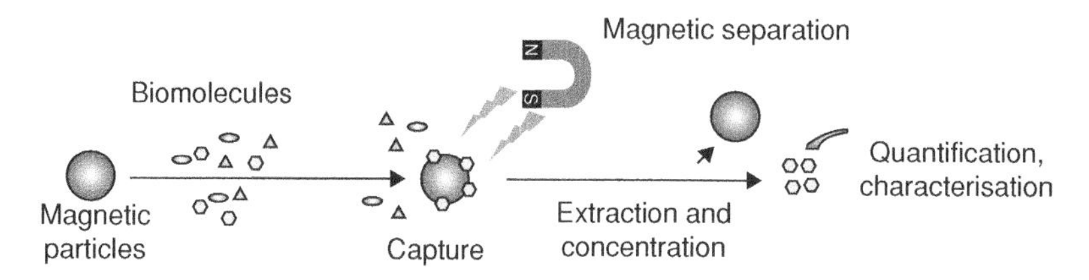

Figure 11.5 Magnetic separation of biomolecules from a sample mixture (schematic picture).

industry. Hence, it is essential to be able to extract, purify and concentrate the biological molecule of interest via a fast, efficient (i.e. with minimal loss (or maximal recovery) of the compound of interest) and a simple to set up process.

For this type of application, magnetic particles are particularly well suited, as a magnet is enough to perform the essential separation steps. This is faster than having to use ultra-centrifugation for instance and, furthermore, particles can almost be tailor-made to a particular application.

The capture of a molecule of interest, schematically represented in Figure 11.5, can either be generic or specific. Generic means that in a sample, one single family of biomolecules will bind onto the particles (e.g. only proteins or only nucleic acids or only viruses). In that case, the physico-chemical properties of the interface should match those of the molecules of interest. For instance, if the goal is to extract DNA from a sample, positively charged particles will be particularly well suited, as electrostatic interactions will allow the binding of the negatively charged DNA (Elaissari *et al.*, 2001). But, for purification or concentration purposes, the interactions between the particles and the molecules of interest should be reversible, to allow the release of the target products, as shown in Figure 11.5. In other words, the end-user should be able "to switch-off" the particle–biomolecule interactions when needed. This is possible when the functional groups at the surface of the particles can be modified on changing the physico-chemical properties of the medium, for instance pH. If the carrier bears amine groups, on increasing the pH, because of the deprotonation of the surface amino moieties, electrostatic interactions with DNA are no longer possible and so the nucleic acids can be released in the medium. The adsorption/desorption of proteins can be controlled using smart polymers whose properties vary under the influence of an external stimulus. poly(*N*-isopropylacrylamide) (PNIPAM) gels loose a part of their hydration water above 32 °C, a critical temperature called the volume phase transition temperature (T_{vpt}). Hence, a particle containing PNIPAM shrinks on heating (see Figure 11.6) inducing a change in the interfacial properties. Other parameters, like pH, salinity, can impact on the properties of the carriers and so be used in the reversible binding of biomolecules (Kawaguchi, Duracher and Elaissari, 2003).

Specific capture means that only one type of biological species should interact with the colloids, excluding any other contaminants. For instance, the DNA of a particular virus of interest, or one single protein, or only tumor cells, or one type of bacteria (and so on) has to be removed from the sample for characterization, purification and whatever other purpose the end-user is interested in. To allow these specific interactions, the particles should bear certain types of molecules or ligands, which specifically recognize the target, at their surface. Antibodies, for instance, have been developed for many different fields

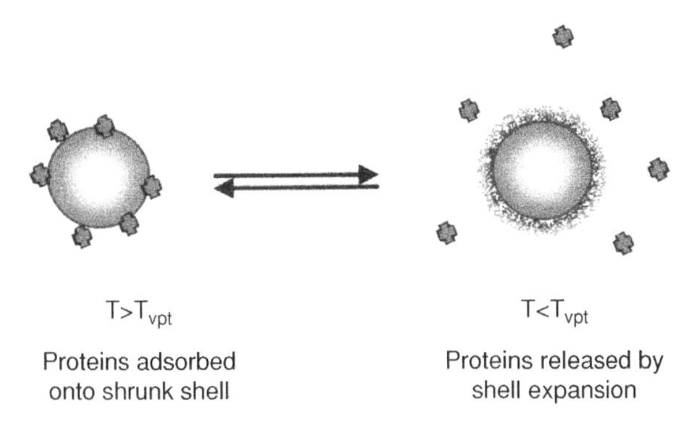

Figure 11.6 *Protein adsorption and release onto smart particles.*

and these proteins can be made to recognize proteins, small molecules and fragments of bacteria, viruses or cells. Hence, binding antibodies onto particles led to new tools to perform immunoseparation, which has become quite popular in biology. A typical example of immunoseparation can be found in the applications for *in vitro* diagnostics described below. Other types of ligands can be used, such as peptides, or short single-stranded DNA fragments (Delair *et al.*, 1999). In this latter case, the well-known base-pairing process brings the specificity. Even drugs were also used as ligands (Shimizu *et al.*, 2000) where a protein, specific to the drug, was detected from the cell extract. In all systems the crucial condition required from the particles is to reject any non-specific adsorption of biomolecules in order to get highly pure target material. Development of particles having non-specific adsorption is still going ongoing (Yuan, Yoshimoto and Nagasaki, 2009).

11.4.2.2 Biomedical Applications

Diagnostics applications of latex particles have long been investigated because latexes have many advantages as an immuno-specific solid phase, for instance, high specific area (allowing an increase in rate). As a consequence, the overall process is faster and often sensitivity is enhanced in comparison with non-dispersed immuno-specific solid phases (like test tubes, for instance). Moreover, they can be in the form of a composite (for instance with iron oxides) and may contain dyes so the particles are coloured or fluorescent.

The aim of *in vitro* diagnostics is to detect and quantify the eventual presence of an antigen in samples from patients, if we consider the case of infectious diseases.

Tests are performed on patients' sera, which correspond to whole blood minus the cellular components. This type of medium contains salts and proteins and the goal is to detect a specific protein (antigen) eventually present in minute amounts in this complex mixture. To reach the required standards of specificity and sensitivity, the tests are often performed in two steps, as shown in Figure 11.7. The first step consists in the capture of the biomolecule to be assayed on the particles. If the antigen is indeed present in the sample, after several washes, it will remain bound onto the particle surface via antibody–antigen interactions. In the second step, an antibody covalently bound to an enzyme (detection conjugate) is incubated and will specifically bind to the captured antigen. After several washes to remove poorly immobilized materials (which could affect the efficacy of the test) the enzyme substrate is added. The enzymatic reaction can lead to the development of a

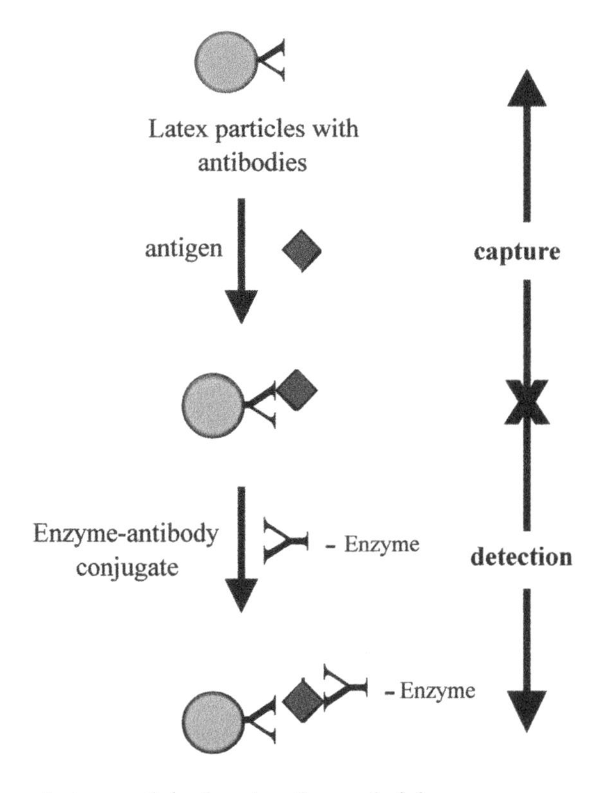

Figure 11.7 *Schematic representation of an immuno-assay according to the enzyme-linked immunosorbent assay (ELISA) method.*

colour or fluorescence, whose intensity is proportional to the amount of enzyme bound on the particles at the end of the detection step, hence proportional to the amount of antigen bound at the end of the capture step. High-throughput, automated ELISA systems have come into wide use at most hospitals.

From this short description, it clearly appears that the efficiency of such an immunoassay will require a close control of the particle characteristics in terms of morphology and surface chemistry, and also the physico-chemical parameters that control the interactions between the colloids and the biomolecules. In particular, the binding of the capture antibody onto the particle surface is critical as this aspect should take place with no, or at least minimized, loss of immunoreactivity. A suitable spacer is often used to secure this point.

In the same field of applications, latexes could be used to visualize and to detect the antigen–antibody interactions. The antigen-induced agglutination of latex particles, leading to the formation of aggregates detectable to the naked eye, was the first generation of rapid tests, cheap to produce and simple to use (Bangs, 1999). The test is rather simple to perform. On a dark plate (to ease the detection of white latexes) a drop of latex suspension is mixed

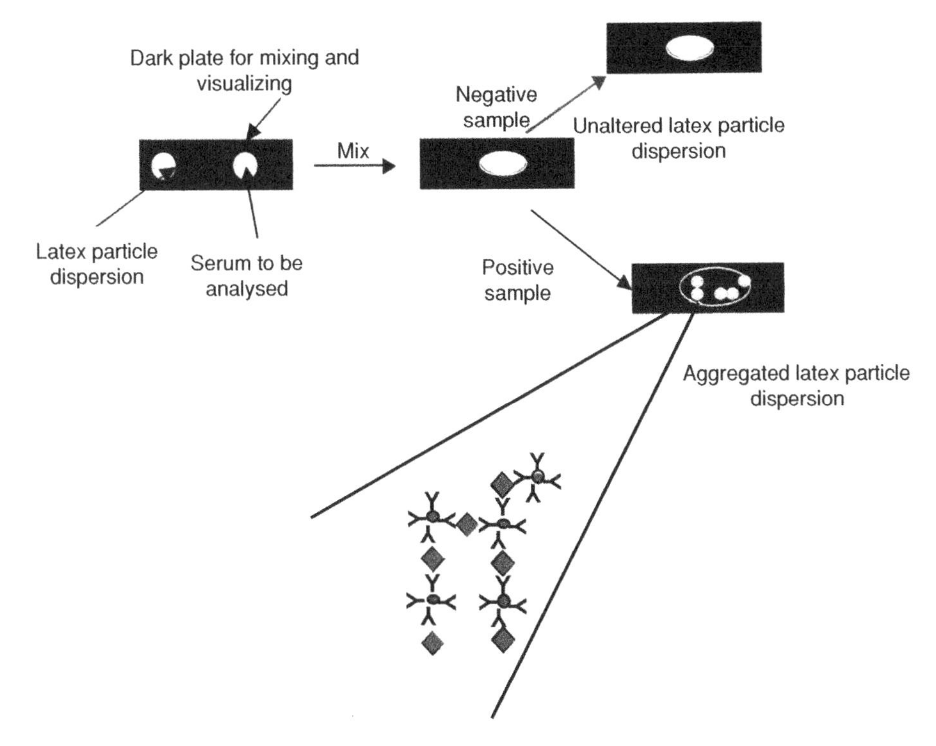

Figure 11.8 *Schematic representation of an agglutination assay.*

with a drop of sample serum to be analyzed. The latex particles bear on their surface antibodies capable of recognition of a specific antigen. If this antigen is present in the sample, it will induce the cross-linking of the particles into aggregates detectable to the naked eye. If the antigen is absent in the sample, the latex will remain in the form of a homogenous dispersion (Figure 11.8). Molecular diagnostics appearing in the middle of the twentieth century have been in great development as they should allow an early detection of a pathogen via its genome, that is, the methodology was evolved by replacing proteins with nucleic acids (Delair *et al.*, 1999).

The above-mentioned method allowed the detection of one single antigen at a time, it is now possible to achieve a multiplexed coding of biomolecules by incorporation of a variety of quantum dots within the latex particles (Han *et al.*, 2001). Hence, latexes prepare the future of modern diagnostics, by allowing the simultaneous detection of several different antigens.

In the last ten years, a strong tendency toward miniaturization of biological analysis has merged with the purpose to analyse smaller and smaller volumes (down to submicrolitres). This generated the development of microsystems consisting of inorganic or organic plain supports on which micrometer size channels are integrated so that all the different steps of a biological sample analysis are carried out fully automated. These systems, the so-called *lab on a chip* systems generate many physico-chemical challenges regarding the handling of fluids in microsized channels. It appeares that latex particles, due to their colloidal properties (control of their size and surface functionalities), should be suitable tools and well adapted for microfluidics systems. They were already employed in various applications such as in packed bed columns for chromatography or as two- or three-dimensional self-assembled systems. One interesting application was regarding the property of magnetic latex particles to self-organize in a network of long chains perpendicular to the microsize channel (Doyle *et al.*, 2002). This network was successfully used for a rapid separation of DNA molecules.

11.4.2.3 Pharmaceutical Applications

In pharmaceutical sciences, colloids are used as carriers of active substances in order to improve the bioavailability of poorly water-soluble drugs (Schmidt and Boodmeier, 1999). Furthermore, they are used to reduce adverse effects of toxic drugs by controlled release, hence the molecule remains in low concentration in the body; to protect fragile substances like peptides, proteins or DNA from enzymatic degradation (Allémann, Leroux and Gurny, 1998; Janes, Calvo and Alonso, 2001); and to improve cellular uptake and target cell compartments (Panyam and Labhasetwar, 2003). Drugs are released from a polymer particle either by leaching-out or by degradation of the particle, depending on the nature and morphology of the particle. Comprehensive studies on the design of particles as drug carriers have been reviewed (Oh, 2008a; Oh *et al.*, 2008b).

Another aspect of the role of particles in pharmaceutical sciences is to investigate them as vaccine adjuvant alternatives to the currently established alum, an aluminium sulfate or phosphate derivative. New generations of vaccines, if they are less toxic than the former ones, are seldom active by themselves and need a helper substance to trigger an appropriate immune response. This is the role of the adjuvant, but though alum is currently used in humans there is a need for a more efficient and biodegradable adjuvant. Kreuter and Speiser originally demonstrated in 1976 that viruses when adsorbed onto poly(methyl methacrylate) particles allowed a better immune response (Kreuter and Speiser, 1976). Using a variety of particles, it was shown that soluble antigens were more efficient when adsorbed or covalently bound onto colloids of various types (polystyrene, iron, silica) (Kavacsovics-Bankowski *et al.*, 1993). Later, it was shown that resorbable lamellar particles of poly(lactic acid) were also efficient adjuvants for the influenza virus (Coombes *et al.*, 1998). Finally, poly(lactic acid-co-glycolic acid) (PLGA) particles were shown to be a potent adjuvant for HIV-1 recombinant protein p55 (Kassaz *et al.*, 2000). The second approach, to date probably the most widely investigated, consists of the encapsulation of the antigen within the polymer matrix via the multiple emulsion method, for instance. This strategy, involving the sustained delivery of the antigen, should allow the reduction of shots (Preis and Langer, 1979). Moreover, it was shown to elicit an immune response using soluble recombinant HIV proteins (Moore *et al.*, 1995).

Both strategies, either based on antigen entrapment (McKeever *et al.*, 2002) or surface adsorption (Singh *et al.*, 2000) were also applied to DNA.

11.5 Conclusions

An increasing number of polymer latex particles with a wide variety of shape, size, morphologies and functionalities can be currently produced by emulsion polymerisation techniques and also by applying physico-chemical phase separation methods on preformed polymers. Due to their outstanding properties, such particles offer a panel of fine applications in many domains where surface and colloidal characteristics play a very important role. They can be used as such in data storage, electronic and optical devices or in catalysis and complexation of metals after adequate surface modification. They can provide nanostructured materials, like colloidal crystals or when they can be self-assembled on flat surfaces to be used in high value applications (e.g., biosensors). However, there is no doubt that colloid polymers have found a large field of applications in all domains dealing with the life sciences where they can serve as suitable tracers, supports or carriers.

The development of novel polymer latex particles, especially in the domain size lower than 100 nm, makes them very attractive, for instance in nanotechnologies and bionanotechnology (which combines biosystems and nanofabrication). Innovating heterogeneous polymerisation techniques, such as those taking place in surfactant assemblies (microemulsions, vesicles, etc.) or using polymerisable amphiphiles should provide polymer colloids with interesting properties. In addition, the possibility to associate organic and inorganic materials at the nanoscale or even at molecular levels should also offer new opportunities. The research is very active and productive toward the synthesis of nanospheres better adapted for diagnostics and in therapy or to conceive systems mimicking biological assemblies. A substantial amount of research is dedicated to the preparation of two or three-dimensional assembled nanostructures with the use of well-characterized polymer colloids which can act as useful templates.

In addition, well-characterized polymer colloids will continue to serve as tools and models in various optical techniques (atomic force microscopy, optical tweezers, photonic force microscopy, etc.) for investigating direct measurements of colloidal forces and studying physical properties of biological membranes and vesicles. All these techniques should be useful in different fields of applications, especially in microfluidics systems.

Finally, it is obvious that advanced research programmes gathering chemists, physicists, biologists and pharmacologists should allow the design of polymer colloids better adapted to the envisioned applications.

References

Abad, C., de la Cal, J.C. and Asua J.M. (1997) The loop process, in *Polymeric Dispersions: Principles and Applications* (ed. J.M. Asua), Kluwer Academic Publishers, Dordrecht, p. 338.

Abad, C., de la Cal, J.C. and Asua, J.M. (1994) *Chem. Eng, Sci.*, **49**, 5025.

ACS (2001) United States Synthetic Rubber Program, 1939-1945, http://portal.acs.org/portal/acs/corg/.

Adrian, J., Esser, E., Hellmann, G. and Pasch, H. (2000) *Polymer*, **41**, 2439.

Aerdts, A.M., Boei, M.M.W.A. and German, A.L. (1993) *Polymer*, **34**, 574.

Aguirre, M., Paulis, M. and Leiza, J.R. (2013) *J. Mat. Chem. A.*, **1**, 3155.

Ahmed, N., Heathly, F. and Lovell, P. (1998) *Macromolecules*, **31**, 2822.

Ahmed, S.M., El-Aasser, M.S., Pauli, G.H. *et al.* (1980) *J. Colloid Interface Sci.*, **73**, 388.

Akhmatskaya, E. and Asua, J.M. (2012) *J. Polym. Sci. Part A. Polym. Chem.*, **50**, 1383.

Akhmatskaya, E. and Asua, J.M. (2013) *Colloid Polym. Sci.*, **291**, 87.

Alam, M.N., Zetterlund, P.B. and Okubo, M. (2008) *Polymer*, **49**, 3428.

Alfrey, T. and Goldfinger, G. (1944) *J. Chem. Phys.*, **12**, 205.

Ali, S.I., Heuts, J.P.A. and van Herk, A.M. (2011) *Soft Matter*, **7**, 5382.

Allémann, E., Leroux, J.C. and Gurny, R. (1998) *Adv. Drug Del. Rev.*, **34**, 171.

Allen, T. (1968) *Particle Size Measurement*, Chapman and Hall Ltd, London.

Alteheld, A., Gourevich, I., Field, L. *et al.* (2005) *Macromolecules*, **38**, 3301.

Anderson, C.D. and Daniels, E.S. (2003) Emulsion Polymerization and Latex Applications. Rapra Review Reports, Shroshire.

Ando, I., Kobayashi, M., Kanekiyo, M. *et al.* (2000) In *Experimental Methods in Polymer Science*, Academic Press, San Diego, p. 261.

Anton, N., Benoit, J.P. and Saulnier, P. (2008) *J. Controlled Release*, **128**, 185.

Antonietti, M. and Landfester, K. (2002) *Prog. Polym. Sci.*, **27**, 689.

Armes, S.P. (1998) Colloidal dispersions of conducting polymer colloids, in *Handbook of Conducting Polymers*, 2nd edn, (eds T.A. Skotheim, R.L. Enselbaumer and J.R. Reynolds), Marcel Dekker, New York, p. 423.

Arora, A., Daniels, E.S., El-Asser, M.S. *et al.* (1995) *J. Appl. Polym. Sci.*, **58**, 313.

Arshady, R. (1993) *Biomaterials*, **14**, 5.

Arshady, R. (1999) In *Microspheres, Microcapsules and Liposomes Vol. 1 Preparation and Chemical Applications* (ed. R. Arshady), Citus Book.

Arzamendi, G. and Asua, J.M. (1989) *J. Appl. Polym. Sci.*, **38**, 2019.

Chemistry and Technology of Emulsion Polymerisation, Second Edition. Edited by A.M. van Herk.

Arzamendi, G. and Asua, J.M. (1990) *Makromol. Chem., Macromol. Symp.*, **35/36**, 249.
Arzamendi, G. and Asua, J.M. (1991) *Ind. Eng. Chem. Res.*, **30**, 1342.
Arzamendi, G., Leiza, J.R. and Asua, J.M. (1991) *J. Polym. Sci., Part A: Polym. Chem.*, **29**, 1549.
Ashkin, A. (1997) *Proc. Natl. Acad. Sci. USA*, **94**, 4853.
Asnaghi, D., Carineti, M., Giglio, M. and Sozzi, M. (1992) *Phys. Rev. Lett.*, **45**, 1108.
Asua, J.M., Sudol, E.D. and El-Aasser, M.S. (1989) *J. Polym. Sci.*, **27**, 3903.
Asua, J.M. (1997) *Polymeric Dispersions: Principles and Applications*, Kluwer Academic publishers, Dordrecht.
Asua, J.M. (2002) *Prog. Polym. Sci.*, **27**, 1283.
Asua, J.M., Beuermann, S., Buback, M. *et al.* (2004) *Macromol. Chem. Phys.*, **205**, 2151.
AzoM.com (2002) Emulsion Styrene Butadiene Rubber (E-SBR) – History, Development and Applications of Emulsion Styrene Butadiene Rubber, http://www.azom.com/details.asp?ArticleID=1848.
Azukizawa, M., Yamada, B., Hill, D.J.T. and Pomery, P.J. (2000) *Macromol. Chem. Phys.*, **201**, 774.
Balandina, V., Berezan, K., Dobromyslowa, A. *et al.* (1936a) *Bull. Acad. Sci. U.S.S.R.* Ser., **7**, 423.
Balandina, V., Berezan, K., Dobromyslowa, A., Dogadkin, B. and Lapuk, M. (1936b) *Bull. Acad. Sci. U.S.S.R.* Ser., **7**, 397.
Ballauff, M. (2001) *Curr. Opin. Colloid Interface Sci.*, **6**, 132.
Bandermann, F., Tausenfreund, I., Sasic, S. *et al.* (2001) *Macromol. Rapid Commun.*, **22**, 690.
Bangs, L.R. (1999) In *Microspheres Microcapsules & Liposomes*, vol. **2** Medical and Biotechnology applications. (ed. R. Arshady), Citus Books, London, p. 43.
Bar, G., Thomann, Y., Brandsch, R. *et al.* (1997) *Langmuir*, **13**, 3807.
Barner-Kowollik, C., Quinn, J.F., Morsley, D.R. and Davis, T.P. (2001) *J. Polym. Sci., Part A: Polym. Chem.*, **39**, 1353.
Barner-Kowollik, C., Buback, M., Charleux, B. *et al.* (2006) *J. Polym. Sci., Part A: Polym. Chem.*, **44**, 5809.
Barner-Kowollik, C. (ed.) (2008) *Handbook of RAFT Polymerisation*, Wiley-VCH, Weinheim.
Barouch, E., Matijevic, E. and Wright, T.H. (1985) *J. Chem. Soc., Faraday Trans. I*, **81**, 1819.
Barrett, K.E.J. (1975) *Dispersion Polymerisation in Organic Media*, Wiley, New York.
Basinska, T. and Slomkowski, S. (1991) *J. Biomed. Sci.-Polym. Ed.*, **3**, 115.
Basinska, T., Slomkowski, S. and Delamar, M, (1993) *J. Bioact. Compatible Polym.*, **8**, 205.
Bassett, D.R. and Hoy, K.L. (1980) In *Polymer Colloids II* (ed. R.M. Fitch), Plenum Press, New York, p. 1.
Bassett, D.R., Derderian, E.J., Johnston, J.E. and MacRury, T.B. (1981) *ACS Symp. Ser.*, **165**, 263.
Bassett, D.R. and Hoy, K.L. (1981) *ACS Symp. Series*, **165**, 371.
Bassett, D.R. (1983) In *Science and Technology of Polymer Colloids*, vol. **I** (eds G.W. Poehlein, R.H. Ottewil and J.W. Goodwin), Martinus Nijhoff Publishers, The Hague, p. 220.

Baumstark, R., Costa, C. and Schwartz, M. (1999) *Eur. Coating. J.*, **5**, 44.
Beach, E., Keefe, M., Heeschen, W. and Rothe, D. (2005) *Polymer*, **46**, 11195.
Bell, C.A., Smith, S.V., Whittaker, M.R. *et al.* (2006) *Adv. Mater.*, **18**, 582.
Benson, S.W. and North, A.M. (1962) *J. Am. Chem. Soc.*, **84**, 935.
Berek, D., Janco, M., Kitayama, T. and Hatada, K. (1994) *Polym.Bull.*, **32**, 629.
Berezan, K., Dobromyslowa, A. and Dogadkin, B. (1936) *Bull. Acad. Sci. U.S.S.R.* Ser., **7**, 409.
Berg, J., Sundberg, D.C. and Kronberg, B. (1986) *Polym. Mat. Sci. Eng.*, **54**, 367.
Bertin, D., Destarac, M. and Boutevin, B. (1998) *Polym. Surf.*, 47.
Biela, T., Duda, A., Penczek, S. *et al.* (2002) *J. Polym. Sci. Part A: Polym. Chem.*, **40**, 2884.
Bielemann, J. (2000) *Additives for Coatings*, Wiley-VCH, Weinheim.
Bierwagen, G. and Rich, D. (1983) *Prog. Org. Coat.*, **11**, 339.
Binks, B.P. (2002) *Curr. Opin. Colloid Interface Sci.*, **7**, 21.
Blackley, D.C. (1983) In *Science and Technology of Polymer Colloids*, NATO ASI Series E68, vol. **I**, (ed. G.W. Poehlein, R.H. Ottewill and J.W. Goodwin), Martinhus Nijhoff Publishers, The Hague, p. 220.
Blackley, D. C. (1975) *Emulsion Polymerisation*, Applied Science Publishers Ltd, London.
Blackley, D.C. (1997) *Polymer Latices, Science and Technology*, Chapman & Hall, London.
Blanpain, P., (2002) Applications in the carpet industry, in *Polymer Dispersions and Their Industrial Applications* (eds Urban & Takamura), Wiley-VCH, Weinheim.
Boborodea, A.G., Daoust, D., Jonas, A.M. and Bailly, C. (2004) *LC-GC North America*, **22**, 52.
Bogner, A., Guimarães, A., Guimarães, R. *et al.* (2008) *Colloid Polym. Sci.*, **286**, 1049.
Bogner, A., Jouneau, P.H., Thollet, G. *et al.* (2007) *Micron*, **38**, 390.
Bogner, A., Thollet, G., Bassett, D. *et al.* (2005) *Ultramicroscopy*, **104**, 290.
Boissé, S., Rieger, J., Belal, K. *et al.* (2010) *Chem. Commun.*, **46**, 1950.
Boissé, S., Rieger, J., Pembouong, G. *et al.* (2011) *J. Polym. Sci., Part A: Polym. Chem.*, **49**, 3346.
Bolze, J., Ballauff, M., Kijlstra, J. and Rudhardt, D. (2003) *Macromol. Mater. Eng.*, **288**, 495.
Bon, S.A.F., Bosveld, M., Klumperman, B. and German, A.L. (1997) *Macromolecules*, **30**, 324.
Bon, S.A.F. and Colver, P.J.C. (2007) *Langmuir*, **23**, 8316.
Bonnefond, A., Micusik, M., Paulis, M. *et al.* (2012) *Colloid. Polym. Sci.* doi: 10.1007/s00396-012-2649-3
Bonvin, D., Valliere, P. and Rippin, W.T. (1989) *Comput. Chem. Eng.*, **13**(1/2), 1.
Booth, G.L. (1970) *Coating Equipment and Processes*, Lockwood, New York.
Bottle, G.A., Lye, J.E. and Ottewill, R.H. (1990) *Makromol. Chem., Macromol. Symp.*, **35/36**, 291.
Bourgeat-Lami, E, (2002) *J. Nanosci. Nanotechnol.*, **2**, 1.
Bourgeat-Lami, E. and Lansalot, M. (2010) *Adv. Polym. Sci.*, **233**, 53.
Bovey, F.A., Kolthoff, I.M., Medalia, A.I. and Meehan, E.L. (1955) *Emulsion Polymerization*, Intersciences, New York.
Braunecker, W.A. and Matyjaszewski, K. (2007) *Prog. Polym. Sci.*, **32**, 93.
Brindley, A., Davies, M.C., Lynn, R.A.P. *et al.* (1992) *Polymer*, **33**, 1112.
Britton, D., Heatley, F. and Lovell, P.A. (2001) *Macromolecules*, **34**, 817.

Brown, W.E. (1947) *J. Appl. Phys.*, **18**, 273.
Bruchez, M., Moronne, M., Gin, P. *et al.* (1998) *Science*, **281**, 2013.
Bruheim, T., Molander, P., Theodorsen, M. *et al.* (2001) *Chromatographia*, **53**, S266.
Brusseau, S., Belleney, J., Magnet, S. *et al.* (2010) *Polym. Chem.*, **1**, 720.
Brusseau, S., D'Agosto, F., Magnet, S. *et al.* (2011) *Macromolecules*, **44**, 5590.
Buback, M., Gilbert, R.G., Hutchinson, R.A. *et al.* (1995) *Macromol. Chem. Phys.*, **196**, 3267.
Buback, M., Gilbert, R.G., Russell, G.T. *et al.* (1992) *J. Polym. Sci., Polym. Chem. Edn.*, **30**, 851.
Buchhammer, H.M., Mende, M. and Oelmann, M. (2003) *Colloids Surf. A*, **218**, 151.
Burguière, C., Chassenieux, C. and Charleux, B. (2003) *Polymer*, **44**, 509.
Burguière, C., Dourges, M.A., Charleux, B. and Vairon, J.P. (1999) *Macromolecules*, **32**, 3883.
Butt, H.J. and Gerharz, B. (1995) *Langmuir*, **11**, 4735.
Butt, H.J. and Kuropka, R. (1995) *J. Coat. Technol.*, **67**, 101.
Butt, H.J., Kuropka, R. and Christensen, B. (1994) *Colloid Polym. Sci.*, **272**, 1218.
Butté, A., Storti, G. and Morbidelli, M. (2000) *Macromolecules*, **33**, 3485.
Cairns, R., Ottewill, R.H., Osmond, D.W.J. and Wagstaff, I. (1976) *J. Colloid Interface Sci.*, **79** 511.
Canegallo, S., Canu, P., Morbidelli, M. and Storti, G. (1994) *J. Appl. Polym. Sci.*, **54**, 1919.
Canetta, E., Marchal, J., Lei, C.-H. *et al.* (2009) *Langmuir*, **25**, 11021.
Canu, P., Canegallo, S., Morbidelli, M. and Storti, G. (1994) *J. Appl. Polym. Sci.*, **54**, 1889.
Cao, J., He, J., Li, C. and Yang, Y. (2001) *Polym. J.*, **33**, 75.
Caris, C.H.M., Kuijpers, R.P.M., van Herk, A.M. (1990) *Macromol. Symp.*, **35–36**, 535.
Caruso, F., Caruso, R.A. and Möhwald, H. (1999a) *Chem. Mater.*, **11**, 3309.
Caruso, F., Susha, A.S., Giersig, M. and Möhwald. (1999b) *Adv. Mater.*, **11**, 950.
Casey, B.S., Morrison, B.R. and Gilbert, R.G. (1993) *Prog. Polym. Sci.*, **18**, 1041.
Cauvin, S., Sadoun, A., Dos Santos, R. *et al.* (2003) *Macromolecules*, **35**, 7919.
Cayre, O.J., Chagneux, N. and Biggs, S. (2011) *Soft Matter*, **7**, 2211.
Chaduc, I., D'Agosto, F., Lansalot, M. and Charleux, C. (2012) *Macromolecules*, **45**, 1241.
Chambard, G., De Man, P. and Klumperman, B. (2000) *Macromol. Symp.*, **150**, 45.
Chang, T. (2003) *Adv.Polym.Sci.*, **163**, 1.
Chang, T., Lee, H.C., Lee, W. *et al.* (1999) *Macromol. Chem. Phys.*, **200**, 2188.
Chan-Seng, D. and Georges, M.K. (2006) *J. Polym. Sci., Polym. Chem.*, **44**, 4027.
Chan-Seng, D., Rider, D.A., Guérin, G. and Georges, M.K. (2008) *J. Polym. Sci., Polym. Chem.*, **46**, 625.
Charleux B., Nicolas J. and Guerret O. (2005) *Macromolecules*, **38**, 5485.
Charleux, B. (2000) *Macromolecules*, **33**, 5358.
Charleux, B., D'Agosto, F. and Delaittre G. (2010) *Adv. Polym. Sci.*, **233**, 125.
Charleux, B. and Nicolas, J. (2007) *Polymer*, **48**, 5813.
Charleux, B., Delaittre, G., Rieger, J. and D'Agosto, F. (2012) *Macromolecules*. doi: 10.1021/ma300713f
Charleux, B., Fanget, P. and Pichot, C. (1992) *Die Makromol. Chem.*, **193**, 205.
Charmot, D., Corpart, P., Adam, H. *et al.* (2000) *Macromol. Symp.*, **150**, 23.
Chen, X.Y., Armes, S.P., Greaves, S.J. and Watts, J.F. (2004) *Langmuir*, **20**, 587.

Chen, Y.C., Dimonie, V. and El-Aasser, M.S. (1991a) *Macromolecules*, **24**, 3779.
Chen, Y.C., Dimonie, V.L. and El-Aasser, M.S. (1991b) *J. Appl. Polym. Sci.*, **42**, 1049.
Chen, Y.C., Dimonie, V.L. and El-Aasser, M.S. (1992a) *J. Appl. Polym. Sci.*, **45**, 487.
Chen, Y.C., Dimonie, V.L. and El-Aasser, M.S. (1992b) *Pure Appl. Chem.*, **64**, 1691.
Chen, Y.C., Dimonie, V.L., Shaffer, O.L. and El-Aasser, M.S. (1993) *Polym. Int.*, **30**, 185.
Chen, M., Wu, L., Zhou, S. and You, B. (2004) *Macromolecules*, **37**, 9613.
Chen J., Chen, M., Zhou, S. and Wu, L. (2006) *J. Polym. Sci. Part A: Polym. Sci.*, **44**, 3807.
Chen, Y. and Sajjadi, S. (2009) *Polymer*, **50**, 357.
Cheng, C.J., Gong, S.S., Fu, Q.L. *et al.* (2010) *New J. Chem.*, **34**, 63.
Chern, C.-S. and Poehlein, G.W. (1987) *J. Polym. Sci., Polym Chem Ed.*, **25**, 617.
Chern, C.-S. and Poehlein, G.W. (1990a) *J. Polym. Sci., Polym Chem Ed.*, **28**, 3055.
Chern, C.-S. and Poehlein, G.W. (1990b) *J. Polym. Sci., Polym Chem Ed.*, **28**, 3073.
Chiefari, J., Jeffery, J., Mayadunne, R.T.A. *et al.* (1999) *Macromolecules*, **32**, 7700.
Cho, I. and Lee, K. (1985) *Polymer (Korea)*, **9**, 110.
Christie, D.I., Gilbert, R.G., Congalidis, J.P. *et al.* (2001) *Macromolecules*, **34**, 5158.
Clay, P.A. and Gilbert, R.G. (1995) *Macromolecules*, **28**, 552.
Clay, P.A., Christie, D.I. and Gilbert, R.G. (1998) In *Advances in Free-Radical Polymerisation*, vol. **685** (ed. K. Matyjaszewski) A.C.S., Washington D.C., p. 104.
Cobbold, A.J. and Mendelson, A.E. (1971) *Science Tools*, **18**, 10.
Coen, E.M. and Gilbert, R.G. (1997) In *Polymeric Dispersions. Principles and Applications*, vol. NATO Advanced Studies Institute (ed. J.M. Asua), Kluwer Academic Publishers, Dordrecht, pp. 67–78.
Coen, E.M., Gilbert, R.G., Morrison, B.R. *et al.* (1998) *Polymer*, **39**, 7099.
Coen, E.M., Morrison, B.R., Peach, S. and Gilbert, R.G. (2004) *Polymer*, **45**, 3595.
Coen, E., Lyons, R.A. and Gilbert, R.G. (1996) *Macromolecules*, **29**, 5128.
Colard, C.A.L., Teixera, R.F.A. and Bon, S.A.F. (2010) *Langmuir*, **26**, 7915.
Collins, E.A. (1991) Experimental measurement of particle size and particle size distribution. Advances in emulsion polymerization and latex technology, 2nd annual short course, Bethlehem, USA.
Colombini, D., Ljungberg, N., Hassander, H. and Karlsson, O.J. (2005) *Polymer*, **46**, 1295.
Colver, P.J., Colard, C.A.L. and Bon, S.F.A. (2008) *J. Am. Chem. Soc.*, **130**, 16850.
Coombes, A.G.A., Major, D., Wood, J.M. *et al.* (1998) *Biomaterials*, **19**, 1073.
Coote, M.L., Zammit, M.D. and Davis, P.D. (1996) *Trends Polym. Sci.*, **4**, 189.
Cowell, C. and Vincent, B. (1982) *J. Colloid Interface Sci*, **87**, 518.
Crassous, J.J., Ballauff, M., Drechsler, M. *et al.* (2006) *Langmuir*, **22**, 2403.
Cui, Z. and Mumper, R.J. (2001) *J. Control Rel.*, **75**, 409.
Cunningham, M.F., Tortosa, K., Lin, M. *et al.* (2002a) *J. Polym. Sci. Part A: Polym. Chem.*, **40**, 2828.
Cunningham, M.F., Tortosa, K., Ma, J.W. *et al.* (2002b) *Macromol. Symp.*, **182**, 273.
Cunningham, M.F., Xie, M., McAuley, K.B. *et al.* (2002c) *Macromolecules*, **35**, 59.
Cunningham, M.F. (2002d) *Prog. Polym. Sci.*, **27**, 1039.
Cunningham, M.F., Ng, D.C.T., Milton, S.G. and Keoshkerian, B. (2006) *J. Polym. Sci. Part A: Polym. Chem.*, **44**, 232.
Cunningham, M.F. (2008) *Prog. Polym. Sci.*, **33**, 365.

D'Agosto, F., Charreyre, M.T., Pichot, C. and Gilbert, R.G. (2003) *J. Polym. Sci., Polym. Chem.*, **41**, 1188.
Daniel, J-C., Schuppiser, J.L., and Tricot, M. (1982) U.S. Patent 4 358 388.
De Brouwer, H., Monteiro, M.J., Tsavalas, J.G. and Schork, F.J. (2000) *Macromolecules*, **33**, 9239.
De Bruyn, H., Gilbert, R.G. and Ballard, M.J. (1996) *Macromolecules*, **29**, 8666.
De Jaeger, N.C., Trappers, J.L. and Lardon, P. (1986) *Particle Characterization*, **3**, 187.
de la Cal, J.C., Adams, M.E. and Asua, J. (1990a) *Makromol. Chem., Macromol. Symp.*, **35–36**, 23.
de la Cal, J.C., Urzay, R., Zamora, A. *et al.* (1990b) *J. Polym. Sci. Part A Polym. Chem.*, **28**, 1011.
de la Cal, J.C., Echeverria, A., Meira, G.R. and Asua, J.M. (1995) *J. Appl. Polym. Sci.*, **57**, 1063.
Dean, T.W.R. (1997) *The Essential Guide to Aqueous Coating of Paper and Board*, Pita, Lancashire, UK.
Delair, T, Pichot, C. and Mandrand, B. (1993) *Colloïd Polym. Sci*, **272**, 72.
Delair, T. (2003) In *Colloidal Biomolecules, Biomaterials, and Biomedical Applications*, vol. **116** (ed. A. Elaissari), Surfactant Science Series, Marcel Dekker, p. 329.
Delair, T., Marguet, V., Pichot, C. and Mandrand, B. (1994) *Colloid Polym. Sci.*, **272**, 962.
Delair, T., Meunier, F., Elaïssari, A. *et al.* (1999) *Colloid Surf. A*, **153**, 341.
Delaittre, G., Nicolas, J., Lefay, C. *et al.* (2005) *Chem. Commun.* 615.
Delaittre, G., Nicolas, J., Lefay, C. *et al.* (2006) *Soft Matter*, **2**, 223.
Delaittre, G., Dire, C., Rieger, J. *et al.* (2009) *Chem. Commun.* 2887.
Derjaguin, B.V. and Landau, L.D. (1941) *Acta Phys. Chim. URSS*, **14**, 633.
Diaconu, G., Asua, J.M, Paulis, M. and Leiza, J.R. (2007) *Macromol. Symp.*, **259**, 305.
Diaconu, G., Paulis, M. and Leiza, J.R. (2008a) *Macromol. React. Eng.*, **2**, 80.
Diaconu, G., Paulis, M. and Leiza, J.R. (2008b) *Polymer*, **49**, 2444.
Diaconu, G., Micusik, M., Bonnefond, A. *et al.* (2009) *Macromolecules*, **42**, 3316.
Dickinson, E. (1983) In *Specialist Periodical Reports*, vol. **4** (ed. D.H. Everett), Royal Society of Chemistry, London, p. 150.
Dimitratos, J., Georgakis, C., El-Aasser, M.S., and Klein, A. (1989) Control of product composition in emulsion copolymerization, in *Polymer Reaction Engineering* (eds K.H. Reichert and W. Geiseler), VCH, New York, p. 33.
Dimonie, V.L., Daniels, E.S., Shaffer, O.L. and El-aasser, M.S. (1997) Control of particle morphology, in *Emulsion Polymerization and Emulsion Polymers* (eds P.A. Lovell, and M.S. El-Aasser) Wiley & Sons Inc., New York, 293.
Dimonie, V.L., El-Aasser, M.S. and Vanderhoff, J.W. (1988) *Polym. Mater. Sci. Eng.*, **58**, 821.
Dinsmore, R.P. (1927) Synthetic rubber. Patent GB297050.
Dire, C., Magnet, S., Couvreur, L. and Charleux, B. (2009) *Macromolecules*, **42**, 95.
Dire, C., Belleney, J., Nicolas, J. *et al.* (2008) *J. Polym. Sci., Part A: Polym. Chem.*, **46**, 6333.
Dire, C., Charleux, B., Magnet, S. and Couvreur, L. (2007) *Macromolecules*, **40**, 1897.
Discher, D.E., Ortiz, V., Srinivas, G. *et al.* (2007) *Prog. Polym. Sci.*, **32**, 838.
Dobler, F., Pith, T., Holl, Y. and Lambla, M. (1992) *J. Appl. Polym. Sci.*, **44**, 1075.
Dos Ramos, J.G. and Silebi, C.A. (1989) *J. Colloid Interface Sci.*, **130**, 14.

Dos Santos, F.D., Fabre, P., Drujon, X. *et al.* (2000) *J. Polym. Sci., Part B: Polym. Phys.*, **38**, 2989.
Dos Santos, A.M., Le Bris, T. and Graillat, C. (2009) *Macromolecules*, **42**, 946.
Dougherty, E.P. (1986a) *J. Appl. Polym. Sci.*, **32**, 3051.
Dougherty, E.P. (1986b) *J. Appl. Polym. Sci.*, **32**, 3079.
Doyle, P.S., Bibette, J., Bancaud, A. and Viovy, J.-L., (2002) *Science*, **295**, 2237.
Dreyer, J.K., Nylander, T., Karlsson, O.J. and Piculell, L. (2011) *ACS Appl. Mater. Interfaces*, **3**, 167.
Druschke, W. (1987) *Adhäsion*, **5**, 29 and, **6**, 26.
Du, L., Kelly, J.Y., Roberts, G.W. and Desimone, J.M. (2008) *J. Supercritical Fluids*, **47**, 447.
Dukhin, A.S. and Goetz, P.J. (2001) *Colloids Surf., A.*, **192**, 267.
Dukhin, A.S., Goetz, P.J., Wines, T.H. and Somasundaran, P. (2000) *Colloids Surf., A.*, **173**, 127.
Dunn, S.E., Brindley, A., Davis, S.S. *et al.* (1994) *Pharm. Res.*, **11**, 1016.
Durant, Y.G., Carrier, R.H. and Sundberg, D.C. (2003) *Polym. React. Eng.*, **11**, 433.
Echeverria, A., de la Cal, J.C. and Asua, J.M. (1995) *J. Appl. Polym. Sci.*, **57**, 1217.
El-Aasser, M.S., Makgawinata, T., Vanderhoff, J.W. and Pichot, C. (1983a) *J. Polym. Sci., Polym. Chem. Ed.*, **21**, 2363.
El-Aasser, M.S. (1983b) Methods of latex cleaning, in *Science & Technology of Polymer Colloids. Surface Characterization of Latexes. Characterization, Stabilization and Application Properties.* Volume II, NATO ASI Series, vol. **68** (eds G.W. Poehlein, R.H. Ottewill and J.W. Goodwin), Martinus Nijhoff Publishers, The Hague, p. 422.
El-Aasser, M.S. and Fitch, R.M. (eds) (1987) *Future Direction in Polymer Colloids*, NATO ASI Series, Series E, Applied Sciences, Martinus Nijhoff Publishers.
El-Aasser, M.S., Hu, R., Dimonie, V.L. and Sperling, L.H. (1999) *Colloid Surfaces*, **156**, 241.
Elaissari, A., Rodrigue, M., Meunier, F. and Herve, C. (2001) *J. Magn. Magn. Mater.*, **225**, 127.
Elaissari, A., Veyret, R., Mandrand, B. and Chatterjee, J. (2003) In *Colloidal Biomolecules, Biomaterials, and Biomedical Applications*, vol **116** (ed. A. Elaissari), Marcel Dekker Edition Surfactant Science Series, Marcel Dekker, p. 1.
Engqvist, C., Forsberg, S., Norgren, M. *et al.* (2007) *Colloids Surf., A.*, **302**, 197.
Erdem, B., Sudol, E.D., Dimonie, V.L. and El-Aasser, M.S. (2000a) *J. Polym. Sci. Part A*, **38**, 4419.
Erdem, B., Sudol, E.D., Dimonie, V.L. and El-Aasser, M.S. (2000b) *J. Polym. Sci. Part A*, **38**, 4431.
Erdem, B., Sudol, E.D., Dimonie, V.L. and El-Aasser, M.S. (2000c) *J. Polym. Sci. Part A*, **38**, 4441.
Eslami, H. and Zhu, S. (2005) *Polymer*, **46**, 5484.
Eslami, H. and Zhu, S. (2006) *J. Polym. Sci., Polym. Chem.*, **44**, 1914.
Everett, D.H., Gülteppe, M.E. and Wilkinson, M.C. (1979) *J. Colloid Interface Sci.*, **71**, 336.
Falkenhagen, J., Much, H., Stauf, W. and Muller, A.H.E. (2000) *Macromolecules*, **33**, 3687.
Farcet, C., Lansalot, M., Charleux, B. *et al.* (2000) *Macromolecules*, **33**, 8559.
Farcet, C., Charleux, B. and Pirri, R. (2001a) *Macromol. Symp.*, **182**, 249.

Farcet, C., Charleux, B. and Pirri, R. (2001b) *Macromolecules*, **34**, 3823.

Farcet, C., Nicolas, J. and Charleux, B. (2002) *J. Polym. Sci. Part A: Polym. Chem.*, **40**, 4410.

Farcet, C., Burguière, C. and Charleux, B. (2003) In *Colloidal Polymers: Synthesis and Characterization*, (ed. A. Elaissari), Surfactant Science Series, vol. **115**, Marcel Dekker Inc., p. 23.

Farinha, J.P.S., Wu, J., Winnik, M.A. *et al.* (2005) *Macromolecules*, **28**, 4393.

Faucheu, J., Chazeau, L., Gauthier, C. *et al.* (2009) *Langmuir*, **25**, 10251.

Fed Reg (1990) 55 Federal Register 17404, April 24.

Feeney, P.J., Napper, D.H. and Gilbert, R. (1984) *Macromolecules*, **17**, 2520.

Feeney, P.J., Napper, D.H. and Gilbert, R.G. (1987) *Macromolecules*, **20**, 2922.

Ferguson, C.J., Hughes, R.J., Pham, B.T.T. *et al.* (2002a) *Macromolecules*, **35**, 9243.

Ferguson, C.J., Russell, G.T. and Gilbert, R.G. (2002b) *Polymer*, **43**, 6371.

Ferguson, C.J., Hughes, R.J., Nguyen, D. *et al.* (2005) *Macromolecules*, **38**, 2191.

Fernandez-Barbero, A. and Vincent, B. (2000) *Phys. Rev. E*, **63**, 1509.

Fessi, H., Puisieux, F. and Devisaguet, J.P. (1987) Eur. Patent 274 961.

Fikentscher, H. (1931) I.G. Farbenindustrie, U.S.

Fikentscher, H. (1934) *Angew. Chem.*, **51**, 433.

Fischer, H. (2001) *Chem. Rev.*, **101**, 3581.

Fitch, R.M., Prenosil, M.B. and Sprick, K.J. (1969) *J. Polymer Sci., Part C*, **27**, 95.

Fitch, R.M. and Tsai, C.H. (1971) In *Polymer Colloids* (ed. R.M. Fitch), Plenum Press, New York, p. 73.

Fitch, R.M. and Shih, L.B. (1975) *Progr. Colloid Polym. Sci.*, **56**, 1.

Fitch, R.M. and McCarvill, W.T. (1979) *J. Colloid Sci.*, **66**, 20.

Fitch, R.M. and Watson, R.C. (1979) *J. Colloid Interface Sci.*, **68**, 14.

Fitch, R.M., Gajria, C. and Tarcha, P.J. (1979) *J. Colloid Sci.*, **71**, 107.

Fitch, R.M., Mallya, P.K., McCarvill, W.T. and Miller, R.S. (1981) Preprints, Div Colloid Surf. Chem. ACS, 182nd Meeting, 23–28 August, Paper N°64.

Fitch, R.M., Palmgren, T.H., Aoyagi, T. and Zuikov, A. (1984) *Angew. Makromol. Chem.*, **123/124**, 261.

Fitzpatrick, F.P., Ramaker, H.-J., Schoenmakers, P.J. *et al.* (2004) *J. Chromatogr. A*, **1043**, 239.

Fitzpatrick, F.P., Staal, B.B.P. and Schoenmakers, P.J. (2005) *J. Chromatogr. A*, **1065**, 219.

Fleer, G.J., Cohen Stuart, M.A., Scheutjens, J.M.H.M. *et al.* (1993) *Polymers at Interfaces*, Chapman & Hall, London.

Flory, P.J. (1953) In *Principles of Polymer Science*, Cornell University Press, Ithaca, New York.

Fontenot, K. and Schork, F.J. (1992) *Polym. React. Eng.*, **1**, 75.

Forcada, J. and Asua, J.M. (1985) *J. Polym. Sci., Part. A: Polym. Chem.*, **23**, 1955.

Ford, W.T., Badley, R.D., Chandran, R.S. *et al.* (1992) In *Polymer Latexes, Preparation, Characterization and Applications*, (eds E.S. Daniels, E.D. Sudol, M.S. El-Aasser), ACS Symp. Series, vol. **492** Chapter 26, p. 422.

www.freedoniagroup.com (2011).

Fricke, H.J. and Maempel, L. (1990) *Adhäsion*, **7/8**, 13.

Fricke, H.J. and Maempel, L. (1994) *Kleben Dichten, Adhäsion*, **11**, 14.

Friddle, R.W., Lemieux, M.C., Cicero, G. *et al.* (2007) *Nat. Nanotechnol.*, **2**, 692.
Friis, N. and Hamielec, A. (1973a) *J. Polym. Sci., Polym. Chem. Ed.*, **11**, 3321.
Friis, N. and Nyhagen, L. (1973b) *J. Appl. Polym. Sci.*, **17**, 2311.
Friis, N. and Hamielec, A. (1974a) *J. Polym. Sci., Polym. Chem. Ed.*, **12**, 351.
Friis, N., Goosney, D., Wright, J.D. and Hamielec, A. (1974b) *J. Appl. Polym. Sci.*, **18**, 2247.
Frilette, V.L. (1944) 108th ACS Meeting, New York.
Fryling, C.F. (1944) *Ind. Eng. Chem., Anal. Ed.*, **16**, 1.
Fryling, C.F. and Harrington, E.W. (1944) *Ind. Eng. Chem.*, **36**, 114.
Fukuda, T., Goto, A. and Ohno, K. (2000) *Macromol. Rapid Commun.*, **21**, 151.
Fukuda, T., Terauchi, T., Goto, A., Ohno, K., Tsujii, Y., Miyamoto, T., Kobatake, S. and Yamada, B. (1996) *Macromolecules*, **29**, 6393.
Furusawa, K., Nagashima, K. and Anzai, C. (1994) *Colloid Polym. Sci.*, **272**, 1104.
Gaillard, C., Fuchs, G., Plummer, C.J.G. and Stadelmann, P.A. (2007) *Micron*, **38**, 522.
Galian, R.E. and Guardia, M. (2009) *Trends. Anal. Chem.*, **394**, 47.
Ganachaud, F., Mouterde, G., Delair, T. *et al.* (1995) *Polym. Adv. Technol.*, **6**, 480.
Gao, J. and Penlidis, A. (2002) *Prog. Polym. Sci.*, **27**, 403.
Gardon, J.L. (1968) *J. Polym. Sci. Part A-1*, **6**, 2859.
Gardon, J.L. (1970a) *Br. Polym. J.*, **2**, 1.
Gardon, J.L. (1970b) *Rubb. Chem. Techol.*, **43**, 74.
Gardon, J.L. (1977) Interfacial, colloidal and kinetic aspects of emulsion polymerization, in *Interfacial Synthesis*, vol. **I** (eds F. Millich and C.E. Carraher Jr), Marcel Dekker, New York.
Garnier, J., Warnant, J., Lacroix-Demazes, P. *et al.* (2012) *Macromol. Rapid. Commun.*, **33**, 1388.
Gast, A.P., Hall, C.K. and Russel, W.B. (1983) *J. Colloid Interface Sci.*, **96**, 251.
Gaynor S.G., Qiu J. and Matyjaszewski, K. (1998) *Macromolecules*, **31**, 5951.
Gaynor, S.G., Wang, J.S. and Matyjaszewski, K. (1995) *Macromolecules*, **28**, 8051.
Ge, J., Hu, Y., Zhang, T. and Yin, Y. (2007) *J. Am. Chem. Soc.*, **129**, 8974.
Gee, G., Davies, C.B. and Melville, W.H. (1939) *Trans. Faraday Soc.*, **35**, 1298.
Georges, M.K., Veregin, R.P.N., Kazmaier, P.M. and Hamer, G.K. (1993) *Macromolecules*, **26**, 2987.
Georges, M.K., Veregin, R.P.N., Kazmaier, P.M. and Hamer, G.K. (1994) *Trends Polym. Sci.*, **2**(2), 66.
Georges, M.K., Lukkarila, J.L. and Szkurhan, A.R. (2004) *Macromolecules*, **37**, 1297.
Gerharz, B., Butt, H.J. and Momper, B. (1996) *Progr. Colloid Polym. Sci.*, **100**, 91.
German, A.L., Van Herk, A.M., Schoonbrood, H.A.S. and Aerdts, A.M. (1997) Latex polymer characterization, Chapter 11, in *Emulsion Polymers and Emulsion Polymerization* (eds P. Lovell and M.S. El-Aasser), John Wiley & Sons.
Gerrens, H. (1956) *Z. Elektrochem.*, **60**, 400.
Gerschberg, D.B. and Longfield, J.E. (1961) Symp. Polym. Kinet. Katal. Syst., 45th AIChE Meeting, New York, Vol. Preprint 10.
Giannetti, E. (1990) *Macromolecules*, **23**, 4748.
Gilbert, R.G. (1995) *Emulsion Polymerization. A Mechanistic Approach*, Academic Press, London.
Gilbert, R.G. and Napper, D.H. (1974) *J. Chem. Soc., Faraday Trans.*, **I 70**, 391.

Gilbert, R.G. and Napper, D.H. (1983) *J. Macromol. Sci. - Rev. Macromol. Chem. Phys.*, **C23**, 127.
Giskehaug, K. (1965) Symp Chem. Polym. Process. London.
Gloor, G.J., Jackson, G., Blas, F.J. and de Miguel, E. (2005) *J. Chem. Phys.*, **123**, 134703.
Goikoetxea, M., Beristain, I., Minari, R.J. *et al.* (2011) *Chem. Eng. J.*, **170**, 114.
Goikoetxea, M., Minari, R.J., Beristain, I. *et al.* (2009) *J. Polym. Sci. Part A: Polym. Chem.*, **47**, 4871.
Goikoetxea, M., Reyes, Y., de las Heras, C.M. *et al.* (2012) *Polymer*, **53**, 1098.
Gombotz, W.R. and Wee, S.F (1998) *Adv. Drug Del. Rev.*, **31**, 267.
Gonzales-Ortiz, L.J. and Asua, J.M. (1995) *Macromolecules*, **28**, 3135.
González-Ortiz, L.J. and Asua, J.M. (1996a) *Macromolecules*, **29**, 383.
González-Ortíz, L.J. and Asua, J.M. (1996b) *Macromolecules*, **29**, 4520.
Goodall, A.R., Wilkinson, M.C. and Hearn, J. (1977) *J. Polym. Sci.*, **15**, 2193.
Goodall, A.R., Hearn, J. and Wilkinson, M.C. (1979) *J. Polym. Sci., Polym. Chem. Ed.*, **17**, 1019.
Goodwin, J.W., Ottewill, R.H., Pelton, R. *et al.* (1978) *Br. Polym. J.*, **10**, 173.
Goodwin, J.W., Ottewill, R.H., Harris, N.M. and Tabony, J. (1980) *J. Colloid Interface Sci.*, **78**, 253.
Goodwin, J.W., Hughes, R.W., Partridge, S.T. and Zukowski, C.F. (1986) *J. Phys. Chem.*, **85**, 559.
Goodwin, J.W., Markham, G. and Vincent, B. (1997) *J. Phys. Chem. B*, **101** 1961.
Goto, A. and Fukuda T. (2004) *Prog. Polym. Sci.*, **29**, 329.
Goto, A., Ohno, K. and Fukuda, T. (1998) *Macromolecules*, **31**, 2809.
Gottlob, K. (1913) Producing isoprene from turpentine oil, US Patent 1065522.
Gourevich, I., Pham, H., Jonkman, J.E.En and Kumacheva, E. (2004) *Chem. Mater.*, **16**, 1472.
Grancio, M.R. and Williams, D.J. (1970) *J. Polym. Sci., Polym. Chem.*, **8**, 2617.
Granier, V., Sartre, A. and Joanicot, M. (1993) *J. Adhesion*, **42**, 255.
Grcev, S., Schoenmaker, P.J. and Iedema, P. (2004) *Polymer*, **45**, 39.
Green, N.G. and Morgan, H. (1999) *J. Phys. Chem.*, **103**, 41.
Groison, E., Brusseau, S., D'Agosto, F. *et al.* (2012) *ACS Macro Lett.*, **1**, 47.
Gromada, J. and Matyjaszewski, K. (2001) *Macromolecules*, **34**, 7664.
Gu, I., Former, N.A., Geier, M.L. and Alivisatos, A.P. (2005) *Science*, **310**, 462.
Gugliotta, L.M., Arotcarena, M., Leiza, J.R. and Asua, J.M. (1995a) *Polymer*, **36**, 2019.
Gugliotta, L.M., Arzamendi, G. and Asua, J.M. (1995b) *J. Appl. Polym. Sci.*, **55**, 1017.
Gugliotta, L.M., Leiza, J.R., Arotcarena, M. *et al.* (1995c) *Ind. Eng. Chem. Res.*, **34**, 3899.
Guillaume, J.L., Pichot, C. and Guillot, C. (1990) *J. Polym. Sci. Polym. Chem.*, **28**, 119.
Guillot, S., Delsanti, M., Desert, S. and Lanvgevin, D. (2003) *Langmuir*, **19**, 230.
Guo, J.S., Sudol, E.D., Vanderhoff, J.W. and El-Aasser, M.S. (1992) In *Polymer Latexes - Preparation, Characterization and Applications*, (eds E.S. Daniels, E.D. Sudol and M.S. El-Aasser), ACS Symposium Series, vol. **492**, American Chemical Society, Washington D.C., p. 99.
Guo, Y., Liu, J. and Zetterlund, P.B. (2010) *Macromolecules*, **43**, 5914.
Guyot, A. and Tauer, K. (1994) *Adv. Polym. Sci*, **111**, 43.
Guyot, A., Guillot, J., Graillat, C. and Llauro, M.F. (1984) *J. Macromol. Sci. Chem.*, **A21**, 683.

Hagen, R., Salmen, L., Karlsson, O. and Wesslen, B. (1996) *J. Appl. Polym. Sci.*, **62**, 1067.
Hamaker, H.C. (1937) *Physica*, **4**, 457.
Hamielec, A. and MacGregor, J.F. (1982) Latex reactor principles, in *Emulsion Polymerization* (ed. I. Piirma) Academic Press, New York.
Hammouri, H., McKenna, T. and Othman, S. (1999) *Ind. Eng. Chem. Res.*, **38**, 4815.
Han, M., Gao, X., Su, J.Z. and Nie, S. (2001) *Nat. Biotechnol.*, **18**, 631.
Hanarp, P., Duncan, S., Sutherland, S. *et al.* (2003) *Colloids Surf. A: Physicochem. Eng. Aspects*, **214**, 23.
Haneke, K.E. (2002) 4-PCH Review of Toxicological Literature, http://ntp.niehs.nih.gov/ntp/htdocs/Chem_Background/ExSumPdf/ Phenylcyclohexene.pdf.
Hansen, F.K., Baumann Ofstad, E. and Ugelstad, J. (1974) *Theory and Practice of Emulsion Technology*, Academic Press, New York.
Hansen, F.K. and Ugelstad, J. (1978) *J. Polym. Sci.*, **16**, 1953.
Hansen, F.K. and Ugelstad, J. (1979a) *J. Polym. Sci.*, **17**, 3033.
Hansen, F.K. and Ugelstad, J. (1979b) *J. Polym. Sci.*, **17**, 3047.
Hansen, F.K. and Ugelstad, J. (1979c) *J. Polym. Sci.*, **17**, 3069.
Hansen, F.K. and Ugelstad, J. (1982) Particle formation mechanisms, in *Emulsion Polymerization* (ed. I. Piirma), Academic Press, New York, p. 51.
Hansen, F.K. (1992a) *ACS Symp. Ser.*, **492**, 12.
Hansen, F.K. (1992b) *Chem. Eng. Sci.*, **48**, 2, 437.
Harada, M., Nomura, M., Kojima, H. *et al.* (1972) *J. Appl. Polym. Sci.*, **16**, 811.
Harkins, W.D. (1945a) *J. Chem. Phys.*, **13**, 381.
Harkins, W.D. (1945b) *J. Chem. Phys.*, **13**, 534.
Harkins, W.D. (1946) *J. Chem. Phys.*, **14**, 47.
Harkins, W.D. (1947) *J. Am. Chem. Soc.*, **69**, 1428.
Harkins, W.D. (1950) *J. Polymer Sci.*, **5**, 217.
Hassander, H., Karlsson, O. and Wesslen, B. (1994) Proceedings ICEM 13-PARIS, Paris, France, **2B**, p. 1215.
Hatada, K., Ute, K., Okamoto, Y. *et al.* (1988) *Polymer Bull.*, **20**, 317.
Hauser, E.A. (1930) *Latex*, Chemical Catalogue Co., New York.
Hawker, C.J. (2002) In *Handbook of Radical Polymerisation* (eds K. Matyjaszewski and T.P. Davis), Wiley Interscience, John Wiley and Sons Inc., p. 463.
Hawker, C.J., Bosman, A.W. and Harth, E. (2001) *Chem. Rev.*, **101**, 3661.
Hawkett, B.S., Napper, D.H. and Gilbert, R.G. (1980) *J. Chem. Soc., Faraday Trans. 1*, **76**, 1323.
Hazot, P. (2001) Doctorate thesis, University Lyon, France.
Healy, T.W. and White, L.R. (1978) *Adv. Colloid Interface Sci.*, **9**, 303.
Hearn, J., Wilkinson, M.C. and Goodall, A.R. (1981) *Adv. Colloid Interface Sci.*, **14**, 173.
Hellgren, A.C., Weissenborn, P. and Holmberg, K. (1999) *Prog. Org. Coat.*, **35**, 79.
Hergeth, W.D., Schmutzler, K. and Wartewig, S. (1990) *Makromol. Chem., Macromol. Symp.*, **31**, 123.
Herrera, V., Pirri, R., Leiza, J.R. and Asua, J.M. (2006) *Macromolecules*, **39**, 69969.
Herrera, V., Pirri, R., Asua, J.M. and Leiza, J.R. (2007) *J. Polym. Sci. Part A: Polym. Chem.*, **45**, 2484.
Herrera, V., Palmillas, Z., Pirri, R. *et al.* (2010) *Macromolecules*, **43**, 1356.

Herrera-Ordonez, J. and Olayo, R. (2000) *J. Polym. Sci. Part A - Polym. Chem.*, **38**, 2201.
Herrera-Ordonez, J. and Olayo, R. (2001) *J. Polym. Sci. Part A - Polym. Chem.*, **39**, 2547.
Hess, B., Kutzner, C., van der Spoel, D. and Lindahl, E. (2008) *J. Chem. Theory Comput.*, **4**, 435.
Hidalgo, M., Guillot, J., Llauro-Darricades, M.F. *et al.* (1992) *J. Chim. Phys. Physico-Chim. Biol.*, **89**, 505.
Hidalgo-Alvarez, R., Martin, A., Fernandez, A. *et al.* (1996) *Adv. Colloid Interface Sci.*, **67**, 1.
Higuchi, T., Tajima, A., Motoyoshi, K., Yabu, H. and Shinomura, M. (2008) *Angew. Chem.*, **120**, 8164.
Hillenkamp, F. and Karas, M. (2000) *Int. J. Mass Spectrom.*, **200**, 71.
Hiller, W. and Pasch, H. (2001) *Polym. Preprints*, **42**, 66.
Hodgson, M. (2000) Thesis, University of Stellenbosh, South Africa.
Hofman, F. and Delbrück, K. (1909, 1912) Germany.
Hohenstein, W.P., Siggia, S. and Mark, H. (1944a) *India Rubber World*, **111**, 173.
Hohenstein, W.P., Vigniello, F. and Mark, H. (1944b) *India Rubber World*, **110**, 291.
Hohenstein, W.P. (1945) *Polym. Bull.*, **1**, 1.
Hohenstein, W.P. and Mark, H. (1946) *J. Polym. Sci.*, **1**, 127.
Hong, L., Cacciuto, A., Luijten, E. and Granik, S. (2006) *Nano Lett.*, **6**, 2510.
Hoseh, M. (1944) German Patents relating to synthetic rubberlike materials in India Rubber World through 1940 and 1941. *India Rubber World*, **110**, 416.
Hourston, D.J., Zhang, H.X., Song, M. *et al.* (1997) *Thermochim. Acta*, **294**, 23.
Hoy, K.L. (1979) *J. Coat. Technol.*, **51**, 27.
Hu, J., Chen, M., Wu, L. (2011) *Polym. Chem.*, **2**, 760.
Hu, X, Zhang, J. and Yang, W. (2009) *Polymer*, **50**, 141.
Huang, H., Lui, H. (2010) *J. Polym. Sci. Part A: Polym. Chem.*, **48**, 5198.
Huges, E.W., Sawyer, W.M. and Vinograd, R.L. (1945) *J. Chem. Phys.*, **13**, 131.
Hugues, M.P. and Morgan, H. (1998) *J. Phys. D: Appl. Phys.*, **31**, 2205.
Hul, H.J.V.D. and Vanderhoff, J.W. (1972) *J. Elektroanal. Chem.*, **37**, 161.
Hunter, R.J. (1987) *Foundations of Colloid Science*, Clarendon Press, Oxford.
Hunter, R.J. (1998) *Colloids Surf., A.*, **141**, 37.
Iijima, T., Yochioka, N. and Tomoi, N. (1992) *Eur. Polym. J.*, **28**, 573.
Illum, L. and Davis, S.S. (1982) *J. Parenteral Sci. Tech.*, **36**, 242.
Ishida, M., Oshima, J., Yoshinaga, K. and Horii, F. (1999) *Polymer*, **40**, 3323.
Israelachvili, J. (1991) *Intermolecular and Surface Forces*, 2nd edn, Academic Press, London.
Ita, P. (2002) Adhesives and Sealants Industry July 2002, p. 24.
Ito, K., Cao, J. and Kawaguchi, S, (2002) In *Functional Colloids and Microparticles* (eds R. Arshady and A. Guyot), Citus, London, IV Chapter 4, p. 109.
Jaeger, W., Wendler, U., Lieske, A., and Bohrisch, J. (1999) *Langmuir*, **15**, 4026.
Jakubowski, W., Min, K. and Matyjaszewski, K. (2006) *Macromolecules*, **39**, 39.
Janes, K.A., Calvo, P. and Alonso, M.J. (2001) *Adv. Drug Del. Rev.*, **47**, 83.
Janshoff, A., Neitzert, M., Oberdörfer, Y. and Fuchs, H. (2000) *Angew. Chem. Int. Ed.*, **39**, 3212.
Ji, J. Yan, L. and Xie, D. (2008) *J. Polym. Sci. Part A Polym. Chem.*, **46**, 3098.

Jia, Z. and Monteiro, M.J. (2012) In *Progress in Controlled Radical Polymerisation: Mechanisms and Techniques* (eds K. Matyjaszewski, B. Sumerlin and N.V. Tsarevsky), American Chemical Society, p. 1100, 293.
Jiang, X., van der Horst, A. and Schoenmakers, P.J. (2002) *J. Chromatogr. A*, **982**, 55.
Jiang, X.L., Lima, V. and Schoenmakers, P.J. (2003a) *J. Chromatogr. A*, **1018**, 19.
Jiang, X.-L., Schoenmakers, P.J., Lou, X.-W. *et al.* (2003b) *Anal. Chem.*, **75**, 5517.
Jiang, X.-L., Schoenmakers, P.J., van Dongen, J.L.M. *et al.* (2004) *J. Chromatogr. A*, **1055**, 123.
Jiang, X.-L., van der Horst, A., Lima, V. and Schoenmakers, P.J. (2005) *J.Chromatogr. A*, **1076**, 51.
Jones, D.A.R. and Vincent, B. (1989) *Colloids Surf.*, **42** 113.
Jönsson, J.E.L., Hassander, H., Jansson, L.H. and Törnell, B. (1991) *Macromolecules*, **24**, 126.
Jönsson, J.-E., Hassander, H. and Törnell, B. (1994) *Macromolecules*, **27**, 1932.
Jönsson, J.E., Karlsson, O.J., Hassander, H. and Törnell, B. (2007) *Eur. Polym. J.*, **43**, 1322.
Jousset, S., Qiu, J., Matyjaszewski, K. and Granel, C. (2001) *Macromolecules*, **34**, 6641.
Jovanovic, R. and Dube, M.A. (2003) *Polym. React. Eng.*, **11**, 233.
Juang, M.S. and Krieger, I.M. (1976) *J. Polym. Sci. Polym. Chem.*, **14**, 2089.
Juhué, D. and Lang, J. (1994a) *Colloids Surf. A*, **87**, 177.
Juhué, D. and Lang, J. (1994b) *Langmuir*, **9**, 792.
Juhué, D. and Lang, J. (1995) *Macromolecules*, **28**, 1306.
Jung, M., Hubert, D.H.W., Bomans, P. *et al.* (2000) *Adv. Mater.*, **12** 210.
Kagawa, Y., Minami, H., Okubo, M. and Zhou, J. (2005) *Polymer*, **46**, 1045.
Kagawa, Y., Zetterlund, P.B., Minami, H. and Okubo, M. (2007a) *Macromolecules*, **40**, 3062.
Kagawa, Y., Kawasaki, M., Zetterlund, P.B. *et al.* (2007b) *Macromol. Rapid. Commun.*, **28**, 2354.
Kahn, M.A. and Armes, S.P. (2000) *Adv. Mater.*, **12**, 671.
Kamigaito, M., Ando, T. and Sawamoto, M. (2001) *Chem. Rev.*, **101**, 3689.
Karg, M., Hellweg, T. (2009) *J. Mater. Chem.*, **19**, 8714-8727.
Karger, B.L., Foret, F. and Preisler, J. (2001) U.S. Pat. Appl. 6, 175, 112.
Karlsson, O., Hassander, H. and Wesslen, B. (1995) *Colloid Polym. Sci.*, **273**, 496.
Karlsson, O.J., Caldwell, K. and Sundberg, D.C. (2000) *Macromol. Symp.*; "Polymers in Dispersed Media II", **151**, 503.
Karlsson, L.E., Karlsson, O.J. and Sundberg, D.C. (2003a) *J. Appl. Polym. Sci.*, **90**, 905.
Karlsson, O.J., Hassander, H. and Colombini, D. (2003b) *C.R. Chim.*, **6**, 1233.
Karlsson, O.J., Stubbs, J.M., Carrier, R.H. and Sundberg, D.C. (2003c) *Polym. React. Eng.*, **11**, 589.
Kassaz, J., Neidleman, J., Singh, M., Ott, G. and O'Hagan, D.T. (2000) *J. Control. Rel.*, **67**, 347.
Kato, K. (1966) *J. Polym. Sci.: Polym. Lett. Ed.*, **4**, 35.
Kato, M., Kamigaito, M., Sawamoto, M. and Higashimura, T. (1995) *Macromolecules*, **28**, 1721.
Katoh, M. (1979) *J. Electron Microsc.*, **28**, 197.
Kavacsovics-Bankowski, M., Cllark, K., Benacerraf, B. and Rock, K.L. (1993) *Proc. Natl. Acad. Sci. USA*, **90**, 4942.

Kawaguchi, H., Duracher, D. and Elaissari, A. (2003) In *Colloidal Biomolecules, Biomaterials, and Biomedical Applications*, vol **116**. (ed. A. Elaissari), Marcel Dekker Edition Surfactant Science Series, Marcel Dekker, p. 189.
Kawaguchi, K. (2000) *Prog. Polym. Sci.*, **25**, 1170.
Kawaguchi, S., Yekta, A. and Winnik, M.A. (1995) *J. Colloid Interface Sci.*, **176**, 362.
Kawahara, H., Goto, T., Ohnishi, K. *et al.* (2001) *J. Appl. Polym. Sci.*, **81**, 128.
Keddie, J.L. (1997) *Mater. Sci. Eng.*, **21**, 101.
Kemmere, M.F., Meuldijk, J., Drinkenburg, A.A.H. and German, A.L. (1998) J. Appl. Polym. Sci., **69**, 2409.
Keoshkerian, B., McLeod, P.J. and Georges, M.K. (2001a) *Macromolecules*, **34**, 3594.
Keoshkerian, B., Szkurhan, A.R. and Georges, M.K. (2001b) *Macromolecules*, **34**, 6531.
Kessel, S., Urbani, C.N. and Monteiro, M.J. (2011) *Angew. Chem., Int. Ed.*, **50**, 8082.
Kim, J., Serpe, M.J. and Lyon, L.A. (2005) *Angew. Chem. Int. Ed.*, **44**, 1333.
Kirsch, S., Pfau, A., Hadicke, E. and Leuninger, J. (2002) *Prog. Org. Coat.*, **45**, 193.
Kirsch, S., Stubbs, J., Leuninger, J. *et al.* (2004) *J. Appl. Polym. Sci.*, **91**, 2610.
Kitayama, T., Janco, M., Ute, K. *et al.* (2000) *Anal.Chem.*, **72**, 1518.
Kitayama, Y., Yorizane, M., Kagawa, Y. *et al.* (2009) *Polymer*, **50**, 3182.
Knoll, A., Magerle, R. and Krausch, G. (2001) *Macromolecules*, **34**, 4159.
Koinuma, H., Tanabe, T. and Hirai, H. (1982) *Makromol. Chem.*, **183**, 28.
Kok, S.J., Arentsen, N.C., Cools, P.J.C.H. *et al.* (2002) *J. Chromatogr. A*, **948**, 257.
Kok, S.J., (2004) Coupling of liquid chromatography and Fourier-transform infrared spectroscopy for the characterization of polymers. Ph.D.Thesis, University of Amsterdam.
Kolthoff, M. and Dale, W.J. (1945) *J. Am. Chem. Soc.*, **67**, 1672.
Kong, X. Z., Pichot, C. and Guillot, J. (1987) *Colloid Polym. Sci.*, **265**, 791.
Koo, H.Y., Yi, D.K., Yoo, S.J. and Kim, D.-Y. (2004) *Adv. Mater.*, **16**, 274.
Kowalski, A., Vogel, M. and Blankenship, M. (1981) Rohm and Haas Company, European Patent EP 22633.
Krämer, I., Pasch, H., Händel, H. and Albert, K. (1999) *Macromol. Chem. Phys.*, **200**, 1734.
Kreuter, J. and Speiser, P.P. (1976) *Infect. Immun.*, **13**, 204.
Kühle, A., Sorensen, A.H. and Bohr, J. (1997) *J. Appl. Phys.*, **81**, 6562.
Kukulj, D. and Gilbert, R.G. (1997) In *Polymeric Dispersions. Principles and Applications*, Vol. NATO Advanced Studies Institute (ed. J.M. Asua), Kluwer Academic Publishers, Dordrecht, pp. 97–107.
Kumacheva, E. (2004) PCT Int. Appl., WO 2004081072 A2 20040923.
Kwak, Y., Goto, A., Tsujii, Y. *et al.* (2002) *Macromolecules*, **35**, 3026.
Kwak, Y., Goto, A., Fukuda, T. *et al.* (2006) *Macromolecules*, **39**, 4671.
Lam, S., Hellgren, A.-C., Sjöberg, M. *et al.* (1997) *J. Appl. Polym. Sci.*, **66**, 187.
Landfester, K. and Spiess, H.W. (1998) *Acta Polym.*, **49**, 451.
Landfester, K., Bechthold, N. and Antonietti, M. (1999) *Macromolecules*, **32**, 5222.
Landfester, K. (2009) *Angew. Chem.*, **48**, 4488.
Lansalot, M., Farcet, C., Charleux, B. *et al.* (1999) *Macromolecules*, **32**, 7354.
Lansalot, M., Farcet, C., Charleux, B. *et al.* (2000) In *Controlled/Living Radical Polymerisation: Progress in ATRP, NMP and RAFT* (ed. K. Matyjaszewski), ACS Symposium Series, vol. **768**, Washington DC, p. 138.
Lansalot, M., Davis, T.P. and Heuts, J.P.A. (2002) *Macromolecules*, **35**, 7582.
Larpent, C., Amigoni-Gerbier, S. and De Souza Delgado, A.-P. (2003) *C.R. Chim.*, **6**, 1275.

Lau, A.W.C., Portigliatti, M., Raphael, E. and Leger, L. (2002) *Europhys. Lett.*, **60**, 717.

Laureau, C., Vicente, M., Barandiaran, M.J. *et al.* (2001) *J. Appl. Polym. Sci.*, **81**, 1258.

Le, T. P., Moad, G., Rizzardo, E. and Thang, S. H. (1998) PCT Int. Appl. WO 98/01478 Chem Abs, 1998 128: 115390.

Leclere, P., Lazzaroni, R., Bredas, J.L. *et al.* (1996) *Langmuir*, **12**, 4317.

Ledesma-Alonso, R., Legendre, D. and Tordjeman, P. (2012) *Phys. Rev. Lett.*, **108**, 106104.

Lee, D.G. and Vandeelkes, M. (1973) In *Oxidation in Organic Chemistry* (ed. W.S. Trahanovsky), Academic Press, New York, p. 177.

Lee, D.I. (1981) *ACS Symp. Ser.*, **165**, 405.

Lee, H.C. and Chang, T. (1996) *Polymer*, **37**, 5747.

Lee, S. and Rudin, A. (1992) *J. Polym. Sci., Polym. Chem. Ed.*, **30**, 865.

Lee, W., Lee, H.C., Chang, T. and Kim, S.B. (1998) *Macromolecules*, **31**, 344.

Lee, W., Lee, H.C., Park, T. *et al.* (1999) *Polymer*, **40**, 7227.

Lee, W., Lee, H.C., Park, T. *et al.* (2001a) *Macromol. Chem. Phys.*, **201**, 320.

Lee, W., Cho, D., Chang, T. *et al.* (2001b) *Macromolecules*, **34**, 2353.

Leiza, J.R., Arzamendi, G. and Asua, J.M. (1993a) *Polym. Int.*, **30**, 455.

Leiza, J.R., de la Cal, J.C., Meira, G.R. and Asua, J.M. (1993b) *Polym. React. Eng.*, **1**, 461.

Leiza, J.R. and Asua, J.M. (1997) Feedback control of emulsion polymerization reactors: A critical review and future directions, in *Polymeric Dispersions: Principles and Applications* (ed. J.M. Asua), Kluwer Academic Publishers, Dordrecht.

Leswin, J.S.K., Meuldijk, J., Gilbert, R.G. and van Herk, A.M. (2009) *Macromol. Symp.*, **275–276**, 24.

Li, C.Y., Chiu, W.Y. and Don, T.M. (2007) *J. Polym. Sci. Part A: Polym. Chem.*, **45**, 3359.

Li, F., Josephson, D.P. and Stein, A. (2011) *Angew. Chem. Int. Ed.*, **50**, 360-388.

Li, J.T. and Caldwell, K.D. (1991) *Langmuir*, **7**, 2034.

Li, M. and Matyjaszewski, K. (2003a) *J. Polym. Sci. Part A: Polym. Chem.*, **41**, 3606.

Li, M. and Matyjaszewski, K. (2003b) *Macromolecules*, **36**, 6028.

Li, M., Jahed, N.M., Min, K., Matyjaszewski, K. (2004a) *Macromolecules*, **37**, 2434.

Li, M., Min, K. and Matyjaszewski, K. (2004b) *Macromolecules*, **37**, 2106.

Li, W., Min, K., Matyjaszewski, K. *et al.* (2008) *Macromolecules*, **41**, 6387.

Li, W., Matyjaszewski, K. (2011) *Macromolecules*, **44**, 5578.

Li, Y., Armes, S.P., Jin, X. and Zu, S. (2003) *Macromolecules*, **36**, 8268.

Liang, J., He, L., Dong, X. and Zhou, T. (2012) *J. Colloid Interface Sci.*, **369**, 435.

Lichti, G., Gilbert, R.G. and Napper, D.H. (1980) *J. Polym. Sci. A.*, **18**, 1297.

Lichti, G., Gilbert, R.G. and Napper, D.H. (1982) Theoretical predictions of the particle size and molecular weight distributions in emulsion polymerization, in *Emulsion Polymerization* (ed. I. Piirma), Academic, New York.

Lichti, G., Gilbert, R.G. and Napper, D.H. (1983) *J. Polym. Sci., Polymer Chem. Ed.*, **21**, 269.

Lin, M., Chu, F., Guyot, A. *et al.* (2005) *Polymer*, **46**, 1331.

Lin, M., Cunningham, M.F. and Keoshkerian, B. (2004) *Macromol. Symp.*, **206**, 263.

Litvinenko, G. and Müller, A. (1997) *Macromolecules*, **30**, 1253.

Liu, H., Hu, X., Wang, Y. *et al.* (2009) *J. Polym. Sci., Polym. Chem.*, **47**, 2892.

Liu, L.J. and Krieger, I.M., (1978) In *Emulsions Latexes and Dispersions* (eds P. Becher and M.N. Yudenfreund), pp. 41.

Llauro, M.F., Petiaud, R., Hidalgo, M. *et al.* (1995) *Macromol. Symp.*, **92**, 117.
Long, J., Osmond, D.W.J. and Vincent, B. (1973) *J. Colloid Interface Sci.*, **42**, 545.
Lopez, A., Chemtob, A., Milton, J.L. *et al.* (2008) *Ind. Eng. Chem. Res.*, **47**, 6289.
Lopez, A., Degrandi-Contraires, E., Canetta, E. *et al.* (2011) *Langmuir*, **27**, 3878.
López-Redón, R., Reyes, Y. and Orea, P. (2006) *J. Chem. Phys.*, **125**, 084508.
Lousberg, H.A., Hoefsloot, H.C.J., Boelens, H.F.M. *et al.* (2002) *Int. J. Polym. Anal. Char.*, **7**, 76.
Lovell, P.A. (1995) *Macromol. Symp.*, **92**, 71.
Lovell, P.A. and El-Aasser, M.S. (1997) *Emulsion Polymerization and Emulsion Polymers*, Wiley, Chichester.
Luo, Y., Tsavalas, J.G. and Schork, F.J. (2001) *Macromolecules*, **34**, 5501.
Luther, M. and Heuck, C. (1927) Polymerizing butadiene hydrocarbons. Patent DE 558890.
Lutz, H. and Hahner, C. (2002) Applications of redispersible powders, in: *Polymer Dispersions and Their Industrial Applications* (eds D. Urban and K. Takamura), Wiley-VCH, Weinheim.
Lyons, R.A., Hutovic, J., Piton, M.C. *et al.* (1996) *Macromolecules*, **29**, 1918.
MacGregor, J.F. (1986) Online Reactor Energy Balances via Kalman Filtering. IUPAC Conf. on Instrumentation and Automation in Rubber, Plastic and Polymerization Industries, Akron, OH.
Maeder, S. and Gilbert, R.G. (1998) *Macromolecules*, **31**, 4410.
Magenau, A.J.D., Strandwitz, N.C., Gennaro, A. and Matyjaszewski, K. (2011) *Science*, **332**, 81.
Magnet, S., Guillot, J., Guyot, A. and Pichot, C. (1992) *Prog. Org. Coatings*, **20**, 73.
Magonov, S.N., Elings, V. and Whangbo, M.-H. (1997) *Surf. Sci.*, **375**, L385.
Mahl, H. (1964) *Kunstoffe*, **54**, 15.
Makarov, C., Khalfin, R.L., Makarov, V. *et al.* (2007) *Polym. Adv. Technol.*, **18**, 712.
Makino, T., Tokunaga, E. and Hogen-Esch, T.E. (1998) *Polym. Prepr. (Am. Chem. Soc. Div. Polym. Chem.)*, **39**, 288.
Makino, T., Tokunaga, E. and Hogen-Esch, T.E. (1998) *Polym. Prepr. Am. Chem. Soc. Div. Polym. Chem.*, **39**, 288.
Manea, M., Chemtob, A., Paulis, M. *et al.* (2008) *AIChE J.*, **54**, 289.
Manuszrak-Guerrini, M., Charleux, B. and Vairon, J.P. (2000) *Macromol. Rapid Commun.*, **21**, 669.
Marciniec, B. and Malecka, E. (2003) *J. Phys. Org. Chem.*, **16**, 818.
Marestin, C., Noël, C., Guyot, A. and Claverie, J. (1998) *Macromolecules*, **31**, 4041.
Mark, H. and Rafft, R. (1941) *High Polymeric Reactions*, Interscience, New York.
Masuhara, H. (2003) *Single Organic Nanoparticles* (ed. H. Masuhara, H. Nakanishi and K. Sasaki), Springer, pp. 121-132.
Matsumoto, T., Okubo, M. and Imai, T. (1974) *Kobunshi Ronbunshu, English Ed.*, **3**, 1814.
Mattoussi, H., Radzilowski, L.H., Dabbousi, B.O. *et al.* (1998) *J. Appl. Phys.*, **83**, 7965.
Matyjaszewski, K. ed. (1998) *Controlled Radical Polymerisation*. ACS Symposium Series, vol. **685**, ACS, Washington DC
Matyjaszewski, K., Qiu, J., Shipp, D. and Gaynor, S.G. (2000a) *Macromol. Symp.*, **155**, 15.
Matyjaszewski, K., Qiu, J., Tsarevsky, N.V. and Charleux, B. (2000b) *J. Polym. Sci. Part A: Polym. Chem.*, **38**, 4724.

Matyjaszewski, K., Shipp, D.A., Qiu, J. and Gaynor, S.G. (2000c) *Macromolecules*, **33**, 2296.

Matyjaszewski, K. ed. (2000d). *Controlled/Living Radical Polymerisation. Progress in ATRP, NMP and RAFT*. ACS Symposium Series,vol. **768**, ACS, Washington DC.

Matyjaszewski, K. and Xia, J. (2001) *Chem. Rev.*, **101**, 2921.

Matyjaszewski, K., Xia, J. (2002) In *Handbook of Radical Polymerisation* (eds K. Matyjaszewski and T.P. Davis), Wiley Interscience, John Wiley and Sons Inc., p. 523.

Matyjaszewski, K. and Davis, T.P. (eds) (2002) *Handbook of Radical Polymerisation*, John Wiley & Sons.

Matyjaszewski, K. ed. (2003) *Advances in Controlled/Living Radical Polymerisation*. ACS Symposium Series, vol. **854**, ACS, Washington DC.

Matyjaszewski, K., Jakubowski, W., Min, K. *et al.* (2006) *Proc. Natl. Acad. Sci. USA*, **103**, 15309.

Matyjaszewski, K. Gnanou, Y. and Leibler, L. (eds) (2007) *Macromolecular Engineering: Precise Synthesis, Materials Properties, Applications*, vol. **1**, Synthetic Techniques, Wiley-VCH, Weinheim.

Matyjaszewski, K. (2012) *Macromolecules*, **45**, 4015.

Maxwell, I.A., Morrison, B.R., Napper, D.H. and Gilbert, R.G. (1991) *Macromolecules*, **24**, 1629.

Maxwell, I.A., Kurja, J., van Doremaele, G.H.J. and German, A.L. (1992a) *Makromol. Chem.*, **193**, 2065.

Maxwell, I.A., Kurja, J., van Doremaele, G.H.J. *et al.* (1992b) Makromol. Chem., **193**, 2049.

Maxwell, I.A. and Kurja, J. (1995) *Langmuir*, **11**, 1987.

Mayo, F.R. and Lewis, F.M. (1944) J. Am. Chem. Soc., **66**, 1594.

Mballa Mballa, M.A., Ali, S.I., Heuts, J.P.A. and van Herk, A.M. (2012) *Polym. Int.*, **61**, 861.

McBain, J.W. (ed.) (1942) *Solubilization and Other Factors in Detergent Action*, Interscience, New York.

McBain, J.W. and Soldate, A.M. (1944) *J. Am. Chem. Soc.*, **64**, 1556.

McCaffery, T.R. and Durant, Y.G. (2002) *J. Appl. Polym. Sci.*, **86**, 1507.

McCarvill, W.T. and Fitch, R.M. (1978a) *J. Colloid Interface Sci.*, **64**, 403.

McCarvill, W.T. and Fitch, R.M. (1978b) *J. Colloid Interface Sci.*, **67**, 204.

Mcconney, M.E., Singamaneni, S. & Tsukruk, V.V. (2010) *Polym. Rev. (Philadelphia, PA, USA)*, **50**, 235.

McDonald, C.J. and Devon, M.J. (2002) *Adv. Colloid Interface Sci.*, **99**, 181–213.

McHale, R., Aldabbagh, F., Zetterlund, P.B. (2007) *J Polym. Sci., Part A: Polym. Chem.*, **45**, 2194.

McKeever, U., Barman, S., Hao, T *et al.* (2002) *Vaccine*, **20**, 1524.

McLeary, J. B. and Klumperman, B. (2006) *Soft Matter*, **2**, 45.

McLeary, J.B., Tonge, M.P., De Wet Roos, D. *et al.* (2004) *J. Polym. Sci., Part A: Polym. Chem.*, **42**, 960.

McLeod, P.J., Barber, R., Odell, P.G. *et al.* (2000) *Macromol. Symp.*, **155**, 31.

Mellon, V. (2009) PhD Thesis. CNRS/CPE Lyon, France.

Mendichi, R. and Schieroni, A.G. (2001) *Curr. Trends Polym. Sci.*, **6**, 17.

Mendoza, J., de la Cal, J.C. and Asua, J.M. (2000) *J. Polym. Sci., Part A: Polym. Chem.*, **38**, 4490.

Mengerink, Y., Peters, R., de Koster, C.G. *et al.* (2001) *J.Chromatogr. A*, **914**, 131.

Metropolis, N., Rosenbluth, A.W., Rosenbluth, M.N. and Teller, A.H. (1953) *J. Chem. Phys.*, **21**, 1087.

Meuldijk, J. and German, A.L. (1999) *Polym. React. Eng.*, **7**, 207.

Meuldijk, J., Kemmere, M.F., De Lima, S.V.W. *et al.* (2003) *Polym. React. Eng.*, **11**, 259.

Micusik, M., Fond, A., Reyes, Y. *et al.* (2010) *Macromol. React. Eng.*, **4**, 432.

Mills, M.F., Gilbert, R.G. and Napper, D.H. (1990) *Macromolecules*, **23**, 4247.

Min, K. and Matyjaszewski, K. (2005a) *Macromolecules*, **38**, 8131.

Min, K., Gao, H. and Matyjaszewski, K. (2005b) *J. Am. Chem. Soc.*, **127**, 3825.

Min, K., Li, M. and Matyjaszewski, K. (2005c) *J. Polym. Sci., Polym. Chem.*, **43**, 3616.

Min, K., Jakubowski, W. and Matyjaszewski, K. (2006a) *Macromol. Rapid Commun.*, **27**, 594.

Min, K., Gao, H. and Matyjaszewski, K. (2006b) *J. Am. Chem. Soc.*, **128**, 10521.

Min, K., Oh, J.K. and Matyjaszewski, K. (2007) *J. Polym. Sci., Polym. Chem.*, **45**, 1413.

Min, K., Gao, H., Yoon, J.A. *et al.* (2009) *Macromolecules*, **42**, 1597.

Min, K.W. and Ray, W.H. (1974) *J. Macromol. Sci., Rev. Macromol. Chem.*, **C11**, 177.

Min, K.W. and Ray, W.H. (1976a) *Chem. React. Eng., Proc. Int. Symp.*, **4**, 31.

Min, K.W. and Ray, W.H. (1976b) *ACS Symp. Series*, **24**, 359.

Min, K.W. and Ray, W.H. (1978) *J. Appl. Polym. Sci.*, **22**, 89.

Mingozzi, I., Cecchin, G. and Morini, G. (1997) *Int. J. Polym. Anal. Charact.*, **3**, 293.

Moad, G. and Solomon, D.H., (1995) *The Chemistry of Free Radical Polymerization*, Pergamon, Oxford.

Moad, G., Chiefari, J., Chong, Y.K. *et al.* (2000) *Polym. Int.*, **49**, 993.

Moad, G., Rizzardo, E. and Thang, S.H. (2005) *Aust. J. Chem.*, **58**, 379.

Moad, G., Rizzardo, E. and Thang, S.H. (2006) *Aust. J. Chem.*, **59**, 669.

Moad, G., Rizzardo, E. and Thang, S.H. (2009) *Aust. J. Chem.*, **62**, 1402.

Montagne, F. (2002) PhD Thesis. Claude Bernard University, Lyon-France.

Montaudo, G. and Lattimer, R.P. (2002) *Mass Spectrometry of Polymers*, CRC Press.

Monteiro, M.J., Hodgson, M. and de Brouwer, H. (2000a) *J. Polym. Sci. Part A Polym. Chem.*, **38**, 3864.

Monteiro, M.J., Sjoberg, M., Gottgens, C.M. and van der Vlist, J. (2000b) *J. Polym. Sci. Part A Polym. Chem.*, **38**, 4206.

Monteiro, M.J. and de Brouwer, H. (2001a) *Macromolecules*, **34**, 349.

Monteiro, M.J. and de Barbeyrac, J. (2001b) *Macromolecules*, **34**, 4416.

Monteiro, M.J. and de Barbeyrac, J. (2002) *Macromol. Rapid Comm.*, **23**, 370.

Monteiro, M.J. (2005a) *J. Polym. Sci., Part A: Polym. Chem.*, **43**, 3189.

Monteiro, M.J. (2005b) *J. Polym. Sci., Part A: Polym. Chem.*, **43**, 5643.

Monteiro, M.J. (2010) *Macromolecules*, **43**, 1159.

Monteiro, M.J. and Cunningham, M.F. (2012) *Macromolecules*, **45**, 4939.

Montroll, E. (1945) *J. Chem. Phys.*, **13**, 337.

Moore, A., McGuirk, P., Adams, S. *et al.* (1995) *Vaccine*, **13**, 1741.

Morgan, L.W. (1982) *J. Appl. Polym. Sci.*, **27**, 2033.

Moritz, H.-U. (1989) In *Polymerization Calorimetry-A Powerful Tool for Reactor Control* (eds K.-H. Reichert and W. Geiseler), VCHVerlag, Weinheim, Germany, pp. 248-266.
Morrison, B.R. and Gilbert, R.G. (1995) *Macromol. Symp.*, **92**, 13.
Morrison, B.R., Casey, B.S., Lacík, I. *et al.* (1994) *J. Polym. Sci. A: Polym. Chem.*, **32** 631.
Morrison, B.R., Maxwell, I.A., Gilbert, R.G. and Napper, D.H. (1992) *ACS Symp. Ser.*, **492**, 28.
Morton, M., Kaizerman, S. and Altier, M.W. (1954) *J. Colloid Sci.*, **9**, 300.
Moser, A., Takano, K., Margulies, D.T. *et al.* (2002) *J. Phys; D Appl.*, **35**, 157.
Mullens, M.E. and Orr, C. (1979) *Int. J. Multiphase Flow*, **5**, 79.
Müller, A.H.E., Zhuang, R., Yan, D. and Litvinenko, G. (1995) *Macromolecules*, **28**, 4326.
Müller, R.H., Mäder, K. and Gohla, S. (2000) *Eur. J. Pharm. Biopharm*, **50**, 161.
Mura, J-L. and Riess, G. (1995) *Polym. Adv. Technol.*, **6**, 497.
Murgasova, R. and Hercules, D.M. (2002) *Anal. Bioanal. Chem.*, **373**, 481.
Murgasova, R. and Hercules, D.M. (2003) *Int. J. Mass Spectrom.*, **226**, 151.
Muroi, S. (1966) *J. Appl. Polym. Sci.*, **10**, 713.
Muscatello, M.M.W., Stunja, L.E. and Asher, S.A. (2009) *Anal. Chem.*, **81**, 4978.
Napper, D.H. (1983) *Polymeric Stabilisation of Colloidal Dispersions*, Academic Press, London.
Naruse, T., Hishiba, T., Tereda, K. *et al.* (2001) Multiphase Strcutured Polymer Particles, Method of Manufacturing and Uses and Thereof. U.S. Patent 6,348,452, 2002/EP 1092736, 2001.
Nawamawat, K., Sakdapipanich, J.T., Ho, C.C. *et al.* (2011) *Colloids Surf., A.*, **390**, 157.
Negrete-Herrera, N., Putaux, J.L., Laurent, D. and Bourgeat-Lami, E. (2006) *Macromolecules*, **39**, 9177.
Negrete-Herrera, N., Putaux, J.L., David, L. *et al.* (2007) *Macromol. Rapid. Commun.*, **28**, 1567.
Nelliappan, V., El-Aasser, M.S., Klein, A. *et al.* (1995) *J. Appl. Polym. Sci.*, **58**, 323.
Nguyen, D., Zondanos, H.S., Farrugia, J.M. *et al.* (2008) *Langmuir*, **24**, 2140.
Nicolas, J., Charleux, B., Guerret, O. and Magnet, S. (2004a) *Macromolecules*, **37**, 4453.
Nicolas, J., Charleux, B., Guerret, O. and Magnet, S. (2004b) *Angew. Chem. Int. Ed.*, **43**, 6186.
Nicolas, J., Charleux, B., Guerret, O. and Magnet, S. (2005) *Macromolecules*, **38**, 9963.
Nicolas, J., Dire, C., Mueller, L. *et al.* (2006a) *Macromolecules*, **39**, 8274.
Nicolas, J., Charleux, B. and Magnet, S. (2006b) *J. Polym. Sci.: Part A: Polym. Chem.*, **44**, 4142.
Nicolas, J., Ruzette, A.-V., Farcet, C. *et al.* (2007) *Polymer*, **48**, 7029.
Nicolas, J., Brusseau, S. and Charleux, B. (2010) *J. Polym. Sci., Part A: Polym. Chem.*, **48**, 34.
Nicolas, J., Guillaneuf, Y., Lefay, C. *et al.* (2012) *Prog. Polym. Sci.* doi: 10.1016/j.progpolymsci.2012.06.002
Nicoli, D.F. and Toumbas, P. (2004) Sensors and methods for high-sensitivity optical particle counting and sizing. US Patent 6794671.
Nielen, M.W.F. (1998) *Anal. Chem.*, **70**, 1563.
Nielen, M.W.F. (1999) *Mass Spectrom. Rev.*, **18**, 309.
Nielen, M.W.F.and Buijtenhuijs, F.A. (2001) *LC-GC Europe*, **14**(2), 82.

Nishida, S., El-Aasser, M.S., Klein, A. and Vanderhoff, J.W. (1981) In *Emulsion Polymers and Emulsion Polymerisation*, ACS SymposiumSeries, vol. **165**, (eds D.R. Bassett and A.E. Hamielec), American Chemical Society, Washington D.C., p. 291.
Noel, L.F.J., Altveer, J.L., Timmermans, M.D.F. and German, A.L. (1994) *J. Polym. Sci. Polym. Chem. Ed.*, **32**, 2223.
Noel, L.F.J., Maxwell, I.A. and German, A.L. (1993) *Macromolecules*, **26**, 2911.
Noël, L.F.J., Van Altveer, J.L., Timmermans, M.D.F. and German, A.L. (1996) *J. Appl. Polym. Sci.*, **34**, 1763.
Noel, R.J., Gooding, K.M., Regnier, F.E. *et al.* (1978) *J. Chromatogr.*, **166**, 373.
Nomura, M., Harada, M., Eguchi, W. and Nagata, S. (1975) *Polym. Prepr. Am. Chem. Soc. Div. Polym. Chem.*, **15**, 217.
Nomura, M., Harada, M., Nakagawara, K. *et al.* (1971) *J. Chem. Eng. Jpn.*, **4**, 160.
Nomura, M., Kubo, M. and Fujita, K. (1983) *J. Appl. Polym. Sci.*, **28**, 2767.
Odeberg, J., Rassing, J., Jönsson, J.-E. and Wesslen, B. (1996) *J. Appl. Polym. Sci.*, **62**, 435.
Odeberg, J., Rassing, J., Jönsson, J.-E. and Wesslen, B. (1998) *J. Appl. Polym. Sci.*, **70**, 897.
Odian, G. (2004) *Principles of Polymerization*, 4th edn. Wiley-Interscience, New York.
Oh J.K., Tomba, J.P., Rademacher, J. *et al.* (2003) *Macromolecules*, **36**, 8836.
Oh, J.K., Tang, C., Gao, H. *et al.* (2006a) *J. Am. Chem. Soc.*, **128**, 5578.
Oh, J.K., Perineau, F. and Matyjaszewski, K. (2006b) *Macromolecules*, **39**, 8003.
Oh, J.K., Dong, H., Zhang, R. *et al.* (2007) *J. Polym. Sci., Polym. Chem.*, **45**, 4764.
Oh, J.K. (2008a) *J. Polym. Sci., Part A: Polym. Chem.*, **46**, 6983.
Oh, J.K., Dtumright, R., Siewart, D.J. and Matyjazewski, K. (2008b) *Prog. Polym. Sci.*, **33**, 448.
Ohlsson, B. and Törnell, B. (1990) *J. Appl. Polym. Sci.*, **41**, 1189.
Ohnesorge, F. and Binnig, G. (1993) *Science*, **260**, 1451.
Ohno, K. (2010) *Polym. Chem.*, **1**, 1545.
Ohshima, H., Makino, K., Kato, T. *et al.* (1993) *J. Colloid Interface Sci.*, **159**, 512-514.
Ohtsuka, Y., Kawaguchi, H. and Hayashi, S. (1981) *Polymer*, **22**, 658.
Okubo, M., Ichikawa, K. and Fujimura, M. (1991) *Colloid. Polym. Sci.*, **269**, 1257.
Okubo, M. and Ichikawa, K. (1994a) *Colloid Polym. Sci.*, **272**, 933.
Okubo, M. and Nakagawa, T. (1994b) *Colloid Polym. Sci.*, **272**, 530.
Okubo, T. (1993) *Prog. Polym. Sci*, **18**, 481.
Olaj, O.F. and Bitai, I. (1987) *Angew. Makromol. Chem.*, **155**, 177.
Ostromislensky, I.I. (1915) *J. Russ. Chem. Soc.*, **47**, 1928.
Ostromislensky, I.I. (1916) *J. Russ. Chem. Soc.*, **48**, 1071.
O'Toole, J.T. (1965) *J. Appl. Polym. Sci.*, **9**, 1291.
Otsu, T. (2000) *J. Polym. Sci. Part A: Polym. Chem.*, **38**, 2121.
Oudhoff, K. (2004) Capillary electrophoresis for the characterization of synthetic polymers, Ph.D. thesis, University of Amsterdam.
Ouzineb, K., Lord, C., Lesauze, N. *et al.* (2006) *Chem. Eng. Sci.*, **61**, 2994.
Ozgur, S. and Natalia, E. (2008) *Nanotechnology*, **19**, 445717.
Padget, J. (1994) *J. Coatings Technol.*, **66**(839), 89.
Paleos, C.N. (1990) *Rev. Macromol. Chem. Phys.*, **C30**, 137.
Pan, G., Sudol, E.D., Dimonie, V.L. and El-Aasser, M.S. (2001) *Macromolecules*, **34**, 481.
Pan, G., Sudol, E.D., Dimonie, V.L. and El-Aasser, M.S. (2002) *Macromolecules*, **35**, 6915.

Pangonis, W., Heller, W. and Jacobson, A. (1957) *Tables of Light Scattering Functions for Spherical Particles*, Wayne State University Press, Detroit.
Panyam, J. and Labhasetwar, V. (2003) *Adv. Drug Del. Rev.*, **55**, 329.
Paquet, D.A. Jr. and Ray, W.H. (1994) *AIChE J.*, **40**, 73.
Park, S., Cho, D., Ryu, J. *et al.* (2002) *J. Chromatogr. A*, **958**, 183.
Parker D.S., Sue, H.J. Huang, J. and Yee, A.F. (1989) *Polymer*, **31**, 2267.
Parts, A.G., Moore, D.E. and Watterson, J.G. (1965) *Makromol. Chem.*, **89**, 156.
Pasch, H. and Hiller, W. (1999) *Macromolecules*, **29**, 6556.
Pasch, H., Mequanint, K. and Adrian, J. (2002) *e-Polymers*, **5**.
Pasch, H. and Kiltz, P. (2003) *Macromol. Rapid Commun.*, **24**, 104.
Patel, P.D. and Russel, W.B. (1987) *J. Rheology*, **31**, 599.
Patsiga, R., Litt, M. and Stannett, V. (1960) *J. Phys. Chem.*, **64**, 801.
Pattamasattayasonthi, N., Chaochanchaikul, K., Rosarpitak, V. and Sombatsompop, N. (2011) *J. Vinyl Addit. Tech.*, **17**, 9.
Pelton, R. (1999) *Adv. Colloid. Interface Sci.*, **85**(10), 1.
Pelton, R.H. and Chibante, P. (1986) *Colloids Surf.*, **20**, 247.
Peng, H., Cheng, S., Feng, L. and Fan, Z. (2003) *J. Appl. Polym. Sci.*, **89**, 3175.
Pepels, M.P.F., Holdsworth, C.I., Pascual, S. and Monteiro, M.J. (2010) *Macromolecules* **43**, 7565.
Percec, V., Guliashvili, T., Ladislaw, J.S. *et al.* (2006) *J. Am. Chem. Soc.*, **128**, 14156.
Percec, V., Popov, A.V., Ramirez-Castillo, E. *et al.* (2002) *J. Am. Chem. Soc.*, **124**, 4940.
Percy, M.J., Amalvy, J.I., Barthet, C. *et al.* (2002) *J. Mater. Chem.*, **12**, 697.
Peters, R., Mengerink, Y., Langereis, S. *et al.* (2002) *J. Chromatogr. A*, **994**, 327.
Pfau, A., Sander, R. and Kirsch, S. (2002) *Langmuir*, **18**, 2880.
Phillipsen, H.J.A. (2004) *J. Chromatogr. A*, **1037**, 329.
Pichot, C. and Delair, T. (1999a) In *Microspheres, Microcapsules and Liposomes* (ed. R. Arshady), Citus Book, London, Chapter 5.
Pichot, C. (1995) *Polym. Adv. Technol.*, **6**, 427.
Pichot, C., Delair, Th. and Elaissari, A. (1999b) In *Polymeric Dispersions: Principles and Applications* (ed. J.M. Asua), Kluwer Academic Publishers, Dordrecht, The Netherlands, *NATO ASI Series*, **335**, 515.
Pichot, C.H., Llauro, M. and Pham, Q. (1981) *J. Polym. Sci. Polym. Chem. Ed.*, **19**, 2619.
Plessis, C., Arzamendi, G., Leiza, J.R. *et al.* (2000) *Macromolecules*, **33**, 4.
Poehlein, G. (1981) Emulsion polymerization in continuous reactors, in *Emulsion Polymerization* (ed. I. Piirma), Academic Press, New York.
Poehlein, G.W. (1997) Reaction engineering for emulsion polymerization, in: *Polymeric Dispersions: Principles and Applications* (ed. J.M. Asua), Kluwer Academic Publishers, Dordrecht, pp. 305.
Popovici, S.T. (2004) Towards small and fast size-exclusion chromatography. Ph.D.Thesis, University of Amsterdam.
Popovici, S.T., Kok, W.Th. and Schoenmakers, P.J. (2004) *J. Chromatogr. A*, **1060**, 237.
Popovici, S.T. and Schoenmakers, P.J. (2005) *J. Chromatogr. A*, **1099**, 92.
Preis, I. and Langer, R.S. (1979) *J. Immunol. Methods*, **28**, 193.
Prescott, S.W., Ballard, M.J., Rizzardo, E. and Gilbert, R.G.G. (2002a) *Aust. J. Chem.*, **55**, 415.

Prescott, S.W., Ballard, M.J., Rizzardo, E. and Gilbert, R.G.G. (2002b) *Macromolecules*, **35**, 5417.
Prescott, S.W. (2003) *Macromolecules*, **36**, 9608.
Prescott, S.W., Ballard, M.J. Rizzardo, E. and Gilbert, R.G. (2005a) *Macromolecules*, **38**, 4901.
Prescott, S.W., Ballard, M.J. and Gilbert, R.G. (2005b) *J. Polym. Sci. A Polym. Chem. Ed.*, **43**, 1076.
Price, C.C. and Adams, C.E. (1945) *J. Am. Chem. Soc.*, **67**, 1674.
Priest, W.J. (1952) *J. Phys. Chem.*, **56**, 1077.
Prodpan, T., Dimonie, V.L., Sudol, E.D. and El-Aasser, M.S. (2000) *Macromol. Symp.*, **155**, 1.
PSLC (2000) The Story of Rubber, http://www.pslc.ws/macrog/exp/rubber/menu.htm.
Pusch, J. and van Herk, A.M. (2005) *Macromolecules*, **38**, 6909.
Pusey, P.N. and Van Megen, W. (1986) *Nature*, **320**, 340.
Qiu, J., Gaynor, S.G. and Matyjaszewski, K. (1999a) *Macromolecules*, **32**, 2872.
Qiu, J., Shipp, D., Gaynor, S.G. and Matyjaszewski, K. (1999b) *Polym. Prepr. (Am. Chem. Soc. Div. Polym. Chem.)*, **40**, 418.
Qiu, J., Pintauer, T., Gaynor, S.G. *et al.* (2000) *Macromolecules*, **33**, 7310.
Qiu, J., Charleux, B. and Matyjaszewski, K. (2001) *Prog. Polym. Sci.*, **26**, 2083.
Rager, T., Meyer, W.H. and Wegner, G. (1999) *Macromol. Chem. Phys.*, **200**, 1672.
Rajatapiti, P., Dimonie, V.L. and El-Aasser, M.S. (1995) *J. Macromol. Sci. Pure Appl. Chem. A*, **32**, 1445.
Rajatapiti, P., Dimonie, V.L., El-Aasser, M.S. and Vratsanos, M.S. (1997) *J. Appl. Polym. Sci.*, **63**, 205.
Ramirez, L.P. and Landfester, K. (2003) *Macromol. Chem. Phys.*, **204**(4), 22.
Ramos, J. and Forcada, J. (2011) *Langmuir*, **27**, 7222.
Rasmusson, M. and Wall, S. (1999) *J. Colloid Interface Sci.*, **209**, 312.
Rasmusson, M., Routh, A. and Vincent, B. (2004) *Langmuir*, **20**, 3536.
Ratanathanawongs, S.K. and Giddings, J.C. (1993) *Polym. Mater. Sci. Eng.*, **70**, 26.
Reculusa, S., Mingotaud, C., Bourgeat-Lami, E. *et al.* (2004) *Nano Lett.*, **4**, 1677.
Regnault, F. (1838) *Ann. Chim. Phys.*, **69**, 157.
Reyes, Y. and Asua, J.M. (2010) *J. Polym. Sci., A. Polym. Chem.*, **48**, 2579.
Reyes, Y., Paulis, M. and Leiza, J.R. (2010) *J. Colloid Interface Sci.*, **352**, 359.
Richard, J. and Maquet, J. (1992) *Polymer*, **33**, 4164.
Richard, J. and Deschamps, F.S. (2003) In *Colloidal Biomolecules, Biomaterials, and Biomedical Applications*, (ed. A. Elaissari), Marcel Dekker Edition Surfactant Science Series, vol. **116**, Marcel Dekker, pp. 429.
Richards, J.R., Congalidis, J.P. and Gilbert, R.G. (1989) *J. Appl. Polym. Sci.*, **37**, 2727.
Rieger, J., Stoffelbach, F., Bui, C. *et al.* (2008) *Macromolecules*, **41**, 4065.
Rieger, J., Osterwinter, G., Bui, C. *et al.* (2009) *Macromolecules*, **42**, 5518.
Rieger, J., Zhang, W., Stoffelbach, F. and Charleux, B. (2010) *Macromolecules*, **43**, 6302.
Riess, G. (1999) *Colloids Surf. A, Physicochem. Eng. Aspects*, **153**, 99.
Rodriguez, R., Barandiaran, M.J. and Asua, J.M. (2007) *Macromolecules*, **40**, 5735.
Roe, C.P. (1968) *Ind. Eng. Chem.*, **60**, 20.
Rohrbach, R., Florin, E.L. and Stelzer, E.H.K. (2001) *Proc. SPIE*, **4431**, 75.
Rosen, B.M. and Percec, V. (2009) *Chem. Rev.*, **109**, 5069.

Routh, A. and Vincent, B. (2002) *Langmuir*, **18**, 5366.
Rudin, A. (1995) *Macromol. Symp.*, **92**, 53.
Russell, G.T., Gilbert, R.G. and Napper, D.H. (1992) *Macromolecules*, **25**, 2459.
Russell, G.T., Gilbert, R.G. and Napper, D.H. (1993) *Macromolecules*, **26**, 3538.
Rynders, R.M., Hegedus, C.R. and Gilicinski, A.G. (1995) *J. Coat. Technol.*, **67**, 59.
Sacanna, S. and Philipse, A.P. (2006) *Langmuir*, **22**, 10209.
Saenz De Buruaga, I., Arotcarena, M., Armitage, P.D. *et al.* (1996) *Chem. Eng. Sci.*, **51**, 2781.
Saenz De Buruaga, I., Echevarria, A., Armitage, P.D. *et al.* (1997a) *AIChE*, **43**(4), 1069.
Saenz De Buruaga, I., Armitage, P.D., Leiza, J.R. and Asua, J.M. (1997b) *Ind. Eng. Chem. Res.*, **36**(10), 4243.
Saenz De Buruaga, I., Leiza, J.R. and Asua, J.M. (2000) *Polym. React. Eng.*, **8**, 39.
Sajjadi, S. and Jahanzad, F. (2006) *Chem. Eng. Sci.*, **61**, 3001.
Sato, F. and Tateyama, M. (1988) Impact resistant methacrylic resin composition. US 47300023.
Sato, H., Ichieda, N., Tao, H. and Ohtani, H. (2004) *Anal. Sci.*, **20**, 1289.
Sauzedde, F., Elaissari, A. and Pichot, C. (2000) *Macromol. Symp.*, **150**, 617.
Save, M., Manguian, M., Chassenieux, C. and Charleux, B. (2005) *Macromolecules*, **38**, 280.
Save, M., Guillaneuf, Y. and Gilbert, R.G. (2006) *Aust. J. Chem.*, **59**, 693.
Sawyer, L.C., Grubb, D.T. and Meyers, G.F. (2008) *Polymer Microscopy*, 3rd edn. Springer, New York.
Scarlet B. (1982) *Particle Size Analysis (1981)* (eds N. Stanley-Wood and T. Allen), Wiley, Chichester, pp. 219-231.
Schantz, S., Carlsson, H.T., Andersson, T. *et al.* (2007) *Langmuir*, **23**, 3590.
Scheiber, J. (1943) *Chemie und Techologie der künstlichen Hartze*, Wiss. Verl. G., Stuttgart.
Schellenberg, C., Akari, S., Regenbrecht, M. *et al.* (1999) *Langmuir*, **15**, 1283.
Schlüter, H. (1990) *Macromolecules*, **23**, 1618.
Schlüter, H. (1993) *Colloid Polym. Sci.*, **271**, 246.
Schmid, A., Armes, S.P., Leite, C.A.P. and Galembeck, F. (2009) *Macromolecules*, **25**, 2486.
Schmid, A., Tonnar, J. and Armes, S.P. (2008) *Adv. Mater.*, **20**, 3331.
Schmidt, C. and Boodmeier, R. (1999) *J. Control. Rel.*, **57**, 115.
Schmidt-Thuemmes, J., Schwarzenbach, E. and Lee, D.I. (2002) Applications in the paper industry, in: *Polymer Dispersions and Their Industrial Applications* (eds D. Urban and K. Takamura), Wiley-VCH, Weinheim.
Schneider, M., Pith, T. and Lambla, M. (1996) *J. Appl. Polym. Sci.*, **62**, 273.
Schneider, M., Pith, T. and Lambla, M. (1997) *J. Mater. Sci.*, **32**, 6343.
Schoenmakers, P.J., Marriott, P. and Beens, J. (2003) *LC-GC Europe*, **16**, 335.
Schonherr, H., Hruska, Z. and Vancso, G.J. (1999) Toward imaging of functional group distributions in surface-treated polymers by scanning force microscopy using functionalized tips. Book of Abstracts, 218th ACS National Meeting, New Orleans, Aug. 22–26, MSE-106.
Schoonbrood, H.A.S., Van Den Boom, M.A.T., German, A.L. and Hutovic, J. (1994) *J. Polym. Sci. Polym. Chem. Ed.*, **32**, 2311.

Schoonbrood, H.A.S., Van Den Reijen, B., De Kock, J.B.L. *et al.* (1995a) *Makromol. Chem. Rapid Commun.*, **16**, 119.

Schoonbrood, H.A.S., Brouns, H.M.G., Thijssen, H.A. *et al.* (1995b) *Makromol. Symp.*, **92**, 133.

Schoonbrood, H.A.S., Van Eynatten, R.C.P.M., Van Den Reijen, B. *et al.* (1996a) *J. Polym. Sci. Polym. Chem. Ed.*, **34**, 935.

Schoonbrood, H.A.S., Van Eynatten, R.C.P.M., Van Den Reijen, B. *et al.* (1996b) *J. Polym. Sci. Polym. Chem. Ed.*, **34**, 949.

Schoonbrood, H.A.S., Unzue, M.J., Beck, O.J., Asua, J.M., Montoya-Goñi, A. and Sherrington, D.C. (1997) *Macromolecules*, **30**, 6024.

Schork, F.J. (1990) Advances in Emulsion Polymerization and Latex Technology, part of the syllabus of the short course given in Davos, Switzerland, p. 1.

Schriemer, D.C. and Li, L. (1996) *Anal.Chem.*, **68**, 2721.

Schuler, B., Baumstark, R., Kirsch, S. *et al.* (2000) *Prog. Org. Coat.*, **40**, 139.

Schuler, B., Baumstark, R., Kirsch, S. *et al.* (2000) *Prog. Org. Coat.*, **40**, 139.

Schuler, H. and Schmidt, C.U. (1992) *Chem. Eng. Sci.*, **47**, 899.

Scrivens, J.H. and Jackson, A.T. (2000) *Int. J. Mass Spectrom.*, **200**, 261.

Sebakhy, K.O., Kessel, S. and Monteiro, M.J. (2010) *Macromolecules*, **43**, 9598.

Shaffer, O.L., El-Aasser, M.S. and Vanderhoff, J.W. (1983) *Proc. - Annu. Meet., Electron Microsc. Soc. Am.*, **41**, 30.

Shaffer, O.L., El-Aasser, M.S. and Vanderhoff, J.W. (1987) *Proc. - Annu. Meet., Electron Microsc. Soc. Am.*, **45**, 502.

Shimizu, N., Handa, H. *et al.* (2008) *Nature Biotech.*, **18**, 877.

Shu, J., Cheng, C., Zheng, Y. *et al.* (2011) *Polym. Bull.*, **67**, 1185.

Siewing, A., Schierholz, J., Braun, D. *et al.* (2001) *Macromol. Chem. Phys.*, **202**, 2890.

Siewing, A., Lahn, B., Braun, D. and Pasch, H. (2003) *J. Polym. Sci. Part A: Polym. Chem.*, **41**, 3143.

Siggia, S., Hohenstein, W.P. and Mark, H. (1945) *India Rubber World*, **111**, 436.

Simms, R.W. and Cunningham, M.F. (2006) *J. Polym. Sci., Polym. Chem.*, **44**, 1628.

Simms, R.W. and Cunningham, M.F. (2007) *Macromolecules*, **40**, 860.

Simms, R.W. and Cunningham, M.F. (2008) *Macromolecules*, **41**, 5148.

Singer, J.M. and Plotz, C.M. (1956) *Am. Med. J.*, **21**, 888.

Singh, M. and O'Hagan, D. (1998) *Adv. Drug Del. Rev.*, **34**, 285.

Singh, M., Briones, M., Kazzaz, J., Donnelly, J., and O'Hagan, D. (2000) Proceedings of the International Symposium on Controlled Release of Bioactive Materials **27**, 548

Sinha, V.R., Singla, A.K., Wadhawan, S. *et al.* (2004) *Int. J. Pharm.*, **274**, 1.

Siwick, B.J., Kalinina, O., Kumacheva, E. *et al.* (2001) *J. Appl. Phys.*, **90**(10), 5328.

Slawinski, M., Schellekens, M.A.J., Meuldijk, J. *et al.* (2000) *J. Appl. Polym. Sci.*, **76**, 1186.

Slomkowski, S., Alemán, J.V., Gilbert, R.G. *et al.* (2011) *Pure Appl. Chem.*, **83**, 2229.

Small, H. (1974) *J. Colloid Interface Sci.*, **48**, 147.

Smith, W.V. and Ewart, R.H. (1948) *J. Chem. Phys.*, **16**, 592.

Smoluchowski, M. von (1917) *Z. Phys. Chem.*, **92**, 129.

Smulders, W., Gilbert, R.G. and Monteiro, M.J. (2003) *Macromolecules*, **36**, 4309.

Solomon, D.H., Rizzardo, E. and Cacioli, P. (1985) US 4,581,429 Chem. Abstr. 1985 102: 221335q.

Sommer, F., Duc, T.M., Pirri, R. *et al.* (1995) *Langmuir*, **11**, 440.

Spiegel, S., Landfester, K., Lieser, G. *et al.* (1995) *Macromol. Chem. Phys.*, **196**, 985.

Staal, B.B.P. (2005) Matrix-assisted laser-desorption/ionization time-of-light mass spectrometry of synthetic polymers. Ph.D.thesis, Technical University of Eindhoven.

Stegeman, G. (1994) On hydrodynamic chromatography in packed columns. Ph.D.Thesis, University of Amsterdam.

Stenius, P. and Kronberg, B. (1983) Conductometry, potentiometry, electrophoresis, and hydrodynamic chromatography, in *Science & Technology of Polymer Colloids. Surface Characterization of Latexes. Characterization, Stabilization and Application Properties. Volume II*, (eds G.W. Poehlein, R.H. Ottewill and J.W. Goodwin), Martinus Nijhoff Publishers, The Hague, pp. 449.

Stevenson, A.F. and Heller, W. (1961) *Tables of Scattering Functions for Heterodisperse Systems*, Wayne State University Press, Detroit.

Stockmayer, W.H. (1957) *J. Polym. Sci.*, **24**, 314.

Stoffelbach, F., Belardi, B., Santos, J.M.R.C.A. *et al.* (2007) *Macromolecules*, **40**, 8813.

Stoffelbach, F., Griffete, N., Bui, C. and Charleux B. (2008) *Chem. Commun.*, 4807.

Stone, W.E.E. and Stone-Masui, J.H. (1983) XPS study of sulfate groups on polystyrene latexes, in: *Science & Technology of Polymer Colloids. Surface Characterization of Latexes. Characterization, Stabilization and Application Properties. Volume II*, (eds G.W. Poehlein, R.H. Ottewill and J.W. Goodwin), Martinus Nijhoff Publishers, The Hague, pp. 480.

Storsberg, J. and Ritter, H. (2002) *Macromol. Chem. Phys.*, **203**, 812.

Storti, G., Carra, S., Morbidelli, M. and Vita, G. (1989) *J. Appl. Polym. Sci.*, **37**, 2443.

Stoye, D. and Freitag, W. (1998) *Paints, Coatings and Solvents, Chapter 4, Pigments and Extenders*, 2nd edn, Wiley-VCH, Weinheim.

Strauch, J., McDonald, J., Chapman, B.E. *et al.* (2003) *J. Polym. Sci. A Polym. Chem. Ed.*, **41**, 2491.

Stromberg, R.R., Swerdlow, M. and Mandel, J. (1953) *J. Res. Nat. Bur. Stand.*, **50**, 299.

Stubbs, J., Karlsson, O., Jonsson, J.-E., *et al.* (1999a) *Colloids Surf., A.*, **153**, 255.

Stubbs, J.M., Durant, Y.G. and Sundberg, D.C. (1999b) *Langmuir*, **15**, 3250.

Stubbs, J.M., Durand, Y. and Sunberg, D. (2003a) *C.R. Chim.*, **6**, 1217.

Stubbs, J.M., Carrier, R., Karlsson, O.J. and Sundberg, D.C. (2003b) *Prog. Colloid Polym. Sci.*, **124**, 131.

Stubbs, J.M. and Sundberg, D.C. (2005a) *J. Polym. Sci., Part B: Polym. Phys.*, **43**, 2790.

Stubbs, J.M. and Sundberg, D.C. (2005b) *Polymer*, **46**, 1125.

Stubbs, J.M., Roose, P., De, D.P. and Sundberg, D.C. (2006) *Langmuir*, **22**, 2697.

Stubbs, J.M. & Sundberg, D.C. (2008a) *J. Coat. Technol. Res.*, **5**, 169.

Stubbs, J.M. and Sundberg, D.C. (2008b) *Prog. Org. Coat.*, **61**, 156.

Stubbs, J.M., Tsavalas, J., Carrier, R. and Sundberg, D. (2010) *Macromol. React. Eng.*, **4**, 424.

Stubbs, J.M. and Sundberg, D.C. (2011) *J. Polym. Sci. Part B: Polym. Phys.*, **49**, 1583.

Studer, D. and Gnaegi, H. (2000) *J. Microsc.*, **197**, 94.

Sundberg, D.C., Casassa, A.P., Pantazopoulos, J. *et al.* (1990) *J. Appl. Polym. Sci.*, **41**, 1425.

Sundberg, D.C. and Durant, Y.G. (2003) *Polym. React. Eng.*, **11**, 379.

Sundberg, E.J. and Sundberg, D.C. (1993) *J. Appl. Polym Sci.*, **47**, 1277.

Talalay, A. and Magat, M. (1945) *Synthetic Rubber from Alcohol*, Interscience, New York.

Talmon, Y. (1987) *Proc. - Annu. Meet., Electron Microsc. Soc. Am.*, **45**, 496.
Tamai, H., Fujii, A. and Suzawa, T. (1987) *J. Colloid Sci.*, **118**, 176.
Tamayo, J. and García, R. (1998) *Appl. Phys. Lett.*, **73**, 2926.
Tanaka, B., Azukizawa, M., Yamazoe, H. *et al.* (2000) *Polymer*, **41**, 5611.
Tanaka, T., Saito, N. and Okubo, M. (2009) *Macromolecules*, **42**, 7423.
Tanaka, T., Okayama, M., Minami, H. and Okubo, M. (2010) *Langmuir*, **26**, 11732.
Tang, H.-I., Sudol, E.D., Adams, M.E. *et al.* (1992) In *Polymer Latexes - Preparation, Characterization and Applications*, ACS Symposium Series, vol. **492** (eds E.S. Daniels, E.D. Sudol and M.S. El-Aasser), American Chemical Society, Washington D.C., p. 72.
Tauer, K. and Kühn, I. (1995) *Macromolecules*, **28**, 2236.
Tauer, K. and Kühn, I. (1997) *NATO ASI Ser., Ser. E: Appl. Sci.*, **335**, 49.
Tauer, K. and Deckwer, R. (1998) Book of Abstracts, 215th ACS National Meeting, Dallas, March 29-April 2, COLL-010.
Tauer, K. (2001) *Surf. Sci. Ser.*, **100**, 429.
Taylor, M. (2002) Synthesis of polymer dispersions, in *Polymer Dispersions and Their Industrial Applications* (eds D. Urban and K> Takamura), Wiley-VCH, Weinheim.
Teixeira, R.F.A. and Bon, S.A.F. (2010) *Adv. Polym. Sci.*, **233**, 19.
Teixeira, R.F.A., McKenzie, H.S., Boyd, A.A. and Bon, S.A.F. (2011) *Langmuir*, **44**, 7415.
Teixeira-Neto, E. and Galembeck, F. (2002) *Colloids Surf., A.*, **207**, 147.
Templeton-Knight, R.L. (1990) *J. Oil Colour Chem. Assoc.*, **11**, 459.
Texter, J. (2003) *C.R. Chim.*, **6**, 1425.
Thickett, S.C. and Gilbert, R.G. (2007) *Polymer*, **48**, 6965.
Thickett, S.C., Morrison, B. and Gilbert, R.G. (2008) *Macromolecules*, **41**, 3521.
Thomson, M.E., Manley, A.-M., Ness, J.S. *et al.* (2010) *Macromolecules*, **43**, 7958.
Thundat, T., Zheng, X.-Y., Chen, G.Y. & Warmack, R.J. (1993) *Surf. Sci.*, **294**, L939.
Tiarks, F., Landfester, K. and Antonietti, M. (2001) *Langmuir*, **17**, 5775.
Tobita, H., Takada, Y. and Nomura, M. (1994) *Macromolecules*, **27**, 3804.
Tobita, H. (1995) *Macromolecules*, **28**, 5128.
Tobolsky, A.V. and Offenbach, J. (1955) *J. Polym. Sci.*, **16**, 311.
Tomba, J.P., Portinha, D., Schroeder, W.F. *et al.* (2009) *Colloid Polym. Sci.*, **287**, 367.
Tortosa, K., Smith, J.-A. and Cunningham, M.F. (2001) *Macromol. Rapid. Commun.*, **22**, 957.
Torza, S. and Mason, S.G. (1970) *J. Colloid Interface Sci.*, **33**, 67.
Trabelsi, S., Raspaud, E. and Langevin, D. (2007) *Langmuir*, **23**, 10053.
Trent, J.S. (1984) *Macromolecules*, **17**, 2930.
Tsavalas, J.G., Gooch, J.W. and Schork, F.J. (2000) *J. Appl. Polym. Sci.*, **75**, 916.
Tsavalas, J.G., Schork, F.J., de Brouwer, H. and Monteiro, M.J. (2001) *Macromolecules*, **34**, 3938.
Tsavalas, J.G., Schork, F. and Landfester, K. (2004) *J. Coat. Technol.*, **1**, 53.
Tsuji, S. and Kawaguchi, H. (2004) *Langmuir*, **20**, 2449.
Tsukruk, V. (1997) *Rubber Chem. Technol.*, **70**, 430.
Türk, J. (1985) *Papier- und Kunststoffverarbeitung*, **5**, 40.
Türk, J. (1993) *Adhäsion*, **10**, 17.
Ugelstad, J., Mørk, P. and Aasen, J.O. (1967) *J. Polym. Sci., A-1*, **5**, 2281.
Ugelstad, J., Mørk, P., Dahl, P. and Rangnes, P. (1969) *J. Polym. Sci. Part C*, **27**, 49.
Ugelstad, J. and Mørk, P. (1970) *Br. Polym. J.*, **2**, 31.

Ugelstad, J., El-Aasser, M.S. and Vanderhoff, J. (1973) *J. Polym. Sci., Polym. Lett. Ed.*, **11**, 503.

Ugelstad, J., Hansen, F.K. and Lange, S. (1974) *Makromol. Chem.*, **175**, 507.

Ugelstad, J. and Hansen, F.K. (1976) *Rubber Chem. Technol.*, **49**, 536.

Ugelstad, J., Stenstad, P., Kilaas, L. *et al.* (1993) *Prog. Blood Purif.*, **11**, 349.

Urban, D., Wistuba, E., Aydin, O. and Schwerzel, T. (1995) New Properties of Acrylic Dispersions. European Industrial Adhesive Conference. *Organized by Exxon Chemical Europe Inc.*, Brussels.

Urban, D. and Takamura, K. eds (2002a) *Polymer Dispersions and Their Industrial Applications*, Wiley-VCH Verlag GmbH, Weinheim.

Urban, D. and Egan, L. (2002b) Applications in the adhesives and sealants industry, in *Polymer Dispersions and Their Industrial Applications* (eds Urban & Takamura), Wiley-VCH, Weinheim.

Urbani, C.N. and Monteiro, M.J. (2009a) *Aust. J. Chem.*, **62**, 1528.

Urbani, C.N. and Monteiro, M.J. (2009b) *Macromolecules*, **42**, 3884.

Urretabizkaia, A., Leiza, J.R. and Asua, J.M. (1994) *AIChE J.*, **40**, 1850.

Urretabizkaia, A., Sudol, E.D, El-Aasser, M.S. and Asua, J.M. (1993) *J. Polym. Sci., Part A: Polym. Chem.*, **31**, 2907.

Ute, K., Janco, M., Niimi, R. *et al.* (2001b) *Polym. Preprints*, **42**, 67.

Ute, K., Niimi, R., Hongo, S. and Hatada, K. (1998) *Polymer J.*, **30**, 439.

Ute, K., Niimi, R., Matsunaga, M. *et al.* (2001a) *Macromol. Chem. Phys.*, **202**, 3081.

Uwins, P.J.R. (1994) *Mater. Forum*, **18**, 51.

Uzulina, I., Kanagasabapathy, S. and Claverie, J. (2000) *Macromol. Symp.*, **150**, 33.

van Berkel, K.Y., Russell, G.T. and Gilbert, R.G. (2003) *Macromolecules*, **36**, 3921.

van Berkel, K.Y., Russell, G.T. and Gilbert, R.G. (2005) *Macromolecules*, **38**, 3214.

Van Cleef, M., Holt, S.A., Watson, G.S. and Myhra, S. (1996) *J. Microsc. (Oxford)*, **181**, 2.

Van den Brink, M., Pepers, M.L.H., Van Herk, A.M. and German, A.L. (2001) *Polym. React. Eng.*, **9**(2), 101.

Van den Hul, H.J. and Vanderhoff, J.W. (1970) *Br. Polym. J.*, **2**, 121.

Van der Horst, A. and Schoenmakers, P.J. (2003) *J. Chromatogr. A*, **1000**, 693.

Van der Velden, G.P.M. (1983) *Macromolecules*, **16**, 1336.

Van Doremaele, G.H.J. (1990) Model Prediction, Experimental Determination, and Control of Emulsion Copolymer Microstructure. Ph.D. Thesis, Eindhoven University of Technology, The Netherlands.

Van Doremaele, G.H.J., Van Herk, A.M. and German, A.L. (1990) *Makromol. Chem., Macromol. Symp.*, **35/36**, 231.

Van Doremaele, G.H.J., Schoonbrood, H.A.S., Kurja, J. and German, A.L. (1992) *J. Appl. Polym. Sci.*, **42**, 957.

Van Hamersveld, E.M.S., Van Es, J.J.G.S., German, A.L. *et al.* (1999) *Prog. Org. Coat.*, **35**, 235.

Van Herk, A.M., van Streun, K.H., van Welzen, J. and German, A.L. (1989) *Br. Polym. J.*, **21**, 125.

Van Herk, A.M. and German, A.L. (1999) Microencapsulated pigments and fillers, in *Microspheres, Microcapsules & Liposomes, Vol. 1, Preparation and Chemical Applications* (ed. R. Arshady), Citus Books, London.

Van Herk, A.M. (2000) *Macromol. Theor. Simul.*, **9**, 433.

Van Herk, A.M. (2001) *Macromol. Rapid Commun.*, **22**, 687.

Van Herk, A.M. and Landfester, K. (eds) (2010) *Hybrid Latex Particles*, Advances in Polymer Science, Issue no. 233, Springer.

Van Krevelen, D.W. (1992) *Properties of Polymers*, 3rd edn. Elsevier, Amsterdam.

Vanderheyden, Y., Popovici, S.T. and Schoenmakers, P.J. (2002) *J. Chromatogr. A*, **957**, 127.

Vanderhoff, J.W. (1981) *ACS Symp. Ser.*, **165**, 61.

Vanderhoff, J.W., Hul, H.J.V.D., Tausk, R.J.M. and Overbeek, J.T.G. (1970) The preparation of monodisperse latexes with well-characterized surfaces, in *Clean Surfaces: Their Preparation and Characterization for Interfacial Studies* (ed. G. Goldfinger), Marcel Dekker, New York, NY, pp. 15-44.

Vanderhoff, J.W., Park, J.M. and El-Aasser, M.S. (1991) *Polym. Mater. Sci. Eng.*, **64**, 345.

Vandezande, G.A. and Rudin, A. (1994) *J. Coat. Technol.*, **66**, 99.

Vanlandingham, M., Mcknight, S., Palmese, G. *et al.* (1997) *J. Adhesion*, **64**, 31.

Vancso, G., Hillborg, H. and Schönherr, H. (2005) *Adv. Polym. Sci.*, **182**, 55.

Vanzo, E., Marchessault, R.H. and Stannett, V. (1965) *J. Colloid Sci.*, **20**, 62.

Velev, O.D., Lenhoff, A.M. and Kaler, E.W. (2000) *Science*, **287**, 2240.

Verdurmen-Noël, E.F.J. (1994) Monomer Partitioning and Composition Drift in Emulsion Copolymerization. PhD Thesis, Eindhoven University of Technology, Eindhoven, The Netherlands.

Verrier-Charleux, B., Graillat, C., Chevalier, Y. *et al.* (1991) *Colloid. Polym. Sci.*, **269**, 398.

Verwey, E.J.W. and Overbeek, J.Th.G. (1948) *Theory of the Stability Lyophobic Colloids*, Elsevier, Amsterdam.

Vicente, M. (2001) Control Optimo de la Producción dePpolímeros en Emulsión en Base a Medidas Calorimetricas. Ph.D. Thesis, The University of the Basque Country Donosita-San Sebastián.

Vincent, B., Edwards, J., Emmett, S. and Croot, R. (1988) *Colloids Surfaces*, **31**, 267.

Vincent, B. (1992) *Adv. Colloid Interface Sci.*, **42**, 279.

Vincent, B. (1993) *J. Chem. Eng. Sci.*, **48**, 429.

Vinograd, R.L., Fong, L.L. and Sawyer, W.M. (1944) 108th ACS Meeting, New York.

Voorn, D.J., Ming, W. and van Herk, A.M. (2006) *Macromolecules*, **39**, 4654.

Wan, X. and Ying, S. (2000) *J. Appl. Polym. Sci.*, **75**, 802.

Wang, J.-S. and Matyjaszewki, K. (1995a) *Macromolecules*, **28**, 7901.

Wang, J.-S. and Matyjaszewski, K. (1995b) *J. Am. Chem. Soc.*, **117**, 5614.

Wang, W.-J., Kharchenko, S., Migler, K. and Zhu, S. (2004) *Polymer*, **45**, 6495.

Wang, Y., Xu, H., Ma, Y. *et al.* (2011) *Langmuir*, **27**, 7207.

Warson, H. and Finch, C.A. (2001) *Applications of Synthetic Resin Latices*, John Wiley & Sons Ltd, Chichester England.

Watanabe, J., Seibel, G. and Inoue, M. (1984) *J. Polym. Sci.: Polym. Lett. Ed.*, **22**, 39.

Watt, I.M. (1997) *The Principles and Practice of Electron Microscopy*, 2nd edn, University Press, Cambridge, UK.

Weber, N., Tiersch, B., Unterlass, M.M. *et al.* (2011) *Macromol. Rapid Commun.*, **32**, 1925.

Weissman, J.M., Sunkara, H.B., Tse, A.S. and Asher, S.A. (1996) *Science*, **274**, 959.

Whittal, R.M., Russon, L.M. and Li, L. (1998) *J. Chromatogr. A*, **794**, 367.

Wignall, G.D., Ramakrishnan, V.R., Linne, M.A. *et al.* (1990) *Mol. Cryst. Liq. Cryst.*, **180A**, 25.

Willemse, R.X.E., Staal, B.B.P., Donkers, E.H.D. and Van Herk, A.M. (2004) *Macromolecules*, **37**, 5717.
Willenbacher, N., Boerger, L., Urban, D. and Varela de la Rosa, L. (2003) Tailoring PSA-Dispersion Rheology for High-Speed Coating, *Adhesives and Sealants Industry*, November 2003.
Winnik, M.A., Zhao, C.L., Shaffer, O. and Shivers, R.R. (1993) *Langmuir*, **9**, 2053.
Winzor, C.L. and Sundberg, D.C. (1992a) *Polymer*, **33**, 3797.
Winzor, C.L. and Sundberg, D.C. (1992b) *Polymer*, **33**, 4269.
Wittemann, A., Drechsler, M., Talmon, Y. and Ballauff, M. (2005) *J. Am. Chem. Soc.*, **127**, 9688.
Woggon, U., Wannemacher, R., Artemyev, M.V. *et al.* (2003) *Appl. Phys. B*, **77**, 469.
Wu, G.F., Zhao, J.F., Shi, H.T. and Zhang, H.X. (2004b) *Eur. Polym. J.*, **40**, 2451.
Wu, J., Winnik, M.A., Farwaha, R. and Rademacher, J. (2004a) *Macromolecules*, **37**, 4247.
Wunder, S., Lu, Y., Albrecht, M. and Ballauff, M. (2011) *ACS Catal.*, **1**, 908.
Xu, L.Q., Yao, F., Fu, G.-D. and Shen, L. (2009) *Macromolecules*, **42**, 6385.
Xu, Z., Hu, X., Li, X. and Yi, C. (2008) *J. Polym. Sci., Polym. Chem.*, **46**, 481.
Yamago, S., Lida, K. and Yoshida, J. (2002) *J. Am. Chem. Soc.*, **124**, 13666.
Yau, W.W. and Gillespie, D. (2001) *Polymer*, **42**, 8947.
Young, J.L., Spontak, R.J. and DeSimone, J.M. (1999) *Polym. Preprints (ACS, Division of Polym. Chem.)*, **40**, 829.
Yuan, X., Yoshimoto, K. and Nagasaki, Y. (2009) *Anal. Chem.*, **81**, 1549.
Zackrisson, M., Stradner, A., Schurtenberger, P. and Bergenholtz, J. (2005) *Langmuir*, **21**, 10835.
Zetterlund, P.B., Kagawa, Y. and Okubo, M. (2008) *Chem. Rev.*, **108**, 3747.
Zetterlund, P.B. (2011) *Polym. Chem.*, **2**, 534.
Zhan, Q., Gusev, A. and Hercules, D.M. (1999) *Rapid Commun. Mass Spectrom.*, **13**, 2278.
Zhang, J.H., Li, Y.F., Zhang, M. and Yang, B. (2010) *Adv. Mater.*, **22**, 4249.
Zhang, L.F. and Eisenberg, A. (1995) *Science*, **268**, 1728.
Zhang, S.-W., Zhou, S.-X., Weng, Y.-M. and Wu, L.-M. (2006) *Langmuir*, **22**, 4674.
Zhang, W., D'Agosto, F., Boyron, O. *et al.* (2011a) *Macromolecules*, **44**, 7584.
Zhang, W., D'Agosto, F., Boyron, O. *et al.* (2012) *Macromolecules*, **45**, 4075.
Zhang, X., Boissé, S., Zhang, W. *et al.* (2011b) *Macromolecules*, **44**, 4149.
Zhang, Y., Zha, L.S. and Fu, S.K. (2007) *J. Appl. Polym. Sci.*, **92**, 839.
Zhao, K., Sun, P. and Liu, D. (2004) *Huadong Ligong Daxue Xuebao* **30**, 398.
Zhong, Q., Inniss, D., Kjoller, K. and Elings, V.B. (1993) *Surf. Sci.*, **290**, 688.
Zhou, J., van Duijneveldt, J.S. and Vincent, B. (2011) *Phys. Chem. Chem. Phys.*, **13**, 110.
Zhou, J., Wang, G.Nn., Marquez, M. and Hu, Z.B. (2009) *Soft Matter*, **5**, 820.
Zosel, A. (1985) *Colloid Polym. Sci.*, **263**, 541.
Zosel, A. (1986) *J. Adhesion*, **30**, 14.
Zosel, A., Heckmann, W., Ley, G. and Mächtle, W. (1987) *Colloid Polym. Sci.*, **265**, 113.
Zosel, A., Heckmann, W., Ley, G. and Mächtle, W. (1989) *Adv. Org. Coat. Sci. Tecnhol.*, **11**, 15.
Zosel, A. (1991) *J. Adhesion*, **34**, 201.
Zosel, A. (2000) *Adhesives and Sealants Industry*, October, 30.
Zubitur, M. and Asua, J.M. (2001) *Macromol. Mater. Eng.*, **286**, 362.

Index

Chemistry and Technology of Emulsion Polymerisation, Second Edition. Edited by A.M. van Herk.